The Mycota

Edited by
K. Esser and P.A. Lemke

Springer
*Berlin
Heidelberg
New York
Barcelona
Hong Kong
London
Milan
Paris
Singapore
Tokyo*

The Mycota

I *Growth, Differentiation and Sexuality*
   Ed. by J.G.H. Wessels and F. Meinhardt

II *Genetics and Biotechnology*
    Ed. by U. Kück

III *Biochemistry and Molecular Biology*
     Ed. by R. Brambl and G. Marzluf

IV *Environmental and Microbial Relationships*
    Ed. by D. Wicklow and B. Söderström

V *Plant Relationships*
   Ed. by G. Carroll and P. Tudzynski

VI *Human and Animal Relationships*
    Ed. by D.H. Howard and J.D. Miller

VII *Systematics and Evolution*
     Ed. by D.J. McLaughlin, E.G. McLaughlin, and P.A. Lemke

VIII *Biology of the Fungal Cell*
      Ed. by R.J. Howard and N.A.R. Gow

IX *Fungal Associations*
    Ed. by B. Hock

X *Industrial Applications*
   Ed. by H.D. Osiewacz

XI *Agricultural Applications*
    Ed. by F. Kempken

XII *Human Fungal Pathogens*
     Ed. by J.E. Domer and G.S. Kobayashi

The Mycota

A Comprehensive Treatise
on Fungi as Experimental Systems
for Basic and Applied Research

Edited by K. Esser and P.A. Lemke

VII Systematics and Evolution
Part B

Volume Editors:
D.J. McLaughlin, E.G. McLaughlin, and P.A. Lemke

With 120 Figures and 16 Tables

Springer

Series Editors

Professor Dr. Dr. h.c. mult. KARL ESSER
Allgemeine Botanik
Ruhr-Universität
44780 Bochum, Germany
Tel.: +49(234)32-22211
Fax: +49(234)32-14211
e-mail: Karl.Esser@ruhr-uni-bochum.de

Professor Dr. PAUL A. LEMKE †, Auburn, USA

Volume Editors

Professor Dr. David J. McLaughlin
Department of Plant Biology
University of Minnesota
St. Paul, MN 55108-1095, USA
Tel.: +1(612)625-5736
Fax: +1(612)625-1738
e-mail: davem@tc.umn.edu

Professor Dr. Esther G. McLaughlin
Department of Biology
Augsburg College
Minneapolis, MN 55454, USA
Tel.:+1(612)330-1074
Fax: +1(612)330-1649
e-mail: McLaugh@augsburg.edu

Professor Dr. PAUL A. LEMKE †, Auburn, USA

ISBN 3-540-66493-9 Springer-Verlag Berlin Heidelberg New York

Library of Congress Cataloging-in-Publication Data
The Mycota. Includes bibliographical references and index. Contents: 1. Growth, differentiation, and sexuality/editors, J.G.H. Wessels and F. Meinhardt – 2. Genetics and biotechnology. 1. Mycology. 2. Fungi. 3. Mycology – Research. 4. Research. I. Esser, Karl, 1924– . II. Lemke, Paul A., 1937– . QK603.M87 1994 589.2 ISBN 3-540-57781-5 (v. 1: Berlin: alk. paper) ISBN 0-387-57781-5 (v. 1: New York: alk. paper) ISBN 3-540-58003-4 (v. 2: Berlin) ISBN 0-387-58003-4 (v. 2: New York)

This work is subject to copyright. All rights are reserved, whether the whole or part of the material is concerned, specifically the rights of translation, reprinting, reuse of illustrations, recitation, broadcasting, reproduction on microfilm or in any other way, and storage in data banks. Duplication of this publication or parts thereof is permitted only under the provisions of the German Copyright Law of September 9, 1965, in its current version, and permission for use must always be obtained from Springer-Verlag. Violations are liable for prosecution under the German Copyright Law.

Springer-Verlag Berlin Heidelberg New York
a member of BertelsmannSpringer Science + Business Media GmbH

© Springer-Verlag Berlin Heidelberg 2001
Printed in Germany

The use of general descriptive names, registered names, trademarks, etc. in this publication does not imply, even in the absence of a specific statement, that such names are exempt from the relevant protective laws and regulations and therefore free for general use.

Production Editor: PRO EDIT GmbH, Heidelberg, Germany
Cover design: Springer-Verlag, E. Kirchner
Typesetting: Best-set Typesetter Ltd., Hong Kong
Printed on acid-free paper   SPIN: 10743511   31/3130/So 5 4 3 2 1 0

Series Preface

Mycology, the study of fungi, originated as a subdiscipline of botany and was a descriptive discipline, largely neglected as an experimental science until the early years of this century. A seminal paper by Blakeslee in 1904 provided evidence for self-incompatibility, termed "heterothallism", and stimulated interest in studies related to the control of sexual reproduction in fungi by mating-type specificities. Soon to follow was the demonstration that sexually reproducing fungi exhibit Mendelian inheritance and that it was possible to conduct formal genetic analysis with fungi. The names Burgeff, Kniep and Lindegren are all associated with this early period of fungal genetics research.

These studies and the discovery of penicillin by Fleming, who shared a Nobel Prize in 1945, provided further impetus for experimental research with fungi. Thus began a period of interest in mutation induction and analysis of mutants for biochemical traits. Such fundamental research, conducted largely with *Neurospora crassa*, led to the one gene: one enzyme hypothesis and to a second Nobel Prize for fungal research awarded to Beadle and Tatum in 1958. Fundamental research in biochemical genetics was extended to other fungi, especially to *Saccharomyces cerevisiae*, and by the mid-1960s fungal systems were much favored for studies in eukaryotic molecular biology and were soon able to compete with bacterial systems in the molecular arena.

The experimental achievements in research on the genetics and molecular biology of fungi have benefited more generally studies in the related fields of fungal biochemistry, plant pathology, medical mycology, and systematics. Today, there is much interest in the genetic manipulation of fungi for applied research. This current interest in biotechnical genetics has been augmented by the development of DNA-mediated transformation systems in fungi and by an understanding of gene expression and regulation at the molecular level. Applied research initiatives involving fungi extend broadly to areas of interest not only to industry but to agricultural and environmental sciences as well.

It is this burgeoning interest in fungi as experimental systems for applied as well as basic research that has prompted publication of this series of books under the title *The Mycota*. This title knowingly relegates fungi into a separate realm, distinct from that of either plants, animals, or protozoa. For consistency throughout this Series of Volumes the names adopted for major groups of fungi (representative genera in parentheses) are as follows:

Pseudomycota

Division: Oomycota (*Achlya, Phytophthora, Pythium*)
Division: Hyphochytriomycota

Eumycota

Division: Chytridiomycota (*Allomyces*)
Division: Zygomycota (*Mucor, Phycomyces, Blakeslea*)
Division: Dikaryomycota

Subdivision:	Ascomycotina
Class:	Saccharomycetes (*Saccharomyces, Schizosaccharomyces*)
Class:	Ascomycetes (*Neurospora, Podospora, Aspergillus*)
Subdivision:	Basidiomycotina
Class:	Heterobasidiomycetes (*Ustilago, Tremella*)
Class:	Homobasidiomycetes (*Schizophyllum, Coprinus*)

We have made the decision to exclude from *The Mycota* the slime molds which, although they have traditional and strong ties to mycology, truly represent nonfungal forms insofar as they ingest nutrients by phagocytosis, lack a cell wall during the assimilative phase, and clearly show affinities with certain protozoan taxa.

The Series throughout will address three basic questions: what are the fungi, what do they do, and what is their relevance to human affairs? Such a focused and comprehensive treatment of the fungi is long overdue in the opinion of the editors.

A volume devoted to systematics would ordinarily have been the first to appear in this Series. However, the scope of such a volume, coupled with the need to give serious and sustained consideration to any reclassification of major fungal groups, has delayed early publication. We wish, however, to provide a preamble on the nature of fungi, to acquaint readers who are unfamiliar with fungi with certain characteristics that are representative of these organisms and which make them attractive subjects for experimentation.

The fungi represent a heterogeneous assemblage of eukaryotic microorganisms. Fungal metabolism is characteristically heterotrophic or assimilative for organic carbon and some nonelemental source of nitrogen. Fungal cells characteristically imbibe or absorb, rather than ingest, nutrients and they have rigid cell walls. The vast majority of fungi are haploid organisms reproducing either sexually or asexually through spores. The spore forms and details on their method of production have been used to delineate most fungal taxa. Although there is a multitude of spore forms, fungal spores are basically only of two types: (i) asexual spores are formed following mitosis (mitospores) and culminate vegetative growth, and (ii) sexual spores are formed following meiosis (meiospores) and are borne in or upon specialized generative structures, the latter frequently clustered in a fruit body. The vegetative forms of fungi are either unicellular, yeasts are an example, or hyphal; the latter may be branched to form an extensive mycelium.

Regardless of these details, it is the accessibility of spores, especially the direct recovery of meiospores coupled with extended vegetative haploidy, that have made fungi especially attractive as objects for experimental research.

The ability of fungi, especially the saprobic fungi, to absorb and grow on rather simple and defined substrates and to convert these substances, not only into essential metabolites but into important secondary metabolites, is also noteworthy. The metabolic capacities of fungi have attracted much interest in natural products chemistry and in the production of antibiotics and other bioactive compounds. Fungi, especially yeasts, are important in fermentation processes. Other fungi are important in the production of enzymes, citric acid and other organic compounds as well as in the fermentation of foods.

Fungi have invaded every conceivable ecological niche. Saprobic forms abound, especially in the decay of organic debris. Pathogenic forms exist with both plant and animal hosts. Fungi even grow on other fungi. They are found in aquatic as well as soil environments, and their spores may pollute the air. Some are edible; others are poisonous. Many are variously associated with plants as copartners in the formation of lichens and mycorrhizae, as symbiotic endophytes or as overt pathogens. Association with animal systems varies; examples include the predaceous fungi that trap nematodes, the microfungi that grow in the anaerobic environment of the rumen, the many insectassociated fungi and the medically important pathogens afflicting humans. Yes, fungi are ubiquitous and important.

There are many fungi, conservative estimates are in the order of 100 000 species, and there are many ways to study them, from descriptive accounts of organisms found in nature to laboratory experimentation at the cellular and molecular level. All such studies expand our knowledge of fungi and of fungal processes and improve our ability to utilize and to control fungi for the benefit of humankind.

We have invited leading research specialists in the field of mycology to contribute to this Series. We are especially indebted and grateful for the initiative and leadership shown by the Volume Editors in selecting topics and assembling the experts. We have all been a bit ambitious in producing these Volumes on a timely basis and therein lies the possibility of mistakes and oversights in this first edition. We encourage the readership to draw our attention to any error, omission or inconsistency in this Series in order that improvements can be made in any subsequent edition.

Finally, we wish to acknowledge the willingness of Springer-Verlag to host this project, which is envisioned to require more than 5 years of effort and the publication of at least nine Volumes.

Bochum, Germany
Auburn, AL, USA
April 1994

KARL ESSER
PAUL A. LEMKE
Series Editors

Addendum to the Series Preface

In early 1989, encouraged by Dieter Czeschlik, Springer-Verlag, Paul A. Lemke and I began to plan *The Mycota*. The first volume was released in 1994, five other volumes followed in the subsequent years. Also on behalf of Paul A. Lemke, I would like to take this opportunity to thank Dieter Czeschlik, his colleague Andrea Schlitzberger, and Springer-Verlag for their help in realizing the enterprise and for their excellent cooperation for many years.

Unfortunately, after a long and serious illness, *Paul A. Lemke* died in November 1995. Without his expertise, his talent for organization and his capability to grasp the essentials, we would not have been able to work out a concept for the volumes of the series and to acquire the current team of competent volume editors. He also knew how to cope with unexpected problems which occurred after the completion of the manuscripts. His particular concern was directed at Volume VII; in this volume, a posthumous publication of his is included.

Paul A. Lemke was an outstanding scientist interested in many fields. He was extremely wise, dedicated to his profession and a preeminent teacher and researcher. Together with the volume editors, authors, and Springer-Verlag, I mourn the loss of a very good and reliable friend and colleague.

Bochum, Germany
April 2000

KARL ESSER

Volume Preface

This is an exciting time to produce an overview of the systematics and evolution of the fungi. Homoplasy is evident in all lineages, e.g., those based on the gross morphology of the chytrid zoospore, the perithecium and apothecium, the smut teliospore and the agaric fruiting body, and some classifications based on light microscope morphology have been shown to be unsound. Molecular and subcellular characters, aided by new methods of phylogenetic analysis, have allowed us to see through the conflicts between various phenetic classification schemes and have given us some confidence that we are beginning to achieve a true phylogeny of the fungi. Molecular data have both supported ultrastructural characters that first began to unravel the homoplasies unrecognized at the light microscopic level, and have also revealed the relationships of fungi to other eukaryotes. They continue to enlarge the scope of the fungi, e.g., with the recent addition of the Microsporidia (see Cavalier-Smith, Chap. 1, Vol. VII, Part A), and they have shown the need for more detailed chemical, subcellular, and developmental studies for a fuller understanding of these organisms and their relationships.

This volume is a mixture of phylogenetic and more classical systematics. Progress in knowledge of species and development of taxonomic characters is mixed. Groups with few species have been studied in great detail, while in groups with large numbers of species much effort is still needed to find and determine the taxa. Classical systematics groups organisms on a phenetic basis, then sets up a classification; phylogeny is a secondary consideration. Phylogenetic systematics first determines organism relationships, then constructs a systematic classification that reflects the phylogeny. Molecular characters have made possible the establishment of a monophyletic and hopefully more permanent classification for the fungi. Thus, Volume VII of *The Mycota* contains both classical and phylogenetic classifications, reflecting the available data and the orientation of different authors. The incompleteness of some classifications, e.g., those for the Urediniomycetes (Swann, Frieders and McLaughlin, Chap. 2, Vol. VII, Part B) and Homobasidiomycetes (Hibbett and Thorn, Chap. 5, Vol. VII, Part B), demonstrates that we are in the early stages of a phylogenetic systematics for these groups.

The taxonomic outline used in *The Mycota*, Vol. VII, differs somewhat from that of other volumes in the series (Table 1), reflecting current mycological systematics. There is a lack of agreement on the naming of higher taxa, and the rules of nomenclature permit more than one name for these taxa. Cavalier-Smith (Chap. 1, Vol. VII, Part A) presents an alternative view to the taxonomic outline used for the remainder of the volume (Table 2). Some of the nomenclatural problems stem from a lack of resolution of deep branches in molecular evolutionary trees, a problem that appears likely to be resolved only with additional data from multiple genes and the addition of missing taxa to the analysis. Problems also arise from a difference of opinion among authors. The term *fungi* has assumed an ecological meaning for all organisms with a similar nutritional mode, and, therefore, Eumycota, rather than Fungi, is less confusing for the members of the phylum that encompasses a monophyletic group of these organisms. Pseudofungi (Cavalier-Smith, Chap. 1, Vol. VII, Part A) implies that organisms that lie outside the Eumycota but possess the fungal lifestyle are not fungi, but

Table 1. Taxonomic outline at the kingdom, phylum, and class levels as used in other volumes in the series and in this Volume. The classification in this volume is necessarily confusing at this time because authors are using their own classifications, rather than an imposed classification

Mycota, Vol. I	Mycota, Vol. VII
PSEUDOMYCOTA	PSEUDOMYCOTA[a,b]
Oomycota	Oomycota[c]
	Peronosporomycetes
Hyphochytriomycota	Hyphochytriomycota
	Hyphochytriomycetes
	Plasmodiophoromycota
	Plasmodiophoromycetes
EUMYCOTA	EUMYCOTA
Chytridiomycota	Chytridiomycota[d]
	Chytridiomycetes
Zygomycota	Zygomycota[d]
	Zygomycetes
	Trichomycetes
Dikaryomycota	
Ascomycotina	Ascomycota[e]
Saccharomycetes	Saccharomycetes
Ascomycetes	Plectomycetes
	Hymenoascomycetes[a]
	Loculoascomycetes[a]
Basidiomycotina	Basidiomycota
Heterobasidiomycetes	Urediniomycetes
	Ustilaginomycetes
	Heterobasidiomycetes[a,f]
Homobasidiomycetes	Homobasidiomycetes[a,f]

[a] Artificial taxon.
[b] For a natural classification for Oomycota and Hyphochytriomycota, kingdom Stramenopila (Stramenipila, Dick, Chap. 2, Vol. VII, Part A) or Chromista have been proposed, and for Plasmodiophoromycota, kingdom Protozoa (see Cavalier-Smith, Chap. 1, Vol. VII, Part A).
[c] Or Heterokonta (see Cavalier-Smith, Chap. 1, and Dick, Chap. 2, Vol. VII, Part A).
[d] Probably paraphyletic (see Cavalier-Smith, Chap. 1, Vol. VII, Part A, and Berbee and Taylor, Chap. 10, Vol. VII, Part B).
[e] A phylogenetic classification for Ascomycota is not available. Current thinking among ascomycete scholars is that three classes should be recognized, as follows: "Archiascomycetes", which may not be monophyletic, Hemiascomycetes (see Kurtzman and Sugiyama, Chap. 9, Vol. VII, Part A), and a filamentous group, Euascomycetes, that eventually will be subdividable, perhaps at the subclass level [M.E. Berbee and J.W. Taylor, 1995, Can J Bot 73 (Suppl. 1):S677, and Chap. 10, Vol. VII, Part B; J.W. Spatafora, 1995, Can J Bot 73 (Suppl. 1):S811]. Saccharomycetes as used here (see Barr, Chap. 8, Vol. VII, Part A) includes "Archiascomycetes" and Hemiascomycetes. See the relevant chapters for further speculation on the ultimate disposition of these groups.
[f] Heterobasidiomycetes as used in Vol. VIIB cannot be separated from Homobasidiomycetes. Hymenomycetes [E.C. Swann and J.W. Taylor, 1995, Can J Bot 73 (Suppl. 1):S862] has been proposed as a class for these groups (see Berbee and Taylor, Chap. 10, Vol. VII, Part B).

Table 2. Taxonomic outline at the kingdom, phylum, and class levels as used in the rest of this volume compared with that of Cavalier-Smith, Chap. 1, Vol. VII, Part A

Mycota, Vol. VII	Chapter 1, Vol. VII, Part A
PSEUDOMYCOTA[a]	CHROMISTA
Oomycota	Bigyra
Peronosporomycetes	Oomycetes
Hyphochytriomycota	
Hyphochytriomycetes	Hyphochytrea
	PROTOZOA
Plasmodiophoromycota	Cercozoa
Plasmodiophoromycetes	Phytomyxea
EUMYCOTA	FUNGI
Chytridiomycota	Archemycota
Chytridiomycetes	Chytridiomycetes
	Allomycetes
Zygomycota	
Zygomycetes	Zygomycetes
	Bolomycetes
	Glomomycetes[b]
Trichomycetes	Enteromycetes
	Zoomycetes[c]
	Microsporidia
	Minisporea
	Microsporea
Ascomycota	Ascomycota
Saccharomycetes	Taphrinomycetes
	Geomycetes
	Endomycetes
Plectomycetes	Plectomycetes
Hymenoascomycetes	Discomycetes
	Pyrenomycetes
Loculoascomycetes	Loculomycetes
Basidiomycota	Basidiomycota
Urediniomycetes	Septomycetes
Ustilaginomycetes	Ustomycetes
Heterobasidiomycetes	Gelimycetes[b]
Homobasidiomycetes	Homobasidiomycetes

[a] Artificial taxon.
[b] Probably paraphyletic.
[c] Includes Zygomycetes, Ascomycetes, and Trichomycetes.

in an ecological sense they are fungi. Pseudomycota is therefore used in this series for these fungal organisms that lie outside the Eumycota.

The Mycota, Vol. VII, includes treatments of the systematics and related topics of the Eumycota and Pseudomycota as well as specialized chapters on nomenclature, techniques, and evolution. Certain groups are not treated in this volume: the Labyrinthulomycetes (Pseudomycota) and the slime molds. The evolutionary position of the slime molds has been controversial. Recent evidence suggests that most slime molds are more closely related to the Eumycota than previously believed (S.L. Baldauf and W.F. Doolittle, 1997, Proc Natl Acad Sci USA 94:12007) and they should continue to be of interest to those who study fungi for both ecological and phylogenetic reasons.

Chapters 2 to 4, Vol. VII, Part A, cover the Pseudomycota and Chapters 5-14, Vol. VIIA, and Chapters 1-5, Vol. VII, Part B, the Eumycota. The Pseudomycota contains distantly related groups of fungi (Table 1). The Chytridiomycota and Zygomycota are treated in one and two chapters, respectively, while the Ascomycota and Basidiomycota are treated in five or six chapters each, with separate chapters for yeasts in each phylum, although the yeasts are not monophyletic groups. Chapter 14, Vol. VII, Part A, discusses the special problems of anamorphic genera and their relationships to the teleomorphic genera, and describes the attempts being made to incorporate anamorphs into modern phylogenetic systematics. In Chapter 6, Vol. VII, Part B, Hawksworth discusses the development of a unified system of biological nomenclature. Chapters 7 and 8, Vol. VII, Part B, deal with techniques for cultivation and data analysis, respectively. The final two chapters in Vol. VII, Part B, consider speciation and molecular evolution.

The Mycota, Vol. VII, was originally intended to have been Volume I in the series. Several changes in editors and the unfortunate death of Paul Lemke delayed its production. Added to these difficulties was the fact that these are tumultuous times in systematics because of the rapid development of molecular and phylogenetic analysis techniques and the explosive accumulation of data. As these techniques and new data are more broadly incorporated into systematics, a more stable and useful classification of the fungi will result.

We thank Heather J. Olson for her substantial efforts in compiling the indices.

St. Paul, Minnesota, USA
Minneapolis, Minnesota, USA
April 2000

DAVID J. MCLAUGHLIN
ESTHER G. MCLAUGHLIN
Volume Editors

Contents Part B

The Fungal Hierarchy

1 Basidiomycetous Yeasts
  J.W. Fell, T. Boekhout, A. Fonseca, and J.P. Sampaio
  (With 24 Figures) 1

2 Urediniomycetes
  E.C. Swann, E.M. Frieders, and D.J. McLaughlin (With 20 Figures) 37

3 Ustilaginomycetes
  R. Bauer, D. Begerow, F. Oberwinkler, M. Piepenbring,
  and M.L. Berbee (With 34 Figures) 57

4 Heterobasidiomycetes
  K. Wells and R.J. Bandoni (With 16 Figures) 85

5 Homobasidiomycetes
  D.S. Hibbett and R.G. Thorn (With 14 Figures) 121

Nomenclature and Documentation

6 The Naming of Fungi
  D.L. Hawksworth (With 4 Figures) 171

7 Cultivation and Preservation of Fungi in Culture
  S.-C. Jong and J.M. Birmingham 193

8 Computer-Assisted Taxonomy and Documentation
  O. Petrini and T.N. Sieber (With 2 Figures) 203

Evolution and Speciation

9 Speciation Phenomena
  P.A. Lemke (With 1 Figure) 219

10 Fungal Molecular Evolution: Gene Trees and Geologic Time
   M.L. Berbee and J.W. Taylor (With 5 Figures) 229

Subject Index 247

Biosystematic Index 251

Contents Part A

The Fungal Hierarchy

1 What Are Fungi?
  T. CAVALIER-SMITH (With 6 Figures)

2 The Peronosporomycetes
  M.W. DICK (With 25 Figures)

3 Hyphochytriomycota
  M.S. FULLER (With 1 Figure)

4 Plasmodiophoromycota
  J.P. BRASELTON (With 8 Figures)

5 Chytridiomycota
  D.J.S. BARR (With 11 Figures)

6 Zygomycota: Zygomycetes
  G.L. BENNY, R.A. HUMBER, and J.B. MORTON (With 37 Figures)

7 Zygomycota: Trichomycetes
  G.L. BENNY (With 8 Figures)

8 Ascomycota
  M.E. BARR (With 2 Figures)

9 Ascomycetous Yeasts and Yeastlike Taxa
  C.P. KURTZMAN and J. SUGIYAMA (With 10 Figures)

10 The Monophyletic Plectomycetes: Ascosphaeriales, Onygenales, Eurotiales
   D.M. GEISER and K.F. LOBUGLIO (With 19 Figures)

11 Pyrenomycetes – Fungi with Perithecia
   G.J. SAMUELS and M. BLACKWELL (With 9 Figures)

12 Discomycetes
   D.H. PFISTER and J.W. KIMBROUGH (With 3 Figures)

13 Loculoascomycetes
   M.E. BARR and S.M. HUHNDORF (With 63 Figures)

14 The Taxonomy of Anamorphic Fungi
   K.A. SEIFERT and W. GAMS (With 15 Figures)

Subject Index

Biosystematic Index

List of Contributors

BANDONI, R.J., Department of Botany, University of British Columbia, Vancouver, British Columbia V6T 2B1, Canada

BAUER, R., Universität Tübingen, Lehrstuhl Spezielle Botanik und Mykologie, Auf der Morgenstelle 1, 72076 Tübingen, Germany

BEGEROW, D., Universität Tübingen, Lehrstuhl Spezielle Botanik und Mykologie, Auf der Morgenstelle 1, 72076 Tübingen, Germany

BERBEE, M.L., Department of Botany, University of British Columbia, 6270 University Boulevard, Vancouver, BC V6T 1Z4, Canada

BIRMINGHAM, J.M., American Type Culture Collection, 10801 University Boulevard, Manassas, Virginia 20110-2209, USA

BOEKHOUT, T., Yeast Division, Centraalbureau voor Schimmelcultures, Julianalaan 67a, 2628 BC Delft, The Netherlands

FELL, J.W., Marine Biology and Fisheries, Rosenstiel School of Marine and Atmospheric Science, 4600 Rickenbacker Causeway, Key Biscayne, Florida, USA

FONSECA, A., Centro de Recursos Microbiológicos, Secção Autónoma de Biotechnologia, Faculdade de Ciências e Tecnologica, Universidade Nova De Lisboa, Portugal

FRIEDERS, E.M., Department of Biology, University of Wisconsin-Platteville, Platteville, Wisconsin 53818, USA

HAWKSWORTH, D.L., MycoNova, 114 Finchley Lane, Hendon, London NW4 1DG, UK

HIBBETT, D.S., Harvard University Herbaria, 22 Divinity Avenue, Cambridge, Massachusetts 02138, USA (Present address: Department of Biology, Clark University, 950 Main Street, Worcester, Massachusetts 01610, USA)

JONG, S.-C., American Type Culture Collection, 10801 University Boulevard, Manassas, Virginia 20110-2209, USA

LEMKE †, P.A., Auburn, USA

MCLAUGHLIN, D.J., Department of Plant Biology, University of Minnesota, 1445 Gortner Ave., St. Paul, Minnesota 55108-1095, USA

OBERWINKLER, F., Universität Tübingen, Lehrstuhl Spezielle Botanik und Mykologie, Auf der Morgenstelle 1, 72076 Tübingen, Germany

PETRINI, O., Pharmaton SA, Via ai Mulini, 6934 Bioggio, Switzerland

PIEPENBRING, M., Universität Tübingen, Lehrstuhl Spezielle Botanik und Mykologie, Auf der Morgenstelle 1, 72076 Tübingen, Germany

SAMPAIO, J.P., Centro de Recursos Microbiológicos, Secção Autónoma de Biotechnologia, Faculdade de Ciênçias e Tecnologica, Universidade Nova de Lisboa, Portugal

SIEBER, T.N., Forest Protection and Dendrology, Swiss Technical Institute of Technology, ETH-Zentrum, 8092 Zürich, Switzerland

SWANN, E.C., Department of Plant Biology, University of Minnesota, 1445 Gortner Ave., St. Paul, Minnesota 55108-1095, USA

TAYLOR, J.W., Department of Plant and Microbial Biology, University of California, Berkeley, California 94720-3102, USA

THORN, R.G., Department of Botany, University of Wyoming, Laramie, Wyoming 82071, USA

WELLS, K., Section of Plant Biology, University of California, Davis, California 95616, USA (Present address: 601 Indian Camp Creek Road, Hot Springs, North Carolina 28743, USA)

The Fungal Hierarchy

1 Basidiomycetous Yeasts

J.W. Fell[1], T. Boekhout[2], A. Fonseca[3], and J.P. Sampaio[3]

CONTENTS

I.	Introduction	3
II.	Importance	3
III.	Methods for Collection, Isolation, and Maintenance	5
IV.	Phenotypic Characterization	6
A.	Vegetative Reproduction	6
B.	Generative Reproduction	9
C.	Ultrastructure	13
D.	Physiological Properties	16
E.	Other Characters	17
V.	Molecular Techniques	17
A.	DNA Typing Methods	17
B.	Molecular Sequence Analysis	18
C.	D1/D2 LrDNA Phylogenetic Analyses	18
	1. Urediniomycetes	18
	2. Hymenomycetes	25
	3. Ustilaginomycetes	27
D.	Species and Strain Distinctions Based on D1/D2, ITS, and IGS Regions	27
VI.	Conclusions	28
	References	28

I. Introduction

Yeast phases were first observed among the smuts and jelly fungi by Brefeld (1881, 1888, 1895a,b) and Möller (1895). Subsequently, the resemblance of the *Sporobolomyces* ballistoconidia to actively discharged basidiospores led Kluyver and van Niel (1924, 1927) to suggest that *Sporobolomyces* is a basidiomycete. The most conclusive evidence of a basidiomycete connection was the discovery of mating and a sexual state in strains of *Rhodotorula glutinis* by Banno (1963, 1967), followed by discoveries of sexual states in numerous other species that resulted in the descriptions of genera such as *Rhodosporidium*, *Leucosporidium*, *Sporidiobolus*, *Filobasidium*, *Filobasidiella*, *Cystofilobasidium*, and *Bulleromyces*.

A dominant feature of basidiomycetous yeasts is a growth phase consisting of round, oval, and elongate cells that reproduce by enteroblastic budding (blastoconidia), fission (arthroconidia), and/or forceful ejection (ballistoconidia). These yeasts are not restricted to one group of basidiomycetes, i.e., class or order; rather, they are polyphyletic; there are approximately 220 recognized species in 34 genera distributed among the Urediniomycetes, Ustilaginomycetes, and Hymenomycetes. Consequently, life cycles and ultrastructure among these yeasts are not uniform; rather, they reflect phylogenetic diversity.

In this chapter we will discuss aspects of the biology, ecology, and systematics of taxa traditionally studied by yeast biologists. These discussions will include comparisons with monomorphic and dimorphic parasitic taxa. However, three points should be considered: most species are known only from their anamorphic state; possibly only 1–5% of the species have been discovered; and our understanding of their phylogeny is emerging as a result of recent methods and studies in molecular biology and ultrastructure.

II. Importance

With the increased incidence of immunocompromised patients and the use of antibacterial antibiotics, basidiomycetous yeasts are increasing in importance as a medical problem (Ahearn 1998). The major pathogenic species is *Filobasidiella neoformans*, which has two varieties (Mitchell and Perfect 1995). Both taxa infect the lungs and produce a pneumococcal-type pneumonia. In addition, the central nervous system may be infected to produce a fatal meningoencephalitis. *Malassezia* spp. are lipophilic yeasts that are usually associated with skin and can cause fatal

---

[1] Rosenstiel School of Marine and Atmospheric Science, University of Miami, Key Biscayne, Florida, USA
[2] Yeast Division, Centraalbureau voor Schimmelcultures, Delft, The Netherlands
[3] S.A. Biotecnologia, Faculdade de Ciencias e Tecnologia, Universidade Nova de Lisboa, Portugal

pulmonary infection of neonates, receiving lipid supplementation via broviac catheters (Ahearn 1998). Other medically important yeasts include *Trichosporon beigelii* and various species of *Rhodotorula* and *Cryptococcus* that act as opportunistic pathogens.

Many yeasts are associated with living plants. Examples include *Sporobolomyces* spp., which are found on leaves in temperate regions, and *Phaffia*, that is isolated from tree exudates. Although the majority of the heterobasidiomycetes are plant pathogens (Bandoni 1995), the direct involvement of yeasts in plant diseases is rarely documented. B. Steffenson (unpubl. observ.) reported infections of barley and oat seeds by *Rhodotorula* species in agricultural fields in North Dakota, USA; *Itersonilia perplexans* is a pathogen on chrysanthemums and parsnip (Channon 1963; Sowell 1953); and *Tilletiopsis* spp. (T. Boekhout, unpubl. observ.) can colonize the surface of postharvest apples stored at low temperatures. In contrast to potential plant pathogenicity, yeasts can play a role in biocontrol of plant diseases. *Cryptococcus laurentii, Cryptococcus humicola, Filobasidium floriforme, Sporobolomyces roseus*, and *Rhodosporidium toruloides* can reduce or prevent necrosis or discoloration of apples caused by the gray mold *Botrytis cinerea* (Roberts 1990; Filonow et al. 1996). In addition, cucumber powdery mildew (*Sphaerotheca fuliginea*) and rose powdery mildew (*S. pannosa* var. *rosae*) are antagonized by *Pseudozyma flocculosa* and *P. rugulosa* (cited as *Stephanoascus* and *Sporothrix*, respectively) (Hajlaoui et al. 1992; Jarvis et al. 1989). Growth of cucumber powdery mildew is reduced by *Tilletiopsis minor* (Hijwegen 1986) and *T. pallescens* (Hoch and Provvidenti 1979; Hijwegen 1988; Urquehart et al. 1994), which is reported to antagonize *Erisyphe graminis* f.sp. *hordei* (Klecan et al. 1990). Other yeast species may be mycoparasitic, as suggested by the presence of haustoria and colacosomes.

Basidiomycetous yeasts are important from several agro-industrial aspects (Boekhout and Fell 1995; Demain et al. 1998). *Phaffia rhodozyma* (teleomorph *Xanthophyllomyces dendrorhous*) is a biological source for astaxanthin, an economically important pigment used in aquaculture. There is a growing market for astaxanthin as marine fish farms account for 10 to 15% of the seafood business (Johnson and Schroeder 1995). Several yeasts are exploited for their enzymes and enzymatic activities. *Cryptococcus albidus* produces xylanases (Biely et al. 1981), *C. cellulolyticus* produces cellulases (Nakase et al. 1996), and amylolytic activity has been demonstrated in *Filobasidium capsuligenum, Cryptococcus curvatus, Pseudozyma (Cryptococcus) tsukubaensis*, and *Trichosporon pullulans* (De Mot et al. 1984; De Mot and Verachtert 1985). Lipid accumulation occurs in *Cryptococcus laurentii, C. curvatus, Rhodotorula glutinis, R. gracilis, R. graminis, R. mucilaginosa, Trichosporon cutaneum, T. pullulans*, and a strain (*Candida* 107) closely related with or identical to *Pseudozyma antarctica* (Ratledge 1978, 1982, 1986; Ratledge and Evans 1989; Rolph et al. 1989). *Candida* 107 accumulates up to 41% of its dry weight as fatty acids (Gill et al. 1977) and *T. pullulans* accumulates more than 65% of its biomass as lipid (Reiser et al. 1996). Mutants of *C. curvatus* are able to produce cocoa butter equivalents (Ykema et al. 1988, 1989). Species of *Pseudozyma* produce mannosylerythritol lipids (Kitamoto et al. 1990a,b) and beta-lipase for the synthesis of glucoside esters exhibiting surfactant properties (Björkling et al. 1991); they also produce organic acids such as itaconic acid (J.P. van Dijken and E. de Hulster, unpubl. observ.), which may have applied uses in polymerization reactions (Mattey 1992).

Phenol is degraded by *Cryptococcus elinovii, Rhodotorula glutinis, R. rubra*, and *Trichosporon pullulans* (Neujahr and Varga 1970; Nei 1971a,b; Shoda and Udaka 1980; Zache and Rehm 1989; Katayama-Hirayama et al. 1991). Aromatic compounds are assimilated by *Cryptococcus diffluens, C. humicolus, C. laurentii, C. terreus, Rhodotorula glutinis, R. gracilis, R. graminis, R. mucilaginosa, Trichosporon cutaneum, T. dulcitum*, and *T. moniliiforme* (Mills et al. 1971; Middelhoven et al. 1992). *Trichosporon cutaneum* assimilates a wide variety of carbon sources, making this species a potential candidate for the efficient conversion of various carbon sources into biomass. Moreover, a transformation system and bioreactor technology for efficient cultivation have been developed for this species (Reiser et al. 1996).

Many studies deal with the distribution and occurrences of yeasts in nature, yet we know little regarding their specific ecological roles. A basic problem that hampers ecological studies is the lack of in-depth knowledge of their systematics. For example, species such as *Rhodotorula glutinis, Cryptococcus albidus*, and *C. laurentii* are reported from a diversity of substrates, yet we know, based on % G + C content and molecular sequence analysis, that strains within these species are genetically diverse. Because these specific genotypes have not been taxonomically characterized, their occurrences and activities have not been addressed.

III. Methods for Collection, Isolation, and Maintenance

Most basidiomycetous yeasts are saprotrophs associated with living and dead organic material. However, these populations are not evenly distributed in nature. Cells on a specific substrate occur in low numbers, often in a low metabolic state, and then act opportunistically, growing and reproducing rapidly with the introduction of organic particulates or soluble compounds; consequently, populations become quite dense. Concentrations of yeasts (both ascomycetous and basidiomycetous) in seawater vary from less than 10 cells l^{-1} in low nutrient open ocean water (Fell 1976) to numbers as high as 17 000 cells l^{-1} in high nutrient inshore mangrove regions (J.W. Fell and A. Statzell-Tallman, unpubl. observ.). Similarly, yeasts in soils vary from a few cells to tens of thousand of cells per gram; numbers as high as 250 000 cells g^{-1} have been recorded in soils under fruit trees and berry bushes (Phaff et al. 1978). Some of the organisms that occur in a habitat may be introduced species, for example, terrestrial runoff into seawater and death of a plant or dropping of a fruit with associated yeasts into soils. The introduced species may or may not survive in the new habitat. The occurrence of individual species is often dependent on specific physiological capabilities, such as the ability to utilize certain compounds or to survive and grow at low or high temperatures. Collection and isolation strategies should be designed to reflect these physiological attributes and environmental conditions. For example, *Rhodosporidium lusitaniae* was isolated from woodland soils using a medium containing lignin-related phenolic compounds as the sole carbon source (Fonseca and Sampaio 1992); similarly, psychrophiles, such as *Leucosporidium* and *Mrakia*, require temperatures of 12 °C or less. Basidiomycetous yeasts with ballistoconidia can be isolated with techniques designed to collect falling or ejected spores. Two points to consider are that all isolation media are selective and that rapidly growing ubiquitous species may dominate on artificial media and mask the occurrence of slow-growing habitat-specific species.

Soil and sediment samples can be manually collected from terrestrial and shallow-water environments using sterile jars, vials, plastic bags, and coring devices. Deep-water sediments can be remotely collected using grabs, gravity and piston cores, or submersibles. Water, plant, and animal samples can be collected using similar devices whose goal is to maintain sterility and appropriate temperature conditions during transportation to the laboratory. Isolation procedures should be undertaken as quickly as possible to avoid changes in yeast community structure due to death or proliferation of individual species. Particulate material can be placed directly on agar medium or diluted by placing the material in a test tube with a given volume of sterile water, vortexed, and a dilution series prepared. Standard spread plates can be made from each of the dilution series. Water may be filtered through nitrocellulose filters, 0.45-μm pore size, using a vacuum filter apparatus and the filter placed face-up on solid nutrient agar. Culture plates are incubated, colonies enumerated and periodically isolated at periods of time dependent on incubation conditions. The culture medium should be tailored to the research question. Several general media have been designed; we favor a medium containing (w/v) 2% glucose, 1% peptone, 0.5% yeast extract, and 2% agar. Antibiotics can be added to the medium; chloramphenicol (0.02%) has the advantage that it can be added prior to autoclaving. An alternative is a mixture of penicillin G and streptomycin sulfate, added dry at the rate of 150–500 mg l^{-1} of autoclaved, cooled (approximately 45 °C) medium. Details of these methods can be found in Yarrow (1998) and Fell and Kurtzman (1996).

Cultures can be maintained on potato dextrose agar slants at 5 °C. Because cultures are often difficult to maintain, transfers must be made frequently; in some cases, monthly. Some *Malassezia* species require transfer every 2 weeks, a complex medium and maintenance at 30 °C. Long-term maintenance of yeasts requires low temperature under liquid nitrogen at −75 °C (for details on this technique, see Yarrow 1998). Type and other representative cultures, particularly important industrial, patent, and type cultures should be deposited in a permanent culture collection. For a list of these collections see Kurtzman and Sugiyama (Chap. 9, Vol. VII, Part A).

IV. Phenotypic Characterization

Genera of basidiomycetous yeasts are traditionally characterized by morphological and physiological parameters (Tables 1, 2). However, molecular data indicate that genera such as *Cryptococcus, Bensingtonia, Sporobolomyces,* and *Rhodotorula* are polyphyletic (Fell et al. 1992, 1995, 2000; Nakase et al. 1993). Due to their artificial delineation, anamorphic (mitosporic) genera are often larger and more heterogenous than teleomorphic (meiosporic) genera. However, it seems appropriate to accept traditional generic concepts as working hypotheses until substantial phenotypic and molecular data provide evidence for changes in generic boundaries.

Characteristic morphological features for basidiomycetous yeasts include the presence of ballistoconidia, dikaryotic hyphae, clamp connections, haustorial branches, teliospores, and basidia. In addition, anamorphic basidiomycetous yeasts can be differentiated from ascomycetous yeasts by features such as enteroblastic budding, staining with Diazonium Blue B salt, presence of pyrophosphatidic acid, urease activity, cell wall composition, lamellate cell walls, and septal pore morphology (Kreger-van Rij and Veenhuis 1971a; van der Walt and Hopsu-Havu 1976; von Arx and Weijman 1979; Weijman 1979; Khan and Kimbrough 1980; Hagler and Ahearn 1981; Booth and Vishniac 1987; Moore 1987; Weijman and Golubev 1987; Goto et al. 1988; Prillinger et al. 1993; Roeijmans et al. 1997). Eighty-eight percent of the basidiomycetous yeasts have mol% G + C values above 50, in contrast with only about 8% of the ascomycetous yeasts (Boekhout and Kurtzman 1996). These criteria make it possible to assign anamorphic yeasts to either the Basidiomycota or the Ascomycota.

A. Vegetative Reproduction

Budding (blastoconidiogenesis) in basidiomycetous yeasts is enteroblastic as opposed to holoblastic in most ascomycetous yeasts (von Arx and Weijman 1979). During enteroblastic budding the growing bud emerges through the mother cell wall to leave a scar. In most basidiomycetous yeasts budding is limited to a restricted area at the poles of the cells, and frequently occurs on short denticles. This mode of budding, which is referred to as polar budding, is different from bipolar budding that occurs in ascomycetous genera such as *Hanseniaspora*; specifically, the basidiomycetous yeast buds are formed on a narrow base, but the cells do not become lemon-shaped. Monopolar budding occurs in *Malassezia*, where the bud is formed on a broad base (comparable to bud fission in some bipolar ascomycetes), which gives the parent cell and adhering bud a flask-shaped appearance. Sympodial budding occurs in *Bensingtonia ingoldii, B. intermedia,* some *Malassezia* spp., and *Sympodiomycopsis* (von Arx 1979; Simmons and Guého 1990; Boekhout 1991a; Guého et al. 1996). Percurrent enteroblastic budding may result in annellations, eventually forming a collarette. These annellations, found, for example, in *Udeniomyces* (Boekhout 1991a), are best observed by electron microscopy. Ballistoconidia are a type of blastoconidia that are actively discharged by a mechanism involving Buller's drop (Kluyver and van Niel 1924; Nakase 1989; Boekhout 1991a; Boekhout et al. 1993). We prefer the use of the term ballistoconidia for these mitotically produced cells, contrary to ballistospores, which are meiotically formed. Ballistoconida may be bilaterally symmetrical, e.g., in the genera *Kockovaella, Sporidiobolus, Sporobolomyces,* and *Udeniomyces,* or rotationally symmetrical in *Bullera* and the teleomorph *Bulleromyces*. The presence or absence of ballistoconidia has been used to differentiate *Sporobolomyces* from *Rhodotorula,* and *Bullera* from *Cryptococcus*. However, molecular data suggest that ballistoconidium-forming yeasts are closely related to nonballistoconidium-forming yeasts (Fell et al. 1992, 1995, 2000; Nakase et al. 1995; Suh and Nakase 1995; Swann and Taylor 1995b). In addition, the formation of ballistoconidia is an easily lost character, which may complicate identifications.

Some basidiomycetous yeasts form blastoconidia on elongated stalks with septa at the distal end (*Fellomyces, Kockovaella,* and *Kurtzmanomyces*), in the mid-region (*Sterigmatomyces, Tsuchiyaea*), or at varying positions (*Ballistosporomyces,* a synonym of *Sporobolomyces*) (Yamada and Banno 1984; Yamada et al. 1988a,b, 1989b; Nakase et al. 1991b). Other methods of conidiogenesis occur: in *Reniforma* budding is at the apex of allantoidal cells and the bud is allantoidal at initial formation (Pore and Sorenson 1990); in *Trichosporon* (Guého et al. 1992) conidiogenesis is arthric (arthroconidia). In addition, endospores form in long-standing cultures of

Table 1. Salient characteristics of teleomorphic basidiomycetous yeasts

Genus	rDNA cluster[a]	Teliospores	Basidia[b] Holo	Basidia[b] Phragmo	Pore	Parenthesome	Xylose	CoQ[c]	Fermentation	D-Glucuronate	Inositol	Starch
Bulleromyces	7	−	−	+	Doli[d]	Cups	+	10	−	+	+	+
Chionosphaera	3	−	+	−	Simple	−	−	10	−	−	−	?
Cystofilobasidium	10	+	+	−	Doli	−	+	8	−(+)	+	+	+
Erythrobasidium	5	−	+	−	Simple	−	−	10 (H_2)	−	+	−	−
Filobasidiella	7	−	+	−	Doli	Cups or −	+	10	−	+(−)	+(−)	+
Filobasidium	9	−	+	−	Doli	Cups or −	+	9,10	+(−)	−	−	−
Kondoa	3	+	−	+	Simple	−	−	9	−	−	−	−
Leucosporidium	1	+	−(+)[e]	+(−)[f]	Simple	−	−	9,10	−	+(−)	−	+
Mrakia	10	+	+	+	Doli	−	+	8	+(−)	+	+(−)	−
Rhodosporidium	2	+	−	+	Simple	−	−	9,10	−	−	−	−
Sakaguchia	5	+	−(+)	+(−)	Simple	−	−	10	−	+	−	−
Sporidiobolus	2	+	?	?	Simple	−	−	10	−	−	−	−
Sterigmatosporidium	7	?[g]	?	?	?	?	+	10	−	+	+	+
Xanthophyllomyces	10	−	+	−	?	?	+	10	+	+	−	+

[a] rDNA cluster – see Fig. 19.
[b] Holo = holometabasidium; phragmo = phragmometabasidium.
[c] CoQ = coenzyme Q.
[d] Doli = dolipore.
[e] −(+) = negative, sometimes positive.
[f] +(−) = positive, sometimes negative.
[g] ? = not examined.

Table 2. Salient characteristics of anamorphic basidiomycetous yeasts

Genus	Morphology			rDNA cluster[a]	Pore	Parenthesome	Xylose	CoQ[b]	Fermentation	D-Glucuronate	Inositol	Starch
	Arthro conidia	Ballisto conidia	Stalked conidia									
Apiotrichum	±	–	–	8	Doli[c,g]	–[g]	+	?[f]	–	+	+	–
Bensingtonia	–	+	–	1, 3	Simple	–	–	9	–	+(–)	–	–
Bullera	–	+	–(+)[d]	7	Doli	Cups	+	10	–	+(–)[e]	+(–)	+(–)
Cryptococcus	–	–	–	1, 7–10	Doli[b]	–[h]	+	8–10	–	+	+(–)	+(–)
Fellomyces	–	–	+	7	?	?	+	10	–	+	+	+
Itersonilia	–	+	–	10	Doli	–	+	9	–	+	–	–
Kockovaella	–	–	+	7	?	?	+	10	–	+	+	–
Kurtzmanomyces	–	–	+	3	?	?	–	10	–	?	?	?
Malassezia	–	–	–	6	*[i]	–	–	9	–	?	–	–
Phaffia	–	–	–	10	?	?	+	10	+	+	+	+
Pseudozyma	–	–	–(+)	6	*	?	–	10	–	–	–	–
Reniforma	–	–	–	2	?	?	–	7[f]	–	–	–(+)	–
Rhodotorula	–	–	–	1–3, 5–6	?	?	–	9, 10, 10 (H$_2$)	–	v	–	–
Sporobolomyces	–	+	–(+)	1–3, 5	?	?	–	10, 10 (H$_2$)	–	+(–)	–	+(–)
Sterigmatomyces	–	–	+	3	?	?	–	9	–	+(–)	–	–
Sympodiomycopsis	–	–	–	6	Simple	–	Trace	10	–	?	+	–
Tilletiopsis	–	+	–	6	*	–	–	10	–	+(–)	+(–)	–
Trichosporon	+	–	–	8, 10	Doli, *	Cups or Tubular	–	9, 10	–	+(–)	+(–)	+(–)
Tsuchiyaea	–	–	+	7	?	?	+	9	–	+	+	+
Udeniomyces	–	+	–	10	?	?	+	10	–	+	+(–)	+(–)

[a] rDNA cluster – see Fig. 19.
[b] CoQ = coenzyme Q.
[c] Doli = Dolipore.
[d] –(+) = negative, sometimes positive.
[e] +(–) = positive, sometimes negative.
[f] ? = not examined.
[g] T. Boekhout, unpublished.
[h] *Cryptococcus laurentii*.
[i] H.J. Roeijmans, unpublished.
[j] Not well-defined "micropore"-like structure.

Filobasidiella, Trichosporon, Cystofilobasidium, and *Leucosporidium.*

Different modes of conidiogenesis have been used to separate genera; however, conidiogenesis may be more variable than previously understood (Nakase et al. 1989, 1991a,b; Boekhout 1991a). In particular, recent data on the molecular phylogeny demonstrated that taxa with and without ballistoconidia or conidia with or without elongated stalks occur in the same cluster (Fell et al. 1992, 1995; Nakase et al. 1995). In contrast, the arthroconidia-forming genus *Trichosporon* (with the exception of *T. pullulans*) is monophyletic as demonstrated by 25S rDNA/rRNA sequence analysis (Guého et al. 1992, 1993). In conclusion, vegetative morphology alone has limited value for separating genera.

B. Generative Reproduction

Many basidiomycetous yeasts have dimorphic life cycles in which haploid monokaryotic yeast phases alternate with dikaryotic hyphal phases (Banno 1963, 1967; Olive 1968; Fell 1969, 1970, 1974; Fell et al. 1969, 1973; Rodrigues de Miranda 1972; Kwon-Chung 1975, 1976a, 1977; Fell and Tallman 1980, 1981, 1982; Boekhout et al. 1991). Similary, yeast stages are known from a variety of heterobasidiomycetes, particularly the Urediniomycetes, in genera such as *Agaricostilbum, Cystobasidium, Colacogloea, Kriegeria, Mycogloea, Microbotryum, Naohidea, Occultifur, Septobasidium,* and *Sphacelotheca*; in some orders of the Ustilaginomycetes, such as Ustilaginales (*Ustilago*), Microstromatales (*Microstroma*), Doassansiales (*Rhamphospora*), and in the Hymenomycetes: particularly the Tremellales (*Fibulobasidium, Holtermannia, Sirobasidium, Sirotrema, Tremella, Trimorphomyces, Xenolachne*) (Bandoni 1995; Bauer et al. Chap. 3, this Vol.; Oberwinkler 1990; Boekhout et al. 1998). In most case, yeast stages are recovered by the ballistospore-fall method from basidiocarps or substrates where the fungi are fruiting. However, this technique is subject to contamination by associated saprobic yeasts, which can lead to false interpretations of the production of yeast phases by these filamentous heterobasidiomycetes.

Both heterothallic and homothallic systems of life cycles occur among basidiomycetous yeasts. In heterothallic mating systems, compatible mating types conjugate and develop a dikaryotic mycelium, usually with clamp connections. Following karyogamy, meiosis takes place in the basidium, which germinates to basidiospores that give rise to a haploid yeast phase. The heterothallic incompatibility systems are regulated by mating-type genes, and may be biallelic unifactorial (bipolar: *Kondoa malvinella, Leucosporidium antarcticum, Rhodosporidium diobovatum, R. paludigenum, R. toruloides, R. sphaerocarpum, Sporidiobolus salmonicolor*), multiple allelic unifactorial (*Cystofilobasidium infirmominiatum,* Fig. 1), bifactorial (tetrapolar: *Leucosporidium scottii, Sakaguchia dacryoidea*), or modified bifactorial, which favor outbreeding (Bandoni 1963; Fell 1974; Fell and Tallman 1980, 1981; Wong 1987). Mating in basidiomycetous yeasts, as found in the ascomycetes, is subject to pheromonal control (Flegel 1981). Molecular studies on the regulation of the mating process in the bipolar species *Filobasidiella neoformans* showed the MATalpha locus is at least 60kb in size and composed of at least two genes (Edman 1996).

The homothallic life cycle is distinguished by dikaryotic hyphal cells, with complete clamp connections, that develop from diploid cells in the absence of conjugation (Nyland 1948, 1949; Laffin and Cutter 1959). Examples of species with homothallic strains are *Leucosporidium fellii, Rhodosporidium lusitaniae, R. toruloides,* and *Sporidiobolus johnsonii*. Absence of meiosis may result in diploid basidiospores, as observed in *Sporidiobolus johnsonii* and *S. ruineniae* (Laffin and Cutter 1959). In homothallic strains of *Rhodosporidium toruloides*, the diploid cells appear to originate from a failure during meiosis and the progeny are either homothallic or heterothallic (Fell and Tallman 1984b). Monokaryotic, haploid, or diploid fruiting (self-sporulation) occurs in *Cystofilobasidium capitatum, Leucosporidium scottii, Mrakia frigidia, Rhodosporidium fluviale, R. kratochvilovae,* and *R. sphaerocarpum* (Fell and Tallman 1984a,b, 1997b,c; Kwon-Chung 1987; Hamamoto et al. 1988). Basidia develop in the absence of mating; the hyphae remain uninucleate and complete clamp connections are absent, although incomplete clamp connections may be present.

The genetic background of life cycles is not clear, as ploidy analyses are available only for a small number of species, although we do know that haploid fruiting of *Filobasidiella neoformans* is linked with the alpha-mating type (Wickes et al. 1996). Some self-sporulating strains may restore the heterothallic cycle by mating with a compati-

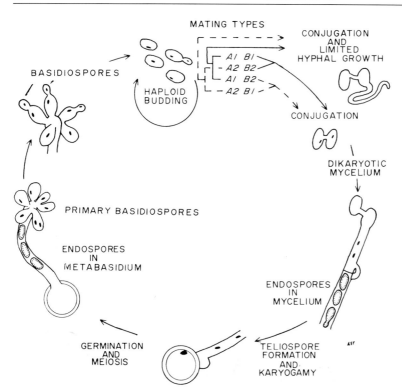

Fig. 1. Life cycle of *Cystofilobasidium infirmominiatum*. (Fell and Tallman 1984b)

ble strain (Schmeding et al. 1981), which suggests the presence of haploid nuclei; alternatively, haploid-diploid mating may take place. In a study of *F. neoformans*, the majority of heterothallic strains and half of asexual strains were haploid, whereas all self-sporulating and the remaining asexual strains were diploid (Takeo et al. 1993), which indicates the presence of a considerable genetic plasticity. These examples indicate that life cycles of basidiomycetous yeasts are complex and poorly understood from a genetic point of view. In principle, a single species may follow different sexual strategies, with different genetic benefits.

Thick-walled probasidia, usually referred to as teliospores, occur in the urediniomycetous genera *Leucosporidium, Mastigobasidium Rhodosporidium, Sakaguchia* and *Sporidiobolus*, in the hymenomycetous yeasts *Cystofilobasidium* and *Mrakia*, and the ustilaginaceous yeastlike fungus *Tilletiaria*. These teliospores may be globose to somewhat angular, solitary or in clusters, intercalary or terminal, and smooth or spiny. The teliospores resemble chlamydospores; however, teliospores differ, as they are the site of karyogamy. Germination occurs by basidia.

Different types of basidia occur: phragmobasidia may be transversely, obliquely to longitudinally septate; nonseptate holobasidia are clavate to cylindrical-capitate. In addition, the basidia may be stalked as in *Sporidiobolus ruineniae* and *S. salmonicolor* (Fell and Tallman 1980). Most teliospore-producing genera have transversely septate basidia (Figs. 2, 3, 4), e.g., *Leucosporidium, Rhodosporidium, S. ruineniae*, although unicellular basidia occur in *Cystofilobasidium* (Fig. 5), *S. johnsonii*, and *S. salmonicolor*. Nonteliospore-forming genera usually have holobasidia, e.g., *Erythrobasidium, Filobasidiella, Filobasidium* (Figs. 6, 7), and *Xanthophyllomyces*; however, longitudinally or obliquely septate basidia (Fig. 8) of the *Tremella* type occur in *Bulleromyces* (Boekhout et al. 1991). The species *Kondoa aeria* and *Occultifur externus* (Fig. 21 *Agaricostilbum* and *Erythrobasidium* clades) produce auricularioid (transversely septate) basidia with ballistospores (Fig. 9), which reproduce by budding or repetition (Fonseca et al. 1999; Sampaio et al. 1999a). The characteristics of these two species are similar to characteristics of simple-pored heterobasidiomycetes such as *Colacogloea peniophorae, Naohidea sebacea* and *Occultifur internus*, namely, the production of auricularioid basidia with forcibly discharged basidiospores that do not arise from teliospores. In contrast to the basidia of the two recently

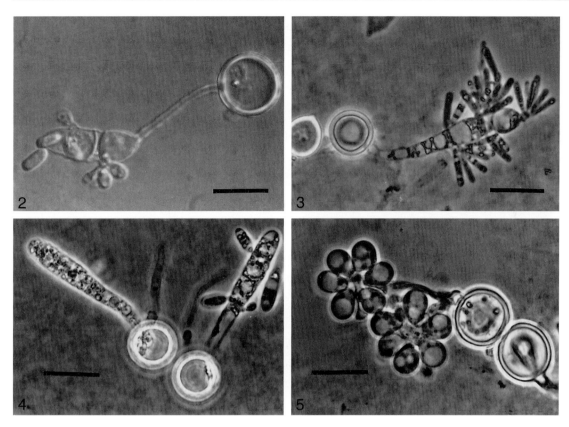

Figs. 2–5. Teliospore and basidial formation. Septate basidia: **Fig. 2**, *Sporidiobolous ruineniae*: **Fig. 3**, *Leucosporidium fellii*: **Fig. 4**, *Leucosporidium scottii*. Holometabasidia: **Fig. 5**, *Cystofilobasidium capitatum*. Bars 10 μm

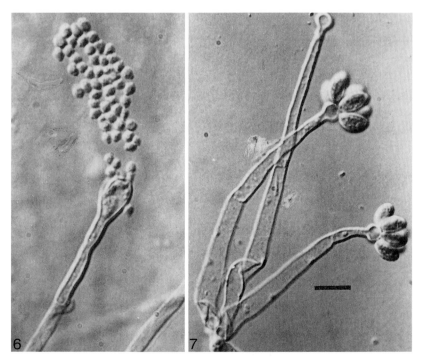

Figs. 6, 7. Holometabasidial formation in the absence of teliospore formation. **Fig. 6.** *Filobasidiella neoformans*. **Fig. 7.** *Filobasidium uniguttulatum*. Bar 10 μm. (Photographs courtesy K.J. Kwon-Chung)

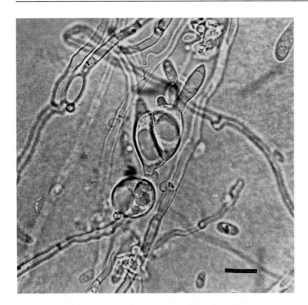

Fig. 8. Basidium of *Bulleromyces albus. Bar* 10 μm

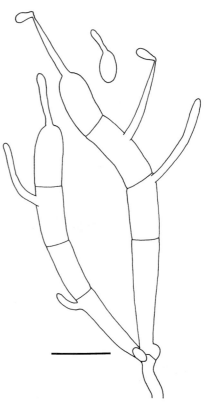

Fig. 10. Transversely septate basidia with ballistospores of *Colacogloea peniophorae. Bar* 10 μm

described species, which developed in artificial culture, the basidia of *C. peniophorae* (Fig. 10), *N. sebacea* and *O. internus* have been observed only under natural conditions in association with the fruiting bodies of other fungi.

Dikaryotic hyphae usually have clamp connections, whereas most monokaryotic hyphae lack these structures. Clamp connections are frequently present in homothallic strains as well, whereas monokaryotic fruiting results in hyphae without clamp connections. Incomplete clamp connections occur in *Itersonilia perplexans* and *Sakaguchia dacryoidea* (Fell and Tallman 1984b; Boekhout 1991b). In *Itersonilia* the monokaryotic yeast phase may give rise to monokaryotic hyphae with incomplete clamp connections, and an otherwise unchanged morphology. In *Tremella mesenterica*, the presence of incomplete clamp connections was reported to be a result of dual mating, with the progeny disomic for the parental B alleles (Wong 1987). Haustorial branches indicate an affinity with Tremellales of Filobasidiales, and the haustoria are thought to play a role in

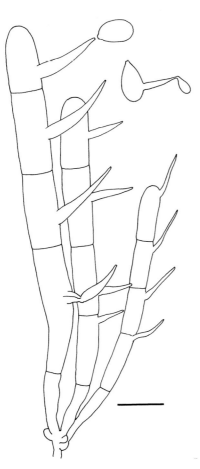

Fig. 9. Transversely septate basidia with ballistospores from the urediniomycetous yeast (Occultifur externus Fig. 21). *Bar* 10 μm

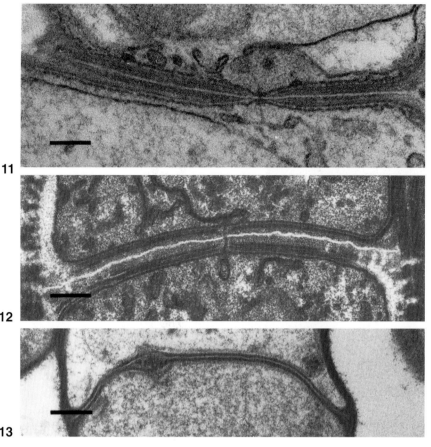

Figs. 11–13. "Simple" septal pores of *Leucosporidium scottii* (**Fig. 11**); and *Bensingtonia yamatoana* (**Fig. 12**); Microporelike structure of *Tilletiaria anomala* (**Fig. 13**). *Bar* 200 nm

mycoparasitic relationships (Bandoni 1987, 1995; Oberwinkler 1987).

Basidiocarps can occur in dimorphic heterobasidiomycetes with a distinct yeast phase and in species that are closely related to yeast taxa. For example, pycnidioid basidiocarps occur in *Heterogastridium pycnidioideum* (Oberwinkler et al. 1990a), synnema-like basidiocarps are present in *Chionosphaera apobasidialis* and *Stilbum vulgare*, and effuse, pustulate or clavarioid basidiocarps occur in some simple-pored taxa and in certain members of the Tremellales (Bandoni 1995).

C. Ultrastructure

Basidiomycetous yeasts are characterized by electron-dense and layered cell walls (Kreger-van Rij and Veenhuis 1971a; Simmons and Ahearn 1987). Septal ultrastructure is an important character complex for understanding phylogenetic relationships (Kreger-van Rij and Veenhuis 1971b; Johnson-Reid and Moore 1972; Moore and Kreger-van Rij 1972; Kwon-Chung and Popkin 1976; Moore 1980, 1987; Rhodes et al. 1981; Khan and Kimbrough 1982; Oberwinkler et al. 1983; Bauer et al. 1989; Boekhout et al. 1992b; Guého et al. 1992; Suh et al. 1993). Simple pored septa, with the cell wall attenuating towards the central pore and without Woronin bodies, occur in urediniomycetous yeasts (Swann and Taylor 1995a,b; E.A. Swann et al. Chap. 2, this Vol.), such as *Leucosporidium scottii* (Fig. 11), *Bensingtonia phyllada*, *B. yamatoana*, *Rhodosporidium sphaerocarpum*, *R. toruloides*, *Sporidiobolus johnsonii* and *S. ruineniae* (Fig. 12; Kreger-van Rij and Veenhuis 1971b; Johnson-Reid and Moore 1972; Moore 1972; Boekhout et al. 1992b). Similar septa were observed in *Agaricostilbum pulcherrimum*, *Helicobasidium mompa*, and *Heterogastridium pycni-*

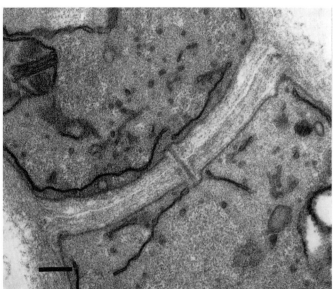

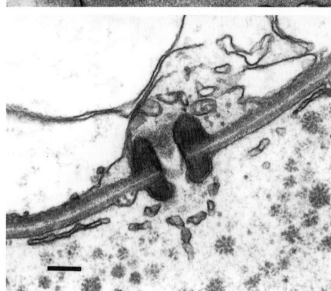

Figs. 14,15. Dolipore-like structure of *Trichosporon inkin* (**Fig. 14**) with inflated margin. Dolipore of *Trichosporon laibachii* (**Fig. 15**) with a vesiculate parenthesome. *Bar* 200 nm

dioideum (Oberwinkler et al. 1990a; McLaughlin et al. 1995). *Kriegeria eriophori* has multiple, but comparable, pores in the septa (McLaughlin et al. 1995 and references therein). Thickened septa with micropore-like structures were observed in species of *Tilletiopsis*, *Tilletiaria anomala* (Fig. 13), and *Ustilago maydis*, which may or may not have an inflated margin (O'Donnell and McLaughlin 1984; Bauer et al. 1989; Boekhout et al. 1992b). We consider these structures to be different from "simple" pores, because they do not have tapering cell walls and probably lack a true pore opening.

Dolipore septa (Figs. 14–17) occur in tremellaceous yeasts, such as *Bulleromyces albus*, *Cryptococcus laurentii*, *Filobasidiella* spp., *Filobasidium* spp., and Trichosporon spp. (except *T. brassicae* and *T. pullulans*) (Moore and Kreger-van Rij 1972; Kwon-Chung 1976b; Kwon-Chung and Popkin 1976; Rhodes et al. 1981; Moore 1987; Boekhout et al. 1991; Guého et al. 1992; Kwon-Chung et al. 1995). A cupulate or tubular parenthesome is present in most of these species, but this may be absent in closely related species, e.g., *Filobasidiella neoformans* versus *F. depauperata* (Kwon-Chung

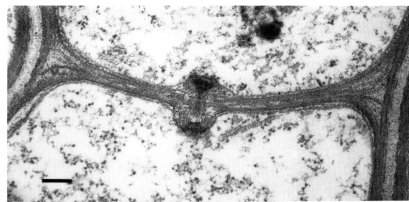

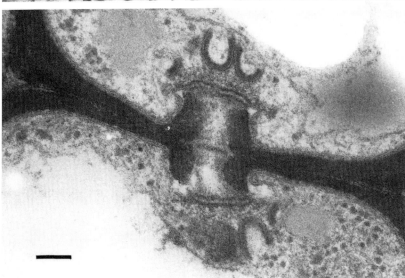

Figs. 16, 17. Dolipore of *Itersonilia perplexans* (**Fig. 16**) without parenthesome and dolipore of *Bulleromyces albus* (**Fig. 17**) with cupulate parenthesome. *Bar* 200 nm

and Popkin 1976; Rhodes et al. 1981; Kwon-Chung et al. 1995). CoQ9 species of *Trichosporon* (e.g., *T. inkin*) lack a parenthesome, but have a dolipore-like structure with a slightly flaired margin (Fig. 14). Different reports show contradictory results on the presence of a cupulate or tubular parenthesome in *F. depauperata* (Khan et al. 1981; Kwon-Chung et al. 1995). The variability in parenthesome type may indicate a fragile structure that is not well preserved by classical chemical fixation procedures or, alternatively, the presence of the parenthesome may depend on different morphogenetic conditions. Therefore, it may be preferable to study these dolipore/parenthesome complexes using techniques such as high-pressure freezing and freeze substitution in order to avoid artifacts which confuse the phylogenetic interpretation of these structures. Some phylogenetically related genera, such as *Cystofilobasidium*, *Mrakia*, and *Itersonilia*, lack a parenthesome (Oberwinkler et al. 1983; Boekhout 1991b; Suh and Sugiyama 1993; Suh et al. 1993; Fell et al. 1995; Swann and Taylor 1995a,b).

Septal ultrastructure correlates with molecular studies of 5S, 18S, and 25S ribosomal RNA/DNA (Walker and Doolittle 1982; Templeton 1983; Blanz and Gottschalk 1984; Walker 1984; Sugiyama and Suh 1993; Suh and Sugiyama 1993; Swann and Taylor 1993, 1995a,b; Fell et al. 1995) and cell wall composition, which can be summarized as follows: "simple" septate species belonging to the Urediniomycetes have a cell wall composition that is dominated by mannose and contains glucose and possibly fucose and rhamnose; however, xylose is usually not present. Species belonging to the Ustilaginomycetes have "micropore-like" structures and contain glucose

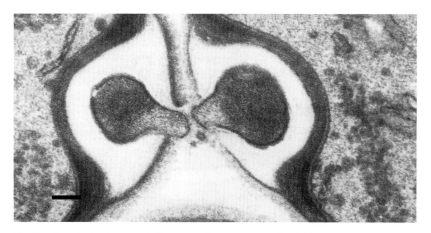

Fig. 18. Lenticular bodies of *Bensingtonia yamatoana*. Bar 200 nm

(as the dominant monomer), galactose, and mannose, but xylose is absent. Dolipore-containing species in the Hymenomycetes (Tremellomycetidae) have predominant levels of glucose, mannose, and xylose (Weijman and Golubev 1987; Prillinger et al. 1990a,b, 1991a,b, 1993; Swann and Taylor 1995a,b; Roeijmans et al. 1997). The literature regarding cell-wall monomers should, however, be viewed with some caution. Some authors analyze whole-cell hydrolysates, in contrast to other researchers who examine hydrolysates of extracellular polysaccharides that contain taxonomically relevant monomers. For example, *Rhodotorula yarrowii* (Fig. 20) was described in *Cryptococcus* because traces of xylose were found in the analysis of whole-cell hydrolysates; however, sequence analysis indicates that The species is a member of the *Urediniomyeetes*. The xylose may have originated in the cytoplasm.

An additional ultrastructural character is the spindle pole body (SPB) that occurs in *Leucosporidium scottii*, *Rhodosporidium toruloides*, and *Sporidiobolus* as small electron-dense disks (McCully and Robinow 1972a,b; Boekhout and Linnemans 1982) with an internal substructure, which is comparable to the SPBs of some smuts that occur on dicotyledons (McLaughlin et al. 1995). The SPBs of *Ustilago maydis* and *U. avenae* are hemispherical with an internal concave electron-dense layer (O'Donnell and McLaughlin 1984; O'Donnell 1992; McLaughlin et al. 1995). SPBs of *Bullera alba*, *Cryptococcus neoformans*, *Itersonilia perplexans* and Tremellales are electron-dense and globular, but without an apparent substructure (Taylor and Wells 1979; McLaughlin et al. 1995).

Electron-dense, lenticular bodies occur in species belonging to the Urediniomycetes, including species of *Bensingtonia* (Fig. 18), *Leucosporidium*, *Rhodosporidium*, *Sporidiobolus* (Kreger-van Rij and Veenhuis 1971b; de Hoog and Boekhout 1982; Boekhout et al. 1992b). Morphologically similar structures referred to as colacosomes, which are visible using high-magnification bright-field microscopy, have been observed in *Colacogloea peniophorae* (Oberwinkler et al. 1990b; Bauer and Oberwinkler 1991), *Cryptomycocolax abnorme* (Oberwinkler and Bauer 1990), and *Heterogastridium pycnidioideum* (T. Boekhout, unpubl. observ.). These structures play a role in cell-cell contact in mycoparasitic relationships, but may also be active within one thallus (T. Boekhout, unpubl. observ.). Their reported presence in species of Ustilaginales and Tilletiales (Moore 1972) needs to be verified, as the published pictures are not clear.

D. Physiological Properties

Examination of physiological properties is the primary method for differentiating species (Barnett et al. 1990; Kurtzman and Fell 1998). Routine tests include fermentation and growth with carbon sources, growth on nitrogen compounds, requirements for vitamins, growth at various temperatures, hydrolysis of urea, and formation of starch-like compounds. Because few basidiomycetes ferment at rates that result in visible reactions, fermentation tests are usually limited to glucose. When positive on that hexose, tests are expanded to galactose, maltose, sucrose, melibiose, and raffinose. Ability to assimilate carbon is routinely tested with 36 compounds. These tests can be made either in liquid or on solid media and the standard compounds are listed in Kurtzman and Sugiyama (Chap. 9, Vol. VII, Part A). Additional compounds, using numerical taxonomic methods, were evaluated by Sampaio and Fonseca (1995) and Sampaio (1999). Among the routinely tested vitamins, p-δ-aminobenzoic acid, biotin, and thiamine appear to be the most taxonomically relevant (Golubev 1989; Boekhout et al. 1993; Sampaio and Fonseca 1995). Most of these physiological properties are used to diagnose species, although some reactions can be characteristic of phylogenetic groups. Formation of extracellular starchlike compounds and assimilation of m-inositol and D-glucuronate occur in the majority of tremellaceous yeasts. These characteristics are absent in most urediniomycetous yeasts, although the ability to utilize D-glucuronate is present among several of the Urediniomycetes. Some ustilaginaceous species assimilate both m-inositol and D-glucuronate, but do not form extracellular starch.

E. Other Characters

Other important systematic characteristics include the number of isoprenoids of the coenzyme Q (Yamada and Kondo 1973; Kuraishi et al. 1985; Yamada et al. 1987) and killer toxin sensitivity patterns (Golubev and Boekhout 1995 and references therein). Coenzymes Q9 and Q10 are prevalent in ustilaginomycetous, urediniomycetous, and tremellaceous yeasts. CoQ7 is present in *Reniforma strues* (H. Roeijmans, unpubl. observ.), CoQ8 occurs in *Cystofilobasidium*, *Mrakia*, and the phylogenetically related species *Cryptococcus aquaticus*, *C. feraegula*, and *C. huempii*.

Killer toxins are produced by species of *Bensingtonia*, *Bullera*, *Cryptococcus*, *Cystofilobasidium*, *Filobasidium*, *Rhodotorula*, *Sporidiobolus*, *Sporobolomyces*, and *Ustilago* (see references in Boekhout et al. 1993). The action spectrum of toxins produced by basidiomycetous yeasts is usually limited to yeasts in the same phylogenetic lineage, such as urediniomycetous or tremellaceous yeasts (Golubev and Boekhout 1995). Species of the Cystofilobasidiales (Fig. 23: *Cystofilobasidium bisporidii*, *C. infirmominiatum*, *Phaffia rhodozyma*, *Udeniomyces* spp.) are sensitive to a narrow spectrum of killer toxins including that of *C. bisporidii* (Golubev and Boekhout 1995).

V. Molecular Techniques

A. DNA Typing Methods

Several DNA-typing methods have been applied to analyze relationships between basidiomycetous yeasts. The most important DNA typing methods are: karyotyping using pulsed field gel electrophoresis (PFGE), genome analysis using random amplified polymorphic DNA (RAPD), restriction fragment analysis (RFLP) of amplified ribosomal DNA, and amplified fragment length polymorphism (AFLP). Karyotype analysis has been applied to a limited number of basidiomycetous yeast species; chromosomal banding patterns vary between most species investigated and in almost all species a considerable intraspecific variation is apparent. For example, the morphologically and physiologically indistinguishable species *Tilletiopsis albescens* and *T. pallescens* differ in their chromosomal makeup (Boekhout et al. 1992a). The number of chromosomes of the astaxanthin producing yeast *Phaffia rhodozyma* was found to vary between 7 and 13 or 9 and 17, depending on the study (Nagy et al. 1994; Adrio et al. 1995). In contrast, such variation is limited or absent in the genus *Malassezia*, although all currently recognized species (Guého et al. 1996) differ in their karyotypes (Gueho et al. 1997), including the recently distinguished lipophilic *Malassezia* species formerly classified as *M. furfur* (Boekhout and Bosboom 1994).

The karyotype patterns of the two varieties of *Cryptococcus neoformans* differ (Boekhout et al. 1997b; Wickes et al. 1994), which supports the hypothesis of two separate species. Most karyotype studies of *C. neoformans* suggested the presence of mitotically stable karyotypes (see, e.g., Perfect et al. 1993; Dromer et al. 1994), but in some recent papers, minor karyotype differences were observed in multiple isolates from one patient or after passage through mice (Fries et al. 1996; Boekhout and van Belkum 1997). In contrast, considerable changes of karyotypes were observed after meiotic recombination, and also after self-sporulation and treatment with mutagens (Boekhout and van Belkum 1997). Restriction analysis of PCR-amplified ribosomal DNA indicated no differentiation between the four serotypes of *C. neoformans*, but varying levels of restriction variation between species of *Cryptococcus* (Vilgalys and Hester 1990).

PCR-typing methods, such as RAPD, have been applied to the medically important basidiomycetous yeasts *Malassezia furfur* and *Cryptococcus neoformans* (van Belkum et al. 1994; Brandt et al. 1995; Sorrell et al. 1996; Boekhout et al. 1997b) and to some nonpathogenic species (Messner et al. 1994). An extensive comparison of environmental, clinical, and veterinary isolates of *C. neoformans* suggested genetic separation of *C. neoformans* var. *neoformans* and *C. neoformans* var. *gattii*. No differences were apparent between environmental or clinical isolates, but veterinary isolates of var. *neoformans* showed a preferential RAPD type. A geographic substructure was present among isolates of var. *gattii*, but not in var. *neoformans* (Boekhout et al. 1997b). Other PCR-based methods include the use of specific primers for species and/or strain identifications (Fell 1993, 1995; Meyer et al. 1993; Mitchell et al. 1994; Haynes et al. 1995). Amplified fragment length polymorphism (AFLP) (Vos et al. 1995) is a universal DNA typing method based on the restriction of DNA by two restriction enzymes, followed by ligation of nucleotide adaptors complementary to the restriction sites. These adaptors and the restriction site sequence act as primer binding sites for PCR. The primers used in PCR are selective as they are complementary to the adaptors/restriction site sequence and to 1–3 additional (selective) nucleotides at the 3′-end. Consequently, the primers amplify only a subset of the ligated primers (Blears et al. 1998). To date, only one paper has been published on the application of AFLP in (ascomycetous) yeasts (de Barros Lopes et al. 1999), however, we are currently applying AFLP in a population structure analysis of *Cryptococcus neoformans*. Our preliminary results suggest considerable genetic distance between the varieties *neoformans* and *gattii* and the presence of serotype hybrids in both varieties (T. Boekhout, unpubl. results).

B. Molecular Sequence Analysis

Significant advancements in our understanding of basidiomycete yeast systematics have resulted from sequence analysis of the large and small subunits of ribosomal RNA and DNA (Blanz and Gottschalk 1984; Guého et al. 1989, 1990, 1992; Yamada and Kawasaki 1989; Yamada et al. 1989a,b, 1990a,b, 1994; Fell and Kurtzman 1990; Yamada and Nakagawa 1990, 1992; Fell et al. 1992, 1995; Fell and Tallman 1992; Van de Peer et al. 1992; Nakase et al. 1993, 1995; Sugiyama and Suh 1993; Suh and Sugiyama 1993, 1994; Boekhout et al. 1995; Nishida et al. 1995; Suh and Nakase 1995; Swann and Taylor 1995a,b; Takashima et al. 1995). Results from these large and small subunit rDNA studies are in general agreement. The following report emphasizes the D1/D2 region of the LrDNA, which constitutes approximately 650 bp. Additional studies employ the internal transcribed spacer (ITS) and intergenic spacer (IGS) regions for further separation of species and strains. The following discussion presents our current views on yeast systematics with the anticipation that studies of additional genes and taxa will provide expanded knowledge.

C. D1/D2 LrDNA Phylogenetic Analyses

There is considerable agreement among phylogenetic studies based on molecular, structural, and other phenotypic studies that the yeasts belong to three classes of the Basidiomycota: Hymenomycetes, Urediniomycetes and Ustilaginomycetes. Accordingly, the yeasts are presented in individual trees that represent those classes. Because of space limitations the individual hymenomycete and urediniomycete trees were separated into sections (Figs. 20–23). The general topology of the two trees and bootstrap values of the major clades is presented in Fig. 19. Yeasts associated with the Ustilaginomycetes are represented in Fig. 24. Strains examined for the following discussion (including Figs. 19–24) were reported in Fell et al. (2000).

1. Urediniomycetes

The Urediniomycetes have unifying characteristics of the presence of diaphragmlike ("simple") septa and the lack of Woronin bodies. Mannose is dominant in whole-cell hydrolysates and xylose is absent, but fucose, galactose, and/or rhamnose may be present, extracellular starchlike compounds are not produced and inositol is not used by the majority of the species. Co-enzymes are

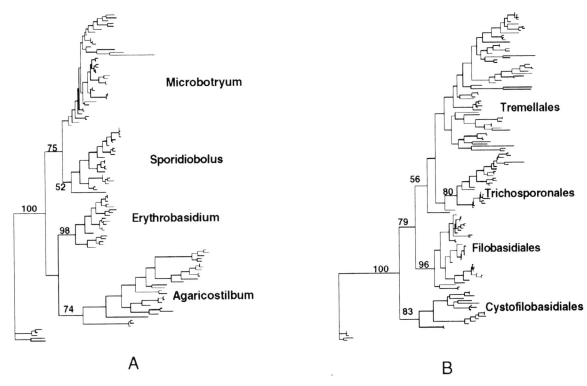

Fig. 19A,B. Topology of phylogenetic trees of **A** Urediniomycetes and **B** Hymenomycetes based on nucleotide sequences of the D1/D2 region of the large subunit rDNA. The Urediniomycete tree was one of 100 equally parsimonious trees. Number of characters = 657, constant characters = 241, parsimony-uninformative characters = 56, parsimony-informative characters = 360. Tree length = 2399, consistency index (CI) = 0.312, retention index (RI) = 0.771, rescaled consistency index (RC) = 0.241, homoplasy index (HI) = 0.688. The Hymenomycete tree was one of 100 equally parsimonious trees. Number of characters = 617, constant characters = 257, parsimony-uninformative characters = 63, parsimony-informative characters = 297. Tree length = 2227, consistency index (CI) = 0.269, retention index (RI) = 0.783, rescaled consistency index (RC) = 0.211, homoplasy index (HI) = 0.731, PAUP*4.0. Bootstrap values from 1000 full heuristic replicates are limited to those >50% (PAUP*4.0). Trees were divided (Figs. 20–23) for presentation purposes

CoQ9 or 10, with the exception of *Reniforma strues*, which has CoQ7.

There are four major clades in the urediniomycete tree (Figs. 20, 21), which are informally labeled as *Microbotryum*, *Sporidiobolus*, *Agaricostilbum*, and *Erythrobasidium* clades. Species distributions within these clades provide an example of the polyphyletic nature of certain genera, particularly some of the anamorphic genera whose taxonomy is based on phenotypic differences such as the presence or absence of ballistoconidia. Three of the anamorphic genera are in two or more clades. *Bensingtonia* occurs in the *Microbotryum* and *Agaricostilbum* clades; *Rhodotorula* is in the *Microbotryum*, *Sporidiobolus*, and *Erythrobasidium* clades; and *Sporobolomyces* is in all four clades. In contrast, some genera, particularly the teleomorphic genera, are limited to single clades. *Leucosporidium* is in the *Microbotryum* clade; *Rhodosporidium* and *Sporidiobolus* are within the *Sporidiobolus* clade; *Kondoa* and the anamorphs *Kurtzmanomyces* and *Sterigmatomyces* are in the *Agaricostilbum* clade; and *Erythrobasidium*, *Sakaguchia*, and *Occultifur* are members of the *Erythrobasidium* clade.

The *Microbotryum* and *Sporidiobolus* clades (Fig. 20) represent the Sporidiobolaceae of Boekhout et al. (1998) and the Microbotryomycetidae of Swann et al. (1999). A characteristic shared by the teleomorphic yeasts in these two clades is the presence of a teliospore (probasidium) and a transversely septate metabasidium. This led Moore (1972, 1980) to describe the order

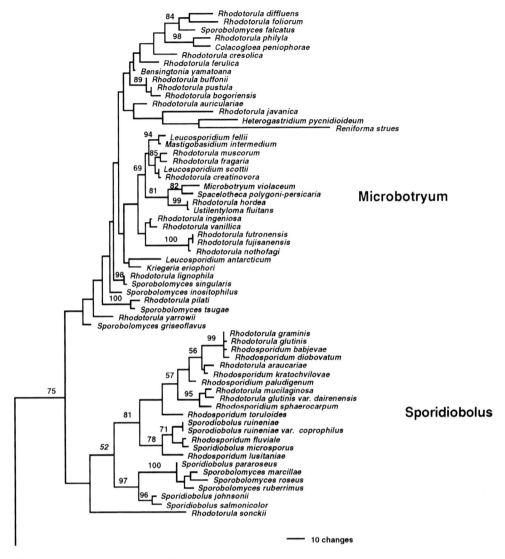

Fig. 20. *Microbotryum* and *Sporidiobolus* clades of the Urediniomycetes, see Fig. 19A

Sporidiales, which he divided into the Sporidiaceae and Sporidiobolaceae for nonballistoconidial and ballistoconidial teliospore-forming yeasts, respectively. However, these two families cannot be discriminated by rRNA/rDNA sequence analysis (Swann and Taylor 1993; Fell et al. 1995, 2000).

The *Microbotryum* clade is phylogenetically diverse as indicated by the presence of plant parasitic fungi in this tree. *Microbotryum violaceum* and *Ustilentyloma fluitans*, which occur in a single cluster (81% bootstrap support), represent the Microbotryaceae and the Ustilentylomataceae of the Microbotryales (Bauer and Oberwinkler 1997). *Colacogloea peniophorae* and *Kriegeria eriophori* are in the Platygloeales and *Heterogastridium pycnidioideum* is in the Heterogastridiales (Bandoni 1995).

Yeast species from six genera are included in the *Microbotryum* clade: the teliospore-forming genera *Leucosporidium* and *Mastigobasidium*, and species of anamorphic genera *Rhodotorula*, *Sporobolomyces*, *Bensingtonia*, and *Reniforma*. Golubev (1999) described *Mastigobasidium*, which is the sexual state of *Bensingtonia intermedia*. *Mastigobasidium intermedium* is closely related to *Leucosporidium fellii*; the two species are morphologically similar as they both produce

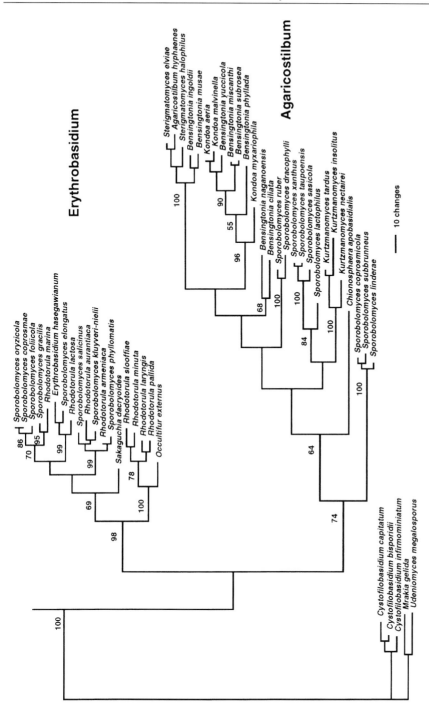

Fig. 21. *Erythrobasidium* and *Agaricostilbum* clades of the Urediniomycetes, see Fig. 19A

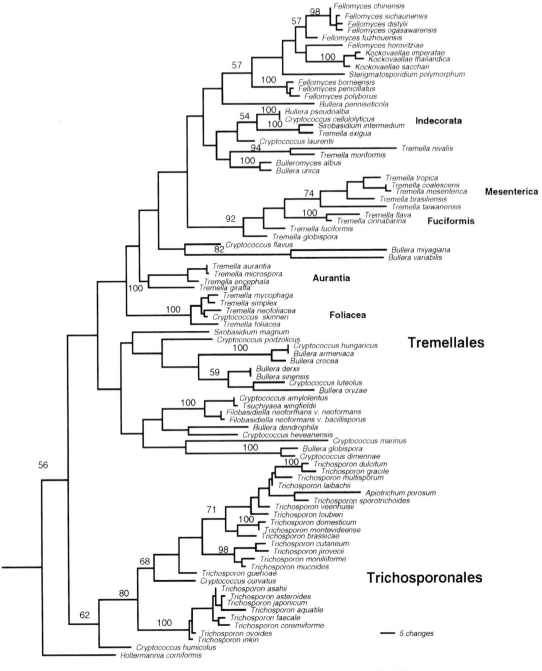

Fig. 22. Tremellales and Trichosporonales clades of the Hymenomycetes, see Fig. 19B

phragmometabasidia with bacilliform basidiospores in clusters on pegs (Statzell-Tallman and Fell 1998). Formation of phragmometabasidia by *Leucosporidium* and *Mastigobasidium* is a characteristic shared with teliospore-forming plant parasites *Microbotryum*, *Sphacelotheca*, and *Kriegeria*. In contrast to the teliosporic members of this clade, teliospores have not been observed in *Colacogloea* or *Kriegeria*.

The majority of the yeasts in the *Microbotryum* clade produce white to cream-colored colonies, an exception is *Rhodotorula fujisanensis* whose colonies may have a light pink color (Johnson and Phaff 1978; Sampaio and Fonseca

Fig. 23. Filobasidiales and Cystofilobasidiales clades of the Hymenomycetes, see Fig. 19B

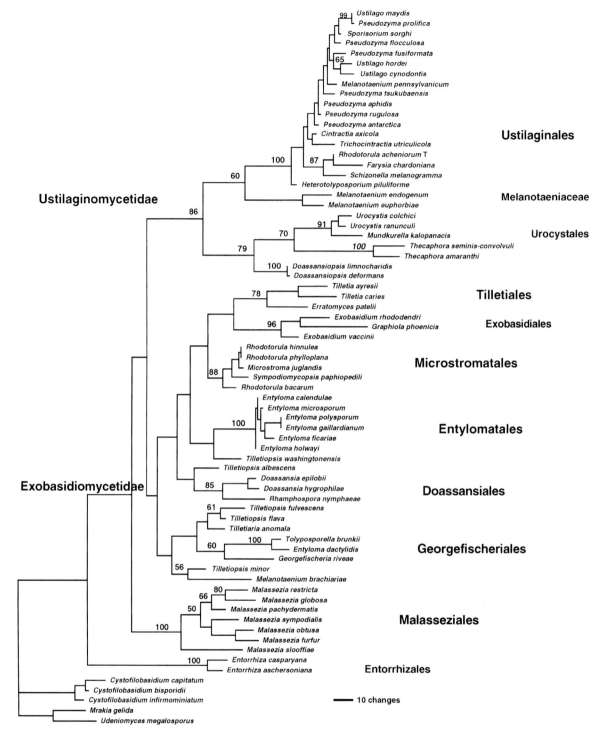

Fig. 24. Yeasts associated with the Ustilaginomycetes. The phylogenetic tree was one of 100 equally parsimonious trees. Number of characters = 526, constant characters = 216, parsimony-uninformative characters = 37, parsimony-informative characters = 271. Tree length = 1483, CI = 0.356 RI = 0.741, RC = 0.264, HI = 0.644. Bootstrap values from 1000 full heuristic replicates are limited to those >50% (PAUP*4.0). Order names are from Begerow et al. (1977)

1995). *Reniforma strues* is a unique species due to the presence of kidney-shaped vegetative cells and the presence of co-enzyme Q7; in contrast, other members of the Urediniomycetes contain CoQ9 or CoQ10. The apparent phylogenetic relationship of *R. strues* to *Heterogastridium pycnidioideum* may be the result of long-branch attraction. Species in the *Microbotryum* clade were isolated from a range of habitats, such as temperate plants, plant debris, polar soils, and marine waters.

The *Sporidiobolus* clade represents the red-pigmented teliosporic yeasts *Rhodosporidium* and *Sporidiobolus* with phragmometabasidia, and their related anamorphs in the genera *Rhodotorula* and *Sporobolomyces*. The presence of carotenoid pigments appears to differentiate the *Sporidiobolus* and *Microbotryales* clades. Species in the *Sporidiobolus* clade are among the most commonly occurring basidiomycetous yeasts and they may be collected from fresh and marine waters, soils, and plants. There are two major clusters in this clade: the *Rhodotorula glutinis* cluster (81% support) and the *Sporidiobolus johnsonii* cluster (97% support). The *R. glutinis* cluster consists of two branches: (1) the nonballistoconidial species of *Rhodotorula* and *Rhodosporidium* (57% support); (2) the *Sporidiobolus ruineniae* branch (78%) includes ballistoconidial positive (*Sporidiobolus*) and negative (*Rhodosporidium*) species. Species in this branch have a unique characteristic among the *Sporidiobolus* clade: the phragmometabasidia are produced on elongate stalks. The *Sporidiobolus johnsonii* cluster consists of ballistoconidial species that are usually associated with plants.

The *Erythrobasidium* clade (Fig. 21), which is strongly supported (98%), consists of the anamorphic genera *Sporobolomyces* and *Rhodotorula* and the teleomorphic genera *Erythrobasidium*, *Sakaguchia*, and *Occultifur*. Sexual mechanisms in the clade are diverse. *Sakaguchia* produces teliospores that germinate to a two- to four-celled metabasidium with repetitively budding basidiospores; in contrast to *Erythrobasidium*, which, according to the original and subsequent descriptions (Hamamoto et al. 1991; Sugiyama and Hamamoto 1998), produces holobasidia directly from dikaryotic hyphae. Sampaio et al. (1999b) questioned the reported life cycle in *Erythrobasidium* by suggesting that the basidiospores are conidiogenous cells. Further studies on the life cycle of *Erythrobasidium* are required to resolve these conflicting observations. Another genus in the clade, *Occultifur* (Sampaio et al. 1999a; Fig. 9) is a nonteliosporic yeast that produces auriculariod basidia with ballistospores.

The *Agaricostilbum* clade (Fig. 21) includes ballistoconidial forming species in the genera *Bensingtonia* and *Sporobolomyces*. There is considerable morphological variability among the other anamorphic genera. The yeast cells of *Sterigmatomyces* form distinct stalks with terminal conidia and a mid-stalk conidial separation. *Kurtzmanomyces* also forms conidia on stalks, however conidial separation is at the distal end of the stalk. Sampaio et al. (1999b) described a species of *Kurtzmanomyces* (*K. insolitus*) that produces ballisto- and stalked-conidia. In addition, some of the *Sporobolomyces* species in this cluster (particularly *S. lactophilus*) can reproduce sympodially with conidia on a long stalk. The only sexually reproducing yeast genus in this clade is *Kondoa*, which forms auriculariod basidia with ballistospores in the absence of teliospore production (Fonseca et al., 2000).

2. Hymenomycetes

The hymenomycetous yeasts have dolipore septa and whole cell hydrolysates that contain glucose, mannose, and xylose. Inositol is usually assimilated and starchlike compounds are produced by the majority of the species. Based on sequence analysis of the small subunit rDNA, Swann and Taylor (1995c) recommended two subclasses among the Hymenomycetes: (1) the Hymenomycetidae, containing the nonyeastlike macrofungi, including the mushrooms and puffballs and (2) the Tremellomycetidae. Analysis of the D1/D2 LrDNA region (Fell et al. 2000) of the Tremellomycetidae resulted in four clades (Fig. 19): the *Tremellales*, *Trichosporonales*, *Filobasidiales*, and *Cystofilobasidiales* (Figs. 22, 23). The genus *Cryptococcus* is polyphyletic and occurs in all four clades. The remaining genera occur in single clades: (1) Tremellales: *Bullera*, *Bulleromyces*, *Fellomyces*, *Filobasidiella*, *Kockovaella*, and *Tsuchiyaea*; (2) Trichosporonales includes all species of *Trichosporon* with the exception of *Trichosporon pullulans*, which occurs in the Cystofilobasidiales; (3) Filobasidiales: *Filobasidium*; (4) Cystofilobasidiales: *Cystofilobasidium*, *Mrakia*, *Phaffia*, *Udeniomyces*, and *Xanthophyllomyces*.

The Tremellales clade is depicted in Fig. 22. For comparative purposes we included filamentous species of the Tremellaceae (*Tremella* spp. and *Holtermannia corniformis*) and two species of the Sirobasidiaceae (*Sirobasidium magnum* and *S. intermedium*). *Tremella* was the major teleomorphic genus represented in this study, however, the survey was limited to 20 of the estimated 120 species of *Tremella* (Bandoni 1995). Chen (1998), in a partial study of the genus *Tremella*, demonstrated five clusters (Indecorata, Mesenterica, Fuciformis, Aurantia, and Foliacea), which are depicted in Fig. 22. Yeast species are present in two of the clusters. *Cryptococcus skinneri* is in the Foliacea cluster, which has strong (100%) statistical support. The Indecorata cluster that lacks statistical support includes *Bulleromyces albus*, *Bullera unica*, *Bullera pseudoalba*, *Cryptococcus cellulolyticus*, and *Cryptococcus laurentii*.

The Tremellales clade and many of the internal clusters lack statistical support, which may reflect the heterogeneity of the order and inadequate sampling of taxa. As more species of *Tremella* are studied and additional species of yeasts discovered, greater resolution within the Tremellales may develop. The genera *Fellomyces* and *Kockovaella* have support, albeit weak (57%), as a single cluster. The distinctive character of the two genera is formation of conidia on stalks with distal conidial separation. *Kockovaella* differs from *Fellomyces* by the formation of ballistoconidia. The close relationship of *Bullera* and *Cryptococcus* species in this tree indicates that the presence or absence of ballistoconidia has little systematic value. In contrast, the three species of *Kockovaella* occur together on a strongly (100%) supported branch. *Sterigmatosporidium*, which occurs in this cluster, has been described as a teleomorph; however, the apparent absence of a tremellaceous basidium suggests that further investigation of the life cycle is required.

The human pathogens *Filobasidiella neoformans* var. *neoformans* and var. *bacillispora* form a statistically supported (100%) cluster within the Tremellales. Previous concepts (Bandoni 1995; Boekhout et al. 1998) placed *Filobasidiella* among the *Filobasidiales*, however, molecular systematics indicates a relationship to the Tremellales. The sexual cycle of *Filobasidiella* is distinct from *Bulleromyces*, which is the only other recognized yeast genus with a teleomorphic state in Tremellales. *Filobasidiella* produces a slender cylindrical capitate holobasidium with basipetally formed chains of basidiospores (Kwon-Chung 1998), whereas *Bulleromyces* produces two- to four-celled basidial morphology similar to many of the Tremellales (Boekhout et al. 1991).

The Trichosporonales clade contains all of the species of *Trichosporon*, except for *T. pullulans*, which is located among the Cystofilobasidiales. The genus *Trichosporon* is characterized by the presence of arthroconidia and the clade has strong statistical support. The order Trichosporonales was proposed (Fell et al. 2000) based on the molecular and morphological characteristics, which demonstrated that the clade is phylogenetically distinct from the other hymenomycetous clades. *Trichosporon* consists of soil-inhabiting species as well as human and animal pathogens. The genus was studied by Guého et al. (1992, 1993) and Middelhoven et al. (1999a,b).

The Filobasidiales clade (Fig. 23) has strong (96%) statistical support and consists of species of *Cryptococcus* and the sole teleomorphic genus *Filobasidium* whose sexual cycle differs from *Filobasidiella* by the formation of petallike whorls of basidiospores at the apex of a slender holobasidium (Kwon-Chung 1998). Previously, the order included the genera *Cystofilobasidium*, *Filobasidiella*, *Filobasidium*, *Mrakia*, and *Xanthophyllomyces* (Wells 1994; Bandoni 1995; Boekhout et al. 1998). Swann and Taylor (1995c) evaluated the *Filobasidiales* and suggested that *Filobasidiella* was more closely related to *Tremella* than to *Filobasidium* and that *Cystofilobasidium* and *Mrakia* did not form a monophyletic group with *Filobasidium*. The data presented in Figs. 22 and 23 concur with Swann and Taylor's recommendations. The majority of the members of the Filobasidiales are species of *Cryptococcus* that originated in a variety of habitats including the Antarctic. Systematics of species in this clade was studied by Fonseca et al. (1999).

The order Cystofilobasidiales (Fell et al. 1999) has teliospores, a feature ascribed to the Urediniomycetes and unique among the Hymenomycetes. Other characteristics include holobasidia and hyphal septa that lack parenthesomes. There are two major clusters among the Cystofilobasidiales that separate the teliosporic genera *Mrakia* and *Cystofilobasidium*. The two genera differ in basidial morphology. *Mrakia* produces a phragmo- or holometabasidium directly attached to the teliospore; in contrast, some, but not all, species of *Cystofilobasidium* produce a long, slender metabasidium with a capitate terminus that bears the basidiospores (Kwon-Chung 1998). The two species of *Mrakia* (*M. frigidia* and *M. gelida*), based on ITS and IGS analyses, have been demonstrated to be distinct species (Diaz and Fell 1999), in contrast to the previously suggested synonymy of the two species (Fell and Statzell-Tallman 1998). Another teleomorph in this clade, *Xanthophyllomyces*, does not produce a teliospore; instead, a slender capitate holobasidium arises from parent–bud cell mating (Golubev 1998). Anamorphic genera in the Cystofilobasidiales include the ballistoconidial genus *Udeniomyces*, the arthroconidial forming species *Trichosporon pullulans*, several species of *Cryptococcus*, and *Phaffia rhodozyma*. The latter is a distinct species from *Xanthophyllomyces dendrorhous* as demonstrated by ITS and IGS analyzes (Fell and Blatt 1999).

3. Ustilaginomycetes

The occurrence of yeasts among the Ustilaginomycetes, as depicted in Fig. 24, is the result of a compilation of sequence data on the plant-associated fungi in the Ustilaginomycetes by Begerow et al. (1997), *Pseudozyma*, *Tilletia*, and *Tilletiopsis* spp. by Boekhout et al. (1995); and yeast data by Fell et al. (2000). Order names in Fig. 24 are adapted from Begerow et al. (1997). Yeast species included among the Ustilaginomycetes are *Pseudozyma* spp., four species of *Rhodotorula*, *Sympodiomycopsis paphiopedili*, and members of the genus *Malassezia*. Species of *Pseudozyma* are in the Ustilaginales cluster. These species were previously classified in *Candida*, *Stephanoascus*, *Trichosporon*, *Sterigmatomyces*, and *Sporobolomyces*. LrDNA analyses indicated that these species belong to the Ustilaginales (Boekhout et al. 1995; Fell et al. 1995); consequently, they were transferred to the genus *Pseudozyma* (Boekhout 1995). These ustilaginaceous yeasts share a relatively uniform morphology of hyphae without clamp connections and the formation of acropetally branched chains of fusiform blastoconidia. *Rhodotorula acheniorum* is also among the Ustilaginales, although that species produces sparse pseudomycelium rather than hyphae. *Rhodotorula hinnulea*, *R. phylloplana*, *Sympodiomycopsis paphiopedili*, and *R. bacarum* are phylogenetically related to *Microstroma juglandis* in the Microstromatales. All of these yeasts occur on plants, however, there is no information to determine whether the yeasts represent free-living saprophytes or life stages of parasitic filamentous fungi.

The genus *Malassezia* (Fig. 24) appears in a clade that is separate from the Ustilaginomycetidae and the Exobasidiomycetidae. The genus is distinct among the Ustilaginomycetes due to a general association with humans and other animals, and a growth requirement, by many of the strains, for fatty acids. The genus was placed in a separate order, Malasseziales, by Begerow et al. (1999; for details for the genus see Guého et al. 1996; Guillot and Guého 1995).

The known characteristics of the ustilaginomycetous yeasts include cell walls containing glucose (dominant), galactose and mannose, no xylose, and usually no rhamnose or fucose. Septa may have microporelike structures, e.g., narrow pore channel with or without flared margins (see McLaughlin et al. 1995). CoQ10 is present, with the exception of *Malassezia*, which has CoQ9. Extracellular starch is not produced.

D. Species and Strain Distinctions Based on D1/D2, ITS, and IGS Regions

The available data (Fell et al. 2000) indicate that strains with identical D1/D2 sequences represent a single species. There are situations where mating genetics and standard phenotypic characteristics indicate that strains with identical D1/D2 sequences represent separate species. A case in point is *Cryptococcus ater*, *C. magnus*, *Filobasidium elegans*, and *F. floriforme* (Fig. 23, Filobasidiales); all four species have identical D1/D2 sequences. *C. ater* and *C. magnus* have similar phenotypic characteristics and possibly represent a single species. *C. magnus*, *F. elegans*, and *F. floriforme* are dissimilar on phenotypic characteristics and the two *Filobasidium* species differ in the morphology of their sexual structures (Kwon-Chung 1998). The problem was resolved (Fell et al. 2000) through sequence analysis of the ITS region, which confirmed the synonymy of the two *Cryptococcus* species and the separation of *C. magnus*, *F. elegans*, and *F. floriforme*. Additional data in our laboratories confirmed that the ITS region is useful for separation of closely related species. Due to the high degree of sequence variability, there may be considerable difficulty to produce adequate ITS alignments for phylogenetic interpretations. Consequently, combined ITS and D1/D2 analyses are useful for species identifications and phylogenetic placements.

The ITS region, which is between the 3' end of the SrDNA and 5' LrDNA genes, has been extensively studied as a method to discriminate species of ascomycetous yeasts and basidiomycetes (e.g., Gardes and Bruns 1993; James et al. 1996). In contrast, the IGS region, which is between the 3'LrDNA and 5'SrDNA, has been of lesser interest to yeast systematics, although the region has been studied for population and strain level differences among a wider range of eukaryotes including *Drosophila*, oats, tomato, rice, nematodes, and *Daphnia* (Tautz et al. 1987; Perry and Palukaitis 1990; Vahidi and Honda 1991; Cordesse et al. 1993; Polanco and Vega 1994; Crease 1995). Of notable exception was the suggested use of the IGS region for typing strains of *Cryptococcus neoformans* (Fan et al. 1995). Because there are often industrial and academic requirements to differentiate strains of yeasts within a species, Fell and Blatt (1999) examined strains of the anamorph *Phaffia rhodozyma* and the teleomorph *Xanthophyllomyces dendrorhous*. These yeasts are commercially important as producers of astaxanthin, which is a component in aquacultural feeds to provide red coloration for marine fish and crustaceans (Johnson and Schroeder 1995). Previous studies (Golubev 1995, 1998) considered *P. rhodozyma* as the anamorphic state of *X. dendrorhous*. ITS studies demonstrated that the two species were distinct and that Pacific rim and European strains differed by one nucleotide. Additional IGS studies confirmed the species distinctions and demonstrated sequence differences between strains. Although the study (Fell and Blatt 1999) used a limited number of strains, the results suggest that the technique could be useful for strain selection and as ecological markers.

VI. Conclusions

Morphological, life history, and molecular studies of basidiomycetous yeasts have rapidly expanded our understanding of yeast systematics and the relationships to other fungi. The present systematic scheme is based on a small data base, considering the estimate that only ~1% of the yeasts in nature have been discovered. With additional taxa and comparative studies between related and unrelated species, a more precise phylogeny should emerge. The informal and formalized clades and current generic concepts that we presented should become clearly defined or rejected with the advancement to new phylogenetic interpretations.

Molecular techniques provide methods for rapid and accurate identifications. Routine identifications using phenotypic markers can be time-consuming, taking as long as 1–2 weeks. In contrast, molecular identifications can be accomplished in a few hours. Data banks for D1/D2 LrDNA regions for the known species of ascomycetes and basidiomycetes are available in GenBank. Therefore, it is possible to identify known species and phylogenetically place unknown species based on computerized searches. Additionally, species-specific oligonucleotide PCR primers and hybridization probes can be developed for rapid identifications and these can be arranged in probe arrays. The problem with molecular techniques is their expense, which renders them unavailable to many laboratories. A goal will be to develop rapid and inexpensive molecular techniques or to devise accurate phenotypic assays.

Possibly the most significant challenge will be to discover and preserve new species. As habitats are altered and disappear, yeast communities are disrupted and species are lost. These losses must be considered in economic and scientific terms due to their potential industrial, medical, and ecological importance.

Acknowledgments. Research funding to JWF was provided by the National Science Foundation (Ocean Sciences Division). TEM plates were made by W.H. Batenburg-van der Vegte.

References

Adrio J, Lopez M, Casqueiro J, Fernandez C, Veiga M (1995) Electrophoretic karyotype of the astaxanthin-producing yeast *Phaffia rhodozyma*. Curr Genet 27:447–450

Ahearn DG (1998) Yeasts pathogenic for humans. In: Kurtzman CP, Fell JW (eds) The yeasts, a taxonomic study, 4th edn. Elsevier, Amsterdam, pp 9–12

Bandoni RJ (1963) Conjugation in *Tremella mesenterica*. Can J Bot 41:468–474

Bandoni RJ (1987) Taxonomic overview of the Tremellales. Stud Mycol 30:87–110

Bandoni RJ (1995) Dimorphic Heterobasidiomycetes: taxonomy and parasitism. Stud Mycol 39:13–28

Banno I (1963) Preliminary report on cell conjugation and mycelial state in *Rhodotorula* yeasts. J Gen Appl Microbiol 9:249–251

Banno I (1967) Studies on the sexuality of *Rhodotorula*. J Gen Appl Microbiol 13:167–196

Barnett JA, Paine RW, Yarrow D (1990) Yeasts characteristics and identification, 2nd edn. Cambridge University Press, Cambridge, 1002 pp

Bauer R, Oberwinkler F (1991) The colacosomes: structures at the host-parasite interface of a mycoparasitic basidiomycete. Bot Acta 104:53–57

Bauer R, Oberwinkler F, Deml G (1989) Ultrastruktur der Basiliensepten phragmobasidialer Brandpilze. Z Mykol 55:163–168

Bauer R, Oberwinkler F, Vanky K (1997) Ultrastructural markers and systematics in smut fungi and allied taxa. Can J Bot 75:1273–1314

Begerow D, Bauer R, Oberwinkler F (1997) Phylogenetic studies on the nuclear large subunit ribosomal DNA of smut fungi and related taxa. Can J Bot 75:2045–2056

Begerow D, Bauer R, Boekhout T (1999) Phylogenetic placements of ustilaginomycetous anamorphs as deduced from nuclear LSU rDNA sequences. Mycol Res (in press)

Biely P, Vrsanska M, Kratky Z (1981) Mechanisms of substrate digestion by endo-1,4-beta-xylanase of *Cryptococcus albidus*. Eur J Biochem 119:565–571

Björkling F, Godtfredsen SE, Kirk O (1991) The future impact of industrial lipases. Tibtech 9:360–363

Blanz P, Gottschalk M (1984) A comparison of 5S ribosomal RNA sequences from smut fungi. Syst Appl Microbiol 5:518–526

Blears MJ, De Grandis SA, Lee H, Trevors JT (1998) Amplified fragment length polymorphism (AFLP): a review of the procedure and its applications. J Ind Microbiol Biotechnol 21:99–114

Boekhout T (1991a) A revision of ballistoconidia-forming yeasts and fungi. Stud Mycol 33:1–194

Boekhout T (1991b) Systematics of the genus *Itersonilia* Derx: a comparative phenetic study. Mycol Res 95:135–146

Boekhout T (1995) *Pseudozyma* Bandoni emend. Boekhout, a genus for yeast-like anamorphs of Ustilaginales. J Gen Appl Microbiol 41:359–366

Boekhout T, Bosboom RW (1994) Karyotyping of *Malassezia* yeasts: taxonomic and epidemiological implications. Syst Appl Microbiol 17:146–153

Boekhout T, Fell JW (1995) Heterobasidiomycetes: systematics and applications. Stud Mycol 38:5–11

Boekhout T, Kurtzman CP (1996) Principles and methods used in yeast classification, and an overview of currently accepted yeast genera. In: Wolf K (ed) Nonconventional yeasts in biotechnology, a handbook. Springer, Berlin Heidelberg New York, pp 1–81

Boekhout T, Linnemans WAM (1982) Ultrastructure of mitosis in *Rhodosporidium toruloides*. Stud Mycol 22:23–38

Boekhout T, van Belkum A (1997) Variability of karyotypes and RAPD types in genetically related strains of *Cryptococcus neoformans*. Curr Gen 32:203–208

Boekhout T, Fonseca A, Batenburg-van der Vegte WH (1991) *Bulleromyces* genus novum (Tremellales), a teleomorph for *Bullera alba*, and the occurrence of mating in *Bullera variabilis*. Antonie van Leeuwenhoek 59:81–93

Boekhout T, Gool J van, Boogert H van den, Jille T (1992a) Karyotyping and G + C composition as taxonomic criteria applied to the systematics of *Tilletiopsis* and related taxa. Mycol Res 95:331–342

Boekhout T, Yamada Y, Weijman ACM, Roeijmans HJ, Batenburg-van der Vegte WH (1992b) The significance of coenzyme Q, carbohydrate composition and septal ultrastructure for the taxonomy of ballistoconidia-forming yeasts and fungi. Syst Appl Microbiol 15:1–10

Boekhout T, Fonseca A, Sampaio J-P, Golubev WI (1993) Classification of heterobasidiomycetous yeasts: characteristics and affiliation of genera to higher taxa of basidiomycetes. Can J Microbiol 39:276–290

Boekhout T, Fell JW, O'Donnell K (1995) Molecular systematics of some yeast-like anamorphs belonging to the Ustilaginales and Tilletiales. Stud Mycol 38:175–183

Boekhout T, Bandoni RJ, Fell JW, Kwon-Chung KJ (1998) Discussion of teleomorphic and anamorphic genera of heterobasidiomycetous yeasts. In: Kurtzman CP, Fell JW (eds) The yeasts, a taxonomic study, 4th edn. Elsevier, Amsterdam, pp 609–626

Boekhout T, Belkum A van, Leenders ACAP, Verbrugh HA, Mukamurangwa P, Swinne D, Scheffers WA (1997) Molecular typing of *Cryptococcus neoformans*: taxonomic and epidemiological aspects. Int J Syst Bact 47:432–442

Booth JL, Vishniac HS (1987) Urease testing and yeast taxonomy. Can J Microbiol 33:396–404

Brandt ME, Hutwagner LC, Kuykendall RJ, Pinner RW, Cryptococcal Disease Active Surveillance Group (1995) Comparison of multilocus enzyme electrophoresis and random amplified polymorphic DNA analysis for molecular subtyping of *Cryptococcus neoformans*. J Clin Microbiol 33:1890–1895

Brefeld O (1881) Die Brandpilze I. In: Untersuchungen über Hefenpilze. Fortsetzung der Schimmelpilze, Heft 4. Arthur Felix, Leipzig

Brefeld O (1888) Basidiomyceten II. Protobasidiomyceten. In: Untersuchungen aus dem Gesammtgebiete der Mykologie, Heft 7. Arthur Felix, Leipzig

Brefeld O (1895a) Die Brandpilze II. In: Untersuchungen aus dem Gesammtgebiete der Mykologie, Heft 11. Heinrich Schöningh, Münster

Brefeld O (1895b) Die Brandpilze III. In: Untersuchungen aus dem Gesammtgebiete der Mykologie, Heft 11. Heinrich Schöningh, Münster

Channon AG (1963) Studies on parsnip canker I. The causes of the diseases. Ann Appl Biol 51:1–15

Chen C (1998) Morphological and molecular studies in the genus *Tremella*. Biblio Mycol 174:1–225

Cordesse F, Cooke R, Tremousaygue D, Grellet F, Delseny M (1993) Fine structure and evolution of the rDNA intergenic spacer in rice and other cereals. J Mol Evol 36:369–379

Crease TJ (1995) Ribosomal DNA evolution at the population level: nucleotide variation in intergenic spacer arrays of *Daphnia pulex*. Genetics 141:1327–1337

De Barros Lopes M, Rainieri S, Henschke PA, Langridge D (1999) AFLP fingerprinting for analysis of yeast genetic variation. Int J Syst Bact 49:915–924

De Hoog GS, Boekhout T (1982) Teliospores, teliospore mimics and chlamydospores. Stud Mycol 22:15–22

De Mot R, Verachtert H (1985) Purification and characterization of extracellular amylolytic enzymes from the yeast *Filobasidium capsuligenum*. Appl Environ Microbiol 50:1474–1482

De Mot R, Demeersman M, Verachtert H (1984) Comparative study of starch degradation and amylase production by non-ascomycetous yeast species. Syst Appl Microbiol 5:421–432

Demain AL, Phaff HJ, Kurtzman CP (1998) The industrial and agricultural significance of yeasts. In: Kurtzman CP, Fell JW (eds) The yeasts, a taxonomic study, 4th edn. Elsevier, Amsterdam, pp 13–20

Diaz MR, Fell JW (2000) Systematics of psychrophilic yeasts in the genus *Mrakia* based on ITS and IGS rDNA sequence analysis. Antonie van Leeuwenhoek 77:7–12

Dromer F, Varma A, Ronin O, Mathoulin S, Dupont B (1994) Molecular typing of *Cryptococcus neoformans* serotype D clinical isolates. J Clin Microbiol 32:2364–2371

Edman JC (1996) Paradoxes of mating in *Cryptococcus neoformans*. In: 3rd Conf on *Cryptococcus* and cryptococcosis. Inst Pasteur de Lyon, Lyon, pp 7–8

Fan M, Currie BP, Gutell RR, Ragan MA, Casadevall A (1994) The 16S-like, 5.8S and 23S-like rRNAs of the two varieties of *Cryptococcus neoformans*: sequence, secondary structure, phylogenetic analysis and restriction fragment polymorphisms. J Med Vet Mycol 32:163–180

Fell JW (1969) Yeasts of the genera *Candida* and *Rhodotorula* with heterobasidiomycetous life cycles. Antonie van Leeuwenhoek 35 (Suppl Yeast Symp 1969):A25–26

Fell JW (1970) Yeasts with heterobasidiomycetous life cycles. In: Ahearn DG (ed) Recent trends in yeast research. Georgia State University Press, Atlanta, pp 49–66

Fell JW (1974) Heterobasidiomycetous yeasts *Leucosporidium* and *Rhodosporidium*. Their systematics and sexual incompatibility systems. Trans Mycol Soc Jpn 15:316–323

Fell JW (1976) Yeasts in oceanic regions. In: Jones EBG (ed) Recent advances in aquatic mycology. Elek Science, London, pp 93–124

Fell JW (1993) Rapid identification of yeast species using three primers in a polymerase chain reaction. Mol Mar Biol Biotechnol 3(2):174–180

Fell JW (1995) rDNA targeted oligonucleotide primers for the identification of pathogenic yeasts in a polymerase chain reaction. J Ind Microbiol 14:475–477

Fell JW, Blatt G (1999) Separation of strains of the yeasts *Xanthophyllomyces dendrorhous* and *Phaffia rhodozyma* based on rDNA IGS and ITS sequence analysis. J Ind Microbiol Biotechnol 21:677–681

Fell JW, Kurtzman CP (1990) Nucleotide sequence analysis of a variable region of the large subunit rRNA for identification of marine occurring yeasts. Curr Microbiol 21:295–300

Fell JW, Kurtzman CP (1996) Yeasts in soils and sediments. In: Hall G (ed) Methods for the examination of organismal diversity in soils and sediments. Cambridge University Press, Cambridge, pp 103–108

Fell JW, Tallman AS (1980) Mating between strains of the yeasts *Aessosporon salmonicolor* and *Sporobolomyces* spp. Int J Syst Bacteriol 30:206–207

Fell JW, Tallman AS (1981) Heterothallism in the basidiomycetous yeast genus *Sporidiobolus*. Curr Microbiol 5:77–82

Fell JW, Tallman AS (1982) Multiple allelic incompatibility factors among bifactorial strains of the yeast *Leucosporidium* (*Candida*) *scottii*. Curr Microbiol 3:213–216

Fell JW, Tallman AS (1984a) Genus *Leucosporidium* Fell, Statzell, Hunter et Phaff. In: Kreger-van Rij NJW (ed) The yeasts, a taxonomic study, 3rd edn. Elsevier, Amsterdam, pp 496–508

Fell JW, Tallman AS (1984b) Genus *Rhodosporidium* Banno. In: Kreger-van Rij NJW (ed) The yeasts, a taxonomic study, 3rd edn. Elsevier, Amsterdam, pp 509–531

Fell JW, Tallman AS (1992) Systematic placement of the basidiomycetous yeasts *Cystofilobasidium lari-marini* comb. nov. as predicted by RNA nucleotide sequence analysis. Antonie van Leeuwenhoek 62:209–213

Fell JW, Tallman AS (1998a) The Genus *Cryptococcus*. In: Kurtzman CP, Fell JW (eds) The yeasts, a taxonomic study, 4th edn. Elsevier, Amsterdam, pp 742–767

Fell JW, Tallman AS (1998b) The Genus *Rhodosporidium*. In: Kurtzman CP, Fell JW (eds) The yeasts, a taxonomic study, 4th edn. Elsevier, Amsterdam, pp 678–692

Fell JW, Tallman AS (1998c) The Genus *Mrakia*. In: Kurtzman CP, Fell JW (eds) The yeasts, a taxonomic study, 4th edn. Elsevier, Amsterdam, pp 676–677

Fell JW, Statzell A, Hunter I, Phaff HJ (1969) *Leucosporidium* gen. nov. The heterobasidiomycetous stage of several yeasts of the genus *Candida*. Antonie van Leeuwenhoek 35:433–462

Fell JW, Hunter I, Tallman AS (1973) Marine basidiomycetous yeasts (*Rhodosporidium* spp. n.) with tetrapolar and multiple allelic bipolar mating systems. Can J Microbiol 19:643–657

Fell JW, Statzell-Tallman A, Lutz MJ, Kurtzman CP (1992) Partial rRNA sequences in marine yeasts; a model for identification of marine eukaryotes. Mol Mar Biol Biotechnol 1:175–186

Fell JW, Boekhout T, Freshwater DW (1995) The role of nucleotide analysis in the systematics of the yeast genera *Cryptococcus* and *Rhodotorula*. Stud Mycol 38:129–146

Fell JW, Roeijmans H, Boekhout T (1999) Cystofilobasidiales, a new order of basidiomycetous yeasts. Int J Syst Bacteriol 49:907–913

Fell JW, Boekhout T, Fonseca A, Scorzetti G, Statzell-Tallman A (2000) Biodiversity and systematics of basidiomycetous yeasts as determined by large subunit rD1/D2 domain sequence analysis. Int J Syst Evol Microbiol 50:1351–1371

Filonow AB, Vishniac H, Anderson JA, Janisiewicz WJ (1996) Biological control of *Botrytis cinerea* in apple by yeasts from various habitats and their putative mechanisms of antagonism. Biol Control 7:212–220

Flegel TW (1981) The pheromonal control of mating in yeasts and its phylogenetic implication: a review. Can J Microbiol 27:373–389

Fonseca A, Sampaio J-P (1992) *Rhodosporidium Iusitaniae* sp. nov., a novel yeast species from Portugal. Antonie van Leeuwenhoek 59:177–181

Fonseca A, Scorzetti G, Fell JW (1999) Diversity in the yeasts *Cryptococcus albidus* and related species as revealed by ribosomal DNA sequence analysis. Can J Microbiol 45:1–21

Fonseca A, Sampaio JP, Inacio J, Fell JW (2000) Emendation of the basidiomycetous yeast genus *Kondoa* and the description of *Kondoa aeria* sp. nov. Antonie van Leeuwenhoek 77:293–302

Fries BC, Chen F, Currie BP, Casadevall A (1996) Karyotype instability in *Cryptococcus neoformans* infections. J Clin Microbiol 34:1531–1534

Gardes M, Bruns TD (1993) ITS primers with enhanced specificity for basidiomycetes-application to the identification of mycorrhizae and rusts. Mol Ecol 2:113–118

Gill CO, Hall MJ, Ratledge C (1977) Lipid accumulation in an oleaginous yeast (*Candida* 107) growing on glucose in single-stage continuous culture. Appl Environ Microbiol 33:231–239

Golubev WI (1989) Catabolism of m-inositol and taxonomic value of D-glucuronate assimilation in yeasts. Mikrobiologiya 58:276–283 (in Russian)

Golubev WI (1995) Perfect state of *Rhodomyces dendrorhous* (*Phaffia rhodozyma*). Yeast 11:101–110

Golubev WI (1998) *Xanthophyllomyces* Golubev. In: Kurtzman CP, Fell JW (eds) The yeasts, a taxonomic study, 4th edn. Elsevier, Amsterdam, pp 718–719

Golubev WI (1999) *Mastigobasidium*, a new teleomorphic genus for the perfect state of ballistosporous yeast *Bensingtonia intermedia*. Int J Syst Bacteriol 49:1301–1305

Golubev WI, Boekhout T (1995) Sensitivity to killer toxins as a taxonomic tool among heterobasidiomycetous yeasts. Stud Mycol 38:47–58

Goto S, Horiguchi S, Kaneko H, Itoh T (1988) Distribution of pyrophosphatidic acid in yeast species. J Gen Appl Microbiol 34:165–182

Guého E, Kurtzman CP, Peterson SW (1989) Evolutionary affinities of heterobasidiomycetous yeasts estimated from 18S and 25S ribosomal RNA sequence divergence. Syst Appl Microbiol 12:230–236

Guého E, Kurtzman CP, Peterson SW (1990) Phylogenetic relationships among species of *Sterigmatomyces* and *Fellomyces* as determined from partial rRNA sequences. Int J Syst Bacteriol 40:60–65

Guého E, Smith MTh, de Hoog GS, Billon-Grand G, Christen R, Batenburg-van der Vegte WH (1992) Contributions to a revision of the genus *Trichosporon*. Antonie van Leeuwenhoek 61:289–316

Guého E, Improvisi L, Christen R, de Hoog GS (1993) Phylogenetic relationships of *Cryptococcus neoformans* and some related basidiomycetous yeasts determined from partial large subunit rRNA sequences. Antonie van Leeuwenhoek 63:175–189

Guého E, Midgley G, Guillot J (1996) The genus *Malassezia* with description of four new species. Antonie van Leeuwenhoek 69:337–355

Guého E, Boekhout T, Ashbee HR, Guillot J, Belkum A van, Faergemann J (1998) The role of *Malassezia* species in the ecology of human skin and as pathogens. Med Mycol 36:220–229

Guillot J, Guého E (1995) The diversity of *Malassezia* yeasts confirmed by rRNA sequence and nuclear DNA comparisons. Antonie van Leeuwenhoek 67:297–314

Hagler A, Ahearn DG (1981) Rapid diazonium blue B test to detect basidiomycetous yeasts. Int J Syst Bacteriol 31:204–208

Hajlaoui MR, Benhamou N, Bélanger RR (1992) Cytochemical study of the antagonistic activity of *Sporothrix flocculosa* on rose powdery mildew, *Sphaerotheca pannosa* var. *rosae*. Phytopathology 82:583–589

Hamamoto M, Sugiyama J, Komagata K (1988a) *Rhodosporidium kratochvilovae* sp. nov., a new basidiomycetous yeast species. J Gen Appl Microbiol 34:119–125

Hamamoto M, Sugiyama J, Komagata K (1988b) Transfer of *Rhodotorula hasegawa* to a new basidiomycetous yeast genus *Erythrobasidium* as *Erythrobasidium hasegawae* comb. nov. J Gen Appl Microbiol 34:279–287

Haynes KA, Westerneng TJ, Fell JW, Moens W (1995) Detection and identification of pathogenic fungi by polymerase chain reaction amplification of large subunit ribosomal DNA. J Med Vet Mycol 33:319–325

Hijwegen T (1986) Biological control of cucumber powdery mildew by *Tilletiopsis minor*. Neth J Plant Pathol 92:93–95

Hijwegen T (1988) Effect of seventeen fungicolous fungi on sporulation of cucumber powdery mildew. Neth J Plant Pathol 94:185–190

Hoch HC, Provvidenti R (1979) Mycoparasitic relationships: cytology of the *Sphaerotheca fuliginea*-*Tilletiopsis* sp. interaction. Phytopathology 69:359–362

James SA, Collins MD, Roberts IN (1996) Use of a rRNA internal transcribed spacer region to distinguish phylogenetically closely related species of the genera *Zygosaccharomyces* and *Torulaspora*. Int J Syst Bacteriol 46:189–194

Jarvis WR, Shaw LA, Traquair JA (1989) Factors affecting antagonism of cucumber powdery mildew by *Stephanoascus flocculosus* and *S. rugulosus*. Mycol Res 92:162–165

Johnson EA, Phaff HJ (1978) *Rhodotorula fujisanensis*, a new taxonomic combination. Curr Microbiol 1:223–225

Johnson EA, Schroeder W (1995) Astaxanthin from the yeast *Phaffia rhodozyma*. Stud Mycol 38:81–90

Johnson-Reid JA, Moore RT (1972) Some ultrastructural features of *Rhodosporidium toruloides*. Antonie van Leeuwenhoek 38:417–435

Katayama-Hirayama K, Tobita S, Hirayama K (1991) Degradation of phenol by yeast *Rhodotorula*. J Gen Appl Microbiol 37:147–156

Khan SR, Kimbrough JW (1980) Septal ultrastructure in some genera of the Tremellaceae. Can J Bot 58:55–60

Khan SR, Kimbrough JW (1982) A reevaluation of the basidiomycetes based upon septal and basidial structures. Mycotaxon 15:103–120

Khan SR, Kimbrough JW, Kwon-Chung KJ (1981) Ultrastructure of *Filobasidiella arachnophila*. Can J Bot 59:893–897

Kitamoto D, Akiba S, Hioki C, Tabuchi T (1990a) Extracellular accumulation of mannosylerythritol lipids by a strain of *Candida antarctica*. Agric Biol Chem 54:31–36

Kitamoto D, Akiba S, Hioki C, Tabuchi T (1990b) Production of mannosylerythritol lipids by *Candida antarctica* from vegetable oils. Agric Biol Chem 54:37–40

Klecan AL, Hippe S, Somerville SC (1990) Reduced growth of *Erysiphe graminis* f.sp. *hordei* induced by *Tilletiopsis pallescens*. Phytopathology 80:325–331

Kluyver AJ, van Niel CB (1924) Ueber Spiegelbilder erzeugende Hefearten und die neue Hefegattung *Sporobolomyces*. Centralbl Bakteriol Ser II 63:3–20

Kluyver AJ, van Niel CB (1927) *Sporobolomyces*: ein Basidiomyzet? Ann Mycol 25:389–394

Kreger-van Rij NJW, Veenhuis M (1971a) A comparative study of the cell wall structure of basidiomycetous and related yeasts. J Gen Microbiol 68:87–95

Kreger-van Rij NJW, Veenhuis M (1971b) Some features of yeasts of the genus *Sporidiobolus* observed by electron microscopy. Antonie van Leeuwenhoek 37:253–255

Kuraishi H, Katayama-Fujimora K, Sugiyama J, Yokoyama T (1985) Ubiquinone systems in fungi I. Distribution of ubiquinones in the major families of Ascomycetes, Basidiomycetes and Deuteromycetes, and their taxonomic implications. Trans Mycol Soc Jpn 26:383–395

Kurtzman CP (1973) Formation of hyphae and chlamydospores by *Cryptococcus laurentii*. Mycologia 65:388–395

Kurtzman CP, Fell JW (1998) The yeasts, a taxonomic study, 4th edn. Elsevier, Amsterdam, 1055pp

Kwon-Chung KJ (1975) A new genus, *Filobasidiella*, the perfect state of *Cryptococcus neoformans*. Mycologia 67:1197–1200

Kwon-Chung KJ (1976a) Morphogenesis of *Filobasidiella neoformans*, the sexual state of *Cryptococcus neoformans*. Mycologia 68:821–833

Kwon-Chung KJ (1976b) Ultrastructure of septal complex in *Filobasidiella neoformans* (*Cryptococcus neoformans*). J Bacteriol 126:524–528

Kwon-Chung KJ (1977) Perfect state of *Cryptococcus uniguttulatus*. Int J Syst Bacteriol 27:293–299

Kwon-Chung KJ (1987) Filobasidiaceae – a taxonomic survey. In: Hoog GS de, Smith MTh, Weijman ACM (eds) The expanding realm of yeast-like fungi. Elsevier, Amsterdam, pp 75–85

Kwon-Chung KJ (1998a) *Filobasidiella* In: Kurtzman CP, Fell JW (eds) The yeasts, a taxonomic study, 4th edn. Elsevier, Amsterdam, pp 656–662

Kwon-Chung KJ (1998b) *Cystofilobasidium* Oberwinkler & Bandoni In: Kurtzman CP, Fell JW (eds) The yeasts, a taxonomic study, 4th edn. Elsevier, Amsterdam, pp 646–653

Kwon-Chung KJ, Popkin TJ (1976) Ultrastructure of septal complex in *Filobasidiella neoformans* (*Cryptococcus neoformans*). J Bacteriol 126:524–528

Kwon-Chung KJ, Chang YC, Bauer R, Swann EC, Taylor JW, Goel R (1995) The characteristics that differentiate *Filobasidiella depauperata* from *Filobasidiella neoformans*. Stud Mycol 38:67–79

Laffin RJ, Cutter VM (1959) Investigations on the life cycle of *Sporidiobolus johnsonii* I. Irradiation and cytological studies. J Elisha Mitchell Sci Soc 75:89–96

Mattey M (1992) The production of organic acids. Crit Rev Biotechnol 12:87–132

McCully EK, Robinow CF (1972a) Mitosis in heterobasidiomycetous yeasts I. *Leucosporidium scottii* (*Candida scottii*). J Cell Sci 10:857–881

McCully EK, Robinow CF (1972b) Mitosis in heterobasidiomycetous yeasts II. *Rhodosporidium* sp. (*Rhodotorula glutinis*) and *Aessosporon salmonicolor* (*Sporobolomyces salmonicolor*). J Cell Sci 11:1–31

McLaughlin DJ, Frieders EM, Lu H (1995) A microscopist's view of heterobasidiomycete phylogeny. Stud Mycol 38:91–110

Messner R, Prillinger H, Altmann F, Lopandic K, Wimmer K, Molnár O, Weigang F (1994) Molecular characterization and application of random amplified polymorphic DNA analysis of *Mrakia* and *Sterigmatomyces* species. Int J Syst Bacteriol 44:694–703

Meyer W, Mitchell TG, Freedman EZ, Vilgalys R (1993) Hybridization probes for conventional DNA fingerprinting used as single primers in the polymerase chain reaction to distinguish strains of *Cryptococcus neoformans*. J Clin Microbiol 31:2274–2280

Middelhoven WJ, Koorevaar M, Schuur GW (1992) Degradation of benzene compounds by yeasts in acidic soils. Plant Soil 145:37–43

Mills SC, Child JJ, Spencer JFT (1971) The utilization of aromatic compounds by yeasts. Antonie van Leeuwenhoek 37:281–287

Mitchell TG, Perfect JR (1995) Cryptococcosis in the era of AIDS – 100 years after the discovery of *Cryptococcus neoformans*. J Clin Microbiol Rev 8:515–548

Mitchell TG, Freedman EZ, White TG, Taylor JW (1994) Unique oligonucleotide primers in PCR for identification of *Cryptococcus neoformans*. J Clin Microbiol 31:253–255

Möller A (1895) Protobasidiomyceten. In: Schimper AFW (ed) Botanische Mitteilungen aus den Tropen 8. Gustav Fischer, Jena, pp 1–179

Moore RT (1972) Ustomycota, a new division of higher fungi. Antonie van Leeuwenhoek 38:567–584

Moore RT (1980) Taxonomic proposals for the classification of marine yeasts and other yeast-like fungi including the smuts. Bot Mar 23:361–373

Moore RT (1987) Micromorphology of yeasts and yeast-like fungi and its taxonomic implications. Antonie van Leeuwenhoek 45:113–118

Moore RT, Kreger-van Rij NJW (1972) Ultrastructure of *Filobasidium* Olive. Can J Microbiol 18:1949–1951

Nagy A, Garamszegi N, Vágvölgyi C, Ferenczy L (1994) Electrophoretic karyotypes of *Phaffia rhodozyma* strains. FEMS Microbiol Lett 123:315–318

Nakase T (1989) Classification of ballistospore-forming yeasts. Yeast 5:S511–516

Nakase T, Okada G, Sugiyama J, Itoh M, Suzuki M (1989) *Ballistosporomyces*, a new ballistospore-forming anamorphic yeast genus. J Gen Appl Microbiol 35:289–309

Nakase T, Hamamoto M, Sugiyama J (1991a) Recent progress in the systematics of basidiomycetous yeasts. Jpn J Med Mycol 32 (Suppl):21–30

Nakase T, Itoh M, Mikata K, Banno I, Yamada Y (1991b) *Kockovaella*, a new ballistospore-forming anamorphic yeast genus. J Gen Appl Microbiol 37:175–197

Nakase T, Takematsu A, Yamada Y (1993) Molecular approaches to the taxonomy of ballistosporous yeasts based on the analysis of the partial nucleotide sequences of 18S ribosomal ribonucleic acids. J Gen Appl Microbiol 39:107–134

Nakase T, Suh S-O, Hamamoto M (1995) Molecular systematics of ballistoconidium-forming yeasts. Stud Mycol 38:163–173

Nakase T, Sukuzi M, Hamamoto M, Takashima M, Hatano T, Fukui S (1996) A taxonomic study on cellulolytic yeasts and yeast-like organisms isolated in Japan. II. The genus *Cryptococcus*. J Gen Appl Microbiol 42:7–15

Nei N (1971a) Microbiological decomposition of phenol I. Isolation and identification of phenol metabolizing yeasts. J Ferment Technol 49:655–660

Nei N (1971b) Microbiological decomposition of phenol II. Decomposition of phenol by *Rhodotorula glutinis*. J Ferment Technol 49:852–860

Neujahr HY, Varga JM (1970) Degradation of phenols by intact cells and cell-free preparations of *Trichosporon cutaneum*. Eur J Biochem 13:37–44

Nishida H, Ando K, Hirata A, Sugiyama J (1995) *Mixia osmundae*: transfer from the Ascomycota to the Basidiomycota based on evidence from molecules and morphology. Can J Bot 73:S660–666

Nyland G (1948) Preliminary observations on the morphology and cytology of an undescribed Heterobasidiomycete from Washington State. Mycologia 40:478–481

Nyland G (1949) Studies on some unusual Heterobasidiomycetes from Washington State. Mycologia 41:686–701

Oberwinkler F (1987) Heterobasidiomycetes with ontogenetic yeast stages – systematic and phylogenetic aspects. Stud Mycol 30:61–74

Oberwinkler F (1990) New genera of auricularioid heterobasidiomycetes. Rep Tottori Mycol Inst 28:113–127

Oberwinkler F, Bauer R (1990) *Cryptomycocolax*: a new mycoparasitic heterobasidiomycete. Mycologia 82:671–692

Oberwinkler F, Bandoni F, Blanz P, Kisimova-Horovitz L (1983) *Cystofilobasidium*: a new genus in the Filobasidiaceae. Syst Appl Microbiol 4:114–122

Oberwinkler F, Bauer R, Bandoni RJ (1990a) Heterogastridiales: a new order of Basidiomycetes. Mycologia 82:48–58

Oberwinkler F, Bauer R, Bandoni RJ (1990b) *Colacogloea*: a new genus in the Heterobasidiomycetes. Can J Bot 68:2531–2536

O'Donnell KL (1992) Ultrastructure of meiosis and the spindle pole body cycle in freeze-substituted basidia of the smut fungi *Ustilago maydis* and *Ustilago avenae*. Can J Bot 70:629–638

O'Donnell KL, McLaughlin DJ (1984) Ultrastructure of meiosis in *Ustilago maydis*. Mycologia 76:468–485

Olive LS (1968) An unusual new heterobasidiomycete with *Tilletia*-like basidia. J Elisha Mitchell Sci Soc 84:261–266

Perfect JR, Ketabchi N, Cox GM, Ingram CW, Beiser CL (1993) Karyotyping of *Cryptococcus* as an epidemiological tool. J Clin Microbiol 31:3305–3309

Perry K, Palukaitis P (1990) Transcription of tomato ribosomal DNA and the organization of the intergenic spacer. Mol Gen Genet 221:102–112

Phaff HJ, Miller MW, Mrak EM (1978) The life of yeasts, 2nd edn. Harvard University Press, Cambridge, 341 pp

Polanco C, Perez de la Vega M (1994) The structure of the rDNA intergenic spacer of *Avena sativa* L.: a comparative study. Plant Mol Biol 25:751–756

Pore RS, Sorenson WG (1990) *Reniforma strues*, a new yeast from wastewater. Mycologia 82:549–553

Prillinger H, Dörfler C, Laaser G, Eckerlein B, Lehle L (1990b) Ein Beitrag zur Systematik und Entwicklungsbiologie Höherer Pilze: Hefe-Typen der Basidiomyceten. Teil I: Schizosaccharomycetales, *Protomyces*-Typ. Z Mykol 56:219–250

Prillinger H, Dörfler C, Laaser G, Hauska G (1990a) Ein Beitrag zur Systematik und Entwicklungsbiologie Höherer Pilze: Hefe-Typen der Basidiomyceten. Teil III: *Ustilago*-Typ. Z Mykol 56:251–278

Prillinger H, Deml G, Dôrfler C, Laaser G, Lockau W (1991b) Ein Beitrag zur Systematik und Entwicklungsbiologie Höherer Pilze: Hefe-Typen der Basidiomyceten. Teil II: *Microbotryum*-Typ. Bot Acta 104:5–17

Prillinger H, Laaser G, Dörfler C, Ziegler K (1991a) Ein Beitrag zur Systematik und Entwicklungsbiologie Höherer Pilze: Hefe-Typen der Basidiomyceten. Teil IV: *Dacrymyces*-Typ, *Tremella*-Typ. Sydowia 53:170–218

Prillinger H, Oberwinkler F, Umile C, Tlachac K, Bauer R, Dörfler C, Taufratzhofer E (1993) Analysis of cell wall carbohydrates (neutral sugars) from ascomycetous and basidiomycetous yeasts with and without derivatization. J Gen Appl Microbiol 39:1–14

Ratledge C (1978) Lipids and fatty acids. In: Rose AH (ed) Economic microbiology, vol 2. Primary products of metabolism. Academic Press, London, pp 263–302

Ratledge C (1982) Microbial oils and fats: an assessment of their commercial potential. Prog Ind Microbiol 16:119–206

Ratledge C (1986) Lipids. In: Pape H, Rehm H-J (eds) Biotechnology, vol 4 1st edn, Microbial products II. VCH, Weinheim, pp 185–213

Ratledge C, Evans CT (1989) Lipids and their metabolism. In: Rose AH, Harrison JS (eds) The yeasts, vol 3. Metabolism and physiology of yeasts, 2nd edn. Academic Press, London, pp 367–455

Reiser J, Ochsner UA, Kälin M, Glumoff V, Fiechter A (1996) *Trichosporon*. In: Wolf K (ed) Nonconventional yeasts in biotechnology. A handbook. Springer, Berlin Heidelberg New York, pp 581–606

Rhodes JC, Kwon-Chung KJ, Popkins TJ (1981) Ultrastructure of the septal complex in hyphae of *Cryptococcus laurentii*. J Bacteriol 145:1410–1412

Roberts RG (1990) Postharvest biological control of gray mold of apple by *Cryptococcus laurentii*. Phytopathology 80:526–530

Rodrigues de Miranda L (1972) *Filobasidium capsuligenum* nov. comb. Antonie van Leeuwenhoek 44:439–450

Roeijmans H, Prillinger H, Umile C, Sugiyama J, Nakase T, Boekhout T (1997) Analysis of carbohydrate composition of cell walls and extracellular carbohydrates. In: Kurtzman CP, Fell JW (eds) The yeasts, a taxonomic study, 4th edn. Elsevier, Amsterdam, pp 99–101

Rolph CE, Moreton RS, Harwood JL (1989) Acyl lipid metabolism in the oleaginous yeast *Rhodotorula gracilis* (CBS 3043). Lipids 24:715–720

Sampaio JP (1999) Utilization of low molecular weight aromatic compounds by heterobasidiomycetous yeasts: taxonomic implications. Can J Microbiol 45:491–512

Sampaio JP, Fonseca A (1995) Physiological aspects in the systematics of heterobasidiomycetous yeasts. Stud Mycol 38:29–46

Sampaio JP, Bauer R, Begerow D, Oberwinkler F (1999a) *Occultifur externus* sp. nov., a new species of simple-pored auricularioid heterobasidiomycete from plant litter in Portugal. Mycologia 91:1094–1104

Sampaio JP, Fell JW, Gadanho M, Bauer R (1999b) *Kurtzmanomyces insolitus* sp. nov., a new anamorphic heterobasidiomycetous yeast species. Syst Appl Microbiol 22:619–625

Schmeding KA, Jong SC, Hugh R (1981) Monokaryotic fruiting and its sexuality in self-fertile strains of *Filobasidiella neoformans* (*Cryptococcus neoformans*). Trans Mycol Soc Jpn 22:1–10

Shoda M, Udaka S (1980) Preferential utilization of phenol rather than glucose by *Trichosporon cutaneum* possessing a partially constitutive catechol 1,2-oxygenase. Appl Environ Microbiol 39:1129–1133

Simmons RB, Ahearn DG (1987) Cell wall structure and diazoneum blue B reaction of *Sporopachydermia quercuum*, *Bullera tsugae*, and *Malassezia* spp. Mycologia 79:38–43

Simmons RB, Guého E (1990) A new species of *Malassezia*. Mycol Res 94:1146–1149

Sorrell TC, Chen SCA, Ruma P, Meyer W, Pfeiffer TJ, Ellis DH, Brownlee AG (1996) Concordance of clinical and environmental isolates of *Cryptococcus neoformans* var. *gattii* by random amplification of polymorphic DNA analysis and PCR fingerprinting. J Clin Microbiol 34:1253–1260

Sowell G (1953) Infection of parsnip by *Itersonilia* sp. Phytopathology 43:485

Sugiyama J, Suh S-O (1993) Phylogenetic analysis of basidiomycetous yeasts inferred from small subunit ribosomal DNA sequence. J Gen Microbiol 139:1595–1598

Sugiyama J, Hamamoto M (1998) *Erythrobasidium* Hammoto, Sugiyama & Komagata. In: Kurtzman CP, Fell JW (eds) The yeasts, a taxonomic study 4th, edn. Elsevier, Amsterdam, pp 654–655

Sugiyama Y, Fukagawa M, Chiu SW, Komagata K (1985) Cellular carbohydrate composition, DNA base composition, ubiquinone systems and diazonium blue B color test in the genera *Rhodosporidium*, *Leucosporidium*, *Rhodotorula* and related basidiomycetous yeasts. J Gen Appl Microbiol 31:519–550

Suh S-O, Nakase T (1995) Phylogenetic analysis of the ballistosporogenous anamorphic genera *Udeniomyces* and *Bullera*, and related basidiomycetous yeasts, based on 18S rDNA sequence. Microbiology 141:901–906

Suh S-O, Sugiyama J (1993) Phylogeny among the basidiomycetous yeasts inferred from small subunit ribosomal DNA sequence. J Gen Microbiol 139:1595–1598

Suh S-O, Sugiyama J (1994) Phylogenetic placement of the basidiomycetous yeasts *Kondoa malvinella* and *Rhodosporidium dacryoideum*, and the anamorphic yeast *Sympodiomycopsis paphiopedili* by means of 18S rRNA gene sequence analysis. Mycoscience 35:367–375

Suh S-O, Hirata A, Sugiyama J, Komagata K (1993) Septal ultrastructure of basidiomycetous yeasts and their taxonomic implications with observations on the ultrastructure of *Erythrobasidium hasegawianum* and *Sympodiomycopsis paphiopedili*. Mycologia 85:30–37

Swann EA, Taylor JW (1993) Higher taxa of basidiomycetes: an 18S rRNA gene perspective. Mycologia 85:923–936

Swann EA, Taylor JW (1995a) Phylogenetic diversity of yeast-producing basidiomycetes. Mycol Res 99:1205–1210

Swann EA, Taylor JW (1995b) Toward a phylogenetic systematics of the Basidiomycota: integrating yeasts and filamentous basidiomycetes using 18S rRNA gene sequences. Stud Mycol 38:147–162

Swann EA, Taylor JW (1995c) Phylogenetic perspectives on basidiomycete systematics: evidence from the 18S rRNA gene. Can J Bot 73:S862–S868

Swann EC, Frieders EM, McLaughlin DJ (1999) *Microbotryum*, *Kriegeria* and the changing paradigm in basidiomycete classification. Mycologia 91:51–66

Takashima M, Suh S-O, Nakase T (1995) *Bensingtonia musae* sp nov. isolated from a dead leaf of *Musa paradisiaca* and its phylogenetic relationship among basidiomycetous yeasts. J Gen Appl Microbiol 41:143–151

Takeo K, Tanaka R, Taguchi H, Nishimura K (1993) Analysis of ploidy and sexual characteristics of natural isolates of *Cryptococcus neoformans*. Can J Microbiol 39:958–963

Tautz D, Tautz C, Webb D, Dover GA (1987) Evolutionary divergence of promoters and spacers in the rDNA family of four *Drosophila* species. Implications for the molecular coevolution in multigene families. J Mol Biol 195:525–542

Taylor JW, Wells K (1979) A light and electron microscopic study of mitosis in *Bullera alba* and the histochemistry of some cytoplasmic substances. Protoplasma 98:31–62

Templeton AR (1983) Systematics of basidiomycetes based on 5S rRNA sequences and other data. Nature 303:731

Urquehart EJ, Menzies JG, Punja ZK (1994) Growth and control activity of *Tilletiopsis* species against powdery mildew (*Sphaerotheca fuliginea*) on greenhouse cucumber. Phytopathology 84:341–351

Vahidi H, Honda BM (1991) Repeats and subrepeats in the intergenic spacer of rDNA from the nematode *Meloidogyne arenaria*. Mol Gen Genet 227:334–336

van Belkum A, Boekhout T, Bosboom R (1994) Monitoring spread of *Malassezia* infections in a neonatal intensive care unit by PCR-mediated genetic typing. J Clin Microbiol 32:2528–2532

Van de Peer Y, Hendriks L, Goris A, Neefs JM, Vancanneyt M, Kersters K, Berny JF, Hennebert GL, De Wachter R (1992) Evolution of basidiomycetous yeasts as deduced from small ribosomal subunit RNA sequences. Syst Appl Microbiol 15:250–258

van der Walt JP, Hopsu-Havu VK (1976) A colour reaction for the differentiation of ascomycetous and hemibasidiomycetous yeasts. Antonie van Leeuwenhoek 42:157–163

Vilgalys R, Hester M (1990) Rapid genetic identification and mapping of enzymatically amplified ribosomal DNA from several *Cryptococcus* species. J Bacteriol 172:4238–4246

von Arx JA (1979) Propagation in the yeasts and yeast-like fungi. In: Kendrick B (ed) The whole fungus. Waterloo, Canada, pp 201–213

von Arx JA, Weijman ACM (1979) Conidiation and carbohydrate composition in some *Candida* and *Torulopsis* species. Antonie van Leeuwenhoek 45:547–555

Vos P, Hogers R, Bleeker M, Reijans M, van de Lee T, Hornes M, Frijters A, Peleman J, Kuiper M, Zabeau M (1995) AFLP: a new technique for DNA fingerprinting. Nucleic Acids Res 23:4407–4414

Walker WF (1984) 5S rRNA sequences from Atractiellales, basidiomycetous yeasts and fungi imperfecti. Syst Appl Microbiol 5:352–359

Walker WF, Doolittle WF (1982) Redividing the basidiomycetes on the basis of 5S rRNA nucleotide sequences. Nature 299:723–724

Weijman ACM (1979) Carbohydrate composition and taxonomy of *Geotrichum*, *Trichosporon* and allied genera. Antonie van Leeuwenhoek 45:119–127

Weijman ACM, Golubev WI (1987) Carbohydrate patterns and taxonomy of yeasts and yeast-like fungi. Stud Mycol 30:361–371

Wells K (1994) Jelly fungi, then and now! Mycologia 86:18–48

Wickes BL, Moore TDE, Kwon-Chung KJ (1994) Comparison of the electrophoretic karyotypes and chromosomal location of ten genes in the two varieties of *Cryptococcus neoformans*. Microbiology 140:543–550

Wickes BL, Mayorga ME, Edman U, Edman JC (1996) Dimorphism and haploid fruiting in *Cryptococcus neoformans*: association with the a-mating type. Proc Natl Acad Sci USA 93:7327–7331

Wong GJ (1987) A comparison of the mating system of *Tremella mesenterica* and other modified bifactorial species. Stud Mycol 30:431–441

Yamada Y, Banno I (1984) *Fellomyces*, a new anamorphic yeast genus for the Q10-equipped organisms whose conidium is freed by an end-break in the sterigma. J Gen Appl Microbiol 30:523–525

Yamada Y, Kawasaki H (1989) The molecular phylogeny of the Q8-equipped basidiomycetous yeasts genera *Mrakia* Yamada et Komagata and *Cystofilobasidium* Oberwinkler et Bandoni based on the partial sequences of 18S and 26S ribosomal ribonucleic acids. J Gen Appl Microbiol 35:173–183

Yamada Y, Kondo K (1973) Coenzyme Q system in the classification of the yeast genera *Rhodotorula* and *Cryptococcus* and the yeast-like genera *Sporobolomyces* and *Rhodosporidium*. J Gen Appl Microbiol 19:59–77

Yamada Y, Nakagawa Y (1990) The molecular phylogeny of the basidiomycetous yeast species, *Leucosporidium scottii* based on the partial sequences of 18S and 26S ribosomal ribonucleic acids. J Gen Appl Microbiol 36:63–68

Yamada Y, Nakagawa Y (1992) The phylogenetic relationships of some heterobasidiomycetous yeast species based on the partial sequences of 18S and 26S ribosomal RNAs. J Gen Appl Microbiol 38:559–565

Yamada Y, Banno I, von Arx JA, van der Walt JP (1987) Taxonomic significance of the coenzyme Q system in yeasts and yeast-like fungi. Stud Mycol 30:299–308

Yamada Y, Itoh M, Kawasaki H, Banno I, Nakase T (1988a) *Kurtzmanomyces* gen. nov., an anamorphic yeast genus for the Q10-equipped organisms whose conidium is freed by an end-break in the sterigma which branches or elongates to produce additional conidia and whose cells contain no xylose. J Gen Appl Microbiol 34:503–506

Yamada Y, Kawasaki H, Itoh M, Banno I, Nakase T (1988b) *Tsuchiyaea* gen. nov., an anamorphic yeast genus for the Q9-equipped organism whose reproduction is either by enteroblastic budding or by the formation of conidia which are disjointed at a septum in the midregion of the sterigmata and whose cells contain xylose. J Gen Appl Microbiol 34:507–510

Yamada Y, Nakagawa Y, Banno I (1989a) The phylogenetic relationship of the Q9-equipped species of the heterobasidiomycetous yeast genera *Rhodosporidium* and *Leucosporidium* based on the partial sequences of 18S and 26S ribosomal ribonucleic acids: proposal of a new genus *Kondoa*. J Gen Appl Microbiol 35:377–385

Yamada Y, Kawasaki H, Nakase T, Banno I (1989b) The phylogenetic relationship of the conidium-forming anamorphic yeast genera *Sterigmatomyces*, *Kurtzmanomyces*, *Tsuchiyaea* and *Fellomyces*, and the teleomorphic yeast genus *Sterigmatosporidium* on the basis of the partial sequnces of 18S and 26S ribosomal ribonucleic acids. Agric Biol Chem 53:2993–3001

Yamada Y, Nagahama T, Kawasaki H, Banno I (1990a) The phylogenetic relationship of the genera *Phaffia* Miller, Yoneyama et Soneda and *Cryptococcus* Kutzing emend. Phaff et Spencer (Cryptococcaceae) based on the partial sequences of 18S and 26S ribosomal ribonucleic acids. J Gen Appl Microbiol 36:403–414

Yamada Y, Nakagawa Y, Banno I (1990b) The molecular phylogeny of the Q10-equipped species of the heterobasidiomycetous yeast genus *Rhodosporidium* Banno based on the partial sequences of 18S and 26S ribosomal ribonucleic acids. J Gen Appl Microbiol 36:435–444

Yamada Y, Maeda K, Mikata K (1994) The phylogenetic relationships of *Rhodosporidium dacryoideum* Fell, Hunter et Tallman based on the partial sequences of 18S and 26S ribosomal RNAs: the proposal of *Sakaguchia* gen. nov., a heterobasidiomycetous yeast genus. Biosci Biotechnol Biochem 58:99–103

Yarrow D (1998) Methods for the isolation, maintenance and identification of yeasts. In: Kurtzman CP, Fell JW (eds) The yeasts, a taxonomic study, 4th edn. Elsevier, Amsterdam, pp 77–100

Ykema A, Verbree EC, Kater MM, Smit H (1988) Optimization of lipid production in the oleaginous yeast *Apiotrichum curvatum* in whey permeate. Appl Microbiol Biotechnol 29:211–218

Ykema A, Verbree EC, Nijkamp HJH, Smit H (1989) Isolation and characterization of fatty acid auxotrophs from the oleaginous yeast *Apiotrichum curvatum*. Appl Microbiol Biotechnol 32:76–84

Zache G, Rehm H-J (1989) Degradation of phenol by a co-immobilized entrapped mixed culture. Appl Microbiol Biotechnol 30:426–432

2 Urediniomycetes

E.C. Swann[1], E.M. Frieders[2], and D.J. McLaughlin[1]

CONTENTS

I. Introduction 37
II. Ecological Diversity 38
A. Plant Associates 38
B. Insect Associates 39
C. Mycoparasites 39
D. Aquatic Fungi and Aquatic Adaptations 39
III. Morphological
and Reproductive Diversity 40
A. Fruitbodies 40
B. Basidia and Basidiospores 41
C. Asexual Reproduction 41
IV. Biochemical Characters 43
V. Ultrastructural Characters 43
A. Septa 43
B. Spindle Pole Bodies 45
C. Colacosomes and Lenticular Bodies 45
D. Microscala 46
E. Other Ultrastructural Characters 46
VI. Phylogeny and Systematics 46
A. Phylogenetic Position of Uredinales 47
B. Dismemberment of Auriculariales 47
C. Teliospore-Forming Yeasts 48
VII. Classification 48
A. Order Atractiellales 49
B. *Mixia osmundae* 49
C. Subclass Microbotryomycetidae 49
D. Subclass Agaricostilbomycetidae 51
E. *Erythrobasidium, Naohidea, Sakaguchia* 51
F. Subclass Urediniomycetidae 51
VIII. Culture 52
IX. Conclusion 52
References 52

I. Introduction

The concept of the monophyletic group of organisms included in the class Urediniomycetes is relatively new (Swann and Taylor 1995b). This taxon contains the rust fungi, the Septobasidiales, and various species that formerly were classified as smuts (e.g., *Microbotryum, Sphacelotheca, Rhodosporidium*), jelly fungi (e.g., *Eocronartium, Helicobasidium, Platygloea, Phleogena*), or with other groups (*Pachnocybe* and *Agaricostilbum* with deuteromycetes; *Mixia osmundae* with ascomycetes). In taking a phylogenetic approach to this taxon, we are explicitly accommodating asexual species in the class, such as the asexual yeasts (see Fell et al., Chap. 1, this Vol.; Swann and Taylor 1995a), and the aquatic hyphomycete *Naiadella fluitans*. In order to provide a name for this diverse group, Swann and Taylor (1995b) enlarged the concept of the existing class name Urediniomycetes, which had previously contained only the Uredinales (Hawksworth et al. 1983).

The circumscription of the Urediniomycetes primarily relies on biochemical and ultrastructural characters that also have been useful in defining subgroups within the class. The use of these characters permits a genuinely phylogenetic concept of the class, in that they are obtainable from sexual and asexual taxa equally. In contrast, reliance on characters associated with sexual morphology suffers from problems caused by homoplasy, and imposes an artificial sexual/asexual duality. The phylogenetic coherence of asexual genera has been improved through the application of biochemical and genomic information to their taxonomy (see Fell et al., Chap. 1, this Vol.). However, until asexual species can be classified with their sexual relatives, neither "anamorph genera" nor "teleomorph genera" will be natural (i.e., monophyletic).

Based on estimates contained in the *Dictionary of the Fungi* (Hawksworth et al. 1995) and other sources (e.g., Vánky 1998), the class Urediniomycetes constitutes about one-third of all teleomorphic basidiomycete species (approximately 7374 spp.), distributed among 214 genera. At least 20% of rust genera, and 60% of nonrust genera, are monotypic. The Uredinales, with over 7000 species, accounts for over 95% of the species in the class (Table 1). The next largest orders are

---

[1] Department of Plant Biology, University of Minnesota, St. Paul, Minnesota 55108 USA
[2] Department of Biology, University of Wisconsin-Platteville, Platteville, Wisconsin 53818 USA

Table 1. Number and percent of genera and species in the three largest orders of Urediniomycetes[a]

	Genera	Species
Uredinales	164 (>75%)	>7000 (>95%)
Septobasidiales	4 (<2%)	175 (2.4%)
Microbotryales	6 (2.8%)	80 (1.1%)
Total	174 (81%)	7255 (>98%)

[a] Total Urediniomycetes genera = 214, species = 7374.

the Septobasidiales and the Microbotryales, a recently circumscribed smut taxon (Table 1). Smaller groups, such as the Atractiellales, Agaricostilbales, Chionosphaeraceae, and other orders and families, represent most of the phylogenetic and ecological diversity within the class but constitute less than 2% of the known species. These numbers are an underestimate, because new genera and species are being added rapidly, and we have not included the large number of asexual (anamorphic) species known to be part of the class.

The most widely known members of the Urediniomycetes are the phytoparasitic rust fungi. These fungi are pathogens on a wide diversity of economically important angiosperms and gymnosperms, potentially causing large losses in yield of agricultural and forest products. In 1998 over 1.7 million m^3 of grain were lost due to rust disease in the USA (D.L. Long, pers. comm.), and about 10% of the crop is lost worldwide annually (Agrios 1997). The rusts also have a great economic impact because of the need for disease control, including basic research on the fungal parasites, development of resistant host cultivars, eradication efforts, and development and application of fungicides. For example, when coffee rust was first introduced into the New World in 1970, the Brazilian government spent $9 million to eradicate infected coffee plants in a futile attempt to control the spread of this disease (Strange 1993). However, a few rusts are being tested for use as biocontrol agents or mycoherbicides. *Puccinia lagenophorae* has been used to successfully control celeriac crop losses due to competition with the annual weed *Senecio vulgaris* (Muller and Rieger 1998). Other phytopathogenic members of the Urediniomycetes are economically important, but to a much lesser extent than the rusts. *Sphacelotheca fagopyri* causes smut on buck-wheat. Several species of *Helicobasidium* are resposible for root rots of minor economic crops such as mulberry and asparagus. The destruction or death of host trees (e.g., *Citrus*, *Magnolia*, *Pyrus*, *Quercus*, *Thea*) has been attributed to insect-associated *Septobasidium* species through the combined effect of both the insect and fungus (Couch 1938). The potential for a role in wood decay exists in some species that are associated with wood and wood products, such as *Pachnocybe ferruginea* (Wang and Zabel 1990) and species of *Atractiella*, *Helicogloea*, *Phleogena*, and *Platygloea*.

II. Ecological Diversity

Urediniomycetes are ubiquitous in nature, present in both aquatic and terrestrial environments. The terrestrial species of this group are better known than aquatic species; however, a number of teliospore-forming yeasts have been collected from estuarine and marine environments (Fell et al. 1988, 1992). Because of their small size and inconspicuous nature, many Urediniomycetes go unnoticed unless one is specifically looking for them, frequently requiring the use of microscopy or culture methods. Modes of nutrition range from parasitism to saprotrophy, with parasitic species constituting the majority (e.g., the rusts).

A. Plant Associates

Plants and plant surfaces are rich sources of Urediniomycetes. The order Uredinales is by far the largest and best studied group of species in the class. Rusts are obligate parasites of a wide diversity of land plants, including the lycophyte *Selaginella* (Hennen 1997), ferns, gymnosperms, and angiosperms. The rusts are characterized by a complex life history, often alternating between two unrelated plant hosts. Other groups of plant parasites include the Microbotryales (a group of smut fungi), the sedge parasite *Kriegeria eriophori*, the moss parasites *Jola* and *Eocronartium*, and the *Lonicera* parasite *Insolibasidium*. A number of presumably saprobic species are often associated with living or dead plant material. Ballistosporic and teliospore-forming yeasts, *Stilbum vulgare*, *Agaricostilbum* spp., *Pachnocybe ferrug-*

inea, and the Atractiellales (*Atractiella*, *Phleogena*, and *Helicogloea*) are variously associated with dead or decaying plant material, or surfaces and exudates of living plants (Nyland 1949; Oberwinkler and Bandoni 1982a; Oberwinkler and Bauer 1989).

B. Insect Associates

There are relatively few Urediniomycetes that are specifically associated with animals; the only widespread zoopathogenic urediniomycete group is the Septobasidiales, all of which are associated with scale insects. This group is most diverse in tropical and subtropical areas, with fewer species in temperate climates (Couch 1938). Species in this order obtain their nutrition from scale insects, which, in turn, feed on living plants, constituting an indirect form of phytoparasitism that in extreme cases can kill the plant host (Couch 1938).

C. Mycoparasites

Mycoparasites are scattered throughout the Urediniomycetes. There is little information available for even the best known mycoparasites in this group regarding molecular phylogenetic data and/or the ultrastructure of the host/parasite interaction. Mycoparasites in the Microbotryomycetidae are generally characterized by the presence of colacosomes (see Sect. V, below). These organelles were first described in *Colacogloea peniophorae*, a parasite of the basidiomycete *Peniophora* (Oberwinkler et al. 1990a), and have been demonstrated in a number of other species (Bauer et al. 1997). Colacosomes also have been found in *Leucosporidium*, *Rhodosporidium*, and *Sporidiobolus* (Bauer et al. 1997), indicating a potential for mycoparasitism in these genera that have been assumed to be saprobic.

Mycoparasites in other groups include *Cystobasidium*, *Naohidea*, *Mycogloea*, members of the Chionosphaeraceae, and some species of *Platygloea* (Bandoni 1995). Some lichen parasites (Diederich and Christiansen 1994) also may belong in the Urediniomycetes. The nature of the host/parasite interface in these fungi is unknown. Some of these species produce haustorial branches that are reminiscent of those found in some Tremellales (Oberwinkler 1990; Roberts 1994, 1996, 1997). The phenomenon of mycoparasitism may be more widespread among Urediniomycetes than is currently suspected. Many of the newly described species of Urediniomycetes are mycoparasites, and species currently thought to be saprobes may be capable of parasitizing fungi.

D. Aquatic Fungi and Aquatic Adaptations

Basidiomycetes adapted to aquatic habitats are uncommon, suggesting that the condition is derived within the phylum, and few of these fungi may be restricted to aquatic situations. Only a few Urediniomycetes appear to be aquatic for part or all of their life cycle. *Naiadella fluitans* (Marvanová and Bandoni 1987) and *Camptobasidium hydrophilum* (Marvanová and Suberkropp 1990) are filamentous species found in freshwater that produce conidia with radiating appendages like other Ingoldian fungi (see Seifert and Gams, Chap. 14, Vol. VII Part A). *Heterogastridium pycnidioideum*, occurring on leaf litter and other habitats on the forest floor (Bandoni and Oberwinkler 1981; Oberwinkler et al 1990b), produces tetraradiate basidiospores, presumably adapted for water dispersal. Marine Urediniomycetes appear to be confined to yeasts, such as *Leucosporidium* and *Rhodosporidium* spp., which may be indigenous to oceans, but other marine species, especially those of estuaries, may, like freshwater yeasts, be derived from terrestrial habitats (Lachance and Starmer 1998; see Fell et al., Chap. 1, this Vol.).

Other Urediniomycetes are potentially aquatic or adapted to water dispersal, but little is known of the biology and ecology of these organisms. Gasteroid taxa, such as *Atractiella*, *Stilbum* (Oberwinkler et al. 1990b), and *Pachnocybe* (Kleven and McLaughlin 1988), are suspected of using water as a means of dispersal. In *Pachnocybe ferruginea* conidium production is vigorous in aerated water cultures, although these conidia show no obvious aquatic adaptation similar to the aquatically produced basidiospores of *Camptobasidium* (Kleven and McLaughlin 1988; Marvanová and Suberkropp 1990). In *Mycogloea* (Olive 1950) and *Kriegeria* (Doublés and McLaughlin 1992) deciduous basidia act as propagules in water, and, when submerged, change from ballistosporic to aballistosporic spore production. The recent finding of freshwater species and the numerous probable adaptations to water dispersal suggest that additional aquatic taxa may be expected.

III. Morphological and Reproductive Diversity

A. Fruitbodies

Because of the tremendous diversity of its members, characterizing the fruitbodies of the Urediniomycetes as a whole is difficult. When present, fruitbodies are usually small in size and range from relatively complex to almost nonexistent. Their texture may be soft and gelatinous to hard and dry. Species in at least four different groups produce stilboid fruitbodies (Fig. 1), Chionosphaeraceae (*Stilbum*, *Chionosphaera*, *Fibulostilbum*), Agaricostilbaceae (*Agaricostilbum*), Pachnocybaceae (*Pachnocybe*), and Atractiellales (*Atractiella*, *Phleogena*). Other fruitbody configurations include resupinate (Fig. 2) (*Helicogloea*, *Helicobasidium*, *Septobasidium*), clavarioid (*Eocronartium*), and pycnidioid (*Heterogastridium*). The fruitbodies of some parasites do not conform easily to standard categories. These species may produce scattered basidia within or on their host (*Kriegeria*, *Herpobasidium*); they may produce basidia in or on relatively

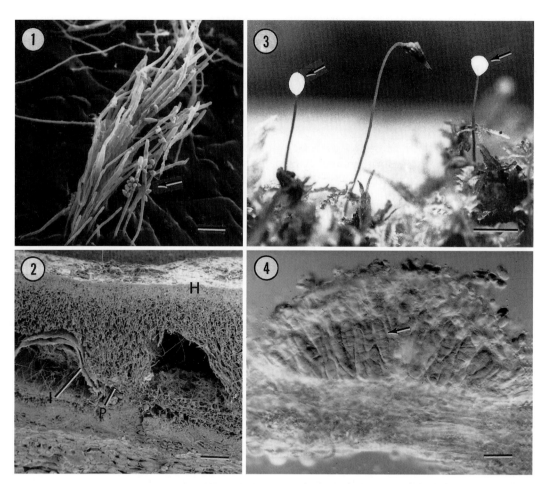

Figs. 1–4. Urediniomycete fruiting bodies. **Fig. 1.** Gasteroid fruiting body of *Agaricostilbum pulcherrimum* produced in culture with clustered sessile basidiospores (*arrow*). (F. Oberwinkler, culture F219). *Bar* 20 µm. **Fig. 2.** Vertical section through *Septobasidium carestianum* fruiting body on branch of *Cornus* sp. with hymenium (*H*), pillars (*P*), insect remains (*I*), and bark of host (below). (Micrograph by J.C. Doublés; McLaughlin, DJM 644). *Bar* 100 µm.

Fig. 3. *Jola* cf. *javensis* fruiting on sporophyte of the moss host (*arrow*) in Costa Rica. Uninfected moss sporophyte (center). (Photograph by J.C. Doublés). *Bar* 2 mm; magnification approximate. **Fig. 4.** Vertical section through telium of *Coleosporium asterum* with septate basidia (*arrow*) and basidiospores on hymenial surface; host tissue (below). (Swann, ECS 502). *Bar* 50 µm

discrete fruitbodies (*Jola*, Fig. 3) or sori (Uredinales, Fig. 4); or they may completely overgrow and incorporate the host structures on which they are fruiting (*Cryptomycocolax*). Urediniomycetous yeasts often produce scattered teliospores in culture but it is unknown where these are produced in nature.

B. Basidia and Basidiospores

Morphological variation is a hallmark of the Urediniomycetes and basidium morphology is no exception (Figs. 5–11). Since the turn of the century, basidium morphology has been an important character in basidiomycete systematics, and has played a central role in the development of taxonomies and evolutionary hypotheses (Rogers 1934, 1971; Talbot 1968; Wells 1994). Terminological debates centered around the interplay between morphological and cytological criteria for defining parts of the basidium. Differing interpretations of the significance of basidial characters and how to define them have contributed to instability in heterobasidiomycete systematics.

The viewpoint taken in this chapter is that the definition should be based on a morphological differentiation of the basidium into distinct parts in association with karyogamy and meiosis. *Probasidia* are specialized regions of the basidium in which karyogamy occurs, but may have different developmental origins. Probasidia may be intercalary (teliospore-forming yeasts), lateral (*Helicogloea*, Fig. 11; *Kriegeria*, not shown), or terminal (*Jola*, Fig. 8). When the probasidium is thick-walled and functions as a diaspore or resistant spore, it is usually called a *teliospore* (Fig. 6); however, the distinction between teliospores and probasidia is murky. By tradition, probasidia of rust fungi are called teliospores whether they are thick-walled resistant structures or not. The distinct morphological compartment in which meiosis occurs is called the *metabasidium*. Metabasidia of the Urediniomycetes are typically transversely segmented with spores borne laterally (Figs. 5–7, 11). Metabasidia are usually persistent, but may be deciduous, and easily separate from the probasidium (*Kriegeria*, Fig. 9; *Mycogloea*). Some taxa have basidia that do not differentiate morphologically into separate compartments in which karyogamy and meiosis occur. These may be holobasidia (*Chionosphaera*; *Pachnocybe*, Fig. 10) or phragmobasidia (*Atractiella*, Fig. 5).

The simple definitions of pro- and metabasidia presented above are not always easily applied, nor is it particularly useful to make such distinctions in all cases. In the rust fungus *Coleosporium asterum*, karyogamy takes place in what is conventionally called a teliospore. The teliospore does not germinate to produce an external, morphologically distinct metabasidium. Instead, meiosis occurs within the teliospore wall, with a transverse septum laid down after each division, resulting in a four-celled metabasidium (Fig. 4). Another example of the difficulties in defining pro- and metabasidia is seen in endocyclic rust species. Endocyclic rusts such as *Endophyllum sempervivi* and *Gymnoconia peckiana* produce external, transversely septate metabasidia; however, they are produced by spores that otherwise fit the morphological description of aeciospores.

Basidiospores are produced in a variety of modes. Spores may be forcibly abstricted from sterigmata (Fig. 11), or may be sessile and released passively (Figs. 5–7, 10). When the basidium or metabasidium is septate, basidiospores may be produced one per segment (Figs. 5, 11), or several may be produced by each segment (*Agaricostilbum*, Fig. 7; *Mycogloea*). Basidiospores may be produced endogenously [*Coleosporium*, *Agaricostilbum*, *Atractiella* (see below)] or exogenously, and sterigmata may be extensions of the basidial wall or arise endogenously. Basidiospores may germinate by repetition, bud as a yeast, produce microconidia, or produce a hyphal germ tube.

C. Asexual Reproduction

Urediniomycetes are frequently dimorphic, with a dikaryotic hyphal phase (+/– clamp connections) and a monokaryotic yeast phase. Groups that lack a yeast phase often have one or more conidial states (Oberwinkler and Bauer 1989). A summary of basidiomycete conidial states can be found in Kendrick and Watling (1979). The most conspicuous elaboration of conidia is in the rust life cycle (Uredinales), where three to four types of asexual spores may be produced by a single species (monokaryotic spermatia, dikaryotic aeciospores, urediniospores, and amphispores). *Insolibasidium deformans* has a distinctive dikaryotic conidial state. The moss parasites *Eocronartium* and *Jola* (Fig. 13) produce a *Sporothrix*-like sympodioconidial state (Frieders 1997). Conidial states are also known in *Spiculogloea occulta*, *Colacogloea penio-*

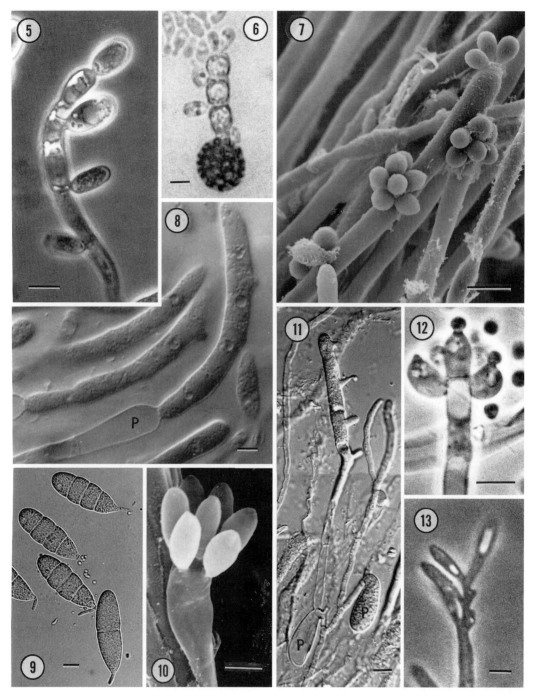

Figs. 5–13. Basidial and conidial morphology in the Urediniomycetes. **Figs. 5–11.** Basidia. **Fig. 5.** Gasteroid auricularioid basidium of *Atractiella* sp. with sessile basidiospores. Differentiated probasidium absent. (Swann, culture from ECS CR27). *Bar* 10 μm. **Fig. 6.** Gasteroid basidium of the urediniomycete smut *Microbotryum reticulatum* with teliospore, sessile basidiospores and yeast stage. (Swann, ECS 698). *Bar* 5 μm. **Fig. 7.** Gasteroid auricularioid basidium of *Agaricostilbum pulcherrimum* with multiple basidiospores on each compartment. (F. Oberwinkler, culture F219). *Bar* 5 μm. **Fig. 8.** Ballistosporic auricularioid basidium of *Jola* cf. *javensis* with differentiated probasidium (*P*). (Frieders, EMF 004). *Bar* 5 μm. **Fig. 9.** Deciduous auricularioid metabasidia of *Kreigeria eriophori* prior to basidiospore production. (Micrograph by J.C. Doublés; see Doublés and McLaughlin 1992). *Bar* 10 μm. **Fig. 10.** Gasteroid holobasidium of *Pachnocybe ferruginea* with apical basidiospores. (Kleven and McLaughlin 1988). *Bar* 2.5 μm. (Kleven and McLaughlin 1988 with permission from Mycologia). **Fig. 11.** Maturing ballistosporic auricularioid basidium of *Helicogloea intermedia* with saccate lateral probasidium (*P*) and an adjacent probasdium prior to metabasidium formation. (Micrograph J.C. Doublés). *Bar* 10 μm. **Figs. 12, 13.** Conidia and conidiophores. **Fig. 12.** Microconidia formation in *Atractiella* sp. (McLaughlin, culture DJM 969). *Bar* 5 μm. **Fig. 13.** Sympodial conidium formation in *Jola* cf. *javensis*. (McLaughlin, culture DJM 739). *Bar* 5 μm. **Figs. 5, 6, 8, 9, 11–13**, bright-field micrographs; **7, 10**, scanning electron micrographs

phorae, *Zygogloea gemellipara*, and *Occultifur internus* (Oberwinkler 1990; Oberwinkler et al. 1990a; Roberts 1994, 1996). Production of microconidia is fairly common among heterobasidiomycetes (Wells 1994). Some Urediniomycetes such as *Platygloea effusa* and members of the Atractiellales produce microconidia on hyphal conidiophores (Fig. 12) or directly from basidiospores (Ingold 1988, 1992; Oberwinkler and Bauer 1989; Oberwinkler et al. 1990a). Some Urediniomycetes are capable of producing endospores (*Naohidea*, Oberwinkler 1990; *Sporidiobolus*, Bandoni 1984; *Atractiella*, Swann, unpubl. observ.). Urediniomycetes that are known primarily or solely as yeasts are discussed in detail by Fell et al. (Chap. 1, this Vol.).

IV. Biochemical Characters

A number of biochemical characters are useful in defining the Urediniomycetes, or for contrasting it with the other two major phylogenetic groups of Basidiomycota. These include the carbohydrates of the cell wall (mono- and polysaccharides), and nucleic acid sequences (and secondary structure).

Prillinger et al. (1990, 1991a,b) examined the monosaccharide composition of isolated cell walls of a diversity of yeast-producing fungi. Using these data, as well as 5S rRNA secondary structure and the production of extracellular amyloid compounds, they recognized four "yeast types" within the basidiomycetes, the *Microbotryum*, *Ustilago*, *Dacrymyces*, and *Tremella* types. Species with a *Microbotryum*-type yeast state belong in the class Urediniomycetes (e.g., Microbotryales, Sporidiobolaceae, *Agaricostilbum*, *Colacogloea*, *Septobasidium*). The key characters that define this yeast type are a predominance of mannose in the cell wall, followed by glucose, galactose, and fucose, and a type A 5S rRNA secondary structure. The *Ustilago* type corresponds to the Ustilaginomycetes, and the *Dacrymyces* and *Tremella* types correspond to the Heterobasidiomycetes (Wells and Bandoni, Chap. 4, this Vol.). The types differ from the *Microbotryum* type in having a predominance of glucose in the cell wall and a type B 5S rRNA secondary structure. Ubiquinone chemistry is a character that may be useful in classifying species within the larger groups defined using cell wall monosaccharides.

V. Ultrastructural Characters

Although the Urediniomycetes is one of the best studied groups of fungi, our knowledge of its ultrastructural characters is fragmentary. Too frequently, conclusions are drawn based on examination of an insufficient number of taxa and/or inadequate quality and quantity of ultrastructural data. For example, few of the 164 genera of Uredinales have been examined ultrastructurally, yet it is assumed that they are all similar at the subcellular level.

Molecular phylogenetics has necessitated a reinterpretation of many ultrastructural characters, e.g., spindle pole body (SPB) morphology and septal pore type (McLaughlin et al. 1995b), as it has with basidial and basidiocarp characters. Ultrastructural characters associated with the septal pore apparatus, the SPB, and nuclear division have been used widely as a basis for phylogenetic hypotheses of basidiomycete taxa (Bandoni 1984; Bauer and Oberwinkler 1990, 1994; Bauer et al. 1995; Boehm and McLaughlin 1989; Boekhout et al. 1992; Bourett and McLaughlin 1986; Khan and Kimbrough 1982; Lü and McLaughlin 1991; McLaughlin et al. 1995a; Moore 1978; Oberwinkler and Bauer 1989; Suh and Sugiyama 1993; Suh et al. 1993; Swann et al. 1999). These characters, and others discussed below, correspond well to phylogenetic lineages of Urediniomycetes generated from rDNA sequence data (McLaughlin et al. 1995a,b; Swann et al. 1999), confirming their value as phylogenetic markers. A complete understanding of relationships within Urediniomycetes requires both molecular and ultrastructural information (Frieders 1997; Swann et al. 1999).

A. Septa

Septa of the Urediniomycetes are simple (Figs. 14, 15). Typically, there is one central septal pore, although multiple pores are found in the sedge parasite *Kriegeria* (Doublés and McLaughlin 1991). The septal wall tapers more or less dramatically toward the pore. Swellings in the septal wall, if present, are proximal to the tapered pore margin (e.g., *Septobasidium* spp., Dykstra 1974; *Eocronartium muscicola*, Boehm and McLaughlin 1989; *Atractogloea stillata*, Oberwinkler and Bauer 1989; *Kondoa malvinella*, Suh et al. 1993; *Uromyces*

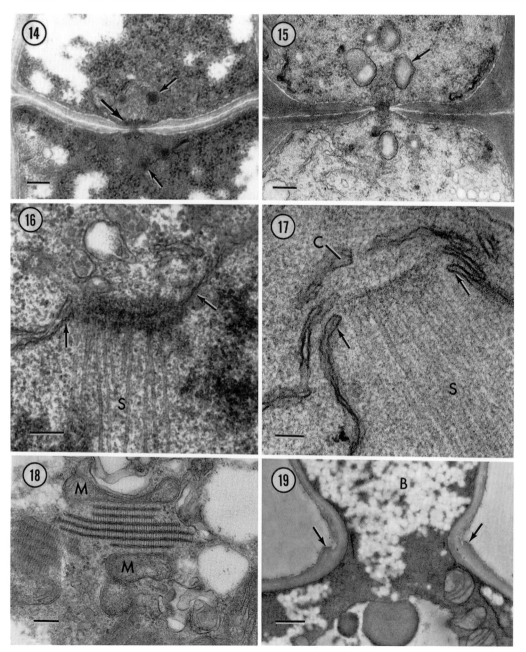

Figs. 14–19. Subcellular structure in the Urediniomycetes. All transmission electron micrographs. **Figs. 14, 15.** Septal pore structure. **Fig. 14.** Pulley-wheel-shaped plug (*large arrow*) in septal pore of *Eocronartium muscicola* and zone of ribosome exclusion surrounded by microbodies (*small arrows*). (McLaughlin, culture DJM 757-5). *Bar* 0.2 μm. **Fig. 15.** Septal pore of *Helicogloea lagerheimii* with adjacent microbodies (*arrow*). (Bandoni, culture RJB 6478-5). *Bar* 0.2 μm. **Figs. 16, 17.** Spindle pole bodies (SPB). **Fig. 16.** Early meiotic metaphase I SPB of *Puccinia malvacearum* inserted in a pore of the nuclear envelope (*arrows*). Spindle (*S*). (Micrograph K. L. O'Donnell; see O'Donnell and McLaughlin 1981a). *Bar* 0.2 μm. **Fig. 17.** Early mitotic metaphase SPB of *Helicobasidium brebissonii* with endoplasmic reticulum cap (*C*). Nuclear envelope (*arrows*); spindle (*S*). (Micrograph T. M. Bourett, culture CBS 324.47). *Bar* 0.1 μm. **Fig. 18.** Microscala in hypha of *Helicogloea variabilis* with rodlets cross-linking ER and mitochondria (*M*). (See McLaughlin 1990). *Bar* 0.2 μm. **Fig. 19.** Break (*arrows*) in outer hyphal wall during branch (*B*) initiation in *Jola* cf. *javensis*. (McLaughlin, culture DJM 739ps). *Bar* 1 μm

dianthi, Jones 1973), rather than being at the pore margin, as is characteristic of dolipore septa. The septal pores may be associated with vesicles, electron-dense globules, or various types of membrane-bound microbodies with or without electron-dense inclusions. Members of the Atractiellales have microbodies consisting of an electron-transparent core bounded by an electron-dense outer layer (Fig. 15). Pore-associated microbodies found in *Agaricostilbum* and *Cryptomycocolax* have been termed Woronin or Woronin-like bodies (Oberwinkler and Bauer 1989, 1990, respectively), but true Woronin bodies are restricted to the Ascomycota. A zone of organelle exclusion bounded by microbodies (Fig. 14) is present in the rust fungi and related taxa (*Herpobasidium*, *Jola*, *Eocronartium*) (Bauer and Oberwinkler 1994; Boehm and McLaughlin 1989; Frieders 1997); the misinterpretation of this region as a reduced dolipore septum (Moore 1985, 1987) has created an incorrect and confusing taxonomic affinity of the Uredinales with the Auriculariales sensu stricto (Wells and Bandoni, Chap. 4, this Vol.) rather than with other members of the Urediniomycetes. Membranous septal pore caps are absent entirely from members of this class. The septal pore may be open or plugged by electron-dense material. In many Urediniomycetes this plugging material is amorphous, but in many members of the Urediniomycetidae the septal pore plug is pulley-wheel-shaped (Bauer and Oberwinkler 1994; Boehm and McLaughlin 1989; Jones 1973; Klevin and McLaughlin 1989; Littlefield and Bracker 1971; Littlefield and Heath 1979; Oberwinkler et al. 1990a), and appears to be diagnostic for this closely related group of plant parasites. Bauer et al. (1995) identified nine septal pore types among members of the Urediniomycetes, based primarily on the various structures associated with the septum, but the phylogenetic significance of these septal categories is doubtful (Frieders 1997).

B. Spindle Pole Bodies

In addition to possessing a simple septum, the Urediniomycetes are characterized by having layered discoid SPBs (Figs. 16, 17), although the SPBs of some taxa, such as members of the Microbotryales and many yeast stages (e.g., those of *Agaricostilbum*, *Leucosporidium*, and *Kriegeria*), tend to be subspherical to subgloboid (Frieders and McLaughlin 1996; McLaughlin et al. 1995b; Swann et al. 1999). The overall size and degree of substructuring of the SPBs is highly variable within the class; additionally, meiotic and mitotic SPBs can differ within a species (McLaughlin et al. 1995b). The SPBs of many Urediniomycetes are more or less internalized within dividing nuclei, surrounded by a continuous or discontinuous nuclear envelope (Bauer et al. 1991, 1992; Berbee et al. 1991; Boekhout and Linnemans 1982; McCully and Robinow 1972a,b; Poon and Day 1976; Swann et al. 1999). However, in the Urediniomycetidae (i.e., *Jola* and *Insolibasidium*, Frieders 1997; *Helicobasidium*, Bourett and McLaughlin 1986; *Eocronartium*, Boehm and McLaughlin 1989; *Herpobasidium*, Bauer and Oberwinkler 1994; *Pachnocybe*, Bauer and Oberwinkler 1990) as well as in *Helicogloea* (McLaughlin, unpubl. results), the SPB is inserted in a pore of the nuclear envelope. In the Uredinales, the nuclear envelope-SPB association is intimate (Harder 1976; Littlefield and Heath 1979; O'Donnell and McLaughlin 1981a–c), but in related taxa the nuclear envelope-SPB association is looser. Most members of the Urediniomycetidae have a more or less fragmented cap of ER over the SPB during nuclear division (Fig. 17), but this feature is lacking and apomorphic in the rusts (Fig. 16; McLaughlin et al. 1995a). In *Eocronartium*, *Jola*, and *Herpobasidium* (Bauer and Oberwinkler 1994; Boehm and McLaughlin 1989; Frieders 1997) the ER cap extends around the entire dividing nucleus, suggesting a close relationship of these moss and fern parasites. Replication of the meiotic SPB has been studied in a few diverse urediniomycete taxa (Bauer and Oberwinkler 1990, 1994; Bauer et al. 1992; Berbee et al. 1991; Oberwinkler and Bauer 1990; O'Donnell and McLaughlin 1981b,c), and replication events were found to be variable in each of these taxa. The significance of these SPB cycle variations remains uncertain until more closely related taxa are studied in detail.

C. Colacosomes and Lenticular Bodies

Colacosomes and lenticular bodies are highly specialized organelles; colacosomes are associated with some, but not all, mycoparasitic taxa while lenticular bodies have been found in supposedly saprobic species (Bauer et al. 1997; Oberwinkler and Bauer 1990; Oberwinkler et al. 1990a). Both consist of an electron-dense central core surrounded by an electron-transparent region and are located along the cell wall of saprobes, or within the parasite adjacent to host cells (see Fell et al., Chap. 1, this Vol.). In mycoparasites, colacosomes penetrate the cell wall of both the parasite and host (Bauer and Oberwinkler 1991b). Cell wall penetration has not been observed for lenticular bodies (Boekhout et al. 1992; Kreger-van Rij and Veenhuis 1971). Although not all mycoparasitic or presumed mycoparasitic taxa have been studied ultrastructurally, colacosomes and lenticular bodies are restricted to members of the Microbotryomycetidae (Bauer et al. 1997; Swann et al. 1999; Sect. VII. C), and these structures appear to be phylogenetically significant.

D. Microscala

An ultrastructural feature that appears to be diagnostic of the Atractiellales is the microscala, or symplechosome (Fig. 18). These membrane complexes, consisting of layers of membranes of endoplasmic reticulum (ER) interconnected by a regular array of rodlets which may also connect to mitochondria, are found in *Helicogloea* (McLaughlin 1990), *Atractiella*, *Phleogena*, and *Saccoblastia* (Bauer and Oberwinkler 1991a; Oberwinkler and Bauer 1989). The microscala is also found in *Pachnocybe*, and has been reported in *Septobasidium* (Bauer and Oberwinkler 1990; Kleven and McLaughlin 1989; Oberwinkler and Bauer 1989), but in *Pachnocybe* it differs from those in the Atractiellales in the degree of ER and mitochondrial cross-linking.

E. Other Ultrastructural Characters

Two other features found among members of the Urediniomycetes, which often are overlooked, include condensed chromatin at interphase and rupture of the outer hyphal cell wall layer at branching (Fig. 19). Interphase condensed chromatin appears to be restricted to the Urediniomycetidae, i.e., *Pachnocybe*, *Eocronartium*, *Jola*, *Herpobasidium*, *Helicobasidium mompa* and the Uredinales (Bauer and Oberwinkler 1994; Boehm and McLaughlin 1989; Bourett and McLaughlin 1986; Frieders 1997; Kleven and McLaughlin 1989; Littlefield and Heath 1979); few micrographs of interphase nuclei have been published for other taxa within the class, however. The functional significance of interphase condensed chromatin in this group of fungi remains unclear. While commonly occurring during basidial germination (Bandoni 1984) and budding in basidiomycetous yeasts, disruption of the outer wall during hyphal branching appears to be rare within the Basidiomycota, although common to many members of the Urediniomycetes. Wall breakage at branching is found in rust fungi (Harder 1984; Littlefield and Heath 1979), and other Urediniomycetidae such as *Jola* (Frieders 1997), *Eocronartium* (Boehm and McLaughlin 1989), *Pachnocybe* (Kleven and McLaughlin 1989), and *Helicobasidium* (Bourett and McLaughlin 1986), as well as in some Agaricostilbomycetidae (*Agaricostilbum*, McLaughlin, unpubl.; *Kondoa*, Suh et al. 1993).

VI. Phylogeny and Systematics

Hypotheses concerning the phylogenetic placement of the diverse groups that compose the Urediniomycetes are varied. The viewpoint offered in this chapter (Fig. 20) is that these species form a monophyletic group (Swann and Taylor 1995b), in opposition to the Ustilagino-

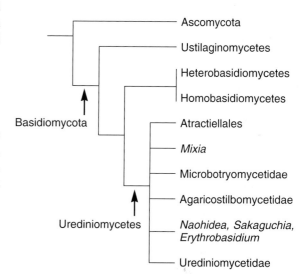

Fig. 20. Phylogenetic tree showing the classes of the Basidiomycota as used in this Volume and the major clades in the Urediniomycetes

mycetes (Bauer et al., Chap. 3, this Vol.) plus the Hymenomycetes sensu Swann and Taylor (1995b), i.e., species usually having a septal pore swelling with pore cap (Heterobasidiomycetes, Wells and Bandoni, Chap. 4, and Homobasidiomycetes, Hibbett and Thorn, Chap. 5, this Vol.). Alternative hypotheses propose that the Urediniomycetes are arranged paraphyletically at the base of the Basidiomycota (Bauer and Oberwinkler 1994; Savile 1955).

The traditional light microscopic characters that have been used to classify basidiomycetes for the past century are not appropriate for inference of monophyletic groups. These characters include basidium and basidiocarp morphology, as well as sexuality itself, i.e., the traditional schizm between classification of sexual and asexual species. Continued uncritical reliance on this homoplasious set of characters, coupled with some highly questionable interpretations of newer character sets (ultrastructural, biochemical) have resulted in taxonomic chaos. R.T. Moore has contributed numerous publications that illustrate this problem [see Bauer and Oberwinkler (1997) for a discussion of Moore's phylum Ustomycota]. These same problems have caused misplacement of even obviously monophyletic taxa such as the rust fungi. Using the principles of phylogenetic systematics, polyphyletic higher taxa (e.g., Ustilaginales sensu lato, Platygloeales, Sporidiales, etc.) are being redistributed into natural, monophyletic taxa.

A. Phylogenetic Position of Uredinales

The Uredinales often have been positioned as an early diverging group, near the base of the Basidiomycota. The characters that have been used to justify this placement include their specialized sexual structures (spermogonia), pleomorphic life history, strict parasitism, lack of clamp connections, bipolar sexual system, and ultrastructural features of the septum and spindle pole body. Insights have been provided recently by ribosomal RNA-based studies regarding the phylogeny of the rust fungi and other Urediniomycetes. Rather than being primitive, the rusts are a highly derived group within the Urediniomycetes. Viewed in this light, alternative interpretations for their putative plesiomorphic characters become rational (McLaughlin et al. 1995a; Swann et al. 1998). No other basidiomycetes have spermogonia; however, other Urediniomycetes have conidial or microconidial states that may possibly function in fertilization (*Atractiella*, Boidin et al. 1979; *Helicogloea* and *Saccoblastia*, Ingold 1992; *Phleogena*, Oberwinkler and Bauer 1989; *Platygloea effusa*, Ingold 1988; *Jola* and *Eocronartium*, Frieders 1997). If the spermatial state of rusts were primitively retained, it would be expected that it might be present in the basal ascomycete groups, but this is not the case. Clamps are found in all of the major lineages of the Basidiomycota, so this character probably arose early in the evolution of the phylum. However, clampless taxa are scattered throughout the basidiomycetes, probably indicating multiple losses of clamp connections. The subclass to which the rust fungi belong is composed solely of clampless taxa. The ultrastructural characters of the rusts (see Sect. V above) are similar to those of other closely related members of the class. Ultrastructural features unique to the rusts, such as the lack of a membrane cap over the large SPB and its association with the nuclear envelope, are better interpreted as synapomorphic for the rusts rather than primitively retained. The same is true for the highly developed biotrophic interaction between rust fungi and their hosts, and their pleomorphic life history.

B. Dismemberment of Auriculariales

For at least the past half century, traditional basidium-based taxon definitions have dominated the systematics of this group (Donk 1972a; Talbot 1968, 1973). These rigid typological concepts began to disintegrate in the 1980's with the application of ultrastructural data combined with a lessened emphasis on basidium morphology. Bandoni (1984) instituted a major realignment of taxa in the orders Auriculariales and Tremellales, using characters associated with septal pore architecture, and the production of a yeast phase. Auriculariales now is restricted to species having septal pore swellings and a complex septal pore apparatus (Wells and Bandoni, Chap. 4, this Vol.). In the absence of a higher taxon to accommodate the phylogenetically diverse simple-septate auricularioid species, Moore (1990) created the order Platygloeales. This was an unfortunate solution, however, giving the impression that these species form a natural group. The polyphyletic nature of such an order has been noted (Bandoni 1984), but the phylogenetic information necessary to redistribute the species into natural groups is lacking for the most part. Ultrastructural and biochemical information is being accumulated that is proving useful in determining natural groupings of auricularioid Urediniomycetes. One feature that has become clear is that the members of the Platygloeales sensu Moore are scattered throughout the Urediniomycetes, and are intermixed with members of other orders.

C. Teliospore-Forming Yeasts

The taxonomy of the teliospore-forming basidiomycetous yeasts is an excellent example of how the uncritical use of morphological criteria in taxonomy can fail in the creation of monophyletic taxa. The order Sporidiales (nomen nudum, see Bauer and Oberwinkler 1997; Boekhout et al. 1998) was created for nonphytopathogenic (i.e., nonsmut), teliospore-forming basidiomycetous yeasts (Moore 1980). Evidence exists to demonstrate that its members (as defined by Moore 1996a) are drawn from at least five different lineages (Fell et al., Chap. 1, this Vol.), and are intermixed with members of other higher taxa (e.g., Platygloeales sensu lato, Tremellales).

VII. Classification

The early history of the classification of some of the fungi comprising the Urediniomycetes is recounted by Ainsworth (1976). During the 19th century the relationship of the Uredinales to the basidiomycetes began to be recognized, but opinion remained divided regarding the Ustilaginales (de Bary 1887; Brefeld 1888, 1889). Patouillard (1900) proposed the family Auriculariaceae to include the taxa now classified in the Uredinales and Ustilaginales, along with *Helicobasidium*, *Helicogloea*, *Jola*, *Platygloea*, *Septobasidium*, and a number of gasteroid genera. With the exclusion of an emended Auriculariales (Bandoni 1984) and Ustilaginomycetes (Bauer et al. 1997), Patouillard's Auriculariaceae is much like our modern concept of the class Urediniomycetes. Patouillard not only realized the close relationships of these taxa but also the adaptiveness of the basidium to ecological conditions, a phenomenon not appreciated by later authors, who proposed classifications based on narrowly defined basidial morphologies (Donk 1972a,b, 1973a,b; Lowy 1968; Talbot 1971; see Swann and Taylor 1993).

One of the most influential systematic treatments for Fungi has been the *Dictionary of the Fungi*. In each of the last three editions (Ainsworth et al. 1971; Hawksworth et al. 1983, 1995) the treatment of the Uredinales and Ustilaginales at the class level has changed significantly, from a single class for these two orders separate from the rest of the basidiomycetes, to a class for each order, to two classes, each of which contains one of these orders plus one or more additional orders. In the latest edition of the *Dictionary*, the elements of a monophyletic Urediniomycetes are dispersed among three classes. The class name Urediniomycetes was introduced for a single order, Uredinales (Hawksworth et al. 1983), and was subsequently emended to provide a name for the monophyletic group containing the Uredinales, part of the Ustilaginales sensu lato (Microbotryales), simple-septate auricularioid taxa, and various gasteroid and yeast genera (Swann and Taylor 1995b). Additional chaos has been introduced into the systematics of the Urediniomycetes by the various phenetic proposals of Moore (1996a,b), which are based on erroneous interpretations of molecular and structural data (Moore 1985, 1987). This confusion is particularly evident in his treatment of the rust fungi, in which the septum has been interpreted as a reduced dolipore/parenthesome complex (Moore 1985, 1987), a claim bolstered by the use of 5S rRNA sequences from yeast cultures that had been misidentified as rusts (Bauer 1987; Prillinger et al. 1991b).

It is difficult to present a complete classification for the Urediniomycetes. Few species have been studied in detail, and the ultrastructural or biochemical evidence required to classify them with confidence is lacking for many species. The classification presented here is intended to reflect the current state of knowledge regarding the phylogenetic relationships among the members of the class (Fig. 20). There are several distinct groups within the class, but the relationships among them remain obscure. We take a conservative approach to the classification of the clades within the Urediniomycetes by recognizing only well supported monophyletic groups.

A. Order Atractiellales

The original circumscription of the order Atractiellales included species with simple septa, gasteroid basidia, and stilboid fruitbodies (*Atractiella*, *Phleogena*, *Agaricostilbum*, *Stilbum*, and *Chionosphaera*) (Oberwinkler and Bandoni 1982a). The order was emended (Oberwinkler and Bauer 1989) using the characteristic septal pore architecture and the presence of microscala (or symplechosomes) to define the group, removing *Agaricostilbum*, *Stilbum*, and *Chionosphaera*. The modern circumscription of the Atractiellales includes genera that are superficially very different, *Atractiella*, *Phleogena*, *Helicogloea*, and *Saccoblastia*, but that are united by septal pore ultrastructure and in having microscala. This circumscription of the Atractiellales has yet to be universally adopted; modified versions of the older circumscription appear in recent publications (Boekhout et al. 1998; Hawksworth et al. 1995; Moore 1996a). Nucleotide sequence analyses indicate that *Atractiella*, *Phleogena*, and *Helicogloea* form a monophyletic group. Conidial states are known from a number of species in this group, but none is known to form a yeast state.

Moore (1996a), using the obsolete concept of the order, placed the mycoparasite *Camptobasidium hydrophilum* in the Atractiellales, following a tentative suggestion made by Marvanová and Suberkropp (1990). This placement is completely unsupported by available data. Microscala were not observed in *Camptobasidium*, and no micrographs of the septal pore have been published. Analyses of nucleotide sequence data (18S) place this species in the Microbotryomycetidae (Swann et al. 1999), where its mycoparasitism makes it a much better fit.

Atractogloea stillata also has been included in this order (Boekhout et al. 1998; Hawksworth et al. 1995; Oberwinkler and Bandoni 1982b). While the phylogenetic affinities of this species are still unclear, it is fairly certain that it does not belong in the Atractiellales. Microscala have not been observed in *A. stillata*, and available electron micrographs of its septa do not suggest a placement in the Atractiellales.

This clade consists of a single order containing two families, Hoehnelomycetaceae and Phleogenaceae, each containing a single genus (Table 2), plus two additional genera which lack family placement. Because the phylogenetic relationships of the species in this clade are still ambiguous, it is premature to change the subordinal classification.

B. *Mixia osmundae*

Mixia osmundae is a recent addition to the Basidiomycota. It was classified in the Ascomycota from the time of its original description as a species of *Taphrina* until 1995, when it was discovered to be a basidiomycete (Nishida et al. 1995). Transmission electron micrographs demonstrate a typically basidiomycetous wall structure and enteroblastic budding. Small subunit rRNA gene sequences (18S) place it in the Urediniomycetes. Thus, what was originally interpreted as an ascus typical of the Taphrinales was discovered to produce exogenous blastospores. It is unknown whether meiosis occurs in the spore-bearing structure. While it is clear that *Mixia* is a basidiomycete, no close relatives have been identified.

C. Subclass Microbotryomycetidae

The subclass Microbotryomycetidae (Swann et al. 1999) is a morphologically and ecologically diverse group. It contains mycoparasites, phytoparasites, and presumed saprobes. Fruitbody morphology ranges from simple teliospores to pustules on leaf surfaces to pycnidioid basidiomes. The namesake of this subclass is the smut fungus order, Microbotryales (Bauer et al. 1997; Bauer et al., Chap. 3, this Vol.). Early nucleotide sequence studies demonstrated a wide gulf between the Microbotryales on the one side and the Ustilaginomycetes on the other (Gottschalk and Blanz 1985; Walker and Doolittle 1982). In the ensuing years, the evidence has become overwhelming that a syndrome of smut characters, including morphology (e.g., pigmented teliospores, gasteroid basidia) and mode of nutrition, have evolved at least twice in the Basidiomycota. Nucleotide sequences (Fell et al. 1995; Swann et al. 1999), ultrastructure (Bauer et al. 1997; McLaughlin et al. 1995b), cell wall monosaccharide composition (Prillinger et al. 1990, 1991a), and fimbrial glycosylation pattern (Celerin et al. 1995) support this conclusion that the Microbotryales are separate from the Ustilaginomycetes and form part of a monophyletic group (Microbotryomycetidae) in the Urediniomycetes.

Other taxa in this group include the teliospore-forming yeasts *Rhodosporidium*, *Leucosporidium*,

Table 2. Phylogenetic outline for classification of Urediniomycetes

A. Atractiellales
   Helicogloea
   Saccoblastia
  Hoehnelomycetaceae
   Atractiella
  Phleogenaceae
   Phleogena

B. Mixiaceae
   Mixia

C. Microbotryomycetidae
   Bensingtonia[a,b]
   Colacogloea
   Kriegeria
   Rhodotorula[a,b]
   Sporobolomyces[a,b]
  Camptobasidiaceae
   Camptobasidium
  Krieglsteineraceae
   Krieglsteinera[c]
  Sporidiobolaceae
   Leucosporidium
   Rhodosporidium
   Sporidiobolus
  Cryptomycocolacales
   Cryptomycocolax[c]
  Heterogastridiales
   Heterogastridium
  Microbotryales
   Microbotryaceae
    Liroa
    Microbotryum
    Sphacelotheca
   Ustilentylomataceae
    Aurantiosporium
    Fulvisporium
    Ustilentyloma

D. Agaricostilbomycetidae
   Bensingtonia[a,b]
   Kondoa
   Kurtzmanomyces[a]
   Mycogloea
   Rhodotorula[a,b]
   Sporobolomyces[a,b]
   Sterigmatomyces[a]
  Agaricostilbales
   Agaricostilbaceae
    Agaricostilbum
   Chionosphaeraceae
    Chionosphaera
    Fibulostilbum
    Stilbum

E. *Erythrobasidium*, *Naohidea*, *Sakaguchia* Clade
   Erythrobasidium
   Naohidea
   Rhodotorula[a,b]
   Sakaguchia
   Sporobolomyces[a,b]

F. Urediniomycetidae
   Helicobasidium[d]
   Platygloea[b,d]
  Platygloeales *sensu stricto*
   Herpobasidium[e]
   Insolibasidium
   Jola
  Eocronartiaceae
   Eocronartium
  Platygloeaceae
   Platygloea[f]
  Septobasidiales
  Pachnocybaceae
   Pachnocybe
  Septobasidiaceae
   Auriculoscypha[g]
   Coccidiodictyon
   Ordonia
   Septobasidium
   Uredinella
  Uredinales[h]

G. Genera *Incertae Sedis*
   Atractogloea
   Cystobasidium
   Kryptastrina
   Naiadella[a]
   Occultifur
   Platycarpa
   Platygloea[b,d]
   Ptechetelium
   Spiculogloea
   Zygogloea

[a] Mitotic genus.
[b] Pro parte.
[c] Classification based on presence of colacosomes.
[d] Not monophyletic.
[e] Classification based on septal and nuclear ultrastructure (Bauer et al. 1997; Oberwinkler and Bauer 1990).
[f] Type species.
[g] Classification based on association with insects (Lalitha and Leelavathy 1990).
[h] See Cummins and Hiratsuka (1983); Laundon (1973).

and *Sporidiobolus* (Sporidiobolaceae) and a variety of less familiar organisms, including *Krieglsteinera*, *Heterogastridium*, *Colacogloea*, *Kriegeria*, and *Camptobasidium*. Asexual species in this group are drawn mostly from *Rhodotorula* and *Sporobolomyces*. One ultrastructural feature shared by several members of this group is the colacosome (Bauer et al. 1997), an organelle functional in mycoparasitism (Oberwinkler et al. 1999a). The presence of colacosomes in *Cryptomycocolax abnorme* (Oberwinkler and Bauer 1990) suggests that it may belong in this group.

The taxonomy of this clade is chaotic. It contains a significant number of monotypic genera that either are not classified in orders or families, or are in monotypic orders and families (Table 2). The Microbotryales is probably monophyletic, but the Sporidiobolaceae (excluding *Sakaguchia*) may be paraphyletic.

D. Subclass Agaricostilbomycetidae

The subclass Agaricostilbomycetidae is not well understood in terms of its biology. *Agaricostilbum* spp. usually are found on dry, dead plant material, particularly from palms (Boekhout et al. 1998; Oberwinkler and Bauer 1989). Members of the Chionosphaeraceae are suspected to be mycoparasitic, due to their frequent association with fungal substrates (Cox 1976; Seifert et al. 1992). *Kondoa*, formerly thought to be a species of *Rhodosporidium*, has been discovered to be very different in terms of molecular phylogenetics (Suh and Sugiyama 1994; Yamada et al. 1989) and morphology.

Agaricostilbum shares an unusual pattern of nuclear division in the yeast phase with the Chionosphaeraceae. In the typical pattern in all other basidiomycetes studied, the nucleus migrates into the bud before dividing (Frieders and McLaughlin 1996; McLaughlin et al. 1995b). In *Agaricostilbum pulcherrimum* and *Stilbum vulgare* the nucleus divides in the parent cell (Swann et al. 1997; McLaughlin, unpubl.). This shared nuclear division pattern supports the conclusion that the Chionosphaeraceae should be a family of Agaricostilbales.

The Agaricostilbomycetidae contains the order Agaricostilbales, with two families, Agaricostilbaceae (*Agaricostilbum* spp.) and Chionosphaeraceae (*Chionosphaera, Stilbum, Fibulostilbum*), the genus *Kondoa*, and a number of asexual species (e.g., *Sterigmatomyces* spp., some *Sporobolomyces* spp.; Table 2). The Agaricostilbales is probably monophyletic and includes asexual species in addition to its teleomorphic members (see Fell et al., Chap. 1, this Vol.). The taxonomy of the remainder of the species in this subclass is undeveloped above the genus level.

E. *Erythrobasidium, Naohidea, Sakaguchia*

This clade, consisting of three sexual genera and a number of asexual species (Table 2), is perhaps the least understood of the major groups discussed in this chapter. Members of this clade are known primarily from culture, and have been isolated from seawater, plant surfaces, and various brewery-associated sources. The species that is best known from nature is the mycoparasite *Naohidea sebacea* (Oberwinkler 1990).

F. Subclass Urediniomycetidae

This clade is composed primarily of phytoparasites (e.g., Uredinales, *Eocronartium, Herpobasidium*). With the exception of some Septobasidiales (Oberwinkler 1987), yeasts are absent in this group; the production of conidia is common. Conidia are especially well developed in the rust fungi, e.g., urediniospores and aeciospores, and are also produced by *Jola* and *Eocronartium* (Frieders 1997), *Helicobasidium* (von Arx 1981), *Pachnocybe* (Kleven and McLaughlin 1988), and *Insolibasidium* (Oberwinkler and Bandoni 1984). The largest, best known, and most economically important group in the subclass is the rust fungi. The closest relatives of the rusts are a group of clampless simple-septate parasites of aerial portions of plants, i.e., the moss and fern parasites *Jola, Eocronartium*, and *Herpobasidium*, as well as *Insolibasidium* and *Platygloea disciformis*. The trophic status of *P. disciformis* is ambiguous. While *Helicobasidium* is closely related to the Platygloeales, it has been provisionally excluded from the order because it does not link closely with the other species in nucleotide studies.

The subclass Urediniomycetidae contains the orders Uredinales, Septobasidiales and Platygloeales sensu stricto, and a number of genera that are not classified in families or orders. Numerous family-level classifications have been used for the rust fungi, ranging from as few as 2 (Arthur 1934) to as many as 14 (Cummins and Hiratsuka 1983), with some authors preferring not to recognize any families within the order (Laundon 1973). The various family-level treatments of the species in the Uredinales have not been subjected to critical phylogenetic evaluation. Despite the obvious differences in trophic mode between the wood-inhabiting genus *Pachnocybe* and the insect parasites of the Septobasidiaceae, we include Pachnocybaceae in the order Septobasidiales. Microscala consisting of interconnected mitochondria with or without connections to endoplasmic reticulum have been illustrated in *Pachnocybe ferruginea* (Kleven and McLaughlin 1989). A similar microscala is reported from *Septobasidium*, but is not illustrated (Bauer and Oberwinkler 1990). A relationship between *Pachnocybe* and *Septobasidium* also has been suggested by nucleotide sequence analyses (Frieders 1997; Gottschalk and Blanz 1985). The concept of a monophyletic group of insect parasites remains unchanged, and is now denoted by a family name (Septobasidiaceae) rather than the now more inclusive ordinal name Septobasidiales. The order Platygloeales sensu stricto is the monophyletic group containing *Platygloea disciformis*, the type of the genus, and four other genera. This concept of the order excludes a large number of genera, many of which are of uncertain taxonomic position within the Urediniomycetes (see Table 2, Genera *Incertae Sedis*).

VIII. Culture

Urediniomycetes are generally isolated on complex media using either the ballistospore discharge method or streak plating on media containing antibiotics. Many ballistosporic yeasts (see Fell et al., Chap. 1, this Vol.) or filamentous species can be isolated by suspending the substrate on which they sporulate above the isolation medium. This technique has been used to separate the yeast-forming *Spiculogloea* from its hypha-producing host based on colony characteristics (Langer and Oberwinkler 1998). Gasteroid species and urediniomycetous smuts can be isolated via streak plating. Obligate parasites in nature, such as the Uredinales and the moss parasites *Jola* and *Eocronartium*, can be axenically cultured from spores, usually urediniospores in the case of the rusts or basidiospores in the moss parasites, or from mycelium in the host or fruitbody. For both groups results were variable and the main obstacle was obtaining the initial isolate; growth was slow and no special nutrient requirements were identified (Frieders 1997; Littlefield 1981; Maclean 1982; Williams 1984).

A limited number of urediniomycete genera and species are available from culture collections, except for basidiomycetous yeasts, which are generally well represented. Some rusts are available from the American Type Culture Collection as urediniospore collections.

IX. Conclusion

The class Urediniomycetes contains elements of historically important higher taxa of basidiomycetes, such as Auriculariales, Heterobasidiomycetes, and Teliomycetes, but it is not fully congruent with any of them. These other taxa were based on the characters that were then available, namely those that could be observed with standard light microscopy. These characters were primarily associated with the basidium, which was thought to be a stable source of taxonomic characters at the time. In contrast, the concept of the Urediniomycetes is a result of phylogenetic analyses of biochemical and ultrastructural characters. Standard morphology will remain an important aspect of the study of fungi in the discovery and recognition of new species, but will probably provide characters that are more useful in separating species than in grouping species into higher taxa. At present, there is only limited basic morphological information for many species. Ultrastructural and some biochemical studies require, or are best performed with, cultured or freshly collected material. There is a need to recollect and culture species that are already described. The problem of relationships among the clades within the Urediniomycetes, and within the clades, may be resolved as data from thorough ultrastructural studies and nucleotide sequencing from more species or additional genes are analyzed. These data are needed to produce a complete systematic treatment for the class.

Acknowledgments. This research was supported by a series of grants from the National Science Foundation, including DEB-9306578 and DEB-9318232. We thank John Haight for printing the figures.

References

Agrios GN (1997) Plant pathology, 4th edn. Academic Press, London
Ainsworth GC (1976) Introduction to the history of mycology. Cambridge University Press, Cambridge

Ainsworth GC, James PW, Hawksworth DL (1971) Ainsworth & Bisby's dictionary of the fungi, 6th edn. Commonwealth Mycological Institute, Kew
Arthur JC (1934) Manual of the rusts in United States and Canada. Purdue Research Foundation, Lafayette, Indiana
Bandoni RJ (1984) The Tremellales and Auriculariales: an alternative classification. Trans Mycol Soc Jpn 25:489–530
Bandoni RJ (1995) Dimorphic heterobasidiomycetes: taxonomy and parasitism. Stud Mycol 38:13–27
Bandoni RJ, Oberwinkler F (1981) *Hyalopycnis blepharistoma*: a pycnidial basidiomycete. Can J Bot 59:1613–1620
Bauer R (1987) Uredinales – germination of basidiospores and pycnospores. Stud Mycol 33:111–125
Bauer R, Oberwinkler F (1990) Meiosis, spindle pole body cycle, and taxonomy of the heterobasidiomycete *Pachnocybe ferruginea*. Plant Syst Evol 172:241–261
Bauer R, Oberwinkler F (1991a) The symplechosome: a unique cell organelle of some basidiomycetes. Bot Acta 104:93–97
Bauer R, Oberwinkler F (1991b) The colacosomes: structures at the host-parasite interface of a mycoparasitic basidiomycete. Bot Acta 104:53–57
Bauer R, Oberwinkler F (1994) Meiosis, septal pore architecture, and systematic position of the heterobasidiomycetous fern parasite *Herpobasidium filicinum*. Can J Bot 72:1229–1242
Bauer R, Oberwinkler F (1997) The Ustomycota: an inventory. Mycotaxon 64:303–319
Bauer R, Berbee ML, Oberwinkler F (1991) An electron microscopic study of meiosis and the spindle pole body cycle in the smut fungus *Sphacelotheca polygoni-serrulati*. Can J Bot 69:245–255
Bauer R, Oberwinkler F, McLaughlin DJ (1992) Meiosis, spindle pole body cycle and basidium ontogeny in the heterobasidiomycete *Agaricostilbum pulcherrimum*. Syst Appl Microbiol 15:259–274
Bauer R, Mendgen K, Oberwinkler F (1995) Cellular interaction of the smut fungus *Ustacystis waldsteiniae*. Can J Bot 73:867–883
Bauer R, Oberwinkler F, Vánky K (1997) Ultrastructural markers and systematics in smut fungi and allied taxa. Can J Bot 75:1273–1314
Berbee ML, Bauer R, Oberwinkler F (1991) The spindle pole body cycle, meiosis, and basidial cytology of the smut fungus *Microbotryum violaceum*. Can J Bot 69:1795–1803
Boehm EWA, McLaughlin DJ (1989) Phylogeny and ultrastructure in *Eocronartium muscicola*: meiosis and basidial development. Mycologia 81:98–114
Boekhout T, Linnemans WAM (1982) Ultrastructure of mitosis in *Rhodosporidium toruloides*. Stud Mycol 22:23–38
Boekhout T, Bandoni RJ, Fell JW, Kwon-Chung KJ (1998) Discussion of teleomorphic and anamorphic genera of heterobasidiomycetous yeasts. In: Kurtzman CP, Fell JW (eds) The yeasts; a taxonomic study, 4th edn. Elsevier, Amsterdam, pp 609–625
Boekhout T, Yamada Y, Weijman ACM, Roeymans HJ, Batenburg-van der Vegte WH (1992) The significance of coenzyme Q, carbohydrate composition and septal ultrastructure for the taxonomy of ballistoconidia-forming yeasts and fungi. Syst Appl Microbiol 15:1–10
Boidin J, Candoussau F, Lanquetin P (1979) Premières récolte d'*Hoehnelomyces* (phragmobasidiomycète auriculariale) en Europe: culture, cycle. Beih Sydowia 8:71–75
Bourett TM, McLaughlin DJ (1986) Mitosis and septum formation in the basidiomycete *Helicobasidium mompa*. Can J Bot 64:130–145
Brefeld O (1888) Untersuchungen aus dem Gesammtgebiete der Mykologie VII. Basidiomyceten II. Protobasidiomyceten. Arthur Felix, Leipzig
Brefeld O (1889) Untersuchungen aus dem Gesammtgebiete der Mykologie VIII. Basidiomyceten II. Autobasidiomyceten. Arthur Felix, Leipzig
Celerin M, Day AW, Castle AJ, Laudenbach DE (1995) A glycosylation pattern that is unique to fimbrae from the taxon Microbotryales. Can J Microbiol 41:452–460
Couch JN (1938) The genus *Septobasidium*. University of North Carolina Press, Chapel Hill, North Carolina
Cox DE (1976) A new homobasidiomycete with anomalous basidia. Mycologia 68:481–510
Cummins GB, Hiratsuka Y (1983) Illustrated genera of rust fungi. American Phytopathological Society, St Paul, Minnesota
de Bary A (1887) Comparative morphology and biology of the fungi, mycetozoa and bacteria. Clarendon Press, Oxford
Diederich P, Christiansen MS (1994) *Biatoropsis usnearum* Räsänen, and other heterobasidiomycetes on *Usnea*. Lichenologist 26:47–66
Donk MA (1972a) The Heterobasidiomycetes: a reconnaissance – I. A restricted emendation. Proc K Ned Akad Wet C 75:365–375
Donk MA (1972b) The Heterobasidiomycetes: a reconnaissance – II. Some problems connected with the restricted emendation. Proc K Ned Akad Wet C 75:376–390
Donk MA (1973a) The Heterobasidiomycetes: a reconnaissance – III^A. How to recognize a basidiomycete? Proc K Ned Akad Wet C 75:1–13
Donk MA (1973b) The Heterobasidiomycetes: a reconnaissance – III^B. How to recognize a basidiomycete? Proc K Ned Akad Wet C 75:14–22
Doublés JC, McLaughlin DJ (1991) A new basidiomycetous septal type: the multiperforate septum in *Kriegeria eriophori*. Am J Bot 87:1542–1548
Doublés JC, McLaughlin DJ (1992) Basidial development, life history, and the anamorph of *Kriegeria eriophori*. Mycologia 84:668–678
Dykstra MJ (1974) Some ultrastructural features in the genus *Septobasidium*. Can J Bot 52:971–972
Fell JW, Kurtzman CP, Tallman AS, Buck JD (1988) *Rhodosporidium fluviale* sp. nov., a homokaryotic red yeast from a subtropical brackish environment. Mycologia 80:560–564
Fell JW, Statzell-Tallman A, Lutz MJ, Kurtzman CP (1992) Partial rRNA sequences in marine yeasts: a model for identification of marine eukaryotes. Mol Mar Biol Biotech 1:175–186
Fell JW, Boekhout T, Freshwater DW (1995) The role of nucleotide sequence analysis in the systematics of the yeast genera *Cryptococcus* and *Rhodotorula*. Stud Mycol 38:129–146
Frieders EM (1997) An integrated approach to understanding the moss parasites and their role in basidiomycete evolution. PhD Thesis, University of Minnesota, St Paul

Frieders EM, McLaughlin DJ (1996) Mitosis in the yeast phase of *Agaricostilbum pulcherrimum* and its evolutionary significance. Can J Bot 74:1392–1406

Gottschalk M, Blanz PA (1985) Untersuchungen an 5S ribosomalen Ribonukleinsäuren als Beitrag zur Klärung von Systematik und Phylogenie der Basidiomyceten. Z Mykol 51:205–243

Harder DE (1976) Mitosis and cell division in some cereal rust fungi. II. The processes of mitosis and cytokinesis. Can J Bot 54:995–1009

Harder DE (1984) Developmental ultrastructure of hyphae and spores. In: Bushnell WR, Roelfs AP (eds) The cereal rusts, vol 1. Academic Press, New York, pp 333–373

Hawksworth DL, Sutton BC, Ainsworth GC (1983) Ainsworth & Bisby's dictionary of the fungi, 7th edn. Commonwealth Mycological Institute, Kew

Hawksworth DL, Kirk PM, Sutton BC, Pegler DN (1995) Ainsworth & Bisby's dictionary of the fungi, 8th edn. CAB International, Egham

Hennen JF (1997) *Uredo vetus* sp. nov., the first record of a rust on *Selaginella*, and the use of the name *Uredo*. Mycologia 89:801–803

Ingold CT (1988) Patterns of basidiospore germination in *Platygloea effusa*. Trans Br Mycol Soc 91:161–162

Ingold CT (1992) Conidia in *Helicogloea* and *Saccoblastia* in relation to taxonomy. Mycol Res 96:734–736

Jones DR (1973) Ultrastructure of the septal pore in *Uromyces dianthi*. Trans Br Mycol Soc 61:227–235

Kendrick B, Watling R (1979) Mitospores in Basidiomycetes. In: Kendrick B (ed) The whole fungus, vol 2. National Museum of Natural Sciences and National Museums of Canada, Ottawa, pp 473–545

Khan SR, Kimbrough JW (1982) A reevaluation of the basidiomycetes based upon septal and basidial structures. Mycotaxon 15:103–120

Kleven NL, McLaughlin DJ (1988) Sporulation of the basidiomycete *Pachnocybe ferruginea* in terrestrial and aquatic environments. Mycologia 80:804–810

Kleven NJ, McLaughlin DJ (1989) A light and electron microscopic study of the developmental cycle in the basiomycete *Pachnocybe ferruginea*. Can J Bot 67:1336–1348

Kreger-van Rij NJW, Veenhuis M (1971) Some features of yeasts of the genus *Sporidiobolus* observed by electron microscopy. Antonie van Leeuwenhoek 37:253–255

Lachance M-A, Starmer WT (1998) Ecology and yeasts. In: Kurtzman CP, Fell JW (eds) The yeasts: a taxonomic study, 4th edn. Elsvier, Amsterdam, pp 21–30

Lalitha CR, Leelavathy KM (1990) A coccoid association in *Auriculoscypha* and its taxonomic significance. Mycol Res 94:571–572

Langer E, Oberwinkler F (1998) *Spiculogloea occulta* (Heterobasidiomycetes) morphology and culture characters. Mycotaxon 69:249–254

Laundon GF (1973) Uredinales. In: Ainsworth GC, Sparrow FK, Sussman AS (eds) The Fungi, vol 4B. Academic Press, New York, pp 247–279

Littlefield LJ (1981) Biology of the plant rusts. Iowa State University Press, Ames, Iowa

Littlefield LJ, Bracker CE (1971) Ultrastructure of septa in *Melampsora lini*. Trans Br Mycol Soc 56:181–188

Littlefield LJ, Heath MC (1979) Ultrastructure of rust fungi. Academic Press, New York

Lowy B (1968) Taxonomic problems in the Heterobasidiomycetes. Taxon 17:118–127

Lü HS, McLaughlin DJ (1991) Ultrastructure of the septal pore apparatus and early septum initiation in *Auricularia auricula-judae*. Mycologia 83:322–334

Maclean DJ (1982) Axenic culture and metabolism of rust fungi. In: Scott KJ, Chakravorty AK (eds) The rust fungi. Academic Press, London, pp 37–120

Marvanová L, Bandoni RJ (1987) *Naiadella fluitans* gen. et sp. nov.: a conidial basidiomycete. Mycologia 79:578–586

Marvanová L, Suberkropp K (1990) *Camptobasidium hydrophilum* and its anamorph *Crucella subtilis*: a new heterobasidiomycete from streams. Mycologia 82:208–217

McCully EK, Robinow CF (1972a) Mitosis in heterobasidiomycetous yeasts. I. *Leucosporidium scottii* (*Candida scottii*). J Cell Sci 10:857–881

McCully EK, Robinow CF (1972b) Mitosis in heterobasidiomycetous yeasts. II. *Rhodosporidium* sp. (*Rhodotorula glutinis*) and *Aessosporon salmonicolor* (*Sporobolomyces salmonicolor*). J Cell Sci 11:1–31

McLaughlin DJ (1990) A new cytoplasmic structure in the basidiomycete *Helicogloea*: the microscala. Exp Mycol 14:331–338

McLaughlin DJ, Berres ME, Szabo LJ (1995a) Molecules and morphology in basidiomycete phylogeny. Can J Bot 73 (Suppl 1):S684–S692

McLaughlin DJ, Frieders EM, Lü HS (1995b) A microscopist's view of heterobasidiomycete phylogeny. Stud Mycol 38:91–109

Moore RT (1978) Taxonomic significance of septal ultrastructure with particular reference to the jelly fungi. Mycologia 70:1007–1024

Moore RT (1980) Taxonomic proposals for the classification of marine yeasts and other yeast-like fungi including the smuts. Bot Mar 23:361–373

Moore RT (1985) The challenge of the dolipore/parenthesome septum. In: Moore D, Casselton LA, Wood DA, Frankland JC (eds) Developmental biology of higher fungi. Cambridge University Press, Cambridge, pp 175–212

Moore RT (1987) Micromorphology of yeasts and yeast-like fungi and its taxonomic implications. Stud Mycol 30:203–226

Moore RT (1990) Platygloeales ord. nov. Mycotaxon 39:245–248

Moore RT (1996a) An inventory of the phylum Ustomycota. Mycotaxon 59:1–31

Moore RT (1996b) Evolutionary advances in the higher fungi. Antonie van Leeuwenhoek Int J Gen Mol Microbiol 72:209–218

Muller SH, Rieger S (1998) Epidemic spread of the rust fungus *Puccinia lagenophorae* and its impact on the competitive ability of *Senecio vulgaris* in celeriac during early development. Biocontrol Sci Technol 8:59–72

Nishida H, Ando K, Ando Y, Hirata A, Sugiyama J (1995) *Mixia osmundae*: transfer from the Ascomycota to the Basidiomycota based on evidence from molecules and morphology. Can J Bot 73 (Suppl 1):S660–S666

Nyland G (1949) Studies on some unusual Heterobasidiomycetes from Washington state. Mycologia 41:686–701

Oberwinkler F (1987) Heterobasidiomycetes with ontogenetic yeast stages – systematic and phylogenetic aspects. Stud Mycol 30:61–74

Oberwinkler F (1990) New genera of auricularioid heterobasidiomycetes. Rep Tottori Mycol Inst 28:113–127

Oberwinkler F, Bandoni RJ (1982a) A taxonomic survey of the gasteroid, auricularioid Heterobasidiomycetes. Can J Bot 60:1726–1750

Oberwinkler F, Bandoni RJ (1982b) *Atractogloea*: a new genus in the Hoehnelomycetaceae (Heterobasidiomycetes). Mycologia 74:634–639

Oberwinkler F, Bandoni RJ (1984) *Herpobasidium* and allied genera. Trans Br Mycol Soc 83:639–658

Oberwinkler F, Bauer R (1989) The systematics of gasteroid, auricularoid Heterobasidiomycetes. Sydowia 41:224–256

Oberwinkler F, Bauer R (1990) *Cryptomycocolax*: a new mycoparasitic heterobasidiomycete. Mycologia 82:671–692

Oberwinkler F, Bauer R, Bandoni RJ (1990a) *Colacogloea*: a new genus in the auricularioid Heterobasidiomycetes. Can J Bot 68:2531–2536

Oberwinkler F, Bauer R, Bandoni RJ (1990b) Heterogastridiales: a new order of basidiomycetes. Mycologia 82:48–58

O'Donnell KL, McLaughlin DJ (1981a) Ultrastructure of meiosis in the hollyhock rust fungus, *Puccinia malvacearum*. I. Prophase I – Prometaphase I. Protoplasma 108:225–244

O'Donnell KL, McLaughlin DJ (1981b) Ultrastructure of meiosis in the hollyhock rust fungus, *Puccinia malvacearum*. II. Metaphase I – Telophase I. Protoplasma 108:245–263

O'Donnell KL, McLaughlin DJ (1981c) Ultrastructure of meiosis in the hollyhock rust fungus, *Puccinia malvacearum*. III. Interphase I – Interphase II. Protoplasma 108:265–288

Olive LS (1950) A new genus of the Tremellales from Louisiana. Mycologia 42:385–390

Patouillard NT (1900) Essai taxonomique sur les familles et les genres de hyménomycètes. (Reprint 1963) A Asher

Poon NH, Day AW (1976) Somatic nuclear division in the sporidia of *Ustilago violacea*. IV. Microtubules and the spindle pole body. Can J Microbiol 22:507–522

Prillinger H, Dörfler C, Laaser G, Hauska G (1990) Ein Beitrag zur Systematik und Entwicklungsbiologie höherer Pilze: Hefe-Typen der Basidiomyceten. Teil III. *Ustilago*-Typ. Z Mykol 56:251–278

Prillinger H, Deml G, Dörfler C, Laaser G, Lockau W (1991a) Ein Beitrag zur Systematik und Entwicklungsbiologie höherer Pilze: Hefe-Typen der Basidiomyceten. Teil II. *Microbotryum*-Typ. Bot Acta 104:5–17

Prillinger H, Laaser G, Dörfler C, Ziegler K (1991b) Ein Beitrag zur Systematik und Entwicklungbiologie höherer Pilze: Hefe-Typen der Basidiomyceten. Teil IV: *Dacrymyces*-Typ, *Tremella*-Typ. Sydowia 43:170–218

Roberts P (1994) *Zygogloea gemellipara*: an auricularioid parasite of *Myxarium nucleatum*. Mycotaxon 52:241–246

Roberts P (1996) Heterobasidiomycetes from Majorca and Cabrera (Balearic Islands). Mycotaxon 60:111–123

Roberts P (1997) New heterobasidiomycetes from Great Britian. Mycotaxon 63:195–216

Rogers DP (1934) The basidium. Univ Iowa Stud Nat Hist 16:106–183

Rogers DP (1971) Patterns of evolution to the homobasidium. In: Peterson RH (ed) Evolution in the higher basidiomycetes. University of Tennesee Press, Knoxville, pp 241–257

Savile DBO (1955) A phylogeny of the basidiomycetes. Can J Bot 33:60–104

Seifert KA, Oberwinkler F, Bandoni R (1992) Notes on *Stilbum vulgare* and *Fibulostilbum phylacicola* gen. et sp. nov. (Atractiellales). Bol Soc Argent Bot 28:213–217

Strange RN (1993) Plant disease control. Towards environmentally acceptable methods. Chapman and Hall, London

Suh S-O, Sugiyama J (1993) Septal pore ultrastructure of *Leucosporidium lari-marini*, a basidiomycetous yeast, and its taxonomic implications. J Gen Appl Microbiol 39:257–260

Suh S-O, Sugiyama J (1994) Phylogenetic placement of the basidiomycetous yeasts *Kondoa malvinella* and *Rhodosporidium dacryoidium*, and the anamorphic yeast *Sympodiomycopsis paphiopedili* by means of 18S rRNA gene sequence analysis. Mycoscience 35:367–375

Suh S-O, Hirata A, Sugiyama J, Komagata K (1993) Septal ultrastructure of basidiomycetous yeasts and their taxonomic implications with observations on the ultrastructure of *Erythrobasidium hasegawianum* and *Sympodiomycopsis paphiopedili*. Mycologia 85:30–37

Swann EC, Taylor JW (1993) Higher taxa of basidiomycetes: an 18S rRNA gene perspective. Mycologia 85:923–936

Swann EC, Taylor JW (1995a) Phylogenetic diversity of yeast-producing basidiomycetes. Mycol Res 99:1205–1210

Swann EC, Taylor JW (1995b) Phylogenetic perspectives on basidiomycete systematics: evidence from the 18S rRNA gene. Can J Bot 73 (Suppl 1):S862–S868

Swann EC, Frieders EM, Hanson RW, McLaughlin DJ (1997) Divide and conquer: the rise and fall of a yeast division paradigm. Inoculum 48(3):37

Swann EC, McLaughlin DJ, Frieders EM, Szabo LJ (1998) The origin of the rust fungi: the Urediniales are not primitive! Inoculum 49(2):51

Swann EC, Frieders EM, McLaughlin DJ (1999) *Microbotryum*, *Kriegeria* and the changing paradigm in basidiomycete classification. Mycologia 91:51–66

Talbot PHB (1968) Fossilized pre-Patouillardian taxonomy? Taxon 17:620–628

Talbot PHB (1971) Principles of fungal taxonomy. St Martin's Press, New York

Talbot PHB (1973) Towards uniformity in basidial terminology. Trans Br Mycol Soc 61:497–512

Vánky K (1998) The genus *Microbotryum* (smut fungi). Mycotaxon 67:33–60

von Arx JA (1981) Genera of fungi sporulating in culture, 3rd edn. Cramer, Vaduz, Lichtenstein

Walker WF, Doolittle WF (1982) Redividing the basidiomycetes on the basis of 5S rRNA sequences. Nature 299:723–724

Wang CJK, Zabel RA (1990) Identification manual for fungi from utility poles in the eastern United States. American Type Culture Collection, Rockville, Maryland

Wells K (1994) Jelly fungi, then and now! Mycologia 86: 18–48

Williams PG (1984) Obligate parasitism and axenic cultures. In: Bushnell WR, Roelfs AP (eds) The cereal rusts. Academic Press, New York, pp 399–430

Yamada Y, Nakagawa Y, Banno I (1989) The phylogenetic relationship of the Q9-equipped species of the heterobasidiomycetous yeast genera *Rhodosporidium* and *Leucosporidium* based on the partial sequences of 18S and 26S ribosomal ribonucleic acids: the proposal of a new genus *Kondoa*. J Gen Appl Microbiol 35:377–385

3 Ustilaginomycetes[1]

R. Bauer[2], D. Begerow[2], F. Oberwinkler[2], M. Piepenbring[2], and M.L. Berbee[3]

CONTENTS

I.	Introduction	57
II.	Diagnosis and Evidence for Monophyletic Origin	57
III.	Nonustilaginomycetous Smut Fungi	58
IV.	Life Cycle	59
A.	Saprobic Phase	60
B.	Parasitic Phase	60
C.	Basidia	62
V.	Hosts, and Their Role in Species Definition	62
VI.	The System	64
A.	Fundamental Characters	65
	1. Cellular Interactions	65
	a) Local Interaction Zones	65
	b) Enlarged Interaction Zones	65
	2. Septation	65
B.	Overview	67
	1. Ustilaginomycetes	67
	2. Taxa Not Ascribed to Any Family	70
C.	Description	70
	1. Entorrhizomycetidae	70
	2. Ustilaginomycetidae	74
	a) Urocystales	74
	b) Ustilaginales	75
	3. Exobasidiomycetidae	76
	a) Malasseziales	76
	b) Georgefischeriales	76
	c) Tilletiales	77
	d) Microstromatales	77
	e) Entylomatales	77
	f) Doassansiales	78
	g) Exobasidiales	78
VII.	Conclusions	79
A.	Evolution of the Basidium	79
B.	Coevolution	79
C.	Evolutionary Trends	80
	References	80

I. Introduction

The Ustilaginomycetes comprises more than 1300 species in ca. 80 genera of basidiomycetous plant parasites. They occur throughout the world, although many species are restricted to tropical, temperate, or arctic regions. Some species of *Ustilago* and *Tilletia*, e.g., the barley, wheat or maize smut fungi, are well known because they are of economic importance (Trione 1982; Thomas 1989; Valverde et al. 1995). For example, from 1983 to 1988, the barley smut fungi reduced annual yields by 0.7% to 1.6% in the prairie provinces in central Canada, causing annual losses of about US$8 000 000 (Thomas 1989). *Tilletia contraversa* Kühn is important in the international wheat trade (Trione 1982) and, 2–5% in a corn field are generally infected by *Ustilago maydis* (DC.) Corda, while up to 80% of a field can be infected if conditions are good for the smut fungus. On the other hand, the galls of *U. maydis* are estimated as a delicacy in the Mesoamerican tradition. They are known in Mexico as huitlacoche and in parts of the USA as maize mushroom, Mexican truffles, or caviar azteca (Valverde et al. 1995).

This chapter focuses on the evolution and suprageneric classification of the Ustilaginomycetes that represents one of the three classes of the Basidiomycota (Fig. 1; Begerow et al. 1997; Swann and Taylor 1993).

II. Diagnosis and Evidence for Monophyletic Origin

The Ustilaginomycetes have a distinctive cell wall carbohydrate composition with a dominance of glucose and absence of xylose that separates them from the Urediniomycetes and Hymenomycetes (Prillinger et al. 1990, 1993). They share the type B secondary structure of the 5S rRNA with the Hymenomycetes (Gottschalk and Blanz 1985) and

---

[1] Part 150 in the series *Studies in Heterobasidiomycetes* from the Botanical Institute, University of Tübingen
[2] Universität Tübingen, Lehrstuhl Spezielle Botanik und Mykologie, Auf der Morgenstelle 1, 72076 Tübingen, Germany
[3] Department of Botany, The University of British Columbia, 6270 University Boulevard, Vancouver, BC V6T 1Z4, Canada

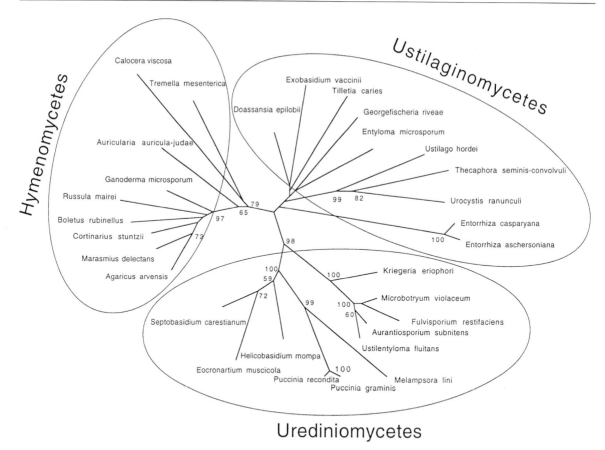

Fig. 1. Topology obtained by neighbor-joining analysis of 518 bp from the 5'-end of the LSU rDNA sequences of 30 basidiomycetes, from Begerow et al. 1997. Percentage bootstrap values of 1000 replicates are given at each furcation. Values smaller than 50% are not shown

the lack of multilayered endoplasmic reticulum elements (parenthesomes) at the pores with the Urediniomycetes (Bauer et al. 1997). An important apomorphy for the Ustilaginomycetes is the presence of zones of host-parasite interaction with fungal deposits, resulting from exocytosis of primary interactive vesicles (Bauer et al. 1997). This feature of the parasitic process is unique among the basidiomycetes.

In using the apomorphic characters discussed above, the Ustilaginomycetes includes all simple-septate (i.e., without multilayered, modified endoplasmic reticulum elements at the pores), holobasidiate phytoparasites of the Basidiomycota, and all simple-septate, phragmobasidiate, teliosporic phytoparasites of the Basidiomycota, producing intracellular organs without penetration necks (Bauer et al. 1997). Sequence analysis supports the monophyly of the Ustilaginomycetes as defined above but with different statistical support in different analyses. Thus, the union of *Tilletia caries* (DC.) L. & C. Tul., *Ustilago hordei* (Pers.) Lagerh. and *Ustilago maydis* is well supported by high bootstrap values in small-subunit (SSU) rDNA sequence analyses (Swann and Taylor 1993, 1995), whereas the bootstrap values for the Ustilaginomycetes are low in large-subunit (LSU) rDNA sequence analyses with an enlarged set of species (Fig. 1; Begerow et al. 1997). In particular, bootstrap support from our LSU data set for the Ustilaginomycetes sank when *Entorrhiza* sequences were included, possibly because *Entorrhiza* is basal in the Ustilaginomycetes and intermediate between the Ustilaginomycetes and the other basidiomycetes.

III. Nonustilaginomycetous Smut Fungi

Like the terms agaric, polypore, lichen, etc., the term smut fungus circumscribes the organization and life strategy of a fungus, but it is not a taxonomic term. Smut fungi evolved in different fungal groups. Most smut fungi are in the Ustilaginomycetes. Other smut fungi, in the Microbotryales, are members of the Urediniomycetes (Fig. 1; Bauer et al. 1997; Begerow et al. 1997). In contrast with the Ustilaginomycetes, available data indicate that the microbotryaceous taxa *Aurantiosporium*, *Bauerago*, *Fulvisporium*, *Liroa*, *Microbotryum*, *Sphacelotheca*, *Ustilentyloma*, and *Zundeliomyces* have a type A 5S rRNA secondary structure (Gottschalk and Blanz 1985; Müller 1989), mannose as the major cell wall carbohydrate (Prillinger et al. 1991, 1993), and cellular interactions without primary interactive vesicles (Bauer et al. 1997). Morphologically, they are distinguishable from the phragmobasidiate members of the Ustilaginomycetes by the lack of intracellular hyphae or haustoria (Bauer et al. 1997). Clustering of the Microbotryales with the Urediniomycetes rather than the Ustilaginomycetes is also supported by sequence analyses (Fig. 1; Begerow et al. 1997; Swann and Taylor 1995). However, there are significant convergences between the microbotryaceous and the ustilaginomycetous phragmobasidiate smut fungi. Certain taxa of both groups are similar with respect to soral morphology, teliosporogenesis, life cycle, basidial morphology, and host range.

The ultrastructural characters reveal the existence of two groups within the Microbotryales. The Ustilentylomataceae possess septa with simple pores, whereas the septa of the Microbotryaceae are poreless. The classification of the Microbotryales is as follows (Bauer et al. 1997); host families are indicated if the host range of the respective genus does not comprise more than two host families; species requiring transfer to other genera are indicated by quotation marks.

    Microbotryales R. Bauer & Oberw.
i. Ustilentylomataceae R. Bauer & Oberw.
    Aurantiosporium M. Piepenbr., K. Vánky & Oberw. on Cyperaceae
    Fulvisporium K. Vánky on Poaceae
    Ustilentyloma Savile on Poaceae
ii. Microbotryaceae R. T. Moore
    Bauerago K. Vánky on Cyperaceae and Juncaceae
    Liroa Ciferri on Polygonaceae
    Microbotryum Léveillé emend. K. Vánky on dicots
[There is some confusion concerning the systematic position of the "*Ustilago*" species on dicots. The "*Ustilago*" species occurring on Asteraceae Caryophyllaceae, Dipsacaceae, Gentianaceae, Lamiaceae, Lentibulariaceae, Onagraceae, Polygonaceae, and Portulacaceae are species of *Microbotryum* (Bauer et al. 1997; Deml and Oberwinkler 1982, Prillinger et al. 1991; Vánky 1998), whereas those occurring on Brassicaceae, Campanulaceae, Haloragaceae, and Oxalidaceae are members of the Ustilaginomycetes (see below)].

Sphacelotheca de Bary emend. Langdon & Fullerton on Polygonaceae
Zundeliomyces K. Vánky on Polygonaceae
Some "*Ustilago*" spp. on Commelinaceae

Even nonbasidiomycetous fungi can cause diseases with the formation of thick-walled propagules similar to those of the smut fungi. Species of *Schroeteria* Winter, for example, look superficially similar to smut fungi (Vánky 1981), but they belong to the ascomycetes (Nagler et al. 1989). Leaf spots similar to sori of *Entyloma* can be formed by representatives of the Protomycetales (Reddy and Kramer 1975).

IV. Life Cycle

The species of the Ustilaginomycetes share an essentially similar life cycle with a saprobic haploid phase and a parasitic dikaryophase (e.g., Sampson 1939; Fig. 2). The haploid phase usually

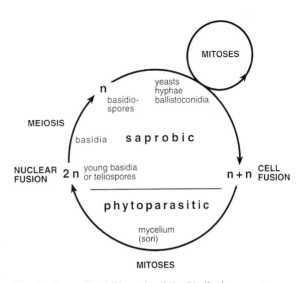

Fig. 2. Generalized life cycle of the Ustilaginomycetes

commences with the formation of basidiospores after meiosis of the diploid nucleus in the basidium and ends with the conjugation of compatible haploid cells to produce dikaryotic, parasitic mycelia. The dikaryotic phase ends with the production of basidia.

Almost all Ustilaginomycetes sporulate on or in parenchymatic tissues of the hosts. In the majority of the Ustilaginomycetes the young basidium becomes thick-walled and at maturity separates from the sorus, thus functioning as a dispersal agent, the teliospore. Dusty teliospores are efficiently dispersed by wind. Sometimes, however, teliospores are also dispersed by water or entire sori or vegetative cells are used as diaspores (Piepenbring et al. 1998a). The sori with the masses of teliospores are usually the most conspicuous stage in the smut's life cycle. Most of the Ustilaginomycetes are dimorphic, producing a yeast or yeastlike phase in the haploid state. However, there are several variations from this generalized life cycle, e.g., homothallism in *Anthracoidea* (Kukkonen and Raudaskoski 1964) or *Exobasidium* (Sundström 1964).

A. Saprobic Phase

Multiplication in the saprobic phase with yeasts or ballistoconidia (i.e., forcibly discharged conidia) may play an essential role in the dispersal of the species and in the infection of host individuals during the vegetation period. Compatible yeast cells are able to conjugate and to infect the respective hosts, e.g., in *Ustilago maydis* (Snetselaar and Mims 1992, 1993) or *Graphiola phoenicis* (Moug.) Poiteau (Cole 1983). Ustilaginomycetous anamorphs appear to be common in nature. Thus, we have isolated numerous ustilaginomycetous anamorphs from different substrata. In addition, the species of *Malassezia*, *Tilletiopsis*, *Sympodiomycopsis*, and *Pseudozyma* sensu Boekhout (1995) represent ustilaginomycetous anamorphs (Figs. 33, 34, Begerow et al. 2000; Boekhout 1987, 1991, 1995; Sugiyama et al. 1991; see also Chap. 7, this Vol.) The species of the Ustilaginomycetes investigated by us produced either a yeast or yeastlike phase, with or without ballistoconidia, or a hyphal anamorph with ballistoconidia from the basidiospores.

The yeast or yeastlike cells of the Ustilaginomycetes are usually elongated and cylindrical in form and tend to produce pseudomycelia with retraction septa (i.e., septa that separate living cells from empty cells) in older cultures. In contrast to this, the yeast cells of the Microbotryales are ellipsoidal and show only yeast growth.

The ballistosporic basidiospores/conidia occurring in the Ustilaginomycetes differ from those of other basidiomycetes in form and ultrastructure. They have the form of a relatively long and narrow curved cylinder with a rounded apex and a slightly constricted base constituting the hilar appendix (e.g., Kollmorgen et al. 1980; Ingold 1987b, 1997). During germination they usually become septate. Ultrastructurally, the hilar septum of the ustilaginomycetous ballistosporic propagules is very characteristic. The median layer of the hilar septum contains electron-opaque, spheroid particles. Fine fibrils extend from the particles towards the ballistosporic basidiospores/conidia. After ejection, the electron-opaque particles remain attached to the ballistosporic basidiospores/conidia (Bauer et al. 1997; Goates and Hoffmann 1986). Thus, morphological and ultrastructural characters can be used to identify a ballistosporic propagule belonging to the Ustilaginomycetes. For example, the form of ballistoconidia indicates that *Tilletiaria anomala* Bandoni & Johri (Bandoni and Johri 1972) is a member of the Ustilaginomycetes.

B. Parasitic Phase

Depending upon the various ustilaginomycetous groups, parasitic hyphae grow either only intercellularly or both intercellularly and intracellularly (Figs. 3–5, 7; Bauer et al. 1997; Luttrell 1987). Haustoria or intracellular hyphae are usually not constricted where they enter or exit the host cells (Figs. 25, 26; Nagler and Oberwinkler 1989; Snetselaar and Tiffany 1990; Mims et al. 1992; Snetselaar and Mims 1994; Bauer et al. 1995b, 1997). Unlike haustoria, intracellular hyphae do not have a consistent characteristic morphology. Except for their intracellular growth and the distinctive matrix, intracellular hyphae are morphologically indistinguishable from intercellular hyphae. As intercellular hyphae, they are branched and septate. Instead of ending within the host cell as haustoria do, they pass completely through host cells, often growing from one host cell into another (Bauer et al. 1997; Luttrell 1987; Mims et al. 1992; Nagler et al. 1990; Snetselaar and Mims 1994; Snetselaar and Tiffany 1990).

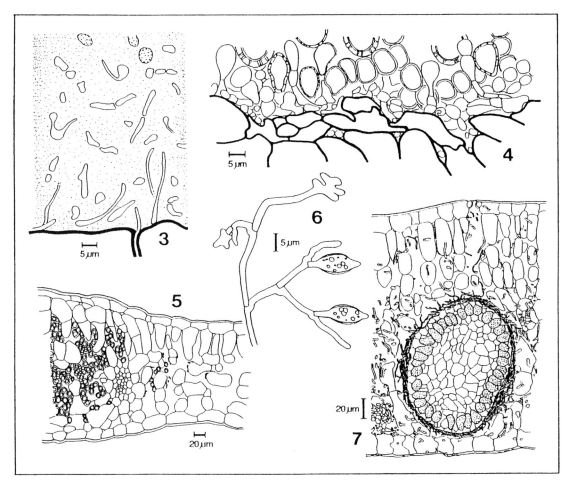

Figs. 3–7. Teliosporogenesis. **Fig. 3.** On the surface of host cells (*bottom*) sporogenous hyphae of *Ustilago maydis* (DC.) Corda disintegrate in a hyaline matrix (*dots*). **Fig. 4.** Sporogenous hyphae of *Conidiosporomyces ayresii* (Berk.) K. Vánky develop teliospores and sterile cells on the host tissue (*bottom*). **Fig. 5.** Transverse section through a part of a leaf spot of *Entyloma microsporum* (Unger) Schröter and normal adjacent leaf tissue. The leaf spot is swollen by the teliospore mass and by additional divisions of the host cells (see palisade layer). **Fig. 6.** Sporogenous hyphae of *Rhamphospora nymphaeae* D.D. Cunn. with clamps, appressoria, and young terminal teliospores. **Fig. 7.** Spore ball of *Doassansiopsis ticonis* M. Piepenbr. with central sterile tissue, a layer of teliospores (*dotted*), peripheral sterile cells, and hyphal sheath in the mesophyll of the host leaf

In many species of the Ustilaginomycetes, hyphal branches arise from structures that look superficially like clamps (Sampson 1939; Nagler 1986; Snetselaar and Mims 1994). Nagler (1986) studied these pseudoclamps in detail and found that they do not function as clamps normally do. They seem to correspond to fusion bridges (Fischer and Holton 1957) for the transport of nuclei and are not involved in conjugate nuclear divisions. However, we found regular clamps in the members of the Doassansiales (Fig. 6), *Graphiola* and in some species of *Exobasidium*.

The teliospores are clustered in sori. Depending on the fungal species, the sori appear in or on different organs of the hosts, e.g., in roots, stems, leaves, inflorescences, flowers, anthers, ovaries, seeds, etc. In contrast with numerous rusts, the teliospores of the Ustilaginomycetes are usually not pediculate (Figs. 4–7). Although the mycelium can be either intercellular or intracellular, teliosporogenesis usually occurs intercellularly either in preformed intercellular spaces or in cavities created by disintegration of host cells (Figs. 3–5, 7; Luttrell 1987). Less commonly, teliospores are produced in host cells. *Entorrhiza* species, for example, sporulate inside living root cells (Deml and Oberwinkler 1981; Fineran 1980) and the species of *Schizonella* use more or less disintegrated epidermal cells for sporulation. Sporulation in disintegrating host cells also occurs in species of *Ustilago* (Langdon and Fullerton 1975; Mims et al. 1992). The species of *Clintamra*, *Exoteliospora* and *Orphanomyces* develop their teliospores externally on the host tissue (Vánky 1987; Bauer et al. 1999a).

Teliosporogenesis varies among the Ustilaginomycetes (Figs. 3–7). One extreme is *Ustilago*, where nearly all hyphae in the sori disarticulate, lose their cell walls, and form teliospores in a matrix resulting from gelatinization of the hyphal walls (Fig. 3; Mims et al. 1992; Snetselaar and Mims 1994). *Rhamphospora* represents another extreme in that the teliospores are formed only terminally on clamped sporogenous hyphae without recognizable gelatinization (Fig. 6; Piepenbring et al. 1998b).

C. Basidia

Historically, basidiomycete classification relied on basidial morphology. However, new phylogenies based on molecular or ultrastructural characters show that basidia are not always the stable markers they were once assumed to be (e.g., Bandoni 1984; Bauer et al. 1997; Begerow et al. 1997; Swann and Taylor 1995). Within each basidiomycete class, basidial types (e.g., Oberwinkler 1977, 1978, 1982, 1985) radiated and converged. Various species of the Ustilaginomycetes, rather than sharing a single, diagnostic basidial type, exhibit a wide range of different types (Figs. 8–21; Ingold 1983, 1987a, 1988, 1989a,b,c; Oberwinkler 1978, 1982, 1985). Ustilaginomycetous basidia may be phragmobasidial, divided into compartments by internal septa (Figs. 8–11, 17), or holobasidial, lacking internal septa (Figs. 13–16, 18–21). The basidiospores of ca. 80 species in the Ustilaginomycetes are ballistosporic (i.e., forcibly discharged from the ends of sterigmata; Figs. 15–17, 21). In the other species, basidiospores are passively released (Figs. 8–14, 18–20). Multiple basidiospore production in yeastlike manner occur in many species having phragmobasidia (Fig. 9).

Some of the combinations of basidial characters common in the Ustilaginomycetes are rare in the other classes. Taxa with ballistosporic holobasidia (Figs. 16, 21) are lacking in the Urediniomycetes, whereas taxa with gastroid phragmobasidia (Figs. 8–11) are extremely rare in the Hymenomycetes. Some specific basidial characters occur only in the Ustilaginomycetes. Fusion of basidial segments, e.g., in species of *Ustilago* or *Cintractia*, or fusion of basidiospores on the basidia, e.g., in species of *Entyloma*, *Urocystis*, or *Tilletia* (Figs. 10, 18; Ingold 1983, 1987a, 1989a, 1989b; Piepenbring 1996) are unknown from basidiomycetes outside the Ustilaginomycetes. In addition, ballistosporic holobasidia with abaxially oriented hilar appendices of the basidiospores (sterigmata turned outwards, basidiospores inwards; Figs. 16, 21; Oberwinkler 1977, 1978, 1982; Ingold 1995; Bauer et al. 1999b; Vánky and Bauer 1996), occur only in the Ustilaginomycetes.

V. Hosts, and Their Role in Species Definition

In contrast with the Urediniomycetes and Hymenomycetes, the Ustilaginomycetes are ecologically well characterized by their plant parasitism. The two species of *Melaniella* occur on spikemosses, *Exoteliospora osmundae* R. Bauer et al. on ferns, the two species of *Uleiella* on conifers, whereas all other parasitic Ustilaginomycetes parasitize angiosperms with a high proportion of species on monocots, especially on Poaceae and Cyperaceae. Thus, of the ca. 1300 species, ca. 42% occurs on Poaceae and ca. 15% on Cyperaceae. In addition, of the 78 ustilaginomycetous genera occurring on angiosperms, 48 genera have species exclusively on monocots, 25 exclusively on dicots and 5 have species on both monocots and dicots. Of the 48 monocotyledonous genera 14 occur exclusively on Poaceae and 15 exclusively on Cyperaceae (see below). Concerning the hosts, two points are noteworthyable: (1) With a few exceptions, the teliospore-forming species of the Ustilaginomycetes parasitize nonwoody herbs, whereas those without teliospores prefer woody trees or bushes. However, almost all species sporulate on parenchymatic tissues of the hosts. (2) Two of the largest angiosperm families the Orchidaceae with about 20000 species and the Poaceae with about 9000 species, play a quite different role for the Ustilaginomycetes. There are no known species on Orchidaceae while the Poaceae are the most important host family of the Ustilaginomycetes. This can be tentatively explained by the completely different ecological strategies of the two families. Orchid species subsist with a few isolated individuals and are highly specialized for

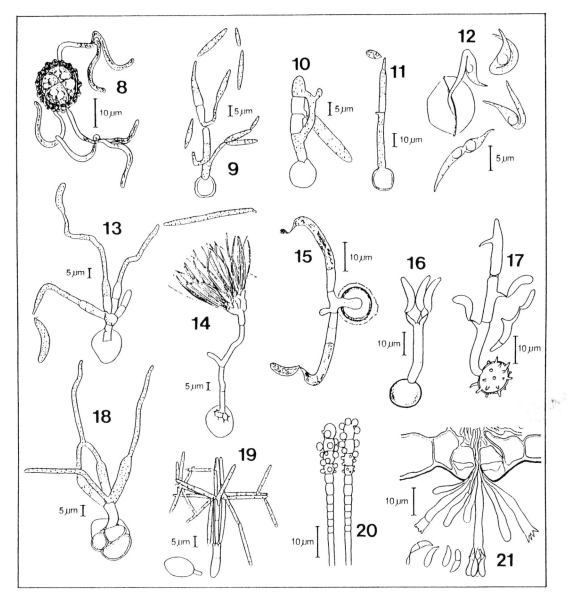

Figs. 8–21. Basidia of the Ustilaginomycetes. **Fig. 8.** *Entorhiza casparyana* (Magnus) Lagerh. **Fig. 9.** *Ustilago maydis* (DC.) Corda. **Fig. 10.** *Cintractia axicola* (Berk.) Cornu. **Fig. 11.** *Anthracoidea altiphila* K. Vánky & Piepenb. **Fig. 12.** *Mycosyrinx cissi* (DC.) G. Beck. **Fig. 13.** *Entyloma guaraniticum* Speg. **Fig. 14.** *Tilletia setaricola* Pavgi & Thirum. **Fig. 15.** *Georgefischeria riveae* Thirum. & Naras. **Fig. 16.** (*"Entyloma"*). *dactylidis* (Pass.) Cif. = *Jamesdicksonia dactylidis* (new combination in prep.). **Fig. 17.** *Tilletiaria anomala* Bandoni & Johri. **Fig. 18.** *Urocystis ranunculi* (Lib.) Moesz. **Fig. 19.** *Rhamphospora nymphaeae* D.D. Cunn. **Fig. 20.** *Graphiola phoenicis* (Moug.) Poiteau. **Fig. 21.** *Exobasidium oxycocci* Rostr.

insect pollination. The Poaceae, however, disperse their dusty pollen by the wind and cover about a third of the land surface with numerous individuals. The ecology of the Ustilaginomycetes, with dusty teliospores or basidiospores dispersed by wind and with the requirement of extensive host populations for successful infection, corresponds well to the ecology of the Poaceae.

Host ranges play an important role in species definition. Many species, e.g., the species of *Entyloma*, *Melanotaenium*, or *Urocystis* (see Vánky 1994), have few defining morphological characters. As a result, species based only on morphology would often be broad and would sometimes include distantly related organisms. For example, Savile (1947) lumped many species of *Entyloma* in two collective species, whereas Vánky (1994) distinguished many different species based on morphology and hosts. To delimit narrower groups closer to "biological species", host ranges are usually considered in species definitions. Fischer and Shaw (1953) argued that morphologically similar smut fungi producing similar symptoms should be considered different species if they parasitize hosts in different families. Vánky (1994) followed a similar approach.

When similar smut fungi parasitize similar hosts, the situation is less clear. Indications can be obtained in the field when a potential host without infection stands close to a systematically related host with heavy infections. Particularly perplexing are the closely related smut species that hybridize under laboratory conditions (Carris and Gray 1994; Fischer and Holton 1957; Huang and Nielsen 1984). For example, *Tilletia contraversa* and *Tilletia caries* (associated mainly with wheat but capable of infecting *Bromus* spp. and other grasses) require different conditions for germination and they differ in spore ornamentation and spore fluorescence. Russell and Mills (1993, 1994) demonstrated overlap in phenotypic characters including karyotype, spore ornamentation, and fluorescence for the two taxa, and Trail and Mills (1990) showed that *Tilletia contraversa* and *Tilletia caries* hybridize when inoculated onto the same plant. Both species can be hybridized with *Tilletia bromi* Brockm. (infecting *Bromus* spp. and other grasses) under laboratory conditions (Carries and Gray 1994). Do these species exchange genes in nature, or do differences in their biology keep them segregated? Determining whether such smut populations belong to the same species will require detailed analysis of population genetic structures to establish the limits of gene exchange (McDermott and McDonald 1993). Only in few cases have fungal species delimitations been subjected to this kind of detailed analysis (McDermott and McDonald 1993; Bucheli and Leuchtmann 1996).

VI. The System

Beginning with Tulasne and Tulasne (1847), the smut fungi have traditionally been divided into the phragmobasidiate Ustilaginaceae or Ustilaginales and the holobasidiate Tilletiaceae or Tilletiales (e.g., Kreisel 1969; Oberwinkler 1987). Durán (1973) and Vánky (1987) discussed difficulties in smut classification in detail but did not list higher taxa in the group. Thus, Vánky (1987) treated all smut fungi in a single order, Ustilaginales, with one family, Ustilaginaceae. The basidiomycetous nature of *Graphiola* was revealed by Oberwinkler et al. (1982). *Microstroma* has traditionally been placed in the Exobasidiales (Hennings 1900).

The classification discussed below is based predominantly on characteristics of host-parasite interactions and septal pore apparatus (Fig. 32; Bauer et al. 1997) and is supported by LSU rDNA sequence analyses (Figs. 33, 34; Begerow et al. 1997), mode of teliosporogenesis and the ultrastructure of teliospore walls (Piepenbring et al. 1998b,c,d). Groups of the Ustilaginomycetes recognized by Bauer et al. (1997) are also evident in both maximum-parsimony and neighbor-joining analyses of LSU rDNA sequence data and are well supported by bootstrap values (Figs. 33, 34; Begerow et al. 1997). In fact, differences between the ultrastructural and the LSU rDNA sequence analyses concern only *Thecaphora* (Glomosporiaceae). *Thecaphora* was interpreted by Bauer et al. (1997) together with *Glomosporium* as basal taxon of the Ustilaginaceae (Fig. 32), whereas in the LSU rDNA analyses *Thecaphora* appears in a position at the base of the Urocystales (Figs. 33–34; Begerow et al. 1997). However, sequence analyses of additional species, such as more species of *Thecaphora* and members of *Glomosporium*, and *Mycosyrinx* (see Fig. 32), are required before any conclusion concerning this discrepancy may be drawn. *Melanotaenium endogenum* (Unger) de Bary was interpreted by Bauer et al. (1997) as a basal taxon of the Urocystales (Fig. 32) and appears in the maximum-parsimony analyses of Begerow et al. (1997) in that position, but not in the analyses of this chapter (Figs. 33, 34). In contrast with the LSU rDNA sequence studies (Figs. 33, 34; Begerow et al. 1997), the ultrastructural analyses at least resolve the relationships among three orders of the Exobasidiomycetidae (Fig. 32, see Exobasidianae; Bauer et al. 1997). Two possible explanations for the lack of resolution in the molecular studies exist: (1) the available sequences of ca. 540 bp are too short to resolve the phylogeny within the Exobasidiomycetidae or (2) the relationships among the orders of the Exobasidiomycetidae cannot be resolved by this kind of sequence analyses because the ancestor of the Exobasidiomycetidae diverged into several groups within a very short time.

The fundamental characters used in classifying the Ustilaginomycetes were discussed in detail by Bauer et al. (1995a, 1995b, 1997) and are therefore only briefly summarized here.

A. Fundamental Characters

1. Cellular Interactions

Hyphae of the Ustilaginomycetes in contact with host plant cells possess zones of host-parasite interaction with fungal deposits resulting from exocytosis of primary interactive vesicles. These zones provide ultrastructural characters diagnostic for higher groups in the Ustilaginomycetes (Fig. 32; Bauer et al. 1997). Initially, primary interactive vesicles with electron-opaque contents accumulate in the fungal cell. Depending on the fungal species, the primary interactive vesicles may fuse with one another before being exocytosed from the fungal cytoplasm. Electron-opaque deposits also appear at the host side, opposite the point of contact with the fungus (Figs. 22–26). Detailed studies indicate that these deposits at the host side originate from the exocytosed fungal material by transfer towards the host plasma membrane (Bauer et al. 1995b, 1997).

The following major types, minor types, and variations were recognized by Bauer et al. (1995b, 1997).

a) Local Interaction Zones (Figs. 22–25)

Short-term production of primary interactive vesicles per interaction site results in local interaction zones.

1. **Local interaction zones without interaction apparatus** (Fig. 22). Primary interactive vesicles fuse individually with the fungal plasma membrane. Depending upon the species, local interaction zones without interaction apparatus are present in intercellular hyphae or haustoria.
2. **Local interaction zones with interaction apparatus** (Figs. 23–25). Fusion of the primary interactive vesicles precedes exocytosis.
   a. **Local interaction zones with simple interaction apparatus** (Fig. 23). Primary interactive vesicles fuse to form one large secondary interactive vesicle per interaction site. Interaction zones of this type are located only in intercellular hyphae.
   b. **Local interaction zones with complex interaction apparatus** (Figs. 24–25). Numerous primary interactive vesicles fuse to form several secondary interactive vesicles per interaction site. Fusion of the secondary interactive vesicles then results in the formation of a complex cisternal net.
      i. **Local interaction zones with complex interaction apparatus containing cytoplasmic compartments** (Fig. 24). The intercisternal space of the cisternal net finally becomes integrated in the interaction apparatus. Depending upon the species, interaction zones of this type are formed by intercellular hyphae or haustoria.
      ii. **Local interaction zones with complex interaction apparatus producing interaction rings** (Fig. 25). The intercisternal space does not become integrated in the interaction apparatus. Transfer of fungal material towards the host plasma membrane occurs in two or three steps. The first transfer results in the deposition of a ring at the host plasma membrane. Depending upon the species, interaction zones of this type are located in intercellular hyphae or haustoria.

b) Enlarged Interaction Zones (Fig. 26)

Continuous production and exocytosis of primary interactive vesicles results in the continuous deposition of fungal material at the whole contact area with the host cell. Depending upon the species, this type of interaction zones is located in intercellular hyphae, intracellular hyphae or haustoria.

2. Septation

Septal pore architecture plays an important role in the arrangement of basidiomycetes (Oberwinkler 1985; Wells 1994). In general, the pores of the Ustilaginomycetes are not associated with differentiated, multilayered caps or sacs derived from the endoplasmic reticulum. The septa produced in the saprobic phase of the dimorphic species of the Ustilaginomycetes are devoid of distinct septal pores. Five types of septation of soral hyphae were recognized by Bauer et al. (1997): (1) Presence of simple pores with two tripartite membrane caps (Fig. 27), (2) presence of simple pores with two outer tripartite membrane caps and two inner non-

66   R. Bauer et al.

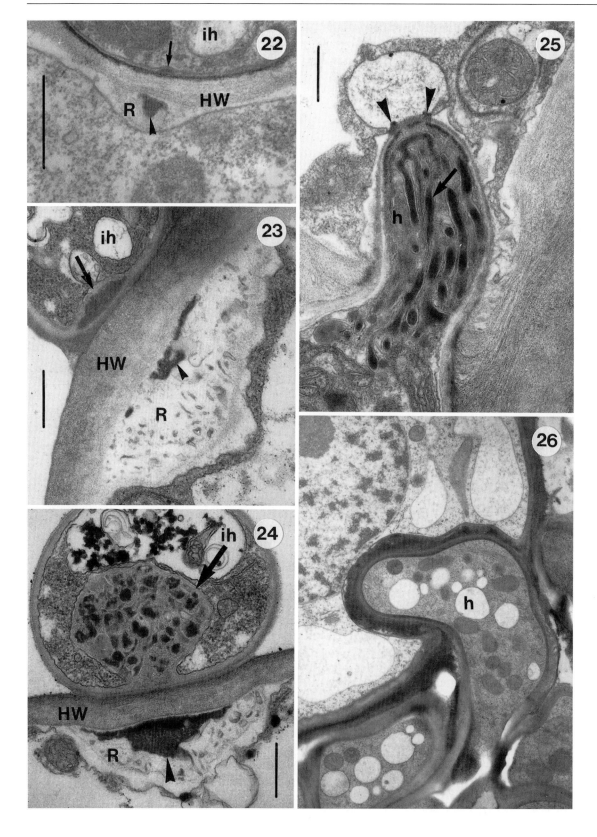

membranous bands (Fig. 28; see also Bauer et al. 1995a), (3) presence of dolipores without membrane caps or bands (Figs. 29; see also Deml and Oberwinkler 1981), (4) presence of dolipores with tripartite membrane bands (Fig. 30; see also Roberson and Luttrell 1989), and (5) septa without distinct pores (Fig. 31), designated poreless septa.

B. Overview

In the following, an overview of the taxa included in the Ustilaginomycetes is given. Host families are indicated if the host range of the respective genus does not comprise more than two families. On the basis of morphology and LSU rDNA sequence analyses (Figs. 33, 34), the anamorphic species are ascribed to teleomorphic taxa. The data indicate that several species require transfer to new genera or families. Species requiring transfer to new genera are indicated by quotation marks.

1. Ustilaginomycetes
R. Bauer, Oberw. & K. Vánky
  I. Entorrhizomycetidae R. Bauer & Oberw.
    1. Entorrhizales R. Bauer & Oberw.
      i. Entorrhizaceae R. Bauer & Oberw.
        Entorrhiza Weber on Cyperaceae and Juncaceae
  II. Ustilaginomycetidae Jülich emend. R. Bauer & Oberw.
    1. Urocystales R. Bauer & Oberw.
      i. Melanotaeniaceae Begerow, R. Bauer & Oberw.
        Exoteliospora R. Bauer, Oberw. & K. Vánky on Osmundaceae
        Melanotaenium de Bary on dicots
        "*Ustilago*" *Speculariae* Stevenson on Campanulaceae
      ii. Doassansiopsaceae Begerow, R. Bauer & Oberw.
        Doassansiopsis (Setchell) Dietel on monocots and dicots
      iii. Urocystaceae Begerow, R. Bauer & Oberw.
        "*Melanotaenium*" *ari* (Cooke) Lagerh. on Araceae
        Mundkurella Thirum. on Araliaceae
        Urocystis Rebenh. ex Fuckel on monocots and dicots
        Ustacystis Zundel on Rosaceae
        Some "*Ustilago*" spp. on Liliaceae
    2. Ustilaginales Clinton emend. R. Bauer & Oberw.
      i. Mycosyringaceae R. Bauer & Oberw.
        Mycosyrinx Beck on Vitaceae
      ii. Glomosporiaceae Cifferi emend. Begerow, R. Bauer & Oberw.
        Glomosporium Kochman on Chenopodiaceae
        Thecaphora Fingerh. emend. K. Vánky on dicots
        Tothiella K. Vánky on Brassicaceae
        "*Ustilago*" *oxalidis* Ellis & Tracy on Brassicaceae
      iii. Ustilaginaceae L. Tulasne & C. Tulasne emend. R. Bauer & Oberw.
        Anthracoidea Brefeld on Cyperaceae
        Cintractia Cornu on Cyperaceae
        Clintamra Cordas & Durán on Liliaceae
        Dermatosorus Sawada ex Ling on Cyperaceae
        Farysia Racib. on Cyperaceae
        Farysporium K. Vánky on Cyperaceae

Figs. 22–26. Cellular interactions in the Ustilaginomycetes. Material illustrated in **Fig. 26** was prepared using freeze substitution. Bars 0.5 µm. **Fig. 22.** Local interaction zone without interaction apparatus, representative for the Entorrhizomycetidae, Georgefischeriales, Tilletiales, and Microstromatales. Intercellular hypha (*ih*) of *Conidiosporomyces ayresii* (Berk.) K. Vánky with the secretion profile of one primary interactive vesicle (*arrow*) in contract with host cell wall (*HW*). Note the electron-opaque deposit at the host side (*arrowhead*). Host response to infection is visible at *R*. **Fig. 23.** Local interaction zone with simple interaction apparatus, representative for the Entylomatales. Intercellular hypha (*ih*) of *Entyloma hieracii* H. & P. Sydow in contact with host cell wall showing the exocytosis profile of the simple interaction apparatus (*arrow*). Note the electron-opaque deposit at the host side (*arrowhead*). Host response to infection is visible at *R*. **Fig. 24.** Local interaction zone with complex interaction apparatus containing cytoplasmic compartments, representative for the Doassansiales. Intercellular hypha (*ih*) of *Doassinga callitrichis* (Liro) K. Vánky et al. in contact with host cell wall (*HW*) showing the exocytosis profile of one complex interaction apparatus (*arrow*). The interaction apparatus and its intercisternal space is excluded from the cytoplasm. Note the electron-opaque deposit at the host side (*arrowhead*). Host response to infection is visible at *R*. **Fig. 25.** Local interaction zone with complex interaction apparatus producing interaction ring, representative for the Exobasidiales. Haustorium (*h*) of *Exobasidium* sp. with interaction apparatus (*arrow*). Note the sectioned interaction ring (*arrowheads*) at the top of the haustorium. **Fig. 26.** Enlarged interaction zone, representative for the Ustilaginomycetidae. Haustorium (*h*) of *Ustacystis waldsteiniae* (Peck) Zundel is surrounded by an electron-opaque matrix

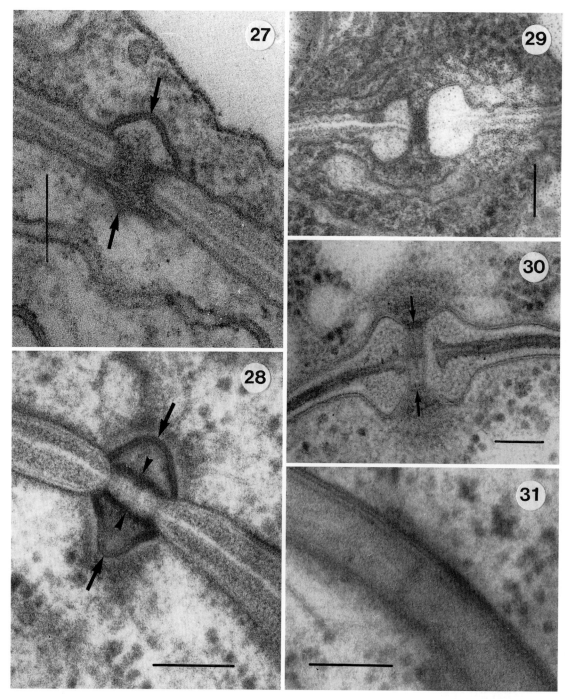

Figs. 27–31. Septation of soral hyphae in the Ustilaginomycetes. Material illustrated in **Figs. 28, 30, 31** was prepared using freeze substitution. *Bars* 0.1 μm. **Fig. 27.** Simple pore with two membrane caps (*arrows*) of *Doassinga callitrichis* (Liro) K. Vánky et al. representative for the Melanotaeniaceae, Microstromatales and the Exobasidianae. **Fig. 28.** Simple pore with two outer membrane caps (*arrows*) and two inner nonmembranous bands (*arrowheads*) of *Ustacystis waldsteiniae* (Peck) Zundel, representative for the Urocystaceae and Doassansiopsaceae. **Fig. 29.** Dolipore without membrane bands of *Entorrhiza casparyana* (Magnus) Lagerh., representative for the Entorrhizales. **Fig. 30.** Dolipore with membrane bands (*arrows*) of *Tilletia barclayana* (Bref.) Sacc. & P. Sydow, representative for the Tilletiales. **Fig. 31.** Poroid structure in a septum of *Mycosyrinx cissi* (DC.) G. Beck, representative for the Ustilaginales and Georgefischeriales

Franzpetrakia Thirum. & Pavgi emend. Guo, K. Vánky & Mordue on Poaceae
Geminago K. Vánky & R. Bauer on Sterculiaceae
Heterotolyposporium K. Vánky on Cyperaceae
Kuntzeomyces P. Henn. ex Sacc. & P. Sydow on Cyperaceae
Leucocintractia M. Piepenbr., Begerow & Oberw. on Cyperaceae
Macalpinomyces Langdon & Full. emend. K. Vánky on Poaceae
Melanopsichium G. Beck on Polygonaceae
Moesziomyces K. Vánky on Poaceae and Eriocaulaceae
Moreaua T.N. Liou & H.C. Cheng on Cyperaceae
Orphanomyces Savile on Cyperaceae
Pericladium Pass. on Tiliaceae
Planetella Savile on Cyperaceae
Schizonella Schröter on Cyperaceae
Sporisorium Ehrenb. on Poaceae
Stegocintractia M. Piepenbr., Begerow & Oberw. on Juncaceae
Tolyposporium Woronin ex Schröter on Juncaceae
Trichocintractia M. Piepenbr. on Cyperaceae
Ustanciosporium K. Vánky emend. M. Piepenbr. on Cyperaceae
Ustilago (Pers.) Roussel on Poaceae (There is some confusion concerning the systematic position of the species of *Ustilago*. The host range of *Ustilago* is restricted to Poaceae. Non-graminicolous "*Ustilago*" species belong to other genera, mostly to *Microbotryum*).
Websdanea K. Vánky on Restionaceae
Anamorphs:
Pseudozyma Bandoni emend. Boekhout
Probably in this family:
Testicularia Klotzsch on Cyperaceae
Tranzscheliella Lavrov on Poaceae
Uleiella Schröter on Auraucariaceae
III. Exobasidiomycetidae Jülich emend. R. Bauer & Oberw.
  1. Malasseziales Moore emend. Begerow, R. Bauer & Boekhout
    Malassezia Baillon

2. Georgefischeriales R. Bauer, Begerow & Oberw.
  i. Georgefischeriaceae R. Bauer, Begerow & Oberw.
    Georgefischeria Thirum. & Narash. emend. Gandhe on Convolvulaceae
    Jamesdicksonia Thirum., Pavgi & Payak emend. R. Bauer Begerow, A. Naglér & Oberw. on Cyperaceae and Poaceae
  ii. Tilletiariaceae Moore
    Phragmotaenium R. Bauer, Begerow, on Poaceae
    Tilletiaria Bandoni & Johri
    Tolyposporella Atkinson on Poaceae
    Anamorphs:
    Tilletiopsis flava (Tubaki) Beokhout
    Tilletiopsis fulvescens Gokhale
  iii. Eballistraceae (in prep.)
    Eballistra A. Nagler & Oberw. on Poaceae
3. Tilletiales Kreisel ex R. Bauer & Oberw.
  i. Tilletiaceae L. & C. Tul. emend. R. Bauer & Oberw.
    Conidiosporomyces K. Vánky on Poaceae
    Erratomyces Piepenb. & R. Bauer on Fabaceae
    Ingoldiomyces K. Vánky on Poaceae
    Neovossia Körn. on Poaceae
    Oberwinkleria K. Vánky & R. Bauer on Poaceae
    Tilletia L. & C. Tul. on Poaceae
4. Microstromatales R. Bauer & Oberw.
  i. Microstromataceae Jülich
    Microstroma Niessl on Juglandaceae and Fagaceae
  ii. Volvocisporiaceae Begerow, R. Bauer & Oberw.
    Volvocisporium Begerow, R. Bauer & Oberw. on Tiliaceae
a. Exobasidianae R. Bauer & Oberw.
5. Entylomatales R. Bauer & Oberw.
  i. Entylomataceae R. Bauer & Oberw.
    Entyloma de Bary on dicots
6. Doassansiales R. Bauer & Oberw.
  i. Melaniellaceae R. Bauer, K. Vánky, Begerow & Oberw.
    Melaniella R. Bauer, K. Vánky, Begerow & Oberw. on Seldginellaceae
  ii. Doassansiaceae (Azb. & Karat.) Moore emend. R. Bauer & Oberw.

Burrillia Setchell on monocots and dicots
Doassansia Cornu on monocots and dicots
Doassinga K. Vánky, R. Bauer & Begerow on Callitrichaceae
Heterodoassansia K. Vánky on monocots and dicots
Nannfeldtiomyces K. Vánky on Sparganiaceae
Narasimhania Thirum. & Pavgi emend K. Vánky on Alismataceae
Pseudodoassania (Setchell) K. Vánky on Alismataceae
Pseudodermatosorus K. Vanky on Alismataceae
Pseudotracya K. Vanky on Hydrocharitaceae
Tracya H. & P. Sydow on Hydrocharitaceae and Lemnaceae
    iii. Rhamphosporaceae R. Bauer & Oberw.
Rhamphospora Cunn. on Nymphaeaceae
7. Exobasidiales P. Henn. emend. R. Bauer & Oberw.
    i. Brachybasidiaceae Gäum.
Brachybasidium Gäumann on Arecaceae
Dicellomyces L.S. Olive on monocots
Exobasidiellum Donk on Poaceae
Kordyana Racib. on Commelinaceae
Proliferobasidium Cunningh. on Heliconiaceae
    ii. Exobasidiaceae P. Henn.
Exobasidium Woronin on dicots
Muribasidiospora Kamat & Rajendren on Anacardiaceae and Ulmaceae
    iii. Cryptobasidiaceae Malencon ex Donk
Botryoconis H. & P. Sydow on Lauraceae
Clinoconidium Pat. on Lauraceae
Coniodictyum Har. & Pat. on Rhamnaceae
Drepanoconis Schröter & P. Henn. on Lauraceae
Laurobasidium Jülich on Lauraceae
"*Sphacelotheca*" *cinnamomi* Hirata on Lauraceae
"*Ustilago*" *onumae* (Shirai) S. Ito on Lauraceae
    iv. Graphiolaceae E. Fischer
Graphiola Poiteau on Arecaceae
Stylina H. Sydow on Arecaceae

2. Taxa Not Ascribed to Any Family
    i. *Ceraceosorus bombacis* (Bakshi) Bakshi on Bombacaceae
    This species is a member of the Exobasidiomycetidae, but of uncertain systematic position within this group.
    ii. *Sympodiomycopsis paphiopedili* Sugiyama, Tokuoka & Komagata
    This anamorphic species is a member of the Microstromatales but of uncertain systematic position within this group.
    iii. *Tilletiopsis albescens* Gokhale
    This anamorphic species is a member of the Exobasidiomycetidae, but of uncertain systematic position within this group (Figs. 33, 34).
    iv. *Tilletiopsis cremea* Tubaki, *Tilletiopsis lilacina* Tubaki, *Tilletiopsis washingtonensis* Nyland
    These three anamorphic species are members of the Entylomatales and appear to be anamorphic representatives of a currently unknown family of the Entylomatales (Figs. 33, 34).
    v. *Tilletiopsis minor* Nyland
    This anamorphic species is a member of the Georgefischeriales, but of uncertain systematic position within this group (Figs. 33, 34).
    vi. *Tilletiopsis pallescens* Gokhale
    This anamorphic species is a member of the Exobasidiomycetidae, but of uncertain systematic position within this group (Figs. 33, 34).

C. Description

Within the Ustilaginomycetes three major groups are evident in the dendrograms resulting from ultrastructural and LSU rDNA sequence analyses (Figs. 32–34; Bauer et al. 1997; Begerow et al. 1997).

1. Entorrhizomycetidae

The Entorrhizomycetidae is the basal group of the Ustilaginomycetes (Figs. 1, 32; Bauer et al. 1997; Begerow et al. 1997). Lack of membrane bands or caps at the pores (Fig. 29) and the presence of local interaction zones without interaction apparatus characterize this group (Bauer et al. 1997). The species of *Entorrhiza*, the single genus currently identified in this group, have dolipores (Fig. 29; Bauer et al. 1997; Deml and Oberwinkler 1981),

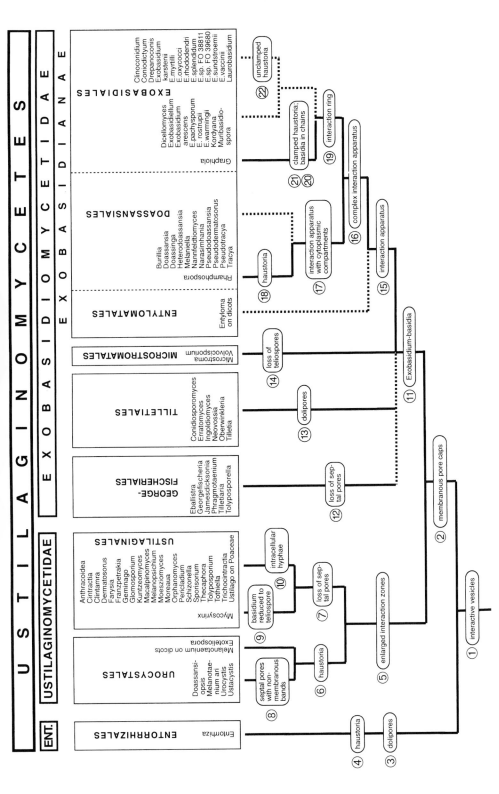

Fig. 32. Phylogenetic tree of the Ustilaginomycetes based predominantly on ultrastructural apomorphies, after Bauer et al. (1997). Apomorphies are illustrated. Their plesiomorphic states occur in the respective sister groups. Clusters on interrupted lines must be regarded as potentially paraphyletic. The apomorphies and their respective plesiomorphic states used in the tree are as follows: *1* interaction with primary interactive vesicles/without primary interactive vesicles; *2* membranous pore caps or bands/without membranous pore caps or bands; *3* dolipores/simple pores; *4* haustoria/intercellular hyphae; *5* enlarged interaction zones/local interaction zones; *6* haustoria/intercellular hyphae; *7* loss of septal pores/simple pores; *8* septal pores with nonmembranous bands/without membranous bands; *9* basidia reduced to teliospores/teliospores only function as probasidia; *10* intracellular hyphae/intercellular hyphae; *11 Exobasidium* basidia/ballistoconidia (repeated loss of ballistosporic basidiospores not labeled); *12* loss of septal pores/simple pores; *13* dolipores/simple pores; *14* loss of teliospores/teliospores; *15* interaction apparatus/primary interactive vesicles; *16* complex interaction apparatus/simple interaction apparatus; *17* interaction apparatus with cytoplasmic compartments/without cytoplasmic compartments; *18* haustoria/intercellular hyphae; *19* interaction ring/without interaction ring; *20* basidia in chains/not in chains; *21* clamped haustoria/intercellular hyphae; *22* unclamped haustoria/intercellular hyphae (this feature is only illustrated to show the taxa of the Exobasidiales having haustoria; it is not considered as apomorphic)

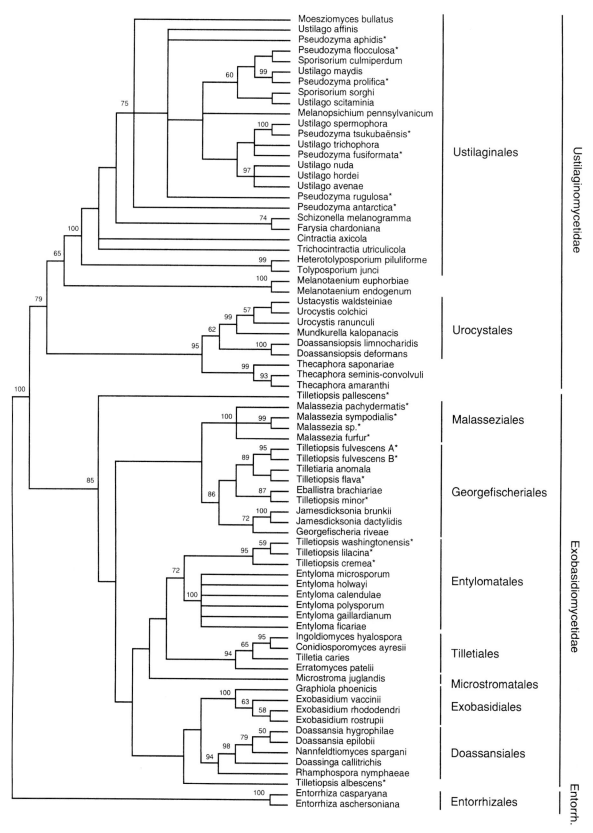

Fig. 33. Strict consensus tree of 71 most parsimonious trees (1516 steps) of 540 bp from the 5′-end of the LSU rDNA sequences of teleomorphic or anamorphic (indicated by asteriks) species of the Ustilaginomycetes, after Begerow et al. (2000). Topology was rooted with *Entorrhiza*. Percentage bootstrap values of 100 replications are given at each furcation. Percentages under 50% are not shown

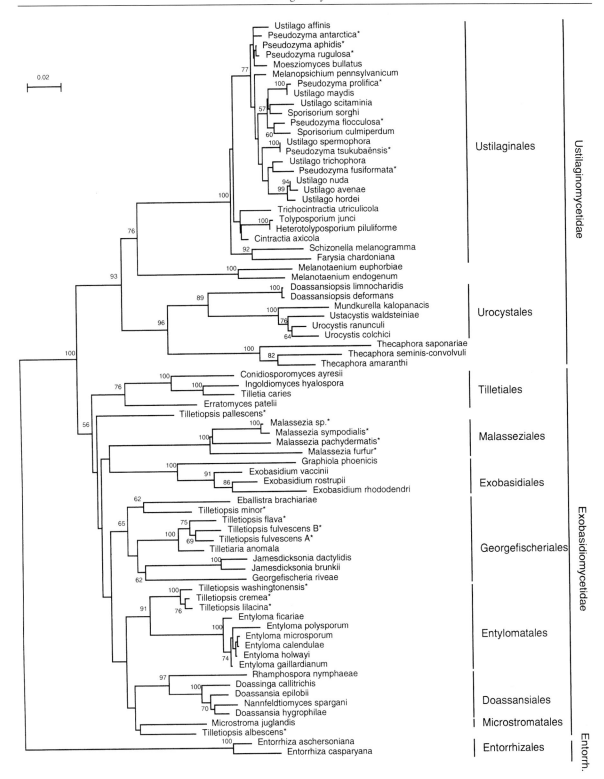

Fig. 34. Topology obtained by neighbor joining analysis of 540 bp from the 5′-end of the LSU rDNA sequences of teleomorphic or anamorphic (indicated by *asterisks*) species of the Ustilaginomycetes, after Begerow et al. (2000). Topology was rooted with *Entorrhiza*. Percentage bootstrap values of 1000 replicates are given at each furcation. Percentages under 50% are not shown

they form teliospores in living host cells (Fineran 1980; Deml and Oberwinkler 1981) in which the exosporium of the teliospores is probably formed by the host (Piepenbring et al. 1998b), and they cause galls on roots of members of the Juncaceae and Cyperaceae. The teliospores germinate internally by becoming four-celled phragmobasidia with cruciform septation (Fig. 8; Weber 1884; Fineran 1982). Although meiosis has not been studied in *Entorrhiza*, the changes in nuclear numbers from binucleate to mononucleate to four-nucleate (Bauer et al. 1997; Fineran 1982) support interpreting the teliospore germlings as basidia. The basidiospores (Fig. 8) resemble conidia of aquatic hyphomycetes (Ingold 1979). Thus, the basidia of *Entorrhiza* show some adaptations to water dispersal in the soil. The haploid phase of *Entorrhiza* is unknown.

2. Ustilaginomycetidae

Presence of enlarged interaction zones (Fig. 26) characterizes this group (Baucr ct al. 1997). The members of the Ustilaginomycetidae are teliosporic, gastroid, and dimorphic. Morphologically and ecologically, they are diverse (Figs. 9–12, 18; see Vánky 1987, 1994), but both ultrastructural and LSU rDNA sequence analyses unite them (Figs. 32–34; Bauer et al. 1997; Begerow et al. 1997). Two groups are recognized.

a) Urocystales

Among the Ustilaginomycetidae, the Urocystales are characterized by the presence of haustoria and of pores in the septa of soral hyphae (Figs. 26, 28; Bauer et al. 1997). Most of the Urocystales sporulate in vegetative parts of the hosts. Three groups are evident by the interpretation of the ultrastructural and LSU rDNA sequence analyses and correlated characters (Figs. 32–34; Bauer et al. 1997; Begerow et al. 1997).

The Melanotaeniaceae represent the basal group of the Urocystales. Lack of the nonmembranous bands in the pores characterizes this group. This family comprises at least *Melanotaenium endogenum* and *Melanotaenium euphorbiae* (Lenz) M.D. Whitehead & Thirum., *Exoteliospora osmundae*, "*Ustilago*" *speculariae* and probably all described *Melanotaenium* species occurring on dicots. *Melanotaenium endogenum* and *M. euphorbiae* form gastroid holobasidia and a *Tilletiopsis*-like pseudohyphal anamorph with ballistoconidia

(e.g., Ingold 1988). Basidia of *Exoteliospora osmundae* have not been observed. Among the Ustilaginomycetidae, formation of ballistoconidia is currently restricted to the Melanotaeniaceae. The ultrastructural as well as the LSU rDNA sequence analyses revealed that smut fungi morphologically similar to *Melanotaenium endogenum*, the type of the genus, evolved in different ustilaginomycetous groups (Figs. 32–34; Bauer et al. 1997; Begerow et al. 1997). Thus, the species of "*Melanotaenium*" occurring on monocots are not closely related with *Melanotaenium endogenum*, belonging to the Urocystaceae (see below) or Georgefischeriales (designated in Figs. 32–34 as members of *Eballistra*, *Jamesdicksonia* or *Phragmotaenium*).

The Doassansiopsaceae represents the sister group of the Urocystaceae. The Doassansiopsaceae share with the Urocystaceae, but not with the Melanotaeniaceae, an essentially identical septal pore apparatus (Bauer et al. 1997). It is composed of a simple pore with two outer tripartite membrane caps and two inner nonmembranous bands (Fig. 28; Bauer et al. 1995a). The species of *Doassansiopsis*, the single genus currently placed in the Doassansiopsaceae, possess complex teliospore balls with a central mass of pseudoparenchymateous cells surrounded by a layer of firmly adhering, lightly colored, teliospores and an external cortex of sterile cells (Fig. 7; Vánky 1987). *Doassansiopsis* species form gastroid holobasidia and a yeast anamorph without ballistoconidia. The position of *Doassansiopsis* in the Urocystales is surprising. Based on teliospore ball morphology and the parasitism of aquatic plants, *Doassansiopsis* was grouped with *Burillia*, *Doassansia*, *Heterodoassansia*, *Nannfeldtiomyces*, *Narasimhania*, *Pseudodoassansia*, and *Tracya* (Vánky 1987, 1994). However, both the ultrastructural as well the LSU rDNA sequence analyses show that *Doassaniopsis* is not closely related to the other complex teliospore ball-forming taxa (Figs. 32–34; Bauer et al. 1997; Begerow et al. 1997).

The Urocystaceae comprises morphologically diverse species with colored teliospores. "*Melanotaenium*" *ari* produces single teliospores, *Mundkurella* is characterized by one- to four-celled teliospores, *Urocystis* (Fig. 18) and *Ustacystis* by teliospores united in balls (Vánky 1987, 1994). In addition, *Mundkurella* and *Ustacystis* (Zundel 1945) are phragmobasidiate, whereas *Urocystis* is holobasidiate (Fig. 18). Basidia of "*Melanotae*-

nium" ari have not been observed. The species in *Urocystis* living on monocots and dicots appear to form a monophyletic group (Figs. 33, 34). The Urocystaceae form a yeastlike anamorph without ballistoconidia.

The evolutionary separation of the Doassansiopsaceae from the Urocystaceae and their respective evolutionary radiations may have an ecological basis. Host species belong to a variety of monocots and dicots in both families, but the hosts of *Doassansiopsis* are exclusively paludal or aquatic plants, whereas those of the Urocystaceae are terrestrial.

b) Ustilaginales

Poreless septa (Fig. 31) characterize the Ustilaginales. Most of the species of this group sporulate in the reproductive parts of their hosts and possess the disarticulating type of teliosporogenesis. A prominent gelatinization of hyphal walls usually precedes teliospore formation (Fig. 3; Luttrell 1987; Snetselaar and Tiffany 1990; Mims and Snetselaar 1991; Mims et al. 1992; Snetselaar and Mims 1994; Piepenbring et al. 1998b). The anamorphs fit the concept of *Pseudozyma* sensu Boekhout (1995). Based on ultrastructural and LSU rDNA sequence analyses the Ustilaginales are divided into three families (Figs. 32–34; Bauer et al. 1997; Begerow et al. 1997; for the discrepancy concerning the systematic position of *Thecaphora* see above).

The Mycosyringaceae may represent the basal group of the Ustilaginales distinguished from the other ustilaginaceous groups by the lack of intracellular hyphae (Fig. 32; Bauer et al. 1997). *Mycosyrinx* is the only genus in the Mycosyringaceae. *Mycosyrinx* species produce teliospores in pairs and their host range is restricted to members of the Vitaceae (Vánky 1996). Basidia, known only from *M. cissi* (DC.) G. Beck, have a unique morphology, being reduced to the teliospores. The basidiospores are sigmoid in shape (Fig. 12; Piepenbring and Bauer 1995), indicating that they may be dispersed by water. The haploid phase is unknown.

The Glomosporiaceae share with the Ustilaginaceae the formation of intracellular hyphae (Bauer et al. 1997). The species of the three genera currently classified in the Glomosporiaceae possess light brown teliospore balls (Vánky 1987) and differ from those of the Ustilaginaceae by the formation of holobasidia. *Glomosporium leptideum* (H. & P. Sydow) Kochmann has true holobasidia (Kochmann 1939), whereas the teliospores of *Tothiella thlaspeos* (Beck) K. Vánky usually germinate with long hyphae in which the cytoplasm is located at the apex (Vanky 1999). Teliospore germination among the species of *Thecaphora* is variable, ranging from true holobasidia to aseptate or septate hyphae that sometimes bear basidiospores (Nagler 1986; Ingold 1987c; Piepenbring and Bauer 1995). We interpret these hyphal germinations as atypical germinations resulting possibly from nonoptimal environmental conditions. For example, for *Thecaphora haumanii* Speg. both germination types have been reported (Piepenbring and Bauer 1995). The hosts of the Glomoporiaceae are dicots.

In contrast with the Glomosporiaceae, the numerous species of the Ustilaginaceae form phragmobasidia (Figs. 9–11) and darkly colored teliospores. Depending upon the species and sometimes also on the environmental conditions, the phragmobasidia vary in morphology (Figs. 9–11; Ingold 1983, 1987a, 1989a, 1989c). The basidia in *Anthracoidea* species are generally two-celled (Fig. 11), whereas those in the other species are usually four-celled (Figs. 9, 10). This family contains many genera (Figs. 32–34; Bauer et al. 1997; Begerow et al. 1997) and comprises, except for 17 species, all phragmobasidiate members of the Ustilaginomycetes. Some of the *Ustilago* species, e.g., *Ustilago hordei*, *U. maydis* or *Ustilago tritici* (Pers.) Rostrup, cause economically important plant diseases on cultivated Poaceae (Thomas 1989; Valverde et al. 1995). The genera are based mainly on characteristics of teliospores and sori, e.g., teliospores free or united in balls, presence or absence of peridia, columellae, sterile cells or sterile hyphae, etc. (see Vánky 1987, 1994). *Anthracoidea* species, for example, form sori with free, black teliospores around ovaries of Cyperaceae. The sori are initially covered by peridia of fungal hyphae. The sori of *Sporisorium* species are also covered by peridia but these can be composed of host tissue as well as of fungal hyphae. The teliospores are free or arranged in balls. Teliospore balls and special soral structures are lacking in *Ustilago* species whose simple teliospores develop by destroying host tissue. We agree with Bandoni (1995) that in many cases (e.g., *Ustilago*) features, upon which ustilaginaceous genera are based, are in need of reevaluation (e.g., see Figs. 33, 34).

Thus, based on morphological and sequence data the *Cintractia* complex was divided in several genera (Piepenbring et al. 1999). With a few exceptions, the Ustilaginaceae occur on monocots, especially on Poaceae (ca. 65%) and Cyperaceae (ca. 30%).

3. Exobasidiomycetidae

The Exobasidiomycetidae represents the sister group of the Ustilaginomycetidae (Figs. 32–34; Bauer et al. 1997; Begerow et al. 1997). The teleomorphic members of the Exobasidiomycetidae share with those of the Ustilaginomycetidae the presence of membrane caps or bands at the septal pores (Figs. 27, 28, 30). However, in both groups, taxa with poreless septa evolved. The Exobasidiomycetidae differ from the Ustilaginomycetidae by forming local interaction zones. Except for the Tilletiariaceae (Fig. 17), the Exobasidiomycetidae are holobasidiate (Figs. 13–16, 18–21). Among the basidiomycetes, formation of ballistosporic holobasidia, in which the hilar appendices of the basidiospores are oriented abaxially (sterigmata turned outwards, basidiospores inwards; Figs. 16, 21), is restricted to the Exobasidiomycetidae. In addition, basidia of this type are common in the Exobasidiales, but occur also in species of the Doassansiales, Georgefischeriales and Tilletiales (Figs. 16, 21; Oberwinkler 1977, 1978, 1982; Ingold 1995; Bauer et al. 1999b; Vánky and Bauer 1996). Therefore, the *Exobasidium* basidium with the specific orientation of the ballistosporic basidiospores may represent an apomorphy for the Exobasidiomycetidae (see Fig. 32).

Teliospores are absent or present within the Exobasidiomycetidae. Formation of teliospore balls occurs only in the Doassansiaceae and in *Tolyposporella*. The smut fungi among the Exobasidiomycetidae show terminal or intercalary teliospore formation (Figs. 4–6; Piepenbring et al. 1998b,d; Roberson and Luttrell 1987; Trione et al. 1989). A gelatinization of hyphal walls preceding teliospore formation is either lacking or not clearly recognizable.

Six groups are recognized on the basis of the ultrastructural characters within the Exobasidiomycetidae (Fig. 32; Bauer et al. 1997). These groups are also evident in the phylogeny resulting from LSU rDNA sequence analyses (Figs. 33, 34; Begerow et al. 1997). In addition, LSU rDNA sequence analyses indicate that the species of the anamorphic genus *Malassezia* represent a group of its own within the Exobasidiomycetidae (Figs. 33, 34; Begerow et al. 2000).

a) Malasseziales

The anamorphic genus *Malassezia* comprises medically important, lipophylic yeasts that constitute part of the fungal microflora of the skin of warm-blooded animals. They have been found associated with a variety of pathological conditions in humans including pityriasis versicolor, seborrheic dermatitis, folliculitis and systemic infections (see Boekhout and Bosboom 1994; Howard and Kwon-Chung 1995; Guého et al. 1998; Begerow et al. 2000; and the references therein). The cell wall of the *Malassezia* yeasts is thick, multilamellate and reveals a unique substructure with an electron-opaque, helicoidal band that corresponds to a helicoidal evagination of the plasma membrane (Takeo and Nakai 1986; Guillot et al. 1995). The sexual phase of *Malassezia* is unknown.

The position of *Malassezia* in the Exobasidiomycetidae is surprising and suggests that either the *Malassezia* species are phytoparasitic in the dikaryophase, or that they originated at least from plant parasites.

b) Georgefischeriales

Among the Exobasidiomycetidae, presence of poreless septa in the soral hyphae characterizes this group. The Georgefischeriales are teliosporic and dimorphic. They interact with their respective hosts by forming local interaction zones without interaction apparatus (Bauer et al. 1997). Haustoria or intracellular hyphae are lacking. The Georgefischeriales sporulate in vegetative parts of the hosts, predominantly in leaves (for *Tilletiaria* see below). Teliospores are yellow to brown in the species of *Georgefischeria* and darkly colored in the other taxa.

Except for *Georgefischeria* with its two species on Convolvulaceae and a few species on Cyperaceae, the Georgefischeriales occur on Poaceae. Because *Tilletiaria anomala* appeared in a plate over which a polypore growing on decaying wood had been suspended (Bandoni and Johri 1972), nothing is known of its life strategy. Like *Tilletiaria anomala*, smut fungi occasionally form teliospores and basidia in culture (Bauer et al. 1997). It is, therefore, conceivable that *Tilletiaria anomala* is a phytoparasite, probably on grasses.

The data resulting from neighbor-joining LSU rDNA analyses (Figs. 33, 34; Begerow et al. 1997) correlate well with the family conception proposed by Bauer et al. (1997). The species of the

Georgefischeriaceae are characterized by holobasidia (Figs. 15, 16) and a *Tilletiopsis*-like pseudohyphal anamorph that produces ballistoconidia. The Tilletiariaceae also form a *Tilletiopsis*-like pseudohyphal anamorph that produces ballistoconidia, but they are phragmobasidiate (Fig. 17; Bandoni and Johri 1972). The members of the Eballistraceae are holobasidiate and characterized by forming a budding yeast phase without ballistoconidia from the basidiospores (Singh and Pavgi 1973). The yeasts produced by the members of the Eballistraceae are spherical to ellipsoidal in form, do not form pseudomycelia, and grow very slowly on agar media. These features are unusual among the Ustilaginomycetes.

c) *Tilletiales*

Presence of dolipores in the septa (Fig. 30) characterizes the Tilletiales among the Exobasidiomycetidae (Fig. 32; Bauer et al. 1997). In contrast with all other groups of the Exobasidiomycetidae, the Tilletiales are not dimorphic. They form local interaction zones without interaction apparatus (Fig. 22) and hyphal anamorphs with ballistoconidia (e.g., Ingold 1987b, 1997). Among all the smut fungi we have studied from cultures, only the Tilletiales presented distinct pores in the septa of the saprobic hyphae. The septa in cultured hyphae of the other taxa were poreless even when distinct pores occurred in septa from sori, e.g., in species of *Entyloma, Rhamphospora, Urocystis,* and *Ustacystis*. This might explain why Moore (1972) and Boekhout et al. (1992) did not find distinct pores in some of these taxa.

The members of the Tilletiales are homogeneous, morphologically and ecologically (Figs. 4, 14). Haustoria and intracellular hyphae are lacking. The teliospores of this group are usually much larger than those of other groups of the Ustilaginomycetes and they are not arranged in balls. Except for *Erratomyces* on Leguminosae, they parasitize grasses and the sori, with the exception of *Erratomyces*, and a few species of *Tilletia* with teliospores in vegetative host organs, appear in ovaries (Piepenbring and Bauer 1997; Vánky 1994; Vánky and Bauer 1992, 1995, 1996). Some tilletian species are economically important. *Tilletia caries* and *T. contraversa*, for example, can cause heavy losses in production of wheat grains (Trione 1982; Mathre 1996). In India and the American tropics the angular black spot disease on leaves of beans is caused by *Erratomyces patelii* (Pavgi & Thirum.) M. Piepenbr. & R. Bauer (Piepenbring and Bauer 1997).

d) *Microstromatales*

Among the Exobasidiomycetidae, the Microstromatales are characterized by the presence of simple pores and local interaction zones without interaction apparatus (Fig. 32; Bauer et al. 1997). Currently, only four teleomorphic species are known in this group: *Microstroma* with its three species on hamamelids and *Volvocisporium triumfetticola* (M.S. Patil) Begerow et al. on Tiliaceae (Hennings 1900; Patil 1977). They are not teliosporic. The young basidia protrude through the stomata and sporulate on the leaf surface (Oberwinkler 1978; Patil 1977). *Volvocisporium* differs mainly from *Microstroma* by the formation of highly septate basidiospores. The Microstromatales form a budding yeast phase without ballistoconidia and pseudomycelia. In contrast with most of the Ustilaginomycetes, the yeast cells are more or less spherical in form.

The identified species of the Microstromatales may represent only the "tip of the iceberg" of this group. Except for *Microstroma juglandis* (Bereng.) Sacc., the known species of the Microstromatales are difficult to detect in nature. In addition, we have isolated some yeasts belonging to this group.

The following orders of the Exobasidiomycetidae are grouped by Bauer et al. (1997) into the superorder Exobasidianae. The members of the Exobasidianae possess local interaction zones with interaction apparatus.

e) *Entylomatales*

The Entylomatales are characterized by the presence of simple interaction apparatus at the interaction sites (Figs. 23, 32; Bauer et al. 1997). Currently, this group comprises only species of *Entyloma* occurring on dicots with the type of *Entyloma, Entyloma microsporum* (Unger) Schröter (Fig. 5). The ultrastructural as well as the LSU rDNA sequence analyses revealed that the genus *Entyloma* is polyphyletic and that the "*Entyloma*" species occurring on monocots belong to the Georgefischeriales (designated in Figs. 32–34 as members of *Eballistra* or *Jamesdicksonia*; Bauer et al. 1997; Begerow et al. 1997). The teliospores of *Entyloma* species assigned to the Entylomatales are lightly colored, whereas those of "*Entyloma*" species belonging to the Georgefischeriales are darkly colored.

The members of the Entylomatales are morphologically very similar (Figs. 5, 13). In fact, the species are not easy to distinguish from each other (Vánky 1994). They form a *Tilletiopsis*-like pseudohyphal anamorph that usually produces ballistoconidia (Boekhout 1991).

f) Doassansiales

A complex interaction apparatus including cytoplasmic compartments (Fig. 24) characterizes this order (Fig. 32; Bauer et al. 1997). The known species of this group have parasitic hyphae with clamps, they are teliosporic and dimorphic and they do not form ballistoconidia in the haploid phase. Although they differ morphologically, they are ecologically uniform. The members of *Burrillia, Doassansia, Heterodoassansia, Nannfeldtiomyces, Narasimhania, Pseudodoassansia, Pseudodermatosorus, Pseudotracya,* and *Tracya* have complex spore balls (Vánky 1987), whereas *Doassinga, Melaniella* and *Rhamphospora* produce single spores (Bauer et al. 1999b; Vánky et al. 1998). In addition, teliospores are darkly colored in *Melaniella*, and lightly colored in *Doassinga, Rhamphospora,* and the genera with complex teliospore balls. The hosts of the Doassansiales are systematically diverse, comprising spikemosses and various angiosperms.

However, the Doassansiales are ecologically well characterized by their occurrence on paludal or aquatic plants. They apparently evolved in the ecological niche of aquatic plants and developed complex spore balls and more or less sigmoid basidiospores in adaptation to water dispersal (Bauer et al. 1997). Interestingly, the species of *Doassansiopsis* in the Urocystales likewise parasitize aquatic plants and possess similar complex spore balls (see above). Thus, *Doassansiopsis* and the Doassansiales are an excellent example for the independent evolution of similar structures under the same environmental pressure.

The morphological and the LSU rDNA sequence analyses reveal a basal dichotomy between *Melaniella* and the other taxa of the Doassansiales (Bauer et al. 1999b). The next dichotomy is between *Rhamphospora* and the Doassansiaceae (Figs. 32–34; Bauer et al. 1977, 1999b; Begerow et al. 1997). In contrast with the members of the Doassansiaceae, *Rhamphospora nymphaeae* D. Cunn., the single species currently placed in the Rhamphosporaceae, forms highly branched haustoria (Bauer et al. 1997).

g) Exobasidiales

The Exobasidiales are characterized by the presence of interaction rings produced by complex interaction apparatus (Figs. 25, 32; Bauer et al. 1997). The union between the species of this group is well supported by LSU rDNA sequences (Figs. 33, 34; Begerow et al. 1997). The members of the Exobasidiales are holobasidiate and dimorphic. They do not form teliospores in the parasitic phase and ballistoconidia in the saprobic phase. In most of the species, the basidiospores become septate during germination. Hosts are monocots and dicots. The sori predominantly appear on leaves. We currently recognize four families in this order.

The Brachybasidiaceae sporulate on the surface of the host organs. The basidia protrude through stomata or emerge from the disintegrated epidermis. The basidia are elongate, ballistosporic, and two-sterigmate. The basidiospores are thin-walled. Available data indicate that the hilar appendices of the basidiospores are oriented adaxially at the apex of the basidia (see Figs. 2 and 6 in Cunningham et al. 1976; Figs. 1.10–2, 1.10–3 in Oberwinkler 1982; Fig. 4 in Oberwinkler 1993, Fig. 1-G in Ingold 1985). *Brachybasidium pinangae* Gäumann, *Dicellomyces gloeosporus* Olive, and *Proliferobasidium heliconiae* Cunningham form persistent probasidia which are arranged in delimited fructifications. In contrast with the other taxa, the basidiospores of the *Kordyana* species usually remain aseptate during germination. The species of the Brachybasidiaceae live on monocots (Cunningham et al. 1976; Gäumann 1922; Oberwinkler 1978, 1982, 1993; Olive 1945).

The Exobasidiaceae are morphologically similar to the Brachybasidiaceae. Like the Brachybasidiaceae, the Exobasidiaceae sporulate through stomata (Fig. 21) or from the disintegrated epidermis, the basidia are elongate and ballistosporic, and the basidiospores are thin-walled. In contrast with the Brachybasidiaceae, however, the hilar appendices of the basidiospores are oriented abaxially at the apex of the basidia (Fig. 21; Oberwinkler 1977, 1978, 1982). In most Exobasidiaceae species, the number of sterigmata per basidium is not fixed, varying from two to eight, with four as the most frequent number. Only a few species form generally two-sterigmate basidia. This family comprises *Exobasidium* and *Muribasidiospora*. However, our analysis revealed that the genus *Muribasidiospora* is polyphyletic (see above). The Exobasidiaceae occur on dicots, predominantly on Ericaceae (Rajendren 1968; Hennings 1900; Mims et al. 1987; Nannfeldt 1981; Oberwinkler 1977, 1978, 1982, 1993).

In contrast with the Brachybasidiaceae and Exobasidiaceae, except for *Laurobasidium* the Cryptobasidiaceae sporulate internally by producing holobasidia in peripheral lacunae of the host galls. During maturation, the galls rupture and liberate the basidiospore mass. The basidia are gastroid and lack sterigmata. The basidiospores are usually thick-walled, resembling the uredospores of rust fungi or the teliospores of smut fungi. In addition, old fructifications often resemble smut sori. These characters may explain why some members of this group were described as smut fungi (see above), while others were originally described as rusts (e.g., *Clinoconidium farinosum* (P. Henn.) Pat. as *Uredo farinosa* P. Henn.).

In contrast to the other members of the Cryptobasidiaceae, however, *Laurobasidium lauri* (Geyler) Jülich sporulates on the surface of the host organ. In addition, the basidia of this species resemble those of *Exobasidium*, but they are gastroid as in the other members of the Cryptobasidiaceae. Thus, *Laurobasidium*, may occupy a systematic position at the base of the Cryptobasidiaceae and intermediate between the Cryptobasidiaceae and the other Exobasidiales.

Except for *Coniodictyum*, the host range of the Cryptobasidiaceae is restricted to laurels. Cryptobasidiaceae species are mainly known from Japan, Africa, South and Middle America (Donk 1956; Lendner 1920; Malençon 1953; Maublanc 1914; Oberwinkler 1977, 1978, 1982, 1993; Sydow 1926).

The Graphiolaceae are parasites of palms. Fructification of the Graphiolaceae starts between the chlorenchyma and hypodermal tissue (Cole 1983). During differentiation of the cylindrical basidiocarp, the epidermis ruptures and globose basidia are produced in chains by disarticulation of sporogenous hyphae within the basidiocarps. The passively released basidiospores arise laterally on the basidia. (Fig. 20; Fischer 1921, 1922; Oberwinkler et al. 1982). The haustoria are constricted at the point of penetration and consist of a clamped basal body (Bauer et al. 1997).

VII. Conclusions

A. Evolution of the Basidium

The sequence of events occurring in the evolution of the basidium of the Ustilaginomycetes is unknown. Nevertheless, a tentative sequence can be outlined from the distribution of the basidial types among the different groups. The presence of simple pores in many members of the Ustilaginomycetes suggests that the first members of this line diverged from simple-septate ancestors. Among the simple-septate basidiomycetes, phragmobasidia are the rule and holobasidia are known only from *Pachnocybe* Berk. and *Chionosphaera* Cox (Oberwinkler and Bauer 1989), two taxa that are phylogenetically distant from the Ustilaginomycetes. *Entorrhiza*, the basal taxon of the Ustilaginomycetes, is phragmobasidiate. From this situation it can be postulated that the common ancestor of the Ustilaginomycetes probably was phragmobasidiate. In the Ustilaginomycetidae and Exobasidiomycetidae, however, phragmobasidia occur only in the Ustilaginaceae, Tilletiariaceae, *Mundkurella*, and *Ustacystis waldsteiniae* (C.H. Beck) Zundel. In other words, except for eight species, the phragmobasidial taxa of the Ustilaginomycetidae and Exobasidiomycetidae are concentrated in a single family, whereas the holobasidial taxa are distributed throughout all orders of the Ustilaginomycetidae and Exobasidiomycetidae. In addition, the basal groups of the Ustilaginomycetidae and Exobasidiomycetidae are holobasidiate. This distribution of basidial types suggests that the common ancestor of the Ustilaginomycetidae and Exobasidiomycetidae was holobasidiate. Consequently, the septation of the basidia of the Ustilaginaceae, Tilletiariaceae *Mundkurella*, and *Ustacystis* must be interpreted as the result of convergent evolution.

B. Coevolution

Except for five species, the host range of the Ustilaginomycetes is restricted to angiosperms. As discussed by Bauer et al. (1997, 1999b), the occurrence of these five species on nonangiosperms may be the result of jumps. Most of the Ustilaginomycetes are parasites of monocots, especially the members of the Poaceae and Cyperaceae. This host distribution suggests that the Ustilaginomycetes may have evolved as pathogens on either early angiosperms or on early monocots with subsequent jumps to the dicots. Thus, the Ustilaginomycetes may have appeared later in the phylogeny of the basidiomycetes than the rusts. Rusts apparently arose as parasites of early vascular plants (Savile 1955). As partly discussed above, within the Ustilaginomycetes there are evident examples of evolution with angiosperm relationships (e.g., the Tilletiales, Georgefischeriales and the *Ustilago-Sporisorium* complex with Poaceae, *Entorrhiza* with Juncaceae and Cyperaceae, *Graphiola* with palms, *Anthracoidea* with Cyperaceae, *Mycosyrinx* with Vitaceae, *Exobasidium* with Ericaceae, etc.). On the other hand, the Doassansiales and *Doassansiopsis* are two excellent examples for evolution with ecosystems. In general, the host ranges of the different groups (e.g., the Georgefischeriales on Poaceae with a few species on Convolvulaceae and Cyperaceae, the Tilletiales on Poaceae with five species on Leguminosae, the Ustilaginaceae on monocots with a few genera on dicots, etc.) reveal that the Ustilaginomycetes ancestors have not only undergone periods of parallel evolution with their hosts; jumps to new hosts may have stimulated the evolution of a large number of taxa.

C. Evolutionary Trends

Our analyses suggest the following evolutionary trends within the Ustilaginomycetes:

- cellular interactions from simple to complex forms,
- multiple evolution of intracellular fungal elements,
- multiple evolution of spore balls,
- repeated loss of septal pores,
- repeated change from simple pores to dolipores,
- primary change from septate to aseptate basidia,
- repeated secondary change from aseptate to septate basidia,
- repeated loss of teliospores as propagules,
- multiple evolution of gastroid taxa,
- repeated loss of the ballistosporic mechanism,
- evolution with host groups, but also with ecosystems,
- repeated jumps to unrelated hosts.

Acknowledgments. We thank Dr. José P. Sampaio, Dr. Micheal Weiß and Dr. Kálmán Vánky for critically reading the manuscript, Dr. Jack W. Fell for the sequences of *Malassezia*, Dr. Teun Boekhout for the sequences of *Tilletiopsis* and *Pseudozyma*, Dr. Kálmán Vánky for many specimens and the Deutsche Forschungsgemeinschaft for financial support.

References

Bandoni RJ (1984) The Tremellales and Auriculariales: an alternative classification. Trans Mycol Soc Jpn 25:489–530

Bandoni RJ (1995) Dimorphic heterobasidiomycetes: taxonomy and parasitism. Stud Mycol 38:13–27

Bandoni RJ, Johri BN (1972) *Tilletiaria*: a new genus in the Ustilaginales. Can J Bot 50:39–43

Bauer R, Mendgen K, Oberwinkler F (1995a) Septal pore apparatus of the smut *Ustacystis waldsteiniae*. Mycologia 87:18–24

Bauer R, Mendgen K, Oberwinkler F (1995b) Cellular interaction of the smut fungus *Ustacystis waldsteiniae*. Can J Bot 73:867–883

Bauer R, Oberwinkler F, Vánky K (1997) Ultrastructural markers and systematics in smut fungi and allied taxa. Can J Bot 75:1273–1314

Bauer R, Oberwinkler F, Vánky K (1999a) Ustilaginomycetes on *Osmunda*. Mycologia 91:699–675

Bauer R, Vánky K, Begerow D, Oberwinkler F (1999b) Ustilaginomycetes on *Selaginella*. Mycologia 91:475–484

Begerow D, Bauer R, Oberwinkler F (1997) Phylogenetic studies on nuclear large subunit ribosomal DNA sequences of smut fungi and related taxa. Can J Bot 75:2045–2056

Begerow D, Bauer R, Boekhout T (2000) Phylogenetic placements of ustilaginomycetous anamorphs as deduced from nuclear LSU rDNA sequences. Mycol Res 104:53–60

Boekhout T (1987) Systematics of anamorphs of Ustilaginales (smut fungi) – a preliminary survey. Stud Mycol 30:137–149

Boekhout T (1991) A revision of ballistoconidia-forming yeasts and fungi. Stud Mycol 33:1–194

Boekhout T (1995) *Pseudozyma* Bandoni emend. Boekhout, a genus for yeast-like anamorphs of Ustilaginales. J Gen Appl Microbiol 41:359–366

Boekhout T, Bosboom R (1994) Karyotyping of *Malassezia* yeasts: taxonomic and epidemiological implications. Syst Appl Microbiol 17:146–153

Boekhout T, Yamada Y, Weijman ACM, Roeijmans HJ, Batenburg-van der Vegte WH (1992) The significance of coenzyme Q, carbohydrate composition and septal ultrastructure for the taxonomy of ballistoconidia-forming yeasts and fungi. Syst Appl Microbiol 15:1–10

Bucheli E, Leuchtmann A (1996) Evidence for genetic differentiation between choke-inducing and asymptomatic strains of the *Epichloe* grass endophyte from *Brachypodium sylvaticum*. Evolution 50:1879–1887

Carris LM, Gray PM (1994) The ability of *Tilletia fusca* to hybridize with the wheat bunt species under axenic conditions. Mycologia 86:157–163

Cole GT (1983) *Graphiola phoenicis*: a taxonomic enigma. Mycologia 75:93–116

Cunningham JL, Bakshi BK, Lentz PL (1976) Two new genera of leaf-parasitic fungi (Basidiomycetidae: Brachybasidiaceae). Mycologia 68:640–654

Deml G, Oberwinkler F (1981) Investigations on *Entorrhiza casparyana* by light and electron microscopy. Mycologia 73:392–398

Deml G, Oberwinkler F (1982) Studies in heterobasidiomycetes, part 24. On *Ustilago violacea* (Pers.,) Rouss, from *Saponaria officinalis*. Phytopathol Z 104:345–356

Donk MA (1956) The generic names proposed for hymenomycetes – VI. Brachybasidiaceae, Cryptobasidiaceae, Exobasidiaceae. Reinwardtia 4:113–118

Durán R (1973) Ustilaginales. In: Ainsworth GC, Sparrow FK, Sussman AS (eds) The Fungi, vol 4B. Academic Press, New York, pp 281–300

Fineran JM (1980) The structure of galls induced by *Entorrhiza* C. Weber (Ustilaginales) on roots of the Cyperaceae and Juncaceae. Nova Hedwigia 32:265–284

Fineran JM (1982) Teliospore germination in *Entorrhiza casparyana* (Ustilaginales). Can J Bot 60:2903–2913

Fischer E (1921) Zur Kenntnis von *Graphiola* and *Farysia*. Ann Mycol 18:118–197

Fischer E (1922) Weitere Beiträge zur Kenntnis der Gattung *Graphiola*. Ann Mycol 20:228–237

Fischer GW, Holton CS (1957) Biology and control of the smut fungi. Ronald Press, New York

Fischer GW, Shaw CG (1953) A proposed species concept in the smut fungi with application to North American species. Phytopathology 43:181–188

Gäumann E (1922) Über die Gattung *Kordyana* Rac. Ann Mycol 20:257–271

Goates BJ, Hoffmann JA (1986) Formation and discharge of secondary sporidia of the bunt fungus, *Tilletia foetida*. Mycologia 78:371–379

Gottschalk M, Blanz PA (1985) Untersuchungen an 5S ribosomalen Ribonucleinsäuren als Beitrag zur Klärung von Systematik und Phylogenie der Basidiomyceten. Z Mycol 51:205–243

Guého E, Boekhout T, Ashbee HR, Guillot J, van Belkhum A, Faergemann J (1998) The role of *Malassezia* species in the ecology of human skin and as pathogens. Med Mycol 36 (Suppl I):220–229

Guillot J, Guého E, Prévost MC (1995) Ultrastructural features of the dimorphic yeast *Malassezia furtur*. J Méd 5:86–91

Hennings P (1900) Exobasidiineae. In: Engler A, Prantl K (eds) Die natürlichen Pflanzenfamilien 1,1. Verlag von Wilhelm Engelmann, Leipzig, pp 103–105

Howard DH, Kwon-Chung KJ (1995) Zoopathogenic heterobasidiomycetous yeasts. Stud Mycol 38:59–64

Huang H-Q, Nielsen J (1984) Hybridization of the seedling-infecting *Ustilago* spp. pathogenic on barley and oats, and a study of the genotypes conditioning the morphology of their spore walls. Can J Bot 62:603–608

Ingold CT (1979) Advances in the study of so-called aquatic hyphomycetes. Am J Bot 66:218–226

Ingold CT (1983) The basidium in *Ustilago*. Trans Br Mycol Soc 81:573–584

Ingold CT (1985) *Dicellomyces scirpi*: its conidial stage and taxonomic position. Trans Br Mycol Soc 84:542–545

Ingold CT (1987a) Germination of teliospores in certain smuts. Trans Br Mycol Soc 88:355–363

Ingold CT (1987b) Ballistospores and blastic conidia of *Tilletia ayresii*, and comparison with those of *T. tritici* and *Entyloma ficariae*. Trans Br Mycol Soc 88:75–82

Ingold CT (1987c) Aerial sporidia of *Ustilago hypodytes* and *Sorosporium saponariae*. Trans Br Mycol Soc 89:471–475

Ingold CT (1988) Ballistospores in *Melanotaenium endogenum*. Trans Br Mycol Soc 91:712–714

Ingold CT (1989a) Basidium development in some species of *Ustilago*. Mycol Res 93:405–412

Ingold CT (1989b) Note on the basidium in the Tilletiaceae. Mycol Res 93:387–389

Ingold CT (1989c) The basidium of *Anthracoidea inclusa* in relation to smut taxonomy. Mycol Res 92:245–246

Ingold CT (1995) Products of teliospore germination in *Tilletia hyalospora*. Mycol Res 99:1247–1248

Ingold CT (1997) Teliospore germination in *Tilletia opaca* and *T. sumatii* and the nature of the tilletiaceous basidium. Mycol Res 101:281–284

Kochman J (1939) Beitrag zur Kenntnis der Brandpilzflora Polens II. Acta Soc Bot Pol 16:53–67

Kollmorgen JF, Owczarzak A, Trione EJ (1980) Morphology and timing of secondary sporidial mating in a wheat-bunt fungus *Tilletia caries*. Trans Br Mycol Soc 75:461–471

Kreisel H (1969) Grundzüge eines natürlichen Systems der Pilze. Cramer, Lehre

Kukkonen I, Raudaskoski M (1964) Studies on the probable homothallism and pseudohomothallism in the genus *Anthracoidea*. Ann Bot Fenn 1:257–271

Langdon RFN, Fullerton RA (1975) Sorus ontogeny and sporogenesis in some smut fungi. Aust J Bot 23:915–930

Lendner A (1920) Un champignon parasite sur une lauracée du genre *Ocotea*. Bull Soc Bot Genève II 12:122–128

Luttrell ES (1987) Relations of hyphae to host cells in smut galls caused by species of *Tilletia*, *Tolyposporium*, and *Ustilago*. Can J Bot 65:2581–2591

Malençon G (1953) Le *Coniodyctium chevalieri* Har. et Pat., sa nature et ses affinités. Bull Soc Mycol Fr 69:77–100

Mathre DE (1996) Dwarf bunt: politics, identification, and biology. Annu Rev Phytopathol 34:67–85

Maublanc MA (1914) Les genres *Drepanoconis* Schr. et Henn. et *Clinoconidium* Pat.: leur structure et leur place dans la classification. Bull Soc Mycol Fr 30: 444–446

McDermott JM, McDonald BA (1993) Gene flow in plant pathosystems. Annu Rev Phytopathol 31:353–373

Mims CW, Snetselaar KM (1991) Teliospore maturation in the smut fungus *Sporisorium sorghi*: an ultrastructural study using freeze substitution fixation. Bot Gaz 152:1–7

Mims CW, Richardson EA, Roberson RW (1987) Ultrastructure of basidium and basidiospore development in three species of the fungus *Exobasidium*. Can J Bot 65:1236–1244

Mims CW, Snetselaar KM, Richardson EA (1992) Ultrastructure of the leaf stripe smut fungus *Ustilago striiformis*: host-pathogen relationship and teliospore development. Int J Plant Sci 153:289–290

Moore RT (1972) Ustomycota, a new division of higher fungi. Antonie van Leeuwenhoek 38:567–584

Müller B (1989) Chemotaxonomische Untersuchungen an Basidiomycetenhefen. Thesis, Universität Tübingen, Tübingen, Germany

Nagler A (1986) Untersuchungen zur Gattungsabgrenzung von *Ginanniella* Ciferri und *Urocystis* Rabenhorst sowie zur Ontogenie von *Thecaphora seminis-convolvuli* (Desm.) Ito. Thesis, Universität Tübingen, Tübingen, Germany

Nagler A, Oberwinkler F (1989) Haustoria in *Urocystis* (Tilletiales). Plant Syst Evol 165:17–28

Nagler A, Bauer R, Berbee M, Vánky K, Oberwinkler F (1989) Light and electron microscopic studies of *Schroeteria delastrina* and *S. poeltii*. Mycologia 81:884–895

Nagler A, Bauer R, Oberwinkler F, Tschen J (1990) Basidial development, spindle pole body, septal pore, and host-parasite interaction in *Ustilago esculenta*. Nord J Bot 10:457–464

Nannfeldt JA (1981) *Exobasidium*, a taxonomic reassessment applied to the European species. Symb Bot Ups 23/2:1–71

Oberwinkler F (1977) Das neue System der Basidiomyceten. In: Frey W, Hurka H, Oberwinkler F (eds) Beiträge zur Biologie der niederen Pflanzen. Gustav Fischer, Stuttgart, pp 59–105

Oberwinkler F (1978) Was ist ein Basidiomycet? Z Mykol 44:13–29

Oberwinkler F (1982) The significance of the morphology of the basidium in the phylogeny of basidiomycetes. In: Wells K, Wells EK (eds) Basidium and basidiocarp. Springer Verlag, Berlin Heidelberg New York pp 9–35

Oberwinkler F (1985) Zur Evolution und Systematik der Basidiomyceten. Bot Jahrb Syst 107:541–580

Oberwinkler F (1987) Heterobasidiomycetes with ontogenetic yeast stages – systematic and phylogenetic aspects. Stud Mycol 30:61–74

Oberwinkler F (1993) Diversity and phylogenetic importance of tropical heterobasidiomycetes. In: Isaac S, Frankland JC, Watling R, Whalley AJS (eds) Aspects of tropical mycology. Cambridge University Press, Cambridge pp 121–147

Oberwinkler F, Bauer R (1989) The systematics of gastroid, auricularioid Heterobasidiomycetes. Sydowia 41:224–256

Oberwinkler F, Bandoni RJ, Blanz P, Deml G, Kisimova-Horovitz L (1982) Graphiolales: basidiomycetes parasitic on palms. Plant Syst Evol 140:251–277

Olive LS (1945) A new *Dacrymyces*-like parasite of *Arundinaria*. Mycologia 37:543–552

Patil MS (1977) A new species of *Muribasidiospora* from Kolhapur. Kavaka 5:31–33

Piepenbring M (1996) Smut fungi (Ustilaginales and Tilletiales) in Costa Rica. Nova Hedwigia Beih 113:1–155

Piepenbring M, Bauer R (1995) Noteworthy germinations of some Costa Rican Ustilaginales. Mycol Res 99:853–858

Piepenbring M, Bauer R (1997) *Erratomyces*, a new genus of Tilletiales with species on Leguminosae. Mycologia 89:924–936

Piepenbring M, Hagedorn G, Oberwinkler F (1998a) Spore liberation and dispersal in smut fungi. Bot Acta 111:444–460

Piepenbring M, Bauer R, Oberwinkler F (1998b) Teliospores of smut fungi. General aspects of teliospore walls and sporogenesis. Protoplasma 204:155–169

Piepenbring M, Bauer R, Oberwinkler F (1998c) Teliospores of smut fungi. Teliospore walls and the development of ornamentation studied by electron microscopy. Protoplasma 204:170–201

Piepenbring M, Bauer R, Oberwinkler F (1998d) Teliospores of smut fungi. Teliospore connections, appendages, and germ pores studied by electron microscopy; phylogenetic discussion of characteristics of teliospores. Protoplasma 204:202–218

Piepenbring M, Begerow D, Oberwinkler F (1999) Molecular sequence data assess the value of morphological characteristics for a phylogenetic classification of species of *Cintractia*. Mycologia 91:485–498

Prillinger H, Dörfler C, Laaser G, Hauska G (1990) Ein Beitrag zur Systematik und Entwicklungsbiologie höherer Pilze. Hefe-Typen der Basidiomyceten. Teil III. *Ustilago*-Typ. Z Mykol 56:251–278

Prillinger H, Deml G, Dörfler C, Laaser G, Lockau W (1991) Ein Beitrag zur Systematik und Entwicklungsbiologie höherer Pilze: Hefe-Typen der Basidiomyceten. Teil II. *Microbotryum*-Typ. Bot Acta 104:5–17

Prillinger H, Oberwinkler F, Umile C, Tlachac K, Bauer R, Dörfler C, Taufratzhofer E (1993) Analysis of cell wall carbohydrates (neutral sugars) from ascomycetous and basidiomycetous yeasts with and without derivatization. J Gen Appl Microbiol 39:1–34

Rajendren RB (1968) *Muribasidiospora* – a new genus of the Exobasidiaceae. Mycopathologia 36:218–222

Reddy MS, Kramer CL (1975) A taxonomic revision of the Protomycetales. Mycotaxon 3:1–50

Roberson RW, Luttrell ES (1987) Ultrastructure of teliospore ontogeny in *Tilletia indica*. Mycologia 79:753–763

Roberson RW, Luttrell ES (1989) Dolipore septa in *Tilletia*. Mycologia 81:650–652

Russell BW, Mills D (1993) Electrophoretic karyotypes of *Tilletia caries*, *T. controversa*, and their F_1 progeny: further evidence for conspecific status. Mol Plant-Microbe Interact 6:66–74

Russell BW, Mills D (1994) Morphological, physiological and genetic evidence in support of a conspecific status for *Tilletia caries*, *T. controversa* and *T. foetida*. Phytopathology 84:576–582

Sampson K (1939) Life cycles of smut fungi. Trans Br Mycol Soc 23:1–23

Savile DBO (1947) A study of the species of *Entyloma* on North American composites. Can J Res 25:105–120

Savile DBO (1955) A phylogeny of the basidiomycetes. Can J Bot 33:60–104

Singh RA, Pavgi MS (1973) Morphology, cytology and development of *Melanotaenium brachiariae*. Cytologia 38:455–466

Snetselaar KM, Mims CW (1992) Sporidial fusion and infection of maize seedlings by the smut fungus *Ustilago maydis*. Mycologia 84:193–203

Snetselaar KM, Mims CW (1993) Infection of maize stigmas by *Ustilago maydis*: light and electron microscopy. Phytopathology 83:843–850

Snetselaar KM, Mims CW (1994) Light and electron microscopy of *Ustilago maydis* hyphae in maize. Mycol Res 98:347–355

Snetselaar KM, Tiffany LH (1990) Light and electron microscopy of sorus development in *Sorosporium provinciale*, a smut of big bluestem. Mycologia 82:480–492

Sugiyama J, Tokuoka K, Suh SO, Hirata A, Komagata K (1991) *Sympodiomyiopsis*: a new yeast-like anamorph genus with basidiomycetous nature from orchid nectar. Antonie van Leuwenhoek 59:95–108

Sundström KR (1964) Studies of the physiology, morphology and serology of *Exobasidium*. Symb Bot Ups 18:3

Swann EC, Taylor JW (1993) Higher taxa of basidiomycetes: an 18S rRNA gene perspective. Mycologia 85:923–936

Swann EC, Taylor JW (1995) Phylogenetic diversity of yeast-producing basidiomycetes. Mycol Res 99:1205–1210

Sydow H (1926) Fungi in itinere costaricensi collecti. Pars secunda. Ann Mycol 24:283–288

Takeo K, Nakai E (1986) Mode of cell growth of *Malassezia* (*Pityrosporum*) as revealed by using plasma membrane configurations as natural markers. Can J Microbiol 32:389–394

Thomas PL (1989) Barley smuts in the prairie provinces of Canada, 1983–1988. Can J Phytopathol 11:133–136

Trail F, Mills D (1990) Growth of haploid *Tilletia* strains in planta and genetic analysis of a cross of *Tilletia caries* x *Tilletia controversa*. Phytopathology 80:367–370

Trione EJ (1982) Dwarf bunt of wheat and its importance in international wheat trade. Plant Dis 66:1083–1088

Trione EJ, Hess WM, Stockwell VO (1989) Growth and sporulation of the dikaryons of the dwarf bunt fungus in wheat plants and in culture. Can J Bot 67:1671–1680

Tulasne L, Tulasne C (1847) Mémoire sur les ustilaginées comparées uredinées. Ann Sci Nat Bot 3:12–127

Valverde ME, Paredes-Lópes O, Pataky JK, Guevara-Lara F (1995) Huitlacoche (*Ustilago maydis*) as a food source – biology, composition, and production. CRC Crit Rev Food Sci Nutr 35:191–229

Vánky K (1981) The genus *Schroeteria* Winter (Ustilaginales). Sydowia 34:157–166

Vánky K (1987) Illustrated genera of smut fungi. Cryptog Stud 1:1–159

Vánky K (1994) European smut fungi. Gustav Fischer, Stuttgart

Vánky K (1996) *Mycosyrinx* and other pair-spored Ustilaginales. Mycoscience 37:173–185

Vánky K (1998) The genus *Microbotryum* (smut Sungi). Mycotaxon 67:33–60

Vánky K (1999) The new classificatory system for smut fungi, and two new genera. Mycotaxon 70:35–49

Vánky K, Bauer R (1992) *Conidiosporomyces*, a new genus of Ustilaginales. Mycotaxon 43:426–435

Vánky K, Bauer R (1995) *Oberwinkleria*, a new genus of Ustilaginales. Mycotaxon 53:361–368

Vánky K, Bauer R (1996) *Ingoldiomyces*, a new genus of Ustilaginales. Mycotaxon 59:277–287

Vánky K, Bauer R, Begerow D (1998) *Doassinga*, a new genus of Doassansiales. Mycologia 90:964–970

Weber C (1884) Über den Pilz der Wurzelanschwellungen von *Juncus bufonius*. Bot Z (Berlin) 42:369–379

Wells K (1994) Jelly fungi, then and now! Mycologia 86:18–48

Zundel GL (1945) A change in generic name. Mycologia 37:795–796

4 Heterobasidiomycetes

K. Wells[1] and R.J. Bandoni[2]

CONTENTS

I. Introduction	85
A. Historical Concepts	86
B. Taxonomic Synopsis	87
II. Heterobasidiomycetidae	87
A. Characteristics and Major Taxa	87
B. Possible Intra- and Interrelationships	89
C. Ceratobasidiales	89
1. Taxonomic Features	89
a) Basidial Ontogeny and Ultrastructural Features	89
b) Basidiocarp Structure	92
2. Habitat and Life Cycle	92
3. Taxonomic Synopsis	93
D. Tulasnellales	93
1. Taxonomic Features	93
a) Basidial Ontogeny and Ultrastructural Features	93
b) Basidiocarp Structure	95
2. Habitat and Life Cycle	95
3. Taxonomic Synopsis	96
E. Dacrymycetales	96
1. Taxonomic Features	96
a) Basidial Ontogeny and Ultrastructural Features	96
b) Basidiocarp Structure	98
2. Habitat and Life Cycle	98
3. Taxonomic Synopsis	98
F. Auriculariales	99
1. Taxonomic Features	99
a) Ultrastructural Features	99
b) Basidial Ontogeny and Basidiocarp Structure	99
2. Habitat and Life Cycle	104
3. Taxonomic Synopsis	105
III. Tremellomycetidae	106
A. Characteristics and Major Taxa	106
B. Possible Intra- and Interrelationships	106
C. Tremellales	108
1. Taxonomic Features	108
a) Ultrastructural Features	108
b) Basidial Ontogeny and Basidiocarp Structure	109
2. Habitat and Life Cycle	110
D. Christianseniales	111
E. Filobasidiales	112
F. Taxonomic Synopsis	113
IV. Culture Techniques	113
V. Conclusions	115
References	116

[1] 601 Indian Camp Creek Road, Hot Springs, North Carolina 28743, USA
[2] Department of Botany, University of British Columbia, Vancouver, British Columbia V6T 2B1, Canada

I. Introduction

The concept of the Tremellales, or jelly fungi, as defined by Martin (1952) has been radically revised during the past several decades. Some of the families included by Martin in the Tremellales are now treated as separate orders, e.g., Tulasnellales, Dacrymycetales, and Ceratobasidiales. Many of the taxa included by Martin in the Auriculariaceae are placed in a separate subclass and are almost certainly more closely related to the rusts than to the taxa treated here. Electron microscopy has shown that there are fundamental differences in the spindle apparatus and in the structure of the septal pore resulting in the realization that there are at least two lines of phylogeny in the basidiomycetes, i.e., those with a complex septal pores and those with simple pores. Those taxa with simple pores generally possess disc-like spindle pole bodies; whereas those with complex pores generally possess biglobular spindle pole bodies. This is not to imply, however, that all basidiomycetes fall with ease into one of these categories, as some clearly do not. Thus, there may be a third, or even more, major lines of evolution within the basidiomycetes.

Data on the structure of the septal pore apparatus and the spindle pole body have certainly improved our understanding of the basidiomycetes; however, other types of data have also suggested revisions in the taxonomy of this group. Culture studies of species of *Tremella* and related genera have demonstrated that these fungi have a modified bifactorial compatibility system, i.e., one controlled by a single pair of A alleles and multiple B alleles as in some smut species[3]. Further, many of the Tremellales and related orders are dimorphic and most, if not all, seem to be parasitic on other fungi in their natural habitat. For these reasons, and others, the Heterobasidiomycetes are divided into two subclasses, i.e., the Tremellomycetidae and the Heterobasidiomycetidae. Although a number of taxa of the Heterobasidiomycetidae are parasitic on a variety of hosts, many are saprobic and some, possibly, mycorrhizal. Data on the mode of nutrition within the Heterobasidiomycetidae is quite fragmentary, although most are known from collections on decaying wood and other plant parts. The species studied have hyphal haploid and dikaryotic phases. The known heterothallic species are uni- or bifactorial with multiple alleles at the mating loci (Wells 1994).

Within the Heterobasidiomycetidae the Tulasnellales and Dacrymycetales seem to be related and each is well defined; however, the Ceratobasidiales are difficult to separate from some species of the Homobasidiomycetes. We believe that our concept of the Ceratobasidiales serves a useful purpose until such time as the phylogeny of the Homobasidiomycetes is more obvious. The Auriculariales is a large and somewhat diverse order as defined here, but data supporting another organization are not available at this time. The species with consistent sphaeropedunculate basidia are included in the Hyaloriaceae of the Auriculariales. This, in our opinion, is the best disposition at this time. The realization that the Hyaloriaceae includes such a diverse group of species has been a major step in the understanding of the Auriculariales. Martin (1952) placed major emphasis on basidial ontogeny and morphology in his work. Much of his system is still sound; however, it is obvious that holobasidia, i.e., non-segmented basidia, have evolved at least once in the Auriculariales, and very possibly several times in the Tremellomycetidae. Also, the transversely segmented basidia of *Auricularia* appear to be closely related to the basidia of *Exidia*, which are usually longitudinally segmented but may be obliquely or, more rarely, even transversely segmented.

A. Historical Concepts

As techniques of data gathering have changed, so have the systematics of the Heterobasidiomycetes and Homobasidiomycetes. Fries (1874) established the core families of the Basidiomycota on the basis of hymenial configuration and texture of the basidiocarps with little or no emphasis on basidial structure. The studies of the Tulasne brothers (Tulasne 1853; Tulasne and Tulasne 1872, 1873) and those of Brefeld (1888) of basidial morphologies made possible refinements of Fries' classification. Brefeld (1888) divided the Basidiomycetes into the Protobasidiomycetes with segmented basidia and the Autobasidiomycetes with nonsegmented basidia. The Dacrymycetaceae and Tulasnellaceae were included in the Autobasidiomycetes by Brefeld. Brefeld (1888) included the rusts within the Protobasidiomycetes but segregated the smuts in a separated class, the Hemibasidii. Modifications of Brefeld's arrangement of the major taxa have been supported in part by several authors (Donk 1956, 1966; Talbot 1968). Patouillard (1900) offered a more flexible system in which the basidia of the Dacrymycetaceae and Tulasnellaceae were held to be heterobasidia and segregated the basidiomycètes into the basidiomycètes hétérobasidiés and basidiomycètes homobasidiés. Patouillard also proposed that the capacity of the basidiospores to form secondary spores is of prime importance. Martin (1945), following Patouillard (1900), included the "jelly fungi" in the order Tremellales of the Heterobasdiomycetes. Martin proposed that the Tulasnellaceae, Sirobasidiaceae, Tremellaceae, Dacrymycetaceae, Phleogenaceae, Hyaloriaceae, and Auriculariaceae be included in the Tremellales. Subsequently, he added the Ceratobasidiaceae to the order (Martin 1948). Bandoni (1984) offers a more detailed description of the various systems of classification of the Heterobasidiomycetes[4] proposed during the past several decades.

The application of electron microscopy has made possible significant improvements in the knowledge of the biology and systematics of the

---

[3] Modified bifactorial compatibility is also found in some species of *Ustilago* and in *Leucosporidium*; however, the labels of the A and B mating type loci in *Leucosporidium* are reversed.

[4] Moore (1996) has recently proposed a division of the basidiomycetes into two phyla, i.e., Basidiomycota and Ustomycota. He separates the orders that are included here in the Heterobasidiomycetes into four of the five classes of his Basidiomycota. We do not consider that Moore's proposals are correlated with the available macro-, micro-, and ultrastructural information nor are they in conformity with the available DNA sequencing data (Bauer and Oberwinkler 1997).

Basidiomycota. Bandoni (1984, 1987) redefined the Tremellales to include only those taxa of Martin's families Sirobasidiaceae and Tremellaceae plus other related species. The Auriculariales are restricted to those species with segmented basidia, complex pores with continuous caps, and biglobular spindle pole bodies (Wells 1994). We are essentially following Bandoni's (1984) redefinition of the Auriculariales and Tremellales supplemented with additional ultrastructural data.

Spindle pole body terminology follows Wells (1977) and basidial terminology is as recommended by Martin (1957). An alternative basidial terminology has been proposed by Donk (1958) and Talbot (1973). Even though the latter system has been widely accepted, it offers no benefits as compared to the earlier terms suggested by Martin. Although the terms *phragmobasidium*, i.e., segmented basidium, and *holobasidium*, i.e., nonsegmented basidium, are useful descriptive terms, it is evident that hobobasidia have evolved independently in *Tremellodendropsis* of the Auriculariales and perhaps several times in the Tremellomycetidae; therefore, phylogenetic significance should not be attached to these terms. The terminology of complex septal pores proposed by Bracker and Butler (1963) is used here; however, the reader may wish to consult the more widely used terminology of Moore and McAlear (1962). Brefeld (1888), Möller (1895), and Ingold (1982a) designated the small lunate elongate, subglobose, or stellate spores arising from germinating basidiospores or monokaryotic hyphae of many species of the Heterobasidiomycetidae as *conidia*. Because of the distinctive nature of these spores and because their function has not been clearly established, the term *microconidium* (*-ia*) is preferable.

B. Taxonomic Synopsis

The Heterobasidiomycetes are defined here in a modified sense (Wells 1994), i.e., those basidiomycetes with an interphase, biglobular spindle pole body (Figs. 1, 2) and a complex septal pore apparatus with a cap that is imperforate (Figs. 4–8), perforate with large pores (Fig. 3), sacculate (Figs. 9, 10), or absent. The basidiospores of most species are capable, under certain environmental conditions, of developing *secondary spores*, i.e., ballistospores, blastospores, or microconidia of varied morphologies. The basidia are generally deeply divided or segmented with the septa oriented longitudinally, obliquely, or transversely. Partially septate or nonseptate (i.e., holobasidia) basidia are reported in both subclasses, e.g., in the Ceratobasidiales and Auriculariales of the Heterobasidiomycetidae and in the Filobasidiales, Rhynchogastremaceae, and Christianseniales of the Tremellomycetidae.

1. Heterobasidiomycetidae: septal pore apparatus with cap imperforate or perforate with large pores; basidiospores generally capable of forming secondary ballistospores, microconidia, or the somatic phase directly; monokaryotic stage mycelial.
2. Tremellomycetidae: septal pore apparatus with sacculate cap or cap lacking; with dimorphic life cycle, i.e., monokaryotic blastospores developed from basidiospores, secondary ballistospores, or hyphae, and dikaryotic stage mycelial; often, or possibly always, mycoparasitic.

II. Heterobasidiomycetidae

A. Characteristics and Major Taxa

The Heterobasidiomycetidae include four orders, i.e., Ceratobasidiales, Tulasnellales, Dacrymycetales, and Auriculariales. The Ceratobasidiales are distinguished by nonsegmented basidia, i.e., holobasidia, with epibasidia terminating, often imperceptibly, in attenuate sterigmata (Fig. 11). The basidiospores are capably of germinating by repetition (Fig. 11G) and by direct germination. The septal pore cap, where known, may be porous, but the pores are consistently larger than those of the higher Homobasidiomycetes (Fig. 3). The septal pore caps of the remaining orders are continuous (Figs. 4–8). The basidia of the Tulasnellales have distinctive microscopic features (Fig. 12). Following meiosis, four spore-like epibasidia are differentiated, into which the four, haploid nuclei and cytoplasm of the hypobasidium migrate. The epibasidia are delimited from the collapsing hypobasidium by wall formation. The basidiospores are capably of germinating directly or by repetition (Fig. 12E). Furcate (i.e., forked) basidia are present in all species of the Dacrymycetales, with few exceptions (Fig. 13) (Wells 1994). The basidiospores generally become septate and develop thickened walls after discharge and are capable of germinating directly or by the

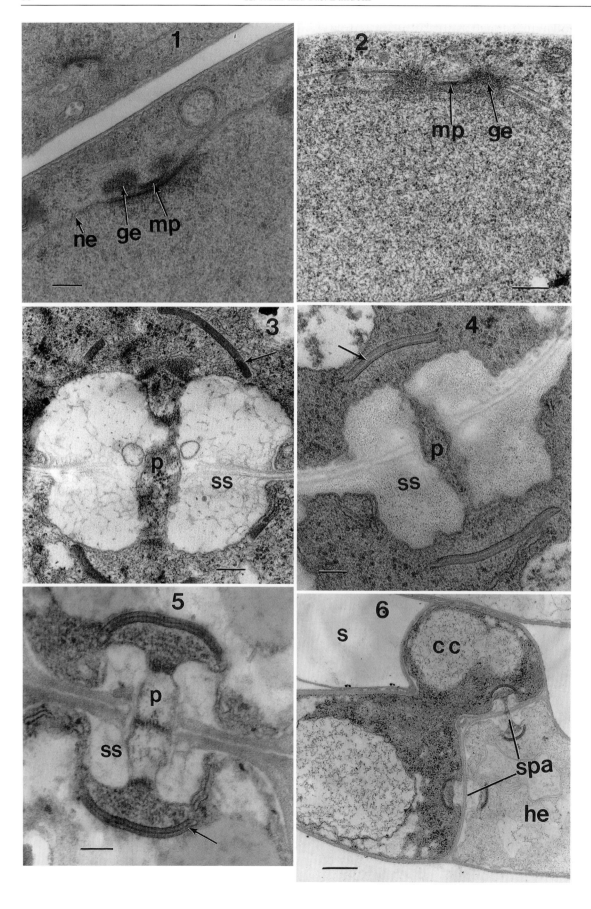

formation of microconidia (Fig. 13I); germination by repetition is unknown in the group. The basidia of the Auriculariales differentiate two to four segments following meiosis and each segment gives rise to a sterigma or an epibasidium and sterigma (Figs. 14, 15). The adjoining walls of the basidia vary from transverse (Fig. 14I,J), oblique (Fig. 14A–C), to longitudinal (Figs. 14G,H,K; 15E–M). The basidiospores germinate directly, by the development of secondary ballistospores (Figs. 14L; 15A,B), or by the production of microconidia.

B. Possible Intra- and Interrelationships

The ultrastructure of the spindle pole bodies and the septal pore apparatus with continuous caps suggest that the Tulasnellales, Dacrymycetales, and Auriculariales, as here defined, are derived from a common ancestor and that these orders are related to the Ceratobasidiales and the Tremellomycetidae (Bandoni 1984; Wells 1994). The available DNA sequencing data (Swann and Taylor 1993, 1995a,b) in general support this view (see Hibbett and Thorn, Chap. 5, this Vol.); however, recent molecular studies (Hibbett et al. 1997; Langer 1998) suggest that the Tulasnellales may be closer to the Homobasidiomycetes and not as closely related to the Dacrymycetales as previously envisioned (Wells 1994). Comparative morphological data, especially the unique basidial morphology of the Tulasnellales, do not support a direct lineage with the primitive Homobasidiomycetes. Unfortunately, sufficient sequencing data are not available for most heterobasidiomycetous and homobasidiomycetous groups to permit a more detailed analysis.

Although many questions of phylogeny in this group of fungi have been answered, several major difficulties in the systematics of the Heterobasidiomycetidae, as here defined, are obvious. The boundary between the Homobasidiomycetes and the Ceratobasidiales is vague, especially in view of the studies of the septal pore apparatus (Langer and Oberwinkler 1993; Langer 1994). Langer and Oberwinkler (1993) reported both continuous and perforate caps in several species of the Corticiaceae. Until such time as more information is available, especially sequencing and ultrastructural data, it is unproductive to ponder the position of the Ceratobasidiales among the larger taxa of the basidiomycetes. There are sequencing data (Swann and Taylor 1993, 1995b) suggesting that the Auriculariales, at least the species of *Auricularia* that have been sampled, are more closely related to the higher basidiomycetes than to the Dacrymycetales and Tremellales. The Dacrymycetales seem to be monophyletic, with no indications of their having given rise to other known taxa.

The Heterobasidiomycetidae, as defined here, is most probably a "horizontal taxon"; however, the available evidence does suggest (Bandoni 1984; Swann and Taylor 1993, 1995a,b; Wells 1994) that the orders treated here have a common ancestor. Rather than introduce additional new taxa into an already confusing scheme, we have redefined the Heterobasidiomycetes and recommend its usage until such time as the phylogeny of the lower basidiomycetes is more clearly documented with reliable information.

C. Ceratobasidiales

1. Taxonomic Features

a) Basidial Ontogeny and Ultrastructural Features

The order Ceratobasidiales is delimited essentially as Martin (1948) defined the family

---

Figs. 1–6. Spindle pole bodies and septal pore apparatuses. **Fig. 1.** *Auricularia auricula-judae*. Section through a prophase, biglobular spindle pole body fixed with chemical fixatives illustrating the globular elements (*ge*) and middle piece (*mp*) in close association with the nuclear envelope (*ne*). *Bar* 0.2 µm. (Lü and McLaughlin 1995 with permission of the Canadian Journal of Botany). **Fig. 2.** *Tremella globospora*. A longitudinal section through a prophase, biglobular spindle pole body showing the middle piece (*mp*) and globular elements (*ge*). Material was fixed by freeze substitution. *Bar* 0.3 µm. (Berbee and Wells 1988 with permission of Mycologia). **Fig. 3.** *Ceratobasidium cornigerum*. Near median section of a septal pore apparatus fixed with chemical fixatives showing the septal swelling (*ss*), pore (*p*), and portions of the large-pored caps (*arrow*). *Bar* 0.2 µm.

Fig. 4. *Tulasnella* sp. Near median section of a septal pore apparatus showing the pore (*p*), septal swelling (*ss*), and caps (*arrow*). The specimen was fixed with chemical fixatives. *Bar* 0.1 µm. **Fig. 5.** *Cerinomyces altaicus*. Median section through a septal pore apparatus showing the pore (*p*), septal swelling (*ss*), and caps (*arrow*). The material was preserved with chemical fixatives. *Bar* 0.1 µm. (Wells 1994 with permission of Mycologia). **Fig. 6.** *Exidiopsis plumbescens*. Section through a clamp connection fixed by chemical fixatives showing two septal pore apparatuses (*spa*). The section through the septal pore apparatus between the clamp cell (*cc*) and the hyphal element (*he*) is a median section. The clamp cell has been walled off by secondary wall development to differentiate a spur (*s*) devoid of contents. *Bar* 0.2 µm. (Bandoni and Wells 1992)

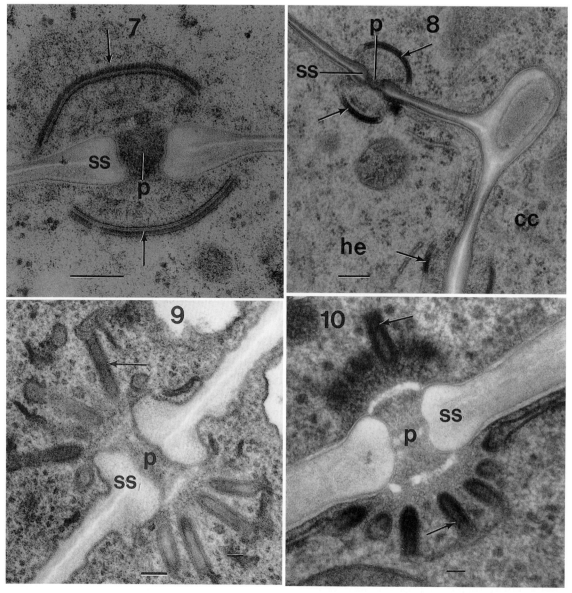

Figs. 7–10. Septal pore apparatuses. **Fig. 7.** *Exidiopsis plumbescens.* Median section through septal pore apparatus prepared by freeze substitution showing the caps (*arrows*), pore (*p*), and septal swellings (*ss*). Bar 0.2 µm. **Fig. 8.** *Auricularia auricula-judae.* Section through a clamp connection fixed by freeze substitution showing a portion of the septum in the hyphal element (*he*) and clamp cell (*cc*). Caps (*arrows*) are visible in both septal pore apparatuses. The median section through the septal pore apparatus in the hyphal element illustrates the septal pore (*p*) and septal swelling (*ss*). Bar 0.2 µm. (Lü and McLaughlin 1991 with permission of Mycologia). **Fig. 9.** *Tremella globospora.* Septal pore apparatus processed with chemical fixatives showing the saccules (*arrows*), septal pore (*p*), and septal swelling (*ss*). Bar 0.1 µm. (Berbee and Wells 1988 with permission of Mycologia). **Fig. 10.** *Tremella globospora.* A median section of a septal pore apparatus prepared by freeze substitution showing the saccules (*arrows*), septal pore (*p*), and septal swelling (*ss*). Bar 0.1 µm. (Berbee and Wells 1988) with permission from Mycologia

Ceratobasidiaceae, i.e., the basidia are without internal segments but develop distinctive epibasidia (Fig. 11A–E), except in the species of *Tofispora*, the basidiospores are capable of secondary ballistopore formation (Fig. 11G), and the septal pores are complex with "large-pored" caps (Fig. 3; Wells 1994; Müller et al. 1998). The hyphae are often broad and branch at wide angles (Fig. 11A–E). These features suggest a common origin with the remaining taxa of the Heterobasid-

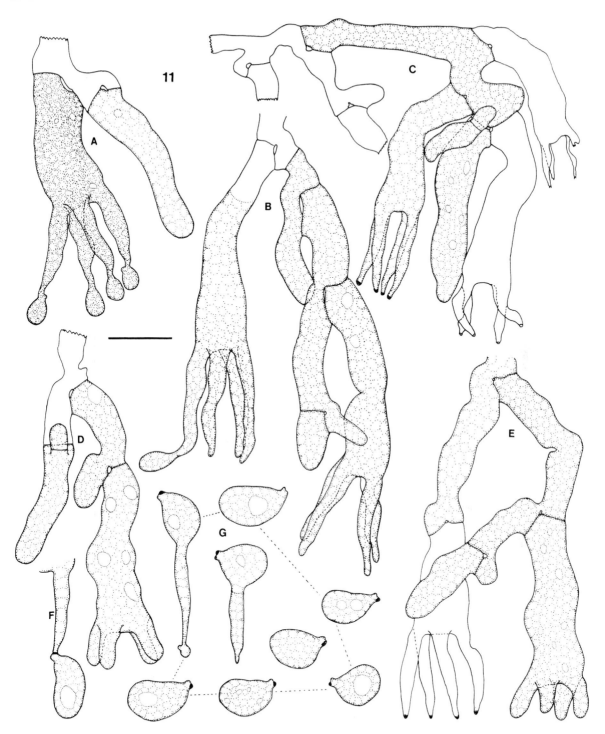

Fig. 11A–G. *Oliveonia atrata*. **A–E** Segments of fertile hyphae with basidia in varying stages of ontogeny. Some basidia, and structures in the basidiocarp are sometimes filled with a granular purplish gray material as depicted in **A**. **F** Sterigma and basidiospore. **G** Basidiospores, two germinating by repetition. *Bar* 10 μm. All figures redrawn from the sketches used by Wells (1994) and reproduced here. (All figs. K. Wells 2596)

iomycetidae; however, the "large-pored" caps are also found in *Galzinia culmigna* (J. Webster et D.A. Reid) B.N. Johri et Bandoni (Wells 1994), *Waitea circinata* Warcup et P.H.B. Talbot (Tu and Kimbrough 1978), and *Sistotrema brinkmannii* (Bres.) J. Erikss. (Langer 1994). Further, comparative studies of septal pore cap structure suggest a gradation in cap pore diameter between the Ceratobasidiales and Corticiaceae (Patton and Marchant 1978). Müller et al. (1998) reported four to six pores in each cap of *Thanatephorus cucumeris* (A.B. Frank) Donk, three to five in *Ceratobasidium cornigerum* (Bourdot) D.P. Rogers, and one to four in *Waitea circinata*. The pores in *T. cucumeris* and *C. cornigerum* are regularly distributed in the cap, whereas those of *W. circinata* are irregularly distributed. Ultrastructural studies of nuclear divisions in taxa of this group (Langer 1994) indicate that a biglobular spindle pole body is present.

In *Thanatephorus cucumeris*, the probasidium is dikaryotic although other hyphal elements in the basidiocarp are generally multinucleate (Tu et al. 1977). Following karyogamy, meiosis with asynchronous disjunction of the chromosomes takes place along an obliquely to longitudinally oriented spindle apparatus in division I, whereas the spindles of division II vary from longitudinal to transverse (Tu et al. 1977). During the late stages of meiosis an inner wall protrudes through the outer wall layer of the distal portion of the basidium. These protrusions become the epibasidia on which the attenuate sterigmata develop. Uninucleate basidiospores develop and following forcible discharge the spores may germinate by giving rise to monokaryotic hyphae or by the formation of a secondary ballistospore (Tu et al. 1977). Basidial ontogeny in other species seems to be similar (Wells 1994); however, some diversity has been noted, e.g., the basidia of *Thanatephorus sterigmaticus* (Bourdot) P.H.B. Talbot consistently bear only two epibasidia and the basidia of the species of *Tofispora* consistently lack epibasidia. The basidia of *Oliveonia atrata* (Bres.) P.H.B. Talbot develop from both prostrate and ascending hyphae (Fig. 11A–E) to develop loosely organized hymenia. While the spore-bearing projections are clearly distinct from those of most Homobasidiomycetes, the epibasidia and sterigmata are difficult to differentiate (Fig. 11A,C,E,F). Germination by repetition is common (Fig. 11G). In *O. atrata* some basidia (Fig. 11A), spores, or hyphae are filled with a granular purplish gray material. Roberts (1998a) proposed the family Oliveoniaceae P. Roberts to include *Oliveonia*; however, insufficient ultrastructural and sequencing data are available to support his proposal at this time.

b) Basidiocarp Structure

Basidiocarps throughout the order are resupinate, generally thin, and waxy-gelatinous to arachnoid to discontinuous (Talbot 1965; Langer 1994). Often sections of the basidiocarps show only a few repent hyphae from which the basidia arise directly or the basidia develop from short, fertile hyphae (Fig. 11C). Older specimens may have erect fertile hyphae bearing terminal clusters of basidia. The hymenium is generally poorly defined and consists of basidia or basidia and cystidioid elements. The basidiocarps of the species of *Uthatobasidium* are generally composed of basal and ascending layers terminating in a reasonably well-defined hymenium (Talbot 1965). Clamp connections are present or absent, and hyphal segments are multinucleate in some taxa and binucleate in other taxa. Characteristically, the tertiary hyphae and those in sclerotia, when formed, are broad and branch at a wide angle. In some species the subhymenial hyphae develop thickened walls, especially those adjacent to the substrate.

2. Habitat and Life Cycle

Many of the species seem to be saprobic on decaying wood, humus, or soil (Rogers 1935; Talbot 1965; Langer 1994), others have been isolated from mycorrhizal roots (Warcup and Talbot 1967), and several species are serious, facultative plant pathogens (Talbot 1965), especially *Thanatephorus cucumeris* (Sneh et al. 1996).

Early reports proposed that *Thanatephorus cucumeris* was homothallic; however, Adams and Butler (1982) demonstrated that anastomosis group 4 was unifactorial. Earlier studies (Whitney and Parmeter 1963) of anastomosis group 1 were suggestive of a unifactorial compatibility system. Evidence of the compatibility system in other anastomosis groups is inconclusive. Presumably the species of this order have a typical basidiomycete life cycle. The basidiospores of most Ceratobasidiales are capable of germination by repetition (Rogers 1935; Talbot 1965; Langer 1994) with the exception of some species of *Tofispora* (Langer 1994). Mitospores are seemingly rare; however, sclerotia are known in *Ceratobasidium* and *Thanatephorus* (Talbot 1965).

3. Taxonomic Synopsis

There are contrasting opinions concerning the taxonomy and nomenclature of a number of taxa included here (Langer 1994; Jülich 1981; Roberts 1998a,c,d; Roberts and Spooner 1998). With the notable exception of *Thanathephorus cucumeris* (Sneh et al. 1996), information on DNA sequencing, ultrastructure, compatibility, and comparative morphology is sparse.

The features of *Ceratobasidium calosporum* D. P. Rogers, the type species of the genus, suggest that it is not a member of this order, and family, as here defined (F. Oberwinkler, pers. comm.). Thus, it may be necessary to select another type species for the genus in order to conserve *Ceratobasidium* as currently circumscribed.

Ceratobasidiales Jülich, Higher taxa basid., p. 344. 1981.
Ceratobasidiaceae G.W. Martin, Lloydia 11:114. 1948.
= Oliveoniaceae P. Roberts, Folia Cryptog. Estonica 33:128. 1998.
Aquathanatephorus C.C. Tu et Kimbr., Bot. Gaz. (Crawfordsville) 139:459. 1978.
Ceratobasidium D.P. Rogers, Stud. Nat. Hist. Iowa Univ. 17:4. 1935.
Oliveonia Donk, Fungus 28:20. 1958.
Oncobasidium P.H.B. Talbot et Keane, Austral. J. Bot. 19:203. 1971.
Sebacinella Hauerslev, Friesia 11:95. 1976.
Thanatephorus Donk, Reinwardtia 3:376. 1956.
Tofispora G. Langer, Biblioth. Mycol. 158:32. 1994.
Uthatobasidium Donk, Reinwardtia 3:376. 1956.

D. Tulasnellales

1. Taxonomic Features

a) Basidial Ontogeny and Ultrastructural Features

Basidial ontogeny and structure are distinguishing characters of the Tulasnellales. The imperforate septal pore cap (Fig. 4; Khan and Talbot 1976; Wells 1994) and biglobular spindle pole bodies (Taylor 1985) point to a common origin with the Heterobasidiomycetidae. In most published electron micrographs[5] of sections cut parallel to the pore, the caps generally appear more flattened, as in the Dacrymycetales, when compared to the dome-shaped caps of the Auriculariales (Khan and Talbot 1976; Moore 1978; Bandoni and Oberwinkler 1982; Wells 1994).

The dikaryotic probasidia arise from prostrate or ascending hyphae either with or without basal clamp connections (Roberts 1994b; Rogers 1932). Karyogamy is followed by meiosis in the hypobasidium (Rogers 1932). The appearance of the subglobose to elongate, incipient epibasidia is approximately commensurate with meiosis. The epibasidia are not precisely distal and often originate from lateral portions of the hypobasidium (Juel 1897; Rogers 1932; Roberts 1994b; Wells 1994). Following the migration of the haploid nuclei into the epibasidia, a postmeiotic mitotic division takes place in some species, prior to the delimitation of the epibasidia from the collapsing hypobasidium (Rogers 1932). The "septum" delimiting the epibasidium may be the result of a secondary wall developed within the maturing epibasidium or localized wall formation (Khan and Talbot 1976). The epibasidium subsequently differentiates a sterigma directly or a tubular extension terminating in a sterigma on which an asymmetrical basidiospore is developed and forcibly discharged (Rogers 1932). Germination of basidiospores is either direct or by secondary ballistospore formation.

Gloeotulasnella cystidiophora (Höhn. et Litsch.) Juel is without clamps (Fig. 12; Roberts 1994b). The obovoid to clavate probasidia arise through proliferations beneath existing basidia (Fig. 12C,D). Four subglobose to elongate epibasidia develop from the distal regions of the basidium and are subsequently delimited from the collapsed hypobasidium (Fig. 12A–D). There seems to be a pause in ontogeny after epibasidial delimitation; subsequently, a sterigma or tubular extension terminating in a sterigma arises from each epibasidium (Fig. 12A,C,D). The subglobose spores develop, with the hilar appendices pointing towards the central axis of the basidium, and are discharged. The basidiospores are capable of germination by repetition (Fig. 12E), the development of unique, stellate microconidia (Ingold 1984b; Roberts 1994b), or direct germination (Ingold 1984b).

A feature of basidial development in this order, and in other taxa forming phragmobasidia and distinct epibasidia, is the lack of synchrony in ontogeny of most basidia. In the Tulasnellales this asynchrony is evident in the development of the

---

[5] The reader is reminded that most studies of basidiomycete septal pores are of chemically fixed material which induces distortion (Figs. 3–6, 9), especially in the septal swellings, when compared to material fixed by freeze substitution (Figs. 7, 8, 10; Hoch and Howard 1981; Berbee and Wells 1988; Lü and McLaughlin 1991).

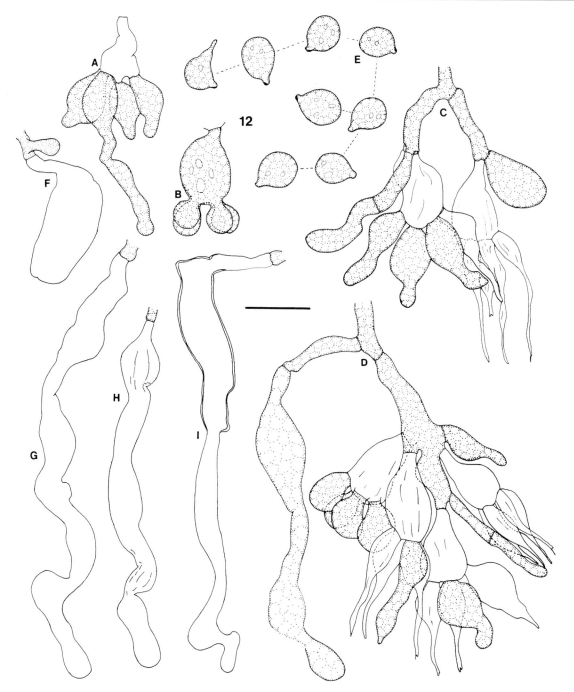

Fig. 12A–I. *Gloeotulasnella cystidiophora.* **A** Basidium with epibasidia delimited from the hypobasidium. **B** Hypobasidium with developing epibasidia. **C, D** Segments of fertile with basidia in varied stages of ontogeny. A gloeocystidium-like structure arising from the fertile hypha is illustrated in **D. E** Basidiospores, one germinating by repetition. **F–I** Gloeocystidia-like elements of varied morphologies. *Bar* 10 μm. (All figs. K. Wells 3582)

epibasidia. It is rare to find the epibasidia of a basidium at the same developmental stage and often one, or more, epibasidium has released the basidiospore before the adjacent ones have initiated spore development (Fig. 12A,C,D; Roberts 1992, 1993a, 1994a,b; Wells 1994). This feature is, evidently, an adaptation to the capacity of the gelatinous basidiocarps of many species of the Heterobasidiomycetes, and other taxa with gelatinous basidiocarps, to withstand repeated periods of high and low humidity (Rockett and Kramer 1974).

The essential features visible by light microscopy of the nuclear cycle in the probasidium were established by the studies of Juel (1897) and Rogers (1932). Karyogamy occurs in the enlarging probasidium followed by two meiotic divisions on a transverse to oblique spindle. Densely staining polar structures are visible and chromosome separation is asynchronous (Rogers 1932). Taylor (1985) examined serial sections of conjugating dividing nuclei in *Tulasnella pruinosa* Bourdot et Galzin (= *T. araneosa* Bourdot et Galzin) by electron microscopy and found that the spindle pole bodies are biglobular during prophase and monoglobular during division. The nuclear envelope remains intact during division except at the poles, and the chromosomes are situated along the pole-to-pole microtubules and separate asynchronously (Taylor 1985). During late anaphase the nuclear envelopes surrounding the pole-to-pole microtubules of the conjugately dividing nuclei are continuous; therefore, at this stage the mitotic spindles of the dikaryon are in a common nucleoplasm. Whether this unique feature is restricted to the Tulasnellales or whether it is a feature of conjugate nuclear division in those basidiomycetes lacking clamps has not been established (Taylor 1985).

b) Basidiocarp Structure

Rogers (1933) expanded the original definition of *Gloeotulasnella* to include species developing waxy gelatinous to mucous basidiocarps with, generally, clavate to capitate basidia embedded in the gelatinous matrix. Generally, the epibasidia develop tubular extensions from the inflated bases. In contrast, in Rogers' (1933) concept of *Tulasnella*, the basal portion of the epibasidia tend to develop subulate projections directly, the basidia tend to be obovoid and are formed in waxy to arid or arachnoid basidiocarps. Olive (1957) rejected these generic concepts and reduced *Gloeotulasnella* to a subgenus; however, Roberts (1994b) notes that basidiocarp structure lends some support to the generic concepts proposed by Rogers (1933). Still, Roberts does not retain *Gloeotulasnella* as a genus. It is obvious that additional study is needed to refine generic concepts in this group. In the interim, the genus *Gloeotulasnella* is tentatively maintained to accommodate those species developing gloeocystidia-like elements in the hymenium (Fig. 12F–I) as at least these species seem to form a phylogenetic unit.

Typically, most species develop basidiocarps that appear as faint arachnoid, mucous, waxy-gelatinous, or gelatinous patches on decaying wood, leaves, or old basidiocarps of other fungi, mainly Aphyllophorales. Upon drying, the basidiocarps are often invisible without magnification. Often, the basidiocarps, fresh or dry, have a pinkish to violaceous tinge. Roberts (1994b) notes that in the majority of species the basidiocarps vary from a few strands of prostrate hyphae, from which the basidia develop, to thicker specimens consisting of both prostrate hyphae and ascending, branching hyphae, from which the basidia arise. In others, clusters of basidia arise from little-branched fertile hyphae with long segments; in such species, prostrate hyphae are absent or poorly defined. Hymenia are, in general, poorly differentiated and are composed only of basidia in *Tulasnella* or of basidia and gloeocystidia-like elements in *Gloeotulasnella*. Basidiocarps are lacking in *Tulasnella zooctonia* Drechsler as the basidia arise from the branching hyphae emanating from the parasitized amebae.

2. Habitat and Life Cycle

Most collections of the species have been made from decaying wood or leaves, as well as senescent basidiocarps of other fungi (Bourdot and Galzin 1928; Rogers 1933; Roberts 1992, 1993a, 1994a,b); however, several species have been isolated from the roots of terrestrial orchids (Warcup and Talbot 1967). One species, *Tulasnella zooctonia*, is evidently parasitic on soil amebae (Drechsler 1969). Worrall et al. (1997) found that the four species of the Tulasnellaceae they tested were unable to decay wood. These observations suggest that at least some species of the Tulasnellales are not primary decay organisms of wood but may be mycoparasitic, parasitic on other organisms, mycorrhizal associates, or scavengers on the byproducts of wood-decaying species.

Little is known of the life cycle of the species of this order. To the best of our knowledge, compatibility studies have not been successfully completed with any species. Other than the typical nuclear cycle in the basidium (Rogers 1932) one can only assume that the usual, basidiomycete, monodikaryotic cycle occurs in these species. Conidia are known in *Stibotulasnella conidiophora*. Sessile or short-stalked synnemata bear clusters of annelidic conidia dispersed in liquid droplets; however, the conidiophores sometimes occur singly or in small clusters in this species. Basidia seem to be most abundant around the bases of the synnemata but are sometimes scattered on the prostrate hyphae (Bandoni and Oberwinkler 1982). Petersen (1968) described allantoid to ovoid conidia arising from a single sporogenous element on the somatic hyphae of *Tulasnella pinicola* Bres. The conidia were formed in clusters and, seemingly, held together by a mucus material. Microconidia are known to develop during basidiospore formation in some species (Ingold 1984b).

3. Taxonomic Synopsis

Tulasnellales Rea, British Basid., p. 739, 1922.
Tulasnellaceae Juel, Bih. Kongl. Svenska Vetensk.-Akad. Handl. 23, Afd. III(12):21. 1897.
Tulasnella J Schröt., apud Cohn, Kryptog.-Flora Schlesien 3(1):397. 1888.
Gloeotulasnella Hohn. et Litsch., Sitzungsber. Kaiserl. Akad. Wiss., Math.-Naturwiss. Kl., Abt. 1, 115:1557. 1906.
Pseudotulasnella Lowy, Mycologia 56:696. 1964[6].
Stibotulasnella Oberw. et Bandoni, Can. J. Bot. 60:1875. 1982.

E. Dacrymycetales

1. Taxonomic Features

a) Basidial Ontogeny and Ultrastructural Features

Furcate (i.e., forked) basidia are formed in all but one recognized species of this order and most taxa develop basidia that are remarkably uniform in morphology (Fig. 13; Kennedy 1959; McNabb 1964, 1965a–e, 1966, 1973; D. Reid 1974). In *Dacrymyces unisporus* (L.S. Olive) K. Wells the basidium differentiates a single, eccentrically attached epibasidium (Wells 1994). Bodman (1938) reported that karyogamy in *Dacryopinax spathularia* (Schwein.) G.W. Martin occurs in the probasidium and that meiosis occurs prior to or during epibasidia development. One haploid nucleus passes into each of the two basidiospores and two degenerate in the epibasidia.

The septal pores are complex with imperforate caps (Fig. 5; Wells 1994). A biglobular spindle

[6] The senior author has been unable to verify partial septation in the type specimen.

pole body has been illustrated in *Dacrymyces stillatus* Nees: Fr. (Oberwinkler et al. 1990).

The basidiospores are usually elliptical to cylindrical and laterally depressed or allantoid (D. Reid 1974) except for the subglobose to broadly ovoid spores of *D. unisporus* and *Dacrymyces ovisporus* Bref. Germination by repetition has not been reported among the species of this order. Transverse septation, or both longitudinal and transverse septation in *D. unisporus*, *D. ovisporus*, and some species of *Guepiniopsis*, of the basidiospores occurs following discharge (Bodman 1938; Ingold 1983). Ingold (1983) considered that segmentation of the discharged basidiospore represented the initial stage of spore germination. Further germination of the basidiospore results in the development of hyphae and/or microconidia. In the five species studied, Ingold (1983) reported that the segmented basidiospores on malt agar gave rise to hyphae, whereas those on glass produced microconidia. Microconidia also arise from the hyphae. They germinate on malt agar, producing hyphae or, less often, secondary microconidia. They arise from a minute denticle projecting from a short, conical conidiophore (Fig. 13I). Additional microconidia arise from the same conidiophore on successive denticles, resulting in the enlargement of the conidiophore and a cluster of minute microconidia (Ingold, 1983).

The number and thickness of basidiospore septa capable of developing is a useful taxonomic character (Kennedy 1959; McNabb 1964, 1965a–e, 1966, 1973; D. Reid 1974). The number of septa is quite varied, even within a single collection; however, the most common number of septa developed is a useful taxonomic character.

Dacrymyces stillatus is a common species, especially in the cool, temperate regions. The probasidia are cylindrical to narrowly clavate and arise as lateral proliferations from beneath older basidia (Fig. 13A,B). Two epibasidia are formed, at the tips of which sterigmata and basidiospores are developed (Fig. 13B,E). The hilar appendices are oriented towards the central axis of the basidium (Fig. 13B). Upon discharge, the cylindrical-curved basidiospores are thin-walled but differentiate one to three transverse septa and thickened walls after release. Each segment is capable of germinating directly or giving rise to subglobose to somewhat elongate microconidia (Fig. 13I). Spores discharged directly on a nutrient surface often germinate directly without wall thickening and septa formation (pers. obs.). Strands of thick-walled arthro-

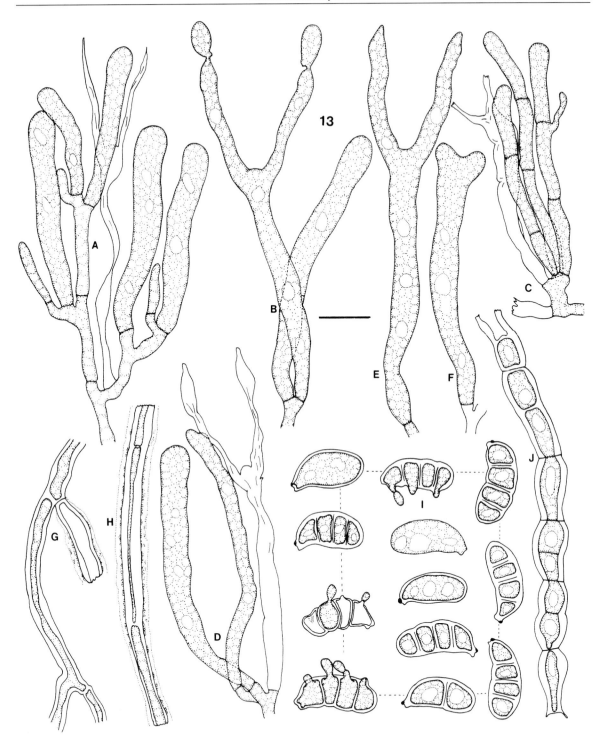

Fig. 13A–J. *Dacrymyces stillatus.* **A, D** Fertile hyphal segments bearing developing and collapsed basidia. **B** Fertile hypha bearing near mature basidium and probasidium. **C** Fertile hypha segment bearing collapsed basidium and dikaryophyses. **E, F** Basidia. **G, H** Segments of thick-walled subhymenial hyphae. Portions of **G** and all of **H** are covered by a gelatinous sheath with an uneven distribution of granule-like structures causing a roughened appearance. **I** Basidiospores. After discharge the spores developed thickened walls and up to three septa. Each segment is capable of developing ovate microconidia or hyphae. **J** Chain of arthrospores. *Bar* 10 μm. All figures redrawn from the same original sketches used by Wells (1994). (All figs K. Wells 3432)

spores (Fig. 13J) develop in separate basidiocarps or are mixed in with the basidia. The basal hyphae are thick-walled and often appear to differentiate a gelatinous sheath that imparts a roughened appearance under the microscope (Fig. 13G,H).

b) Basidiocarp Structure

The basidiocarps range from effused, applanate, through pustulate, conic, obconic, discoid, cupulate, cerebriform, coarsely lobed, or clavate to clavarioid. They are soft to firm gelatinous in texture and generally yellow to orange when fresh, but hyaline to whitish, cream, tan, reddish brown, or red characterize basidiocarps of some taxa. The basidiocarps are composed entirely of thin-walled hyphae, of thick-walled hyphae, or of a combination of thin-walled and thick-walled hyphae (D. Reid 1974). The sterile portions of the basidiocarps of the species of *Heterotextus* are covered with a palisade-like layer of thick-walled, somewhat inflated, terminal hyphal elements. A partially differentiated surface layer is also evident in *Dacrymyces chrysospermus* Berk. et M.A. Curtis. In section, the hyphae of the basidiocarps of most species of Dacrymycetales tend to be more densely arranged than in most other taxa of the Heterobasidiomycetes and often possess thickened, gelatinized walls that are smooth or appear roughened when stained (Fig. 13G,H; D. Reid 1974). Unlike the basidia, which tend to be rather consistent throughout the order, the basidiocarps are exceptionally variable among the various taxa, and often within a single species. This variation has resulted in a lack of agreement as to the limits of the various genera (Kennedy 1959; McNabb 1964, 1965a–e, 1966, 1973; D. Reid 1974).

The ontogeny and structure of the basidiocarps of *Calocera cornea* (Batsch: Fr.) Fr. from culture and the natural habitat were examined by Kennedy (1972). She found that the basidiocarps consisted of a central zone of compact hyphae parallel to the long axis of the simple to sparsely branching, clavarioid fruiting bodies, a median, loosely interwoven zone, and an amphigenous, compact layer of basidia. The developing basidiocarps were very sensitive to moisture and were capable of alternating periods of rapid growth and inactivity.

2. Habitat and Life Cycle

Basidiocarps of all the species in this order have been collected almost exclusively from wood. *Calocera viscosa* (Pers.: Fr.) Fr. basidiocarps often arise from humus around conifer stumps or possibly other buried wood, although they also occur directly on conifer wood. Some species seem to occur predominately on deciduous wood, e.g., *Calocera cornea*; others seem to be restricted to conifer wood, e.g., *C. viscosa*; and some occur on both, e.g., *Dacrymyces capitatus* Schwein., *D. minor* Peck, and *D. stillatus*. Seifert (1983) found that the 16 species he tested resulted in wood decay levels normally associated with the Aphyllophorales. Worrall et al. (1997) conducted decay tests of 10 species and reported similar results. Most species caused a brown rot, as defined by Worrall et al. (1997). Thus, available evidence suggests that the species of the Dacrymycetales are saprobic and capable of degrading a significant amount of wood.

It seems likely that the basic basidiomycete mono- dikaryotic life cycle exists among the Dacrymycetales; however, few phases have been carefully documented. One species, *Cerinomyces aculeatus* N. Maek., is very probably bifactorial but only a single specimen was analyzed (Maekawa 1987). Evidence is also available that *Calocera cornea* and *Ditiola pezizaeformis* (Lév.) D.A. Reid are bifactorial (Yen 1949); however, only isolates from single collections were paired. In *Dacrymyces stillatus* dikaryotic arthrospores are produced in separate basidiocarps or interspersed with the basidia. These are possibly dispersed by water or insects and would seem capable of establishing new colonies. The basidiospores of many species are capable of producing microconidia following segmentation (Fig. 13I; Brefeld 1888; D. Reid 1974; Ingold 1983); however, whether such agents function as fertilizaing agents, dispersal agents, or both, is unknown. Ingold (1983) observed that the microconidia of *D. stillatus* germinated on malt agar to form hyphae or, less often, additional microconidia.

3. Taxonomic Synopsis

Kennedy (1959), McNabb (1964, 1965a–e, 1966, 1973) and D. Reid (1974) were followed in developing the generic concepts. Species concepts have been significantly improved by the studies of McNabb (1964, 1965a–e, 1966, 1973) and D. Reid (1974). While D. Reid's work is restricted to those species known from Great Britain, McNabb considered all described species. The reader should also consult Oberwinkler (1993) for a different perspective on the systematics of this order.

Dacrymycetales Henn., *apud* Engl. et Prantl, Naturl. Pflanzenfam. 1(1**):96. 1897.
Dacrymycetaceae Bref., Untersuch. Gesammtgeb. Mykol. 7:138. 1888.
= Cerinomycetaceae Jülich, Higher Taxa Basid., p. 358. 1981.
Calocera (Fr.) Fr., Syst. Orbis Veg., p. 91. 1825.
Cerinomyces G.W. Martin, Mycologia 41:82. 1949.
Dacrymyces Nees: Fr., Syst. Mycol. 1:lv. 1821.
= *Arrhytidia* Berk. et M.A. Curtis, Hooker's J. Bot. Kew Gard. Misc. 1:235. 1849.
Dacryonaema Nannf., Svensk Bot. Tidskr. 41:336. 1947.
Dacryopinax G.W. Martin, Lloydia 11:116. 1948.
Ditiola Fr., Syst. Mycol. 2:169. 1822.
= *Femsjonia* Fr., Summa Veg. Scand., Sect. 2, p. 341. 1849.
Guepiniopsis Pat., Tabulae Analyticae Fungorum, Fasc. I, Ser. 1, p. 27. 1883.
Heterotextus Lloyd, Mycol. Writings 7:1151. 1922.

F. Auriculariales

1. Taxonomic Features

The Auriculariales include those species with complex septal pores with imperforate caps (Figs. 6, 7, 8) and transversely (Fig. 14I,J), obliquely (Fig. 14A–C), to longitudinally oriented basidial septa (Fig. 14G,H,K), or partially septate basidia. The basidiospores are generally capable of germinating with the development of hyphae, secondary ballistospores (Fig. 14L), or microconidia (Ingold 1982a,b, 1984a,b, 1985, 1992, 1995; Bandoni 1984; Wells 1994).

a) Ultrastructural Features

The meiotic spindle pole bodies in *Auricularia fuscosuccinea* (Mont.) Farl. (McLaughlin 1981) and *Sebacina sublilacina* G.W. Martin[7] (Wells, unpubl. observ.) are biglobular, as in the other studied species of the Heterobasidiomycetes. The mitotic spindle pole bodies during clamp formation in *A. auricula-judae* (Bull.: Fr.) J. Schröt. (Lü and McLauglin 1995) and during the first postmeiotic division in *A. fuscosuccinea* (McLaughlin 1981) are also biglobular (Fig. 1). All species of the order that have been examined possess a septal pore apparatus with an imperforate cap (Khan and Kimbrough 1980; Wells 1994; Müller et al. 1998). Material fixed by freeze substitution results in decidedly less distorted images than by chemical fixation (Lü and McLaughlin 1991; Bandoni and Wells 1992). Freeze substitution yields septal swelling that are smaller and more rounded than those preserved by chemical fixation (Figs. 6–8). In both fixatives an electron-dense plate is often visible in the pore itself, and the imperforate cap is composed of modified endoplasmic reticulum enclosing an electron-dense region (Fig. 8; Lü and McLaughlin 1991). The septal pore apparatus of *Tremellodendropsis tuberosa* (Grev.) D.A. Crawford is typical of those of the other representatives of this order; however, the imperforate cap is generally associated with a strand of unmodified ER (Wells 1994). A similar association of the ER with the cap has been demonstrated by Müller et al. (1998) in *Serendipita vermifera* (Oberw.) P. Roberts.

b) Basidial Ontogeny and Basidiocarp Structure

While considerable variation occurs in the orientation of the septa, basidial segmentation is complete in the Auriculariales, except in several species of *Tremellodendropsis* (Petersen 1985, 1987). Cylindrical, transversely septate basidia are characteristic of the Auriculariaceae (Fig. 14I,J). In *Auricularia fuscosuccinea* the probasidium becomes differentiated into four segments soon after meiosis (McLaughlin 1980). Differentiation of epibasidia, basidiospore development, and nuclear migration into the spore follow, generally, a basipetal sequence. Oberwinkler (pers. comm.) and the senior author have observed a similar pattern of maturation in *Auricularia auricula-judae* (Fig. 14I,J) and *A. mesenterica* (Dickson: Fr.) Pers. A unique basidial feature among the members of the Auriculariales that have been studied ultrastructurally is the formation of a septal pore apparatus in the center of the transverse hypobasidial septa. The septal pores of the basidial septa are very similar to those formed in hyphal septa. During the terminal stages of vacuolation of hypobasidial segment, the pore is occluded by the septal swelling within the viable segment; the septal swelling within the empty

---

[7] D. Reid (1970) and Roberts (1998b) have equated *Sebacina sublilacina* with *S. subhyalina* A. Pearson; however, we disagree with this disposition. Roberts' redefinition of *Stypella* is a significant improvement in the systematics of the Hyaloriaceae, and it would be appropriate to include our concept of *S. sublilacina* in the redefined genus.

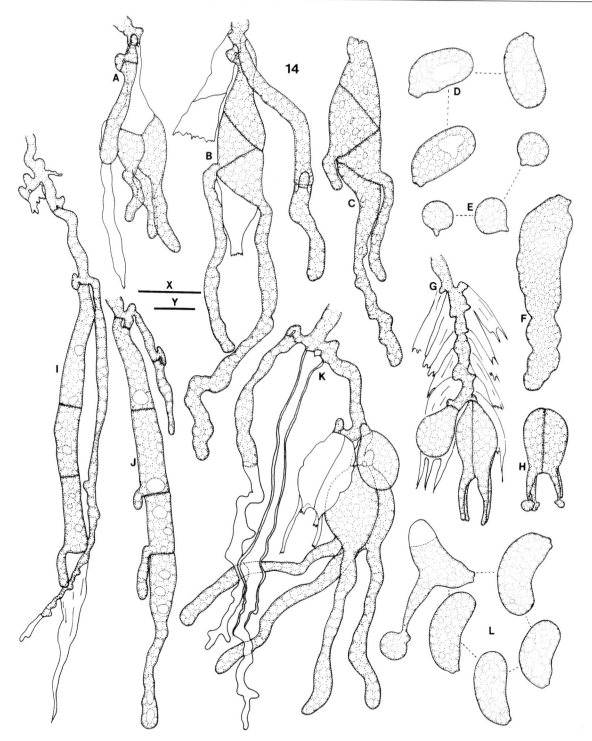

Fig. 14A–L. *Patouillardina cinerea.* **A, B** Short segments of fertile hyphae bearing maturing basidia. **C** Basidium. **D** Basidiospores. **E–H** *Basidiodendron eyrei*. **E** Basidiospores. **F** Gloeocystidium. **G** Fertile hypha bearing developing basidia with involucre-like sheath of collapsed basidia. **H** Basidium. **I, J** *Auricularia auricula-judae*. **I, J** Short segments of fertile hyphae bearing segmented basidia. **K, L** *Exidia glandulosa*. **K** Segment of fertile hypha bearing basidia in several stages of ontogeny and dikaryophyses, one of which has a thickened wall. **L** Basidiospores, one germinating by repetition. *Bars* 10 μm; *bar X* applies to **A–H, K, L**; *bar Y* applies to **I, J** (**A–D** J. Rick, XII-1939; **E–H** Herb. H. Bourdot 18464; **I, J** K. Wells 3695; **K, L** K. Wells 3110). **A–D** redrawn from original sketches used by Wells (1994)

segment breaks down (McLaughlin 1980). The ontogeny of the septal pore apparatuses within the basidia of *A. fuscosuccinea*, and very probably of the other species as well, would seem to be associated with the basipetal development of the hypobasidium that permits cytoplasmic flow within the segmented basidium.

The best-known species of the Patouillardinaceae is *Patouillardina cinerea* Bres (Fig. 14A–D). The adjoining walls of the hypobasidial segments of the often fusiform basidia are most frequently oriented obliquely to the long axis of the basidium; however, transverse and longitudinal septa are often present (Martin 1935; Wells 1994). As in *Auricularia*, basidial maturation generally follows a basipetal pattern. The basidiocarps of *P. cinerea* are waxy-gelatinous and resupinate with a well-defined hymenium of basidia and distinctive dikaryophyses (= hyphidia) (Martin 1935; Wells 1994).

The remaining families of the Auriculariales develop segmented basidia in which the septa are essentially longitudinal, i.e., parallel to the long axis of the basidium (Figs. 14G,H,K; 15E–M); however, a considerable amount of variation in septal orientation often occurs, especially within some species, e.g., *Exidiopsis diversa* K. Wells (Wells 1987). As in the Tulasnellales, there is often a lack of synchrony in the ontogeny of the segments of individual hypobasidia.

The species of the Exidiaceae generally possess subglobose, obovoid, to pyriform basidia that are, typically, with subbasidial clamps (Fig. 14G,H,K). Enucleate hypobasidial stalks are sometimes present in some species but never consistently so. Cytological studies have been made of *Basidiodendron eyrei* (Wakef.) Luck-Allen[8], *Exidia glandulosa* Bull.: Fr., *E. crenata* Schwein: Fr.[9], and *E. pinicola* (Peck) Coker[10] (Furtado 1969; Whelden 1935a,b). These studies show the typical transverse to oblique orientation of the spindle apparatus and sequence of meiosis followed by segmentation and differentiation of the epibasidia. In some species, basidia are scattered over the substrate without an organized hymenium, as in *Serendipita vermifera*, whereas *E. grisea* (Pers.) L. Maire forms a continuous, effused basidiocarp with a defined hymenium of basidia and dikaryophyses. The effused basidiocarps of *Basidiodendron* are frequently encountered in both tropical and temperate regions (Luck-Allen 1963; Wells and Raitviir 1975). Yellow granular gloeocystidia (Fig. 14F) are characteristic of the genus. Several species develop basidia in a close acropetal sequence resulting in an involucre-like sheath of collapsed basidia as the basidiocarps age (Fig. 14G). In several species, the urniform basidia develop attenuate projections bearing the basidiospores, which are difficult, or impossible, to characterize as epibasidia and sterigmata, or as protosterigmata and spicula (Fig. 14G,H).

Sterile spines are characteristic of the hymenia of the speices of *Heterochaete*. The basidiocarps of *Heterochaete* vary from effused, applanate, to pustulate or cerebriform, gelatinous masses (Bodman 1952). Some species seem to have affinities with those of *Eichleriella*, others with *Exidia*, and some with *Exidiopsis*. The species are quite abundant in both numbers of species and numbers of samples in tropical and subtropical regions. Certainly the genus, as currently defined, is not a natural one (Bodman 1952), but sorting out the related elements will require a major effort. The basidiocarps of the species of *Eichleriella* are applanate with reflexed margins on drying. Well-defined hymenia are present, as are distinct basal layers of brownish, thick-walled hyphae. Among the species of *Exidia* the basidiocarps originate as pustules expanding into discoid, obconic, or turbinate forms. In some species of *Exidia* the basidiocarps remain discrete, whereas in others they anastomose, resulting in cerebriform, foliose to pileate basidiocarps. Under especially favorable conditions, *E. glandulosa* can form extensive basidiocarps up to several meters in the longest dimension (per. observ.); these are probably formed by coalescence of numerous basidiocarps. In *Exidia* s. str., the basidiocarps are tough-gelatinous with a well-defined unilateral or inferior hymenium of basidia and an intricate, interwoven layer of dikaryophyses that disintegrate at maturity creating a brownish, granular layer above the developing basidia (Klett 1964).

Oberwinkler (1963, 1964), Hauerslev (1976, 1986, 1993), and Roberts (1993b) have described a number of taxa with very thin basidiocarps that are often invisible without magnification and with poorly defined hymenia. Roberts (1993b) proposed several new genera to accommodate these

---

[8] As *Sebacina deminuta* Bourdot.
[9] As *Exidia recisa* Ditmar: Fr.
[10] As *Exidia saccharina* Alb. et schwein.: Fr. While these specific epithets may represent the same species, authentic specimens are available for *E. pinicola* but evidently not for *E. saccharina*.

taxa. *Serendipita* P. Roberts is composed of species with basidia arising in isolated clusters from the scattered hyphae on the substrate. The species often develop elongated to sigmoid basidiospores, but allantoid to fusiform spores are known. The known specimens are rather consistently associated with corticoid fungi; therefore, it is possible that the species are parasites or scavengers of the by-products of wood-decaying species. The species of *Endoperplexa* P. Roberts develop very thin, resupinate sporocarps bearing cystidiate structures that are septate in *E. septocystidiata* (Hauerslev) P. Roberts. Both *Serendipita* and *Endoperplexa* are composed of species that were initially described in *Sebacina* or *Exidiopsis* and make possible a more homogenous definition of *Exidiopsis*; however, the relationships of both taxa are obscure, although some of the known features of *Endoperplexa* suggest an affiliation with *Basidiodendron*.

While the acceptance of *Ceratosebacina* P. Roberts and *Microsebacina* P. Roberts (Roberts 1993b) would further refine the definition of *Exidiopsis*, and the genus does remain a somewhat heterogeneous taxon, these proposals need further study. *Exidiopsis gloeophora* (Oberw.) Wojewoda is possibly intermediate between *E. calospora* Bourdot et Galzin (= *E. longispora* (Hauerslev) Wojewoda)[11] and the species included by Roberts (1993b) in *Exidiopsis*. We also believe, at this time, that *E. fugacissima* (Bourdot et Galzin) Sacc. et Trotter, the type of *Microsebacina*, should be retained in *Exidiopsis*.

The only known species of *Renatobasidium*, *R. notabile* Hauerslev, was reported as exhibiting percurrent basidial ontogeny (i.e., some probasidia arise as proliferation through the basal septum of depleted basidia), as in the homobasidiomycete genera *Repetobasidium* J. Erikss. and *Repetobasidiellum* J. Erikss. et Hjortstam (Hauerslev 1993). The senior author has examined the type specimen and noted some evidence of internal proliferation; however, definitive observations were not possible. It will be necessary to examine fresh material or, preferably, undertake studies of basidial ontogeny with electron microscopy.

A consistent character of the species of the Hyaloriaceae is the sphaeropedunculate basidium (Wells 1964b). Such basidia, as in *Sebacina crozalsii* Bourdot et Galzin, arise as clavate, spathulate, to capitate probasidia (Fig. 15E,F,M). Following karyogamy and the initiation of meiosis, the stalk becomes vacuolated and is eventually delimited by a wall at approximately the time of the differentiation of the hypobasidial segments (Fig. 15J,K,L). The length of the enucleate stalk varies from a few microns in such species as *Elmerina caryae* (Schwein.) D.A. Reid to up to 25–35 µm in length in *Myxarium nucleatum* Wallr. In all examples examined, an enucleate stalk is present on all basidia of the taxa of this family; however, vacuolated stalks are difficult to distinguish in mature basidia even with phase optics, especially in herbarium specimens. Cytological studies have been made of *Myxarium nucleatum*[12], *Sebacina sublilacina*[13], *Stypella grilletii* (Boud.) P. Roberts[14], and *Stypella subgelatinosa* (P. Karst.) P. Roberts[15] (Whelden 1934, 1935a,b, 1937). Furtado (1968, 1969) followed the chromosome behavior in *Myxarium nucleatum* and demonstrated typical basidiomycete meiotic divisions with asynchronous disjunction during both anaphase I and II. The ontogeny of the sphaeropedunculate basidium of *Myxarium nucleatum* has been followed ultrastructurally (Wells 1964b). In the species examined cytologically, the spindle apparatuses of divisions I and II are oriented transversely to obliquely. Internal wall formation is initiated in *M. nucleatum* following division I of meiosis, and completed following division II (Furtado 1968; Wells 1964b). The wall delimiting the enucleate stalk seems to be laid down in association with walls delimiting the hypobasidial segments. In the species studied ultrastructurally, the epibasidia are apparently initiated as early as division II or during segmentation of the hypobasidium. In *Sebacina crozalsii*, as in most other species of *Stypella* as recently redefined by Roberts (1998b), the basidia are interspersed with branching dikaryophyses (Fig. 15K,M–O) arising from the

---

[11] D. Reid (1970) defined this species as lacking clamps; however, the senior author examined two specimens from PC (i.e., Bourdot 7568 and 14847) collected prior to the publication date and labeled by Bourdot as *Sebacina calospora* and both have abundant clamps. Hauerslev (1976) also found clamps in a specimen studied by Bourdot and collected prior to the publication date. Since clamps were not seen in two of Hauerslev's collections, which he determined as *S. calospora*, some collections do lack clamps; however, whether these represent a different species is a matter for further study. Further, the proper name can only be determined by the careful examination of addition specimens from those studied by Bourdot prior to the publication date.

[12] As *Exidia nucleata* (Schwein.) Burt.
[13] As *Sebacina fugacissima* Bourdot et Galzin.
[14] As *Tremella grilletii* Boud.
[15] As *Protodontia uda* Höhn.

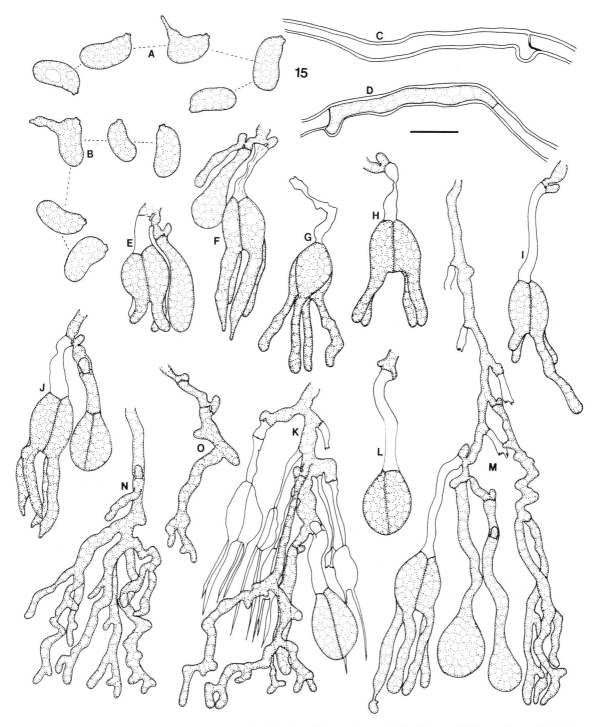

Fig. 15A–O. *Sebacina crozalsii* (Hyaloriaceae). **A, B** Basidiospores, two germinating by repetition. **C, D** Segments of the thick-walled hyphae adjacent to the substrate. **E–J, L** Basidia, some showing portions of fertile hyphae. **K, M** Portions of fertile hyphae bearing basidia and dikaryophyses. **N, O** Apices of dikaryophyses. *Bar* 10 μm. **A, D–F, J, K, N, O** R.J. Bandoni 5297; **B, C, I, L, M** Herb. H. Bourdot, 39633 (A. de Crozals 23); **G, H** H. Bourdot 39065 (A. de Crozals 38)

fertile hyphae. Such dikaryophyses seem to have a high lipid content and become agglutinated in the thinner, more delicate species. *Sebacina crozalsii* is unique among this group in developing a prostrate, basal layer of thick-walled hyphae of varying thickness (Fig. 15C,D).

The basidiocarps of the Hyaloriaceae are remarkably diverse. As examples, those of *Sebacina podlachica* Bres. are resupinate with smooth hymenia, those of *Sebacina sublilacina* are initially pustulate but soon become continuous and effused, both *Stypella subgelatinosa* and the tropical *Protohydnum cartilagineum* A. Möller are resupinate with hydnoid hymenia[16], those of *Elmerina caryae* are resupinate to dimidiate with poroid to almost lamellate hymenia. *Phlogiotis helvelloides* (DC: Fr.) G.W. Martin and *Pseudohydnum gelatinosum* (Scop.: Fr.) P. Karst. both develop distinctly pileate basidiocarps; however, the hymenia of *P. gelatinosum* are hydnoid and those of *P. helvelloides* are smooth to faintly wrinkled. The basidiocarps of the species of *Tremellodendropsis*, which arise from the soil, are generally tough gelatinous to coriaceous and simple to caespitose or clavarioid (Crawford 1954; Petersen 1985, 1987). Throughout the family the basidiocarps are generally waxy to gelatinous and the hymenia revive poorly after drying. A number of species of the Hyaloriaceae develop sterile hymenial elements other than dikaryophyses that have been variously termed cystidia, gloeocystidia, etc.

The basidia of the species of *Tremellodendropsis*, a genus often assigned to a separate family, Tremellodendropsidaceae, arise as clavate to spathulate structures. In some species, such as *T. pusio* Berk. and *T. transpusio* D.A. Crawford, the terminal inflated portion becomes completely segmented following vacuolation of the stalk. In *T. flagelliformis* (Berk.) D.A. Crawford and *T. tuberosa* only partial septation occurs in the distal portion of the hypobasidium (Crawford 1954; Petersen 1985, 1987; Wells 1994). Epibasidia are relatively short and subulate in all species. The basidia of the species of *Tremellodendropsis* seem to be sphaeropedunculate basidia, although some species are only partially segmented. While the tough, clavarioid basidiocarps arising from the soil are unique among the Hyaloriaceae, the basidia and basidiocarp texture are reminiscent of the basidia and the resupinate, toothed basidiocarps of *Protohydnum cartilagineum*. Because the basidia of *Tremellodendropsis* are consistently sphaeropedunculate, the genus is placed in the Hyaloriaceae.

Further, the stalked basidia and finely branched dikaryophyses of *Bourdotia galzinii* (Bres.) Torrend indicate a close relationship with *Myxarium mesomorphum* (Bourdot et Galzin) Hauerslev and *Bourdotia galzinii* f. *microcystidiate* Hauerslev. These three taxa form a coherent unit and are best included in the Hyaloriaceae.

The basidia found in most species of the Sebacinaceae are subglobose, ovoid to pyriform, and lack basal clamps. The subhymenial hyphae are also without clamp connections but with slightly thickened, often brownish walls (Wells and Oberwinkler 1982). Cytological studies of *Sebacina epigaea* (Berk. et Broome) Rea and *Tremellodendron candidum* (Schwein.: Fr.) G.F. Atk. show the characteristic pattern of basidial ontogeny of the Auriculariales (Whelden 1935b, 1937). Simple to branching dikaryophyses are present in the hymenia of the known species. The basidiocarps of *Sebacina epigaea* are initially soft-gelatinous and pustulate but become continuous with age, those of *S. incrustans* (Pers.: Fr.) Tul. are coriaceous and occur on the ground or encrusting debris or bases of trees and shrubs, those of *Tremellodendron candidum* are coriaceous and arise from the soil to form upright, branching basidiocarps. The basidiocarps of *Tremelloscypha gelatinosa* (Murrill) Oberw. et K. Wells arise from decaying wood and are infundibuliform with a poorly defined stipe and a decurrent, inferior hymenium (Wells and Oberwinkler 1982). *Tremellostereum dichroum* (Lloyd) Ryvarden is very similar to *T. gelatinosa*, differing only in being somewhat dimidiate[17].

2. Habitat and Life Cycle

All of the species of the Auriculariales that have been examined are bifactorial, except *Exidiopsis opalea* (Bourdot et Galzin) D.A. Reid, which is unifactorial, and *Basidiodendron eyrei*, which is homothallic (Wells 1994). Secondary ballistospore

---

[16] The teeth in these two species do not seem to respond to gravity as do most hydnaceous homobasidiomycete species.

[17] Since there are only a few known specimens of *T. gelatinosa* and one of *T. dichroum* differing only somewhat in macromorphology, it is quite conceivable that only one species is involved. The known collections are from southern Florida, Jamaica, the Bahamas, and Mexico.

development has been demonstrated in the majority of species (Bourdot and Galzin 1928; Martin 1952; Oberwinkler 1963; Roberts 1993b; Wojewoda 1977). Microconidia are known throughout the order (Barnett 1937; Brefeld 1888; Ingold 1982a,b, 1984a,b, 1985, 1992, 1995). While most of the microconidia are lunate in form, those of *Exidiopsis calospora* are subglobose and those of *Exidia crenata* (= *E. recisa*) are bacilliform.

Although the majority of the Auriculariales have been collected from decaying wood, there are notable exceptions. Specimens of *Tremellodendron* have been collected only from the soil. Whether such taxa are saprobic or mycorrhizal is not clear. While species of *Sebacina* often develop basidiocarps on living or dead wood, leaf debris, etc., the mycelia seem to be closely associated with the adjacent soil. *Serendipita vermifera* has been isolated from the roots of a terrestrial orchid, suggesting a mycorrizal association (Warcup and Talbot 1967).

Several species of the Exidiaceae were tested by Worrall et al. (1997) for their wood-decaying properties. With the exception of *Basidiodendron cinereum* (Bres.) Luck-Allen, all species caused a white rot and a relatively low weight loss resulting in anatomical changes in the wood similar to those induced by some ascomycetes. Wood decay was stimulated by the addition of nutrients, suggesting a primitive wood-decaying capacity as compared to the higher basidiomycetes and the Dacrymycetales. In contrast, *Auricularia auricula-judae* proved to be an efficient wood-decaying species, causing a white rot and resulting in anatomical changes of the substrate similar to those caused by some of the Aphyllophorales.

3. Taxonomic Synopsis

Auriculariales J. Schröt., *apud* Cohn, Kryptog.- Flora Schlesien 3(1):382. 1887.
= Exidiales R.T. Moore, *apud* Sneh, E., et al., *Rhizoctonia* sp.: tax., mol. biol., ecol., path., p. 20. 1996[18].
Auriculariaceae Fr., Epicrisis Syst. Mycol., p. 530. 1838[19].

Auricularia Bull.: Mérat, Nouv. Fl. Env. Paris, ed 2., 1:33. 1821.
= *Hirneola* Fr., Kongl. Vetensk. Acad. Handl. 1848:144. 1848, *nom. cons.*
Mylittopsis Pat., J. Bot. (Morot) 9:247. 1895.
Paraphelaria Corner, Persoonia 4:346. 1966[20].
Exidiaceae R.T. Moore, Mycologia 70:1016. 1978.
Basidiodendron Rick, Brotéria Sér. Trimest. Ciênc. Nat. 7:74. 1938.
= *Metabourdotia* L.S. Olive, Amer. J. Bot. 44:429. 1957.
Craterocolla Bref., Unters. Gesamtgeb. Mykol. 7:98. 1888.
Ductifera Lloyd, Mycol. Writings 5:711. 1917.
Efibulobasidium K. Wells, Mycologia 67:148. 1975.
Eichleriella Bres., Ann. Mycol. 1:115. 1903.
Endoperplexa P. Roberts, Mycol. Res. 97:471. 1993.
Exidia Fr., Syst. Mycol. 2:220. 1822.
Exidiopsis (Johan-Olsen *apud* Bref.) A. Möller, Bot. Mitth. Tropen 8:167. 1895.
= *Ceratosebacina* P. Roberts, Mycol. Res. 97:470. 1993.
= *Microsebacina* P. Roberts, Mycol. Res. 97:473. 1993.
Fibulosebacea K. Wells et Raitv., Trans. Brit. Mycol. Soc. 89:344. 1987.
Heterochaete Pat., *apud* Pat. et Lagerh., Bull. Soc. Mycol. France. 8:120. 1892.
Pseudostypella McNabb, New Zealand J. Bot. 7:259. 1969.
Renatobasidium Hauerslev, Mycotaxon 49:228. 1993.
Serendipita P. Roberts, Mycol. Res. 97:474. 1993.
Hyaloriaceae A. Möller, Bot. Mitth. Tropen 8:173. 1895.
= Myxariaceae Jülich, Higher Taxa Basid., p. 380. 1981.

---

[18] Moore (1996) evidently restricted the Auriculariales to *Auricularia* and *Myllittopsis* and included the order in the "Auriculariomycetes" along with the Uredinales and *Eocronartium*, i.e., with taxa developing simple pores, plate-like spindle pole bodies, and distinctive 18S rRNA gene sequences (Swann and Taylor 1995a,b).

[19] It is conceivable that *Auriculoscypha* D.A. Reid et Manim. is a member of the Auriculariaceae; however, since the nature of the septal pore apparatuses and spindle pole bodies are unknown for the type species, *A. anacardiicola* D.A. Reid et Manim., the affinities of the genus with *Auricularia* and *Mylittopsis* are not established. The parasitic habitat on scale insects, basidial morphology, and evident dimorphic life cycle suggest a possible relationship with *Septobasidium* and allied genera (D. Reid and Manimohan 1985; Lalitha and Leelavathy 1990; Lalitha et al. 1994; see Swann et al., Chap. 2, this Vol.).

[20] This genus is mentioned here simply to draw attention to it. Ultrastructural or DNA sequencing data will be required to properly classify the single included species, *Paraphelaria amboinensis* (Lév.) Corner. Jülich (1982) has described a second genus, *Aphelariopsis* Jülich, that is probably closely related.

= Tremellodendropsidaceae Jülich, Higher Taxa Basid., p. 392. 1981.
Bourdotia (Bres). Trotter, *apud* Sacc., Syll. Fung. 23:571. 1925.
Elmerina Bres., Ann. Mycol. 10:507. 1912.
= *Aporpium* Bondartzev et Singer, *apud* Singer, Mycologia 36:67. 1944.
= (?) *Protodaedalea* Imazeki, Rev. Mycol. (Paris) 20:159. 1955.
Heterochaetella (Bourdot) Bourdot et Galzin, Hymén. France, p. 51. 1928.
Heteroscypha Oberw. et Agerer, *apud* Agerer et Oberw., Beih. Sydowia 8:31. 1979.
Hyaloria A Möller, Bot. Mitth. Tropen 8:173. 1895.
Myxarium Wallr., *apud* Bluff et Fingerh., Compend. Fl. German., Sect. II, 4:260. 1833.
Phlogiotis Quél., Enchiridion Fung., p. 202. 1886.
Protohydnum A Möller, Bot. Mitth. Tropen 8:173. 1895.
Protomerulius A Möller, Bot. Mitth. Tropen 8:172. 1895.
Pseudohydnum (Scop.: Fr.) P. Karst., Not. Sällsk. Fauna Fl. Fenn. Forh. 9:374. 1868.
Stypella A Möller, Bot. Mitth. Tropen 8:166. 1895.
= *Protodontia* Höhn., Sitzungsber. Kaiserl. Akad. Wiss., Math.-Naturwiss. Kl., Abt. 1, 116:83 1907.
Tremellodendropsis (Corner) D.A. Crawford, Trans. Royal. Soc. New Zeal. 82:618. 1954.
Sebacinaceae Oberw. et K. Wells, *apud* K. Wells et Oberw., Mycologia 74:329. 1982.
Sebacina Tul., J. Linn. Soc., Bot. 13:36. 1871.
Tremellodendron G.F. Atk., J. Mycol. 8:106. 1902.
Tremelloscypha D.A. Reid, Beih. Sydowia 8:332. 1979.
Tremellostereum Ryvarden, Mycotaxon 27:321. 1986.
Patouillardinaceae Jülich, Higher Taxa Basid., p. 383. 1981.
Patouillardina Bres., *apud* Rick, Broteria 5:7. 1906.
Taxa of Uncertain Affinities
Heteroacanthella Oberw., Trans. Mycol. Soc. Japan 31:208. 1990[21].

Monosporonella Oberw. et Ryvarden, Mycol. Res. 95:377. 1991.

III. Tremellomycetidae

A. Characteristics and Major Taxa

A distinctive group of characters distinguishes this subclass; however, these traits are not consistently present among all taxa. All of the species of the subclass that have been examined have a biglobular spindle pole body (Fig. 2). The septal pore apparatus consists of the septal swelling and the caps, if present, are sacculate (Figs. 9, 10). Species that have been cultured are dimorphic, with the possible exception of *Tetragoniomyces uliginosus* (Karst.) Oberw. et Bandoni, and many appear to be mycoparasitic. Both phragmobasidia and holobasidia are present. With few exceptions, e.g., *Phragmoxenidium mycophilum* Oberw. et Schneller, which has auricularioid basidia, the adjoining walls of the segments of most phragmobasidiate species of the order are oblique to longitudinal. The inclusion of *P. mycophilum* in this subclass is, at best, an uncertain disposition. The basidia of the Christianseniales and Filobasidiales are aseptate or partially septate distally. Analyses of 18S and 25S rRNA (Fell et al. 1999; Swann and Taylor 1995a,b) support the view that the Tremellomycetidae is monophyletic; however, relationships within the subclass are quite problematical.

B. Possible Intra- and Interrelationships

The presence of biglobular spindle pole bodies (Fig. 2), phragmobasidia (Fig. 16), complex pores (Figs. 9, 10), and gelatinous basidiocarps suggests a relationship with the Ceratobasidiales, Tulasnellales, Dacrymycetales, Auriculariales, and homobasidiomycetes. The findings that most species examined have modified, bifactorial compatibility and a dimorphic life cycle hint at a relationship with the Ustilaginales, but a number of other features are not correlated. Relationships among the families and orders included here are also obscure. Bandoni (1987, 1995) suggested that restricting the Tremellales to the phragmobasidiate taxa, i.e., Tremellaceae, Sirobasidiaceae, Phragmoxenidiaceae, and Tetragoniomycetaceae, would

---

[21] Roberts (1998c) has proposed a new family in the Auriculariales to accommodate this genus; however, we reserve judgment until such time as additional data are available, especially a determination of nuclear behavior within the "basidium" of the sole species.

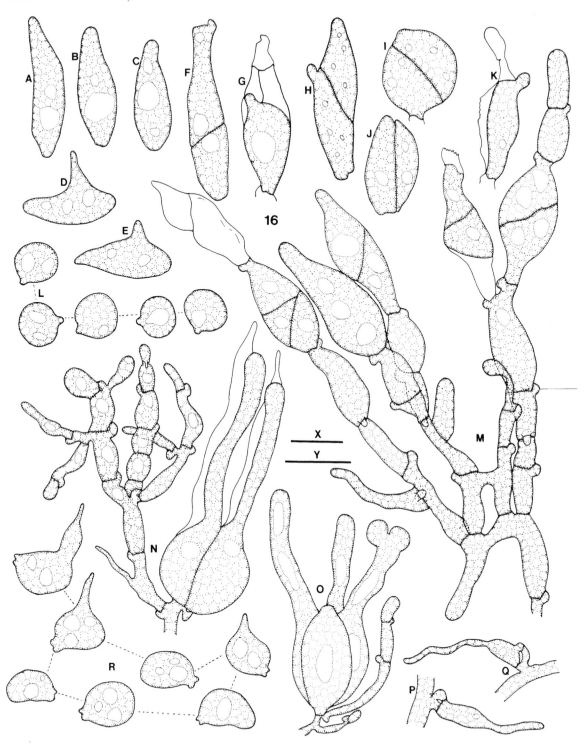

Fig. 16A–R. Tremellales. **A–M** *Sirobasidium magnum*. **A–E** Epibasidia, **D** and **E** forming sterigmata. **F–K** Segmented hypobasidia. **L** Basidiospores. **M** Branching segment of fertile hyphae terminating in single or chains of hypobasidia. **N–R** *Tremella mesenterica*. **N** Segment of fertile hypha bearing segmented basidium (*right*) and dikaryotic conidia (*left*). **O** Segmented basidium with developing epibasidia and subbasidial proliferations. **P, Q** Tremelloid haustoria. **R** Basidiospores, three germinating by repetition. *Bars* 10 μm; *bar X* applies to **N, O, R**; *bar Y* applies to **A–M, P, Q**. **A–M** R.J. Bandoni, 8966; **N, P, Q** N. Strid, 17-VIII-1982 (neotype); **O, R** F. Oberwinkler, 4-I-1983. **A–M** redrawn from original drawings used by Wells (1994); **N, Q, R** redrawn from original drawings used by Wong et al. (1985) and are reproduced here with permission of Mycologia

result in a more coherent order. The remaining families, i.e., Filobasidiaceae, Carcinomycetaceae, and Rhynchogastremaceae, are composed of species bearing cylindrical to urniform holobasidia, or partially, distally septate basidia. Basidiocarps are essentially absent in the Filobasidiaceae, Cystofilobasidiaceae, and Rhynchogastremaceae; the basidiospores are passively released, and most of the species have a haploid, unicellular budding stage. Following Bandoni (1987, 1995), these would be included in the Filobasidiales. Others (Metzler et al. 1989; Wells 1994), however, have proposed including the holobasidiate taxa in the Tremellales as there is some morphological evidence that the holobasidiate taxa may be polyphyletic.

We suggest, at this time, that three orders be recognized in this subclass, i.e., Tremellales, Christianseniales, and Filobasidiales. Such a system segregates, with some exceptions, taxa with segmented basidia and forcibly discharged basidiospores in the Tremellales (Bandoni 1987). The Christianseniales contain those species bearing suburniform to narrowly clavate basidia with or without partial, apical septation. Basidiospores in the Christianseniales arise on distinct sterigmata and are probably passively released, although some species develop asymmetrically attached basidiospores, suggesting forcible discharge. Many of the species develop hymenia on the surfaces of galls of host tissue. The known species of the Filobasidiales have cylindrical basidia with sessile, passively released basidiospores; basidiocarps are unknown in this order. The apices of the basidia of *Filobasidiella lutea* P. Roberts, an intrahymenial parasite, appear partially septate, as are those of some species of the Christianseniales (Roberts 1997).

While the recognition of three orders in this subclass is certainly not ideal, basidial morphology within each of the three orders is more or less uniform. The Christianseniales are probably monophyletic; however, the relationship to the Filobasidiales and Tremellales is certainly not clear. Further, sequencing data (Swann and Taylor 1995a,b; Fell et al. 1999) hint at the possibility that the taxa of the Tremellomycetidae, as interpreted here, do not represent phylogenetic units and that the genus *Tremella*, as currently defined, may have given rise to species of several separate taxa. The Tremellomycetidae also seem to be a monophyletic group but their relationship to other taxa, especially to some smut-like fungi, is still debatable.

C. Tremellales

1. Taxonomic Features

a) Ultrastructural Features

Complex septal pores with sacculate caps have been demonstrated in the species of the Tremellaceae (Figs. 9, 10), Sirobasidiaceae, and Tetragoniomycetaceae examined. Complex pores lacking sacculate caps have been illustrated for those species of Syzygosporaceae and Phragmoxenidiaceae studied (Wells 1994). The saccules are finger- or cup-like projections of modified endoplasmic reticulum arrayed in a hemisphere around the septal swelling (Figs. 9, 10; Khan 1976; Berbee and Wells 1988; Chen 1998). The opening of the saccule is directed away from the pore. The outer limiting membrane of the saccule is continuous with a membrane of the endoplasmic membrane. This outer membrane is associated with an inner electron-dense layer that is similar in appearance to the inner dense layer in the caps of septal pore apparatuses of the species of the Heterobasidiomycetidae possessing continuous or porous caps (Berbee and Wells 1988). *Tremella globospora* D.A. Reid has been shown to possess a biglobular spindle pole body, as in the other species of the Heterobasidiomycetes studied (Fig. 2; Berbee and Wells 1988). Tremelloid haustoria are produced by a number of species throughout the subclass (Bandoni 1984, 1987, 1995; Wells 1994). These structures generally arise as short lateral hyphal branches with a basal clamp connection. The branch consists of a single, binucleate cell, subglobose to pyriform or short clavate, from which one or more narrow (i.e., <1.0μm in diameter) simple to branched filament arises (Fig. 16P,Q; Zugmaier et al. 1994; Chen 1998). It has been shown in several species, including *T. mesenterica* Retz.: Fr., that the apex of the haustorium penetrates the hyphal wall of the presumed host. This is followed by the development of a minute pore connecting the cytoplasm of the host to that of the haustorium, with continuity between the plasma membranes of the host and parasite. In mature areas of contact between the apex of the haustorium and the host the connection is broken by secondary wall formation by the host cytoplasm (Zugmaier et al. 1994; Chen 1998).

b) Basidial Ontogeny and Basidiocarp Structure

Basidial morphology and basidiocarp structure in the Tremellales, as here defined, are quite varied (Bandoni 1987; Wells 1994). In the Tremellaceae the basidia arise singly or in clusters, generally by proliferations through or near subbasidial clamp connections (Fig. 16N,O). Following meiosis, four segments generally form with the walls predominately longitudinal, but also oblique or even transverse. Each hypobasidial segment bears a distinctive epibasidium at the apex of which a sterigma and an asymmetrical basidiospore develops (Whelden 1934). Ontogeny of the hypobasidial segments of a basidium is often not synchronized (Fig. 16N,O). Hymenia consist of dikaryotic conidia (Fig. 16N), basidia, or a mixture of these. In a few species sterile hymenial elements, i.e., dikaryophyses, are present. Forcible discharge of the basidiospores probably occurs in most taxa except in the species of *Xenolachne* (Rogers 1947; Hauerslev 1976); basidiospores are unknown in *Tetragoniomyces* (Oberwinkler and Bandoni 1981). Basidiospore germination is by budding, development of secondary ballistospores (Fig. 16R), or by conjugation tubes if compatible spores are closely situated (Bandoni, 1987).

The endoparasitic species of *Tremella*, e.g., *T. mycophaga* var. *obscura* L.S. Olive, develop basidia within the hymenia of their host or consist of a hymenium formed on the host tissue, as in *T. polyporina* D.A. Reid (Bandoni 1987). The majority of species, however, develop distinct basidiocarps that vary from foliaceous, globose, cerebriform, to irregular in form. The basidiocarps of *T. globospora* are initiated within the perithecia of the host, mainly species of *Valsa* and *Diaporthe*, disrupting the perithecia during enlargement to develop erumpent, tuberculate to pulvinate basidiocarps that may become confluent and cerebriform (Brough 1974). Several large species, e.g., *T. aurantia* Schwein., have a fleshy core composed of the hyphae of the corticiaceous host surrounded mainly by the hymenium and associated hyphae of the parasite. The basidiocarps of *Holtermannia* are tough-gelatinous with simple to branching clavarioid lobes. The basidia, spores, and hymenia are similar to those of *Tremella* (Kobayasi 1937).

Chen's (1998) recent study of a number of species of *Tremella* provides considerable information on comparative morphology and molecular systematics of this large genus. The species of *Holtermannia* have been reported from Japan and the tropical regions of Asia and Central and South America.

The minute, gelatinous, pustulate basidiocarps of the single known species of *Trimorphomyces*, *T. papilionaceus* Bandoni et Oberw., are associated with *Arthrinium* spp. and initially consist solely of conidia (Oberwinkler and Bandoni 1983). More mature basidiocarps bear both conidia and basidia. The basidia are similar to those of *Tremella*. The unusual, dikaryotic conidia arise in pairs at the apices of conidiogenous cells and subsequently fuse prior to release. Following release, these H-shaped, dikaryotic conidia develop additional dikaryotic cells of similar morphology by much the same mechanism as those formed from the initial conidiogenous cells; a dikaryotic yeast state is thus established. The basidiospores germinate either by repetition or by budding to develop a haploid, yeast phase (Oberwinker and Bandoni 1983).

The characteristics of the species of *Sirotrema*, especially those of basidial ontogeny, are intermediate between *Tremella* and *Sirobasidium* (Bandoni 1986). The basidia are generally single but occasionally in chains of two to three, as in the species of *Sirobasidium*. The epibasidia bear both blastic spores or ballistosporic basidiospores, but the epibasidia remain attached as in *Tremella*. The basidiocarps of *Sirotrema* vary from pustulate to subglobose and arise from hysterothecia of hypodermataceous ascomycetes. The known species are dimorphic and bear tremelloid haustoria, often with two filaments (Bandoni 1986).

The basidia of *Bulleromyces albus* Boekhout et A. Fonseca, the sole representative of the genus, have only been observed in compatible pairings of the anamorphic stage, i.e., *Bullera alba* Derx. The basidia develop on clamped, dikaryotic hyphae along with tremelloid haustoria and give rise to epibasidia bearing ballistospores, chains of blastospores, or hyphae. Sacculate caps are associate with the complex septal pores (Boekhout et al. 1991). Whether or not basidiocarps are formed in the natural habitat is unknown.

The basidiocarps of the known species of *Xenolachne* appear as white tomenta on the apothecia of small discomycetes (Rogers 1947; Hauerslev 1976). The two-celled basidia develop in poorly defined hymenia and bear symmetrically attached basidiospores on long, tapering epibasidia/sterigmata. The basidiospores are evidently passively released (Rogers 1947; Hauerslev 1976). Rogers (1947) reports secondary spore formation; however, the mechanism of release of the secondary spores has not been documented.

Tetragoniomyces uliginosus, the only known species of the Tetragoniomycetaceae, develops thick-walled, deciduous hypobasidia. The thick-walled hyphobasidia are not formed in a gelatinous matrix and percurrent proliferation of the fertile hyphae insures the detachment of the mature hypobasidia; therefore, it seems likely that they function as propagules. Conjugation occurs between the segments of the hypobasidium or between germination tubes arising from the segments (Koske 1972; Oberwinkler and Bandoni 1981). Thus, a unique feature of *T. uliginosus* is the elimination of the epibasidium, basidiospore, and, very possibly, the monokaryotic, yeast phase. The basidiocarp of *T. uliginosus* is a smooth to irregular, mucedinous layer over the sclerotium-like structure of the basidiomycetous host[22].

The basidia of the Sirobasidiaceae are distinctive in that the epibasidia become detached and evidently can function as dispersal units (Fig. 16A–M). In *Sirobasidium* the basidia are formed singly or in basipetal, catenulate series (Fig. 16M). The initial basidium differentiates from the terminal dikaryotic segment and subsequent basidia arise basipetally from the subterminal fertile hyphal segments (Fig. 16M). Following meiosis and segmentation, cylindrical to fusiform epibasidia[23] develop from the hypobasidial segments and eventually become detached. Again, ontogeny of the individual hypobasidial segments is often not synchronized. The detached epibasidia can give rise to the monokaryotic yeast phase or differentiate a tubular extension bearing a sterigma and ballistosporic ballistospore (Fig. 16D,E,L; Bandoni 1957, 1987; Ingold 1995). Basidiocarps of *Sirobasidium* vary from small, gelatinous pustules to large, somewhat irregular structures that can anastomose into large continuous patches. In *Fibulobasidium inconspicuum* Bandoni the basidia arise singly or in clusters by the conversion of the clamp cell into the basidium. The epibasidia are deciduous, as in *Sirobasidium*, and germination is by budding or by formation of sterigmata and ballistosporic spores (Bandoni 1979). Thus, the epibasidia in this family can function as agents of dispersal, as can the ballistosporic basidiospores. The basidiocarps of *Fibulobasidium inconspicuum* are small tubercles that often become confluent to develop a linear mass between the bark and wood of recently dead branches and trunks. Large basidiocarps appear applanate to almost effused (Bandoni 1979).

Oberwinkler et al. (1990) tentatively included the Phragmoxenidiaceae, based on *Phragmoxenidium mycophilum*, in the Tremellales. *Phragmoxenidium mycophilum* is an endoparasite of *Uthatobasidium fusisporum* (J. Schöt.) Donk. Unlike most known species of the Tremellales, the phragmobasidia are transversely septate, i.e., auricularioid. Also, since it was not possible to demonstrate a haploid, budding phase, a sacculate septal pore cap, and tremelloid haustoria, the inclusion of this family in the Tremellales is quite provisional. Distinctive basidiocarps are not developed since the basidia of *P. mycophilum* develop within the hymenium of the host (Oberwinkler et al. 1990).

2. Habitat and Life Cycle

Although definitive evidence is lacking for many or most species, growth only on sporocarps or thalli of other fungi or lichens and the common occurrence of tremelloid haustoria (Fig. 16P,Q) on assimilative or basidiocarp hyphae of many species suggests that mycoparasitism is a common feature of the Tremellomycetidae (Bandoni 1995; Diederich 1996).

Brefeld (1888) noted that the basidiospores of *Tremella mesenterica*, and several other species of *Tremella* were capable of germinating by budding resulting in yeastlike colonies. Bandoni (1963) established that such colonies derived from a single basidiospore remained yeastlike; however, when the isolates from a single basidiocarp were paired in all possible combinations, approximately 25% of the resulting colonies gave rise to the dikaryotic stage. He postulated that compatibility in this species was controlled by two alleles at the A locus and multiple alleles at the B locus. Wong and Wells (1985) and Wong et al. (1985) extended these studies of *Tremella mesenterica* to include 46 collections from British Columbia,

---

[22] The septal pore apparatus of the host is characteristic of those studied in the Ceratobasidiales.
[23] See Bandoni (1987) for an alternative interpretation and terminology of basidial ontogeny in the Sirobasidiaceae.

northern California, Sweden, Germany, and Switzerland. All isolates were compatible, as were isolates referable to *T. mesenterica* or *T. lutescens* Pers.: Fr.

Studies of other species in the Tremellaceae, Sirobasidiaceae, and Filobasidiaceae demonstrated that most species are heterothallic and dimorphic, i.e., with a yeastlike haploid phase that persists until appropriate strains are combined to yield the mycelial dikaryotic phase (Wells 1994; Bandoni 1995). Pairings involving several collections were made of *Fibulobasidium inconspicuum* (Bandoni 1979), *Sirobasidium magnum* Boedijn (Flegel 1976), *Tremella fuciformis* Berk. (Fox and Wong 1990), *T. globospora* (Brough 1974; Hanson and Wells 1991), *T. indecorata* Sommerf.: Fr. (Hanson and Wells 1991), and *T. moriformis* (Sm.: Fr.) Berk. (Hanson and Wells 1991). Thus, this type of compatibility, often termed modified bifactorial, is certainly the dominant system in the order. Some isolates of *T. fuciformis* are homothallic, whereas others are heterothallic with a modified bifactorial compatibility system (Fox and Wong 1990). The basidiospores of *Sirobasidium magnum* may be functionally homothallic when isolated from young basidiocarps with a predominance of two-celled basidia (Bandoni and Boekhout 1998).

Control of pairing is bifactorial in the heterothallic species listed above and involves a single allelic pair (A_1, A_2) at one locus and multiple alleles (B_1, B_2, B_3, etc.) at a separate locus on another chromosome (Bandoni 1963; Wong and Wells 1985). Sixty-eight B factors, with a B factor diversity of 74%, were identified in *T. mesenterica* (Wong and Wells 1985), 17 (77% diversity) in *T. indecorata*, 41 (44% diversity) in *T. globospora*, and 19 (15% diversity) in *T. moriformis*. The significantly low B factor diversity in *T. moriformis* is probably due to the fact that all the collections of this species were made in Davis, California, from ornamental hosts, and none was found in surrounding areas (Hanson and Wells 1991).

When haploid compatible strains are mixed, mating is initiated by complementary pheromones (tremerogens) produced by the A_1 and A_2 strains (Bandoni 1965). Budding of the haploid cells ceases, conjugation tubes develop, and conjugation occurs. If the two strains have like B factors, e.g., $B_1 \times B_1$ or $B_2 \times B_2$, conjugation results in the development of an abortive, distorted vesicle with no further development. The dikaryotic mycelium results only from the pairing of strains with unlike A and B factors. When strains with like A factors and unlike B factors are paired, budding continues and there is no apparent interaction (Bandoni 1965; Wong and Wells 1985).

Hormonal initiation of conjugation was initially demonstrated in *Tremella mesenterica* (Bandoni 1965). Similar systems were reported for *T. globospora* by Brough (1970) and for *Sirobasidium magnum* by Flegel (1981). The *T. mesenterica* pheromones were partially characterized by I. Reid (1974), who suggested that they were peptides. The structure of Tremerogen A-10, produced by the "A" strain, was determined as a dodecapeptide (Sakagami et al. 1981). Tremerogen A-13, produced by the "b" strain was judged to be a tridecapeptide (Sakagami et al. 1981; Yoshida et al. 1981). The pheromones of *Tremella brasiliensis* (A. Möller) Lloyd were reported to be different peptides but with similar structural features and similar functions (Ishibashi et al. 1984).

D. Christianseniales

Several significant studies of this group have been made in recent years, resulting in a substantial improvement in our understanding of the Christianseniales. Martin (1937) described *Syzygospora alba* G.W. Martin based on collections from Panama. The basidiocarps are tremelloid in appearance and soft-gelatinous in texture (Oberwinkler and Lowy 1981). Martin had mistaken the unusual conidia as basidia; however, Kao (1956) made a careful cytological study of Martin's specimens and determined that the structures Martin had described as basidia were, in reality, conidia, and that holobasidia were present. She illustrated karyogamy and meiosis in the basidia. Subsequently, Oberwinkler and Lowy (1981) restudied a paratype and additional specimens from Mexico and determined that the basidia were partially septate in the distal region, the basidiospores were asymmetrically attached, tremelloid haustoria were present, the septal pores were complex but without caps, and that *S. alba* was probably a parasite of an undetermined basidiomycete. The conidia develop from a dikaryotic conidiogenous cell in which the dikaryon is delimited by a septum. Each resulting cell forms a lateral outgrowth into which a nucleus from each mother cell migrates following mitosis. The outgrowths fuse to give rise to a dikaryotic zygoconidium that is subsequently released from the conidiogenous branch. The dikaryotic conidia can give rise to haploid cells and hyphae with tremelloid haustoria, but evidently do not develop additional zygoconidia as do those of *Trimorphomyces papilionaceus* (Kao 1956; Oberwinkler and Lowy 1981; Oberwinkler and Bandoni 1982).

Oberwinkler and Bandoni (1982) made an extensive study of the available specimens and proposed the family Carcinomycetaceae to included the genera *Carcinomyces*, *Christiansenia*, and *Syzygospora*. They found that basidiospore discharge was passive in some species and, possibly, forcible in other. Some species develop small, pustulate to large, gyrose basidiocarps on corticoid basidiomycetes. Others induce galls on the basidiocarps of *Collybia* and *Marasmius*. Oberwinkler et al. (1984) made a detailed light and electron microscopic study of living material of *Christiansenia pallida* Hauerslev, which is parasite of *Phanerochaete cremea* (Bres.) Parmasto, and reported that the tremelloid haustoria penetrate the host cells. The zygoconidia seemed to be the reproductive means in the parasitic stage and the basidia developed in the more mature specimens. The zygoconidia are capable of producing a budding, haploid stage.

Ginns' (1986) account of the group recognized nine species, three of which were described as new. Ginns placed all species in the genus *Syzygospora*, reducing *Carcinomyces*, *Christiansenia*, and *Heterocephalacria* Berthier to subgenera. Later Rath (1991) presented a description of the species in Italy and maintained that *Syzygospora* had been applied to the anamorphic stage and that the type did not contain basidia and basidiospores; therefore, he proposed that the correct name for the teleomorph is *Christiansenia*. While there is some question as to the presence of basidia and spores in the holotype, Martin (1937) did not include a description of these structures in the protologue; therefore, *Christiansenia* is the appropriate name for the genus and *Syzygospora* can only be applied to the imperfect stage. However, Carcinomycetaceae was validly published, being based on a teleomorphic genus, and takes precedence over Christianseniaceae.

However confused the nomenclature of the family and genera, this group seems to be monophyletic and is almost certainly related to the Tremellales. Because the known taxa form a natural group and are distinct, they are placed in a separate order, which removes at least some of the heterogeneous elements from the Tremellales.

E. Filobasidiales

(See Fell et al., Chap. 1, this Vol.)

In the Filobasidiaceae saccules appear to absent in the genera *Mrakia*, *Cystofilobasidium*, and *Filobasidiella* but are present in *Filobasidium* (Wells 1994). Homothallism has been reported for *Cystofilobasidium capitatum* (Fell et al.) Oberw. et al.

Kwon-Chung (1987) included in the Filobasidiaceae those taxa with a yeastlike, haploid phase and a mycelial, dikaryotic phase bearing slender, scattered holobasidia with sessile basidiospores. Basidiocarps are evidently not formed, and the septal pores are complex with or without sacculate caps (Wells 1994). *Filobasidiella*, which includes the human pathogen *F. neoformans* Kwon-Chung, develops long, slender holobasidia with an inflated apex from which long chains of sessile spores are developed by basipetal budding. The complex pores of *F. neoformans* also lack saccules. The species of *Filobasidium* produce narrow holobasidia bearing a whorl of sessile basidiospores that are passively released. The dikaryotic hyphae bear clamps, as do the dikaryotic hyphae of *F. neoformans*; however, the complex pores possess saccules. The available sequencing data (Fell et al. 1995; Kwon-Chung et al. 1995; Swann and Taylor 1995a,b) suggest that the Filobasidiaceae is not a monophyletic taxon. In fact, the phylogenetic tree by Fell et al. (1999) suggests that *Filobasidiella neoformans* is not at all closely related to *Filobasidium*.

Cystofilobasidium consists of species developing thick-walled "teliospores" that germinate to form a long, narrow holobasidium bearing terminal, sessile basidiospores. Haustoria are evidently lacking, and the complex pores lack saccules. Fell et al. (1999) erected the order Cystofilobasidiales to include those basidiomycetous yeasts with holobasidia with teliospores and sessile basidiospores. They presented 18S rDNA sequencing data as well as biochemical and physiological characters supporting their concept. Accepting a family Cystofilobasidiaceae, rather than the order, to include these genera is more in keeping with the taxonomy adopted here; however, further studies of the subclass may warrant a change.

The single known species, *Rhynchogastrema coronata* B. Metzler et Oberw., of the Rhynchogastremaceae was isolated from agricultural soil in Germany; therefore, details of its habitat and basidiocarp structure are unknown (Metzler et al. 1989).

The clamped, dikaryotic hyphae bear tremelloid haustoria and the complex septal pores are sacculate. The basidia are distinctly urniform and bear distally four sessile basidiospores with thick, punctate walls which become surrounded by a yellowish exudate. The spores are capable of germinating to form a haploid, yeast phase (Metzler et al. 1989). There are some features suggestive of the Christianseniales, e.g., the partially septate, urniform basidia, but the sessile basidiospores are similar to those of the Filobasidiales. As noted above, distally, partially septate basidia have recently been reported in *Filobasidiella lutea* (Roberts 1997).

F. Taxonomic Synopsis

Tremellales Rea, British Basid., p. 729. 1922.
Tremellaceae Fr., Syst. Mycol. 2:207. 1822.
Bulleromyces Boekhout et A. Fonseca, Stud. Mycol. 33:90. 1991.
Holtermannia Sacc. et Traverso, *apud* Sacc., Syll. Fung. 19:871. 1910.
Sirotrema Bandoni, Can. J. Bot. 64:668. 1986.
Tremella Pers., Neues Mag. Bot. 1:111. 1794, *nom. cons.*
Trimorphomyces Bandoni et Oberw., Syst. Appl. Microbiol. 4:106. 1983.
Xenolachne D.P. Rogers, Mycologia 39:561. 1947.
Sirobasidiaceae A Möller, Bot. Mitth. Tropen 8:165. 1895.
Fibulobasidium Bandoni, Can. J. Bot. 57:264. 1979.
Sirobasidium Lagerh. et Pat., J. Bot. (Morot) 6:468. 1892.
Phragmoxenidiaceae Oberw. et R. Bauer, Syst. Appl. Microbiol. 13:190. 1990.
Phragmoxenidium Oberw., Syst. Appl. Micobiol. 13:187. 1990.
Tetragoniomycetaceae Oberw. et Bandoni, Can. J. Bot. 59:1039. 1981.
Tetragoniomyces Oberw. et Bandoni, Can. J. Bot. 59:1034. 1981.
Christianseniales Rath, Atti Soc. Ital. Sci. Nat. Mus. Civico Storia Nat. Milano 132:17. 1991.
Carcinomycetaceae Oberw. et Bandoni, Nord. J. Bot. 2:502. 1982.
= Christianseniaceae Rath, Atti Soc. Ital. Sci. Nat. Mus. Civico Storia Nat. Milano 132:17. 1991.
Christiansenia Hauerslev, Friesia 9:43. 1969.
= *Carcinomyces* Oberw. et Bandoni, Nord. J. Bot. 2:507. 1982.
= *Heterocephalacria* Berthier, Mycotaxon 12:114. 1980.
Filobasidiales Jülich, Higher Taxa Basid., p. 347. 1981.
Cystofilobasidiaceae K. Wells et Bandoni fam. nov.[24]
Cystofilobasidium Oberw. et Bandoni, Syst. Appl. Microbiol. 4:116. 1983.
Mrakia Y. Yamada et Komag., J. Gen. Appl. Microbiol. 33:456. 1987.
Xanthophyllomyces Golubev, Yeast (ChichesTes) 11:105. 1995.
Filobasidiaceae L.S. Olive, J. Elisha Mitchell Sci. Soc. 84:261. 1968.
Filobasidiella Kwon-Chung, Mycologia 67:1198. 1975.
Filobasidium L.S. Olive, J. Elisha Mitchell Sci. Soc. 84:261. 1968.
Rhynchogastremaceae Oberw. et B. Metzler, Syst. Appl. Microbiol. 12:283. 1989.
Rhynchogastrema B. Metzler et Oberw., Syst. Appl. Microbiol. 12:281. 1989.

IV. Culture Techniques[25]

Brefeld (1888) was able to induce basidiospore germination and hyphal growth on natural media of several species of *Tremella*, *Exidia*, and *Dacrymyces*. He was especially interested in the development of secondary spores, as his definitions of taxa were based in part upon morphology of these spores. Subsequent comparative studies, especially by Möller (1895) and Ingold (e.g.,

---

[24] Heterobasidiomycetes sine carposomatibus, species plerumque zymosae, heterothallicae vel ipsifertiles; hyphae septatae, efibulatae; hypharum septa doliporis praedita parenthesomatibus absentibus sed nonumquam parenthesomatibus imperforatis; probasidia (teliosporae) plerumque formantur, saepe crassitunicata; holobasidia elongata, apice saepe inflato ad subgloboso vel lageniformi; basidiosporae sessiles, tenuitunicatae, hyalinae, blastosporis germinant. Genus Typicum: *Cystofilobasidium* Oberw. et Bandoni.

[25] The techniques discussed here apply primarily to the Heterobasidiomycetidae, Tremellales, and Christinaseniales.

1982a,b, 1983, 1984a,b, 1985, 1992, 1995) have demonstrated that the characteristics of basidiospore germination are reasonably constant and characteristic for many taxa. Since the time of Brefeld, many species have been grown on artificial media, but among the groups discussed here, we have not been able to culture most species of the Sebacinaceae, *Tremellodendropsis*, and many species of the Hyaloriaceae. Some species, such as *Efibulobasidium albescens* (Sacc. et Malbr.) K. Wells and *Exidiopsis calospora*, can be cultivated on artificial media; however, isolates of both species grow vigorously for a time but gradually lose vigor and cease further growth. *Efibulobasidium albescens* is unusual in that monospore cultures yield a *Chaetospermum* anamorph (unpubl. obser.). In reality, few published attempts have been made to culture many species of the Ceratobasidiales, Tulasnellales, Dacrymycetales, and Auriculariales. Most of those which have been cultured have not been studied intensively and our understanding of the behavior of most species on artificial media is rudimentary. In contrast, many of the Filobasidiales are known only from cultural studies.

Isolation of ballistosporic basidiospores is readily performed with fresh specimens by attaching all or a piece of the basidiocarp to the inner surface of a petri dish lid. The lid is then inverted over the agar medium. Air-dried collections, or portions thereof, can be revived in sterile, distilled water by soaking for 15–30min, depending on the size and the rate of rehydration. Specimens can then be rinsed if necessary, preferably with sterile water, blotted to remove excess water, and treated as above. Care must be taken to insure that loose particles of debris or insects on the specimens are brushed or washed away before the lid is inverted over agar. This alone does not insure freedom from contaminating fungi of many kinds, especially species of *Sporobolomyces*, *Tilletiopsis*, *Conidiobolus*, and other fungi with projected spores.

Specimens are usually attached to the petri dish lid using adhesive tape or similar products, or by placing a thick layer of Vaseline at the desired point and pressing the basidiocarp, or part of it, to the lid. For very small structures, however, better results often are obtained using $1 cm^2$ of water agar which, when placed on the inner surface of the lid, usually adheres firmly. A small portion of the fungus is then pressed gently to the agar surface and will adhere to it. With small basidiocarps, drying often occurs too rapidly, resulting in limited sporulation. Desiccation can be largely offset by using folded, sterile paper toweling or filter paper beneath the specimen (Brough 1974).

If monospore isolates are required, the basidiospores can be spread thinly over the agar surface by rotating the disk lid with the attached specimen after appropriate time intervals (Barnett 1937; Brough 1974). Alternatively, the plate can be tilted to about 60–70°. Spore deposits should always be examined under the low power of a compound microscope to insure that there is neither a mixture nor a deposit of a totally different fungus than the one desired. Sporocarps of *Tremella*, for example, often are associated with, or growing on, host fungi. The host may sporulate at the same time as the parasitic *Tremella* species. Species of sporobolomycetaceous fungi as well as those of discomycetes, pyrenomycetes, and others often are present on the basidiocarps or on the substrate.

Isolations from fresh basidiocarps are desirable; however, basidiocarps of some genera, e.g., *Exidia*, *Tremella*, and *Auricularia*, remain viable for several months if properly dried. Members of other genera remain viable for several weeks. The thin, resupinate basidiocarps of *Tulasnella*, *Exidiopsis*, and similar basidiocarps are generally more difficult to revive, especially those from tropical and subtropical regions. Some specimens, or parts of specimens, dried in vials of silica gel in the field remain viable for up to 2 years. The use of silica gel vials is of special value when collecting in tropical and subtropical areas, especially where laboratory facilities are not immediately available.

As in the isolation and growth of many other fungi, numerous media have been used successfully for culture studies of the jelly fungi. We have used a malt extract, yeast extract, and soytone agar (MYP) for both isolation and storage of many Heterobasidiomycetes. To retard bacterial growth in isolation plates, the addition of an antibiotic, e.g., 100mg of tetracycline l^{-1} of medium, is often advisable. MYP agar is also useful as a pairing substrate for the Heterobasidiomycetidae; however, a weaker, conjugation medium of glucose and soytone (CJM) generally gives better results for conjugation studies of the Tremellomycetidae (Bandoni et al. 1975). Pairing studies in *Tremella* and related taxa require freshly prepared media and a vigorous inoculum (Bandoni 1963; Flegel

1968; Brough 1974). Small inocula of the isolates are spotted on the agar surface and mixed with a loop. In studies of *Tremella mesenterica*, 24 crosses are made on a single plate and incubated for 5 days. The crosses are read by scraping off the excess yeast cells and examining with a dissecting microscope for dikaryotic hyphae denoting a fertile pairing with unlike alleles at both loci (A≠B≠), conjugation tubes indicating a common B pairing (A≠B=), or only yeast cells resulting from a common A (A=B≠) or common A and B (A=B=) pairings (Bandoni 1963; Wong and Wells 1985).

Compatibility studies in those species with hyphal monokaryotic stages involve removing plugs of the monokaryon and pairing these in all possible combinations. Since nuclear migration between the paired monokaryon did not occur in the isolates we have studied, it is necessary to examine the contact zone for true clamps, false clamps, or simple septa. As it is difficult to distinguish between false clamps developed in A≠B= pairings and true clamps resulting from A≠B≠ crosses, the technique described by Raper (1976) of removing a plug from the contract zone and incubating the plug on squares of dialysis tubing is favored. Margins of the resulting colonies are then examined with phase optics to determine the presence of the dikaryon in subterminal hyphal segments and the presence of a continuum between the hyphae and clamp cells (Wells and Wong 1985).

Development of basidiocarps in culture is most desirable for genetical analyses, life-cycle studies, or determining basidiocarp variation; however, successful efforts to induce basidiocarp production on artificial media have been limited. Barnett (1937), in one of the first compatibility studies of the Auriculariales, induced basidiocarp formation of *Auricularia auricula-judae* on autoclaved sticks of basswood and hickory. Basidiocarps of *Myxarium nucleatum* developed on a supplemented potato dextrose agar. The basidiocarps of some species of *Auricularia* and *Tremella* are produced for commercial purposes, using, generally, wood products and supplements (Chang et al. 1993). As noted above, Kennedy (1972), in her study of basidiocarp ontogeny and macromorphology variation in *Calocera cornea*, was able to induce basidiocarp development by adding blocks of conifer or poplar wood to Nobles malt agar. *Stilbotulasnella conidiophora* produced both conidia and basidia on common laboratory media (Bandoni and Oberwinkler 1982). Basidiocarps of a strain of *Myxarium nucleatum* developed in a medium of rolled oats supplemented with an extract of decaying wood (Wells 1964a), and fresh multispore isolation of some collections of *Exidia pithya* Fr. will produce clusters of basidia and incipient basidiocarps on MYP agar (pers. obser.). Among the Tremellales, Brough (1974) induced basidiocarp formation of *Tremella globospora* on MYP agar, and Wong and Wells (1985) found that adding a sterile tongue depressor to MYP agar produced occasional basidiocarps of *T. mesenterica*. Flegel (1976) noted basidia and basidiocarps during his studies of *Sirobasidium magnum*; freshly isolated multispore cultures of this species often fruit quickly on MYP agar. Basidiospore development has been induced in some *Stilbotulasnella* isolates by growing the mycelium on MYP agar until the plate is covered. One-cm^2 blocks of mycelium and agar are then removed and transferred to water agar plates with the mycelium side up. Basidia develop overnight at the margins of the block, their location detectable by the presence of basidiospores on the water agar surface adjacent to the block. This "step down" procedure has worked well with some other lower basidiomycetes (R.J.B., pers. obser.) At best, basidiocarps have been produced sporadically; however, their appearance often occurs only with some isolates of very few species. The capacity to sporulate in culture generally diminishes with time in culture, i.e., fresh isolations are more apt to give rise to basidia.

V. Conclusions

Substantial progress has been made in recent years in the understanding of the biology and systematics of the Heterobasidiomycetes. These advances have been based on data from established and new taxa and on macro- and microscopic observations involving phase and electron microscopy. Ultrastructural studies of septal pores and spindle apparatuses have been especially helpful, as have culture studies and DNA sequencing studies. Many new taxa have been described, especially in the Tremellomycetidae, greatly extending the variety of habitats, life cycles, and morphologies recognized. Analyses of the accumulated data have made possible a better understanding of the intra- and interrelationships of the Heterobasidiomycetes. It is now clear that many genera included by Martin (1952) in the Auriculariaceae and Phleogenaceae are almost certainly more closely related to the rusts (i.e., Uredinales and related forms) than to the taxa included here in the Heterobasidiomycetes. Further, the Tremellaceae, as conceived by Martin (1945), have been completely redefined to include only those species with a dimorphic life cycle, and complex pores with or without sacculate caps. The evident widespread mycoparatism within the Tremellomycetidae contributes not only to a better understanding of the biology but also to the systematics of this group, as does the occurrence of a modified bifactorial compatibility system in the vast majority of the species examined. While the available evidence suggests that the Tremellomycetidae is a coherent taxon, the interrelationships within the subclass itself are not clear. Attempts to correlate morphological data and sequencing data in many cases have not been satisfying; therefore, it is unlikely that the taxonomy presented here reflects the actual phylogeny within this subclass.

Within the Heterobasidiomycetidae there are several natural groups. Both the Dacrymycetales and Tulasnellales seem to be coherent taxa, but the relationships of these orders to both the Auriculariales and Ceratobasidiales is not obvious. The Auriculariales consist of several subgroups, such as the Sebacinaceae and Hyaloriaceae, that could possibly be removed from the order, making it more homogeneous; however, more studies using varied techniques are required before such changes could be accepted. The major problem of phylogeny in the Heterobasidiomycetidae is the position of the Ceratobasidiales within any taxonomic scheme. The nature of the relationships between the Ceratobasidiales with each of the remaining orders of the Heterobasidiomycetidae and, perhaps more importantly, the relationships of the order to the Homobasidiomycetes, are uncertain, at best.

Additional studies of all types are needed to further improve the understanding of the biology and taxonomy of the Heterobasidiomycetes. Sequencing studies offer the possibility of solving many of the problems, but data from such studies need to be correlated with culture and morphological studies of all types, plus information on natural habitats and modes of nutrition.

Acknowledgments. Appreciation is due Drs. Mary L. Berbee, David S. Hibbett, Karen Hughes, and D. Lee Taylor for their efforts in aiding us to evaluate DNA sequencing studies, Ronald H. Petersen for his advice on nomenclatural problems, Michael Weiss for correcting the Latin description of Cystofilobasidiaceae, and Jack Fell for much information, some of which was unpublished, on the Filobasidiales. We also wish to express our appreciation to Ellinor K. Wells for the preparation of Figs. 11–16.

References

Adams GC Jr, Butler EE (1982) A re-interpretation of the sexuality of *Thanatephorus cucumeris* anastomosis group four. Mycologia 74:793–800

Bandoni RJ (1957) The spores and basidia of *Sirobasidium*. Mycologia 49:250–255

Bandoni RJ (1963) Conjugation in *Tremella mesenterica*. Can J Bot 41:467–474

Bandoni RJ (1965) Secondary control of conjugation in *Tremella mesenterica*. Can J Bot 43:627–630

Bandoni RJ (1979) *Fibulobasidium*: a new genus in the Sirobasidiaceae. Can J Bot 57:264–268

Bandoni RJ (1984) The Tremellales and Auriculariales: an alternative classification. Trans Mycol Soc Jpn 25:489–530

Bandoni RJ (1986) *Sirotrema*: a new genus in the Tremellaceae. Can J Bot 64:668–676

Bandoni RJ (1987) Taxonomic overview of the Tremellales. Stud Mycol 30:87–110

Bandoni RJ (1995) Dimorphic heterobasidiomycetes: taxonomy and parasitism. Stud Mycol 38:13–27

Bandoni RJ, Boekhout T (1998) Tremelloid genera with yeast phases. In: Kurtzman CP, Fell JW (eds) The yeasts, a taxonomic study, 4th edn. Elsevier, Amsterdam, pp 705–717

Bandoni RJ, Oberwinkler F (1982) *Stilbotulasnella*: a new genus in the Tulasnellaceae. Can J Bot 60:1875–1879

Bandoni RJ, Wells K (1992) Clamp connections and classification of the Auriculariales and Tremellales. Trans Mycol Soc Jpn 33:13–19

Bandoni RJ, Johri BN, Reid SA (1975) Mating among isolates of three species of *Sporobolomyces*. Can J Bot 53:2942–2944

Barnett HL (1937) Studies in the sexuality of the Heterobasidae. Mycologia 29:626–649

Bauer R, Oberwinkler F (1997) The Ustomycota: an inventory. Mycotaxon 64:303–319

Berbee ML, Wells K (1988) Ultrastructural studies of mitosis and the septal pore apparatus in *Tremella globospora*. Mycologia 80:479–492

Bodman MC (1938) Morphology and cytology of *Guepinia spathularia*. Mycologia 30:635–652

Bodman MC (1952) A taxonomic study of the genus *Heterochaete*. Lloydia 15:193–233

Boekhout T, Fonseca A, Batenburg-van Vegte WH (1991) *Bulleromyces* genus novum (Tremellales), a teleomorph for *Bullera alba*, and the occurrence of mating in *Bullera variabilis*. Antonie van Leeuwenhoekro J Microbiol Serol 59:81–93

Bourdot H, Galzin A (1928) Hyménomycètes de France. Marcel Bry, Sceaux

Bracker CE Jr, Butler EE (1963) The ultrastructure and development of septa in hyphae of *Rhizoctonia solani*. Mycologia 55:35–58

Brefeld O (1888) Untersuchungen aus dem Gesammtgebiete der Mykologie. 7 Basidiomyceten. II. Protobasidiomyceten. Arthur Felix, Leipzig

Brough SG (1970) The biology of *Tremella bambusina* Sacc. PhD Thesis, University of British Columbia, Vancouver, British Columbia

Brough SG (1974) *Tremella globospora*, in the field and in culture. Can J Bot 52:1853–1859

Chang S, Buswell JA, Chin S (eds) (1993) Mushroom biology and mushroom porducts. Chinese University Press, Hong Kong

Chen C-J (1998) Morphological and molecular studies in the genus *Tremella*. Bibl Mycol 174:1–225

Crawford DA (1954) Studies on New Zealand Clavariaceae. I. Trans R Soc N Z 82:617–631

Diederich P (1996) The lichenicolous Heterobasidiomycetes. Bibl Lichenol 61:1–198

Donk MA (1956) Notes on resupinate Hymenomycetes – II. The tulasnelloid fungi. Reinwardtia 3:363–379

Donk MA (1958) Notes on the basidium. Blumea (Suppl IV) 4:96–105

Donk MA (1966) Check list of European hymenomycetous Heterobasidiae. Persoonia 4:145–335

Drechsler C (1969) A *Tulasnella* parasitic on *Amoeba terricola*. Am J Bot 56:1217–1220

Fell JW, Boekhout T, Freshwater DW (1995) The role of nucleotide sequence analysis in the systematics of the yeast genera *Cryptococcus* and *Rhodotorula*. Stud Mycol 38:129–146

Fell JW, Roeijmans H, Boekhout T (1999) Cystofilobasidiales, a new order of basidiomycetous yeasts. Int J Syst Bacteriol 49:907–913

Flegel TW (1968) Some aspects of conjugation in the genus *Tremella* Dill. ex Fr. MS Thesis, University of British Columbia, Vancouver, British Columbia

Flegel TW (1976) Conjugation and growth of *Sirobasidium magnum* in laboratory culture. Can J Bot 54:411–418

Flegel TW (1981) The conjugation process in the jelly fungus *Sirobasidium magnum*. Can J Bot 59:929–938

Fox RD, Wong GJ (1990) Homothallism and heterothallism in *Tremella fuciformis* Can J Bot 68:107–111

Fries E (1874) Hymenomycetes Europaei sive epicriseos systematis mycologici, editio altera. E Berling, Uppsala

Furtado JS (1968) Basidial cytology of *Exidia nucleata* Mycologia 60:9–15

Furtado JS (1969) Basidial cytology of *Exidia recisa*. Mycologia 61:415–418

Ginns J (1986) The genus *Syzygospora* (Heterobasidiomycetes: Syzygosporaceae). Mycologia 78:619–636

Hanson LC, Wells K (1991) Compatibility and population studies of three species of *Tremella*. Mycologia 83:273–287

Hauerslev K (1976) New and rare Tremellaceae on record from Denmark. Friesia 11:94–115

Hauerslev K (1986) Three new tremellaceous fungi from Denmark. Windahlia 16:47–48

Hauerslev K (1993) New tremellaceous fungi from Denmark. Mycotaxon 49:217–233

Hibbett DS, Pine EM, Langer E, Langer G, Donoghue J (1997) Evolution of gilled mushrooms and puffballs inferred from ribosomal DNA sequences. Proc Nat Acad Sci USA 94:12002–12006

Hoch HC, Howard RJ (1981) Conventional chemical fixations induce artifactual swelling of dolipore septa. Exp Mycol 5:167–172

Ingold CT (1982a) Basidiospore germination and conidium formation in *Exidia glandulosa* and *Tremella mesenterica*. Trans Br Mycol Soc 79:370–373

Ingold CT (1982b) Basidiospore germination and conidium development in *Auricularia*. Trans Br Mycol Soc 78:161–166

Ingold CT (1983) Basidiospore germination and conidium development in Dacrymycetales. Trans Br Mycol Soc 81:563–571

Ingold CT (1984a) *Myxarium nucleatum* and its conidial state. Trans Br Mycol Soc 83:358–360

Ingold CT (1984b) Patterns of ballistospore germination in *Tilletiopsis*, *Auricularia* and *Tulasnella*. Trans Br Mycol Soc 83:583–591

Ingold CT (1985) Observations on spores and their germination in certain Heterobasidiomycetes. Trans Br Mycol Soc 85:417–423

Ingold CT (1992) The conidial stage in *Exidiopsis effusa* and *E. longispora*. Mycol Res 96:932–934

Ingold CT (1995) Types of reproductive cell in *Exidia recisa* and *Sirobasidium intermediae*. Mycol Res 99:1187–1190

Ishibashi Y, Sakagami Y, Isogai A, Suzuki A (1984) Structures of tremerogens A-9291-I and A-9291-VIII: peptidyl sex hormones of *Tremella brasiliensis*. Biochemistry 23:1399–1404

Juel HO (1897) *Muciporus* und die Familie der Tulasnellaceen. Bih K Sven Vetenskapsakad Handl Bd 23, Afd III 12:1–27

Jülich W (1981) Higher taxa of basidiomycetes. J Cramer, Vaduz

Jülich W (1982) Basidiomycetes of south-east Asia I. The genus *Paraphelaria* (Auriculariales). Persoonia 11:397–406

Kao CJ (1956) The cytology of *Syzygospora alba*. Mycologia 48:677–684

Kennedy LL (1959) The genera of the Dacrymycetaceae. Mycologia 50:874–895

Kennedy LL (1972) Basidiocarp development in *Calocera cornea*. Can J Bot 50:413–417

Khan SR (1976) Ultrastructure of the septal pore apparatus of *Tremella*. J Gen Microbiol 97:339–342

Khan SR, Kimbrough JW (1980) Septal ultrastructure in some genera of the Tremellaceae. Can J Bot 58:55–60

Khan SR, Talbot PHB (1976) Ultrastructure of septa in hyphae and basidia of *Tulasnella*. Mycologia 68:1027–1036

Klett HC (1964) North American species of *Exidia*. PhD Thesis, University of Washington, Seattle, Washington

Kobayasi Y (1937) On the genus *Holtermannia* of Tremellaceae. Sci Rep Tokyo Bunrika Daigaku Sect B 3:75–81

Koske RE (1972) Two unusual tremellas from British Columbia. Can J Bot 50:2565–2567

Kwon-Chung KJ (1987) Filobasidiaceae – *a* taxonomic survey. Stud Mycol 30:75–85

Kwon-Chung KJ, Chang YC, Bauer R, Swan EC, Taylor JW, Goel R (1995) The characteristics that differentiate *Filobasidiella depauperata* from *Filobasidiella neoformans*. Stud Mycol 38:67–79

Lalitha CR, Leelavathy KM (1990) A coccid-association in *Auriculoscypha* and its taxonomic significance. Trans Br Mycol Soc 94:571–572

Lalitha CR, Leelavathy KM, Manimohan P (1994) Patterns of basidiospore germination in *Auriculoscypha anacardiicola*. Trans Br Mycol Soc 98:64–66

Langer E (1998) Evolution of *Hyphodontia* (Corticiaceae, Basidiomycetes) and related Aphyllophorales inferred from ribosomal DNA sequences. Folia Cryptog Estonia 33:57–62

Langer E, Oberwinkler F (1993) Corticioid Basidiomycetes. I. Morphology and ultrastructure. Windahlia 20:1–28

Langer G (1994) Die Gattung *Botryobasidium* Donk (Corticiaceae, Basidiomycetes). Bibl Mycol 158:1–459

Lü H, McLaughlin DJ (1991) Ultrastructure of the septal pore apparatus and early septum initiation in *Auricularia auricula-judae*. Mycologia 83:322–334

Lü H, McLaughlin DJ (1995) A light and electron microscopic study of mitosis in the clamp connection of *Auricularia auricula-judae*. Can J Bot 73:315–332

Luck-Allen ER (1963) The genus *Basidiodendron*. Can J Bot 41:1025–1052

Maekawa N (1987) A new species of the genus *Cerinomyces*. Can J Bot 65:583–588

Martin GW (1935) *Atractobasidium*, a new genus of the Tremellaceae. Bull Torrey Bot Club 62:339–343

Martin GW (1937) A New type of heterobasidiomycete. J Wash Acad Sci 27:112–114

Martin GW (1945) The classification of the Tremellales. Mycologia 37:527–542

Martin GW (1948) New or noteworthy tropical fungi. IV. Lloydia 11:111–122

Martin GW (1952) Revision of the North Central Tremellales. Stud Nat Hist Iowa Univ 19:1–122

Martin GW (1957) The tulasnelloid fungi and their bearing on basidial terminology. Brittonia 9:25–30

McLaughlin DJ (1980) Ultrastructure of the metabasidium of *Auricularia fuscosuccinea*. Am J Bot 67:1225–1235

McLaughlin DJ (1981) The spindle pole body and postmeiotic mitosis in *Auricularia fuscosuccinea*. Can J Bot 59:1196–1206

McNabb RFR (1964) Taxonomic studies in the Dacrymycetaceae I. *Cerinomyces* Martin. N Z J Bot 2:415–424

McNabb RFR (1965a) Taxonomic studies in the Dacrymycetaceae II. *Calocera* (Fries) Fries. N Z J Bot 3:31–58

McNabb RFR (1965b) Taxonomic studies in the Dacrymycetaceae III. *Dacryopinax* Martin. N Z J Bot 3:59–72

McNabb RFR (1965c) Taxonomic studies in the Dacrymycetaceae IV. *Guepiniopsis* Patouillard. N Z J Bot 3;159–169

McNabb RFR (1965d) Taxonomic studies in the Dacrymycetaceae V. *Heterotextus* Lloyd. N Z J Bot 3:215–222

McNabb RFR (1965e) Taxonomic studies in the Dacrymycetaceae VI. *Femsjonia* Fries. N Z J Bot 3:223–228

McNabb RFR (1966) Taxonomic studies in the Dacrymycetaceae VII. *Ditiola* Fries. N Z J Bot 4:546–558

McNabb RFR (1973) Taxonomic studies in the Dacrymycetaceae VIII. *Dacrymyces* Nees ex Fries. N Z J Bot 11:461–524

Metzler B, Oberwinkler F, Petzold H (1989) *Rhynchogastrema* gen. nov. and Rhynchogastremaceae fam. nov. (Tremellales). Syst Appl Microbiol 12:280–287

Möller A (1895) Protobasidiomyceten. Untersuchungen aus Brasilien. Gustav Fischer, Jena

Moore RT (1978) Taxonomic significance of septal ultrastructure with particular reference to the jelly fungi. Mycologia 70:1007–1024

Moore RT (1996) The dolipore/parenthesome septum in modern taxonomy. In: Sneh B, Jabaji-Hare S, Neate S, Dijst G (eds) *Rhizoctonia* species: taxonomy, molecular biology, ecology, pathology and disease control. Kluwer Academic Publishers, Dordrecht, pp 13–35

Moore RT, McAlear JH (1962) Fine structure of Mycota. 7. Observations on septa of ascomycetes and basidiomycetes. Am J Bot 49:86–94

Müller WH, Stalpers JA, van Aelst C, van der Krift TP, Boekhout T (1998) Field emission gun-scanning electron microscopy of septal pore caps of selected species in the *Rhizoctonia* s.l. complex. Mycologia 90:170–179

Oberwinkler F (1963) Niedere Basidiomyceten aus Südbayern III. Die Gattung *Sebacina* Tul. s.l. Ber Bayer Bot Ges 36:41–55

Oberwinkler F (1964) Intrahymeniale Heterobasidiomyceten. Nova Hedwigia 7:489–499

Oberwinkler F (1993) Genera in a monophyletic group: the Dacrymycetales. Mycol Helv 6:35–72

Oberwinkler F, Bandoni RJ (1981) *Tetragoniomyces* gen. nov. and Tetragoniomycetaceae fam. nov. (Tremellales). Can J Bot 59:1034–1040

Oberwinkler F, Bandoni R (1982) Carcinomycetaceae: a new family in the Heterobasidiomycetes. Nord J Bot 2:501–516

Oberwinkler F, Bandoni RJ (1983) *Trimorphomyces*: a new genus in the Tremellaceae. Syst Appl Microbiol 4:105–113

Oberwinkler F, Lowy B (1981) *Syzygospora alba*, a mycoparasitic heterobasidiomycete. Mycologia 73:1108–1115

Oberwinkler F, Bandoni RJ, Bauer R, Deml G, Kisimova-Horovitz L (1984) The life history of *Christiansenia pallida*, a dimorphic, mycoparasitic heterobasidiomycete. Mycologia 76:9–22

Oberwinkler F, Bauer R, Schneller J (1990) *Phragmoxenidium mycophilum* sp. nov., an unusual mycoparastic heterobasidiomycete. Syst Appl Microbiol 13:186–191

Olive LS (1957) Tulasnellaceae of Tahiti. A revision of the family. Mycologia 49:663–679

Patouillard N (1900) Essai taxonomique sur les familles et les genres des hyménomycètes. Lucien Duclume, Lons-le Saunier

Patton AM, Marchant R (1978) A mathematical analysis of dolipore/parenthesome structure in basidiomycetes. J Gen Microbiol 109:335–349

Petersen RH (1968) Hyphal spore germination and asexual spore production in *Gloeotulasnella pinicola*. Mycopathol Mycol Appl 35:145–149

Petersen RH (1985) Type studies in the clavarioid fungi – IX. Persoonia 12:401–413

Petersen RH (1987) Notes on clavarioid fungi. XXI. New Zealand taxa of *Tremellodendropsis*. Mycotaxon 29:45–65

Raper CA (1976) Sexuality and life-cycle of the edible, wild *Agaricus bitorquis*. J Gen Microbiol 95:54–66

Rath F (1991) Christianseniales, Christianseniaceae, ordine e famiglia nuovi (Basidiomycetes). Atti Soc Ital Sci Nat Mus Civ Stor Nat Milano 132:13–24

Reid DA (1970) New or interesting records of British Hymenomycetes, IV. Trans Br Mycol Soc 55:413–441

Reid DA (1974) A monograph of the British Dacrymycetales. Trans Br Mycol Soc 62:433–494

Reid DA, Manimohan P (1985) *Auriculoscypha*, a new genus of Auriculariales (basidiomycetes) from India. Trans Br Mycol Soc 85:532–535

Reid ID (1974) Properties of conjugation hormones (erogens) from the basidiomycete *Tremella mesenterica*. Can J Bot 52:521–524

Roberts P (1992) Spiral-spored *Tulasnella* species from Devon and the New Forest. Mycol Res 96:233–236

Roberts P (1993a) Allantoid-spored *Tulasnella* species from Devon. Mycol Res 97:213–220

Roberts P (1993b) *Exidiopsis* species from Devon, including the new segregate genera *Ceratosebacina*, *Endoperplexa*, *Microsebacina*, and *Serendipita*. Mycol Res 97:467–478

Roberts P (1994a) Long-spored *Tulasnella* species from Devon, with additional notes on allantoid-spored species. Mycol Res 98:1235–1244

Roberts P (1994b) Globose and ellipsoid-spored *Tulasnella* species from Devon and Surrey, with a key to the genus in Europe. Mycol Res 98:1431–1452

Roberts P (1997) New Heterobasidiomycetes from Great Britain. Mycotaxon 63:195–216

Roberts P (1998a) *Oliveonia* and the origin of the holobasidiomycetes. Folia Cryptog Estonica 33:127–132

Roberts P (1998b) A revision of the genera *Heterochaetella*, *Myxarium*, *Protodontia*, and *Stypella* (Heterobasidiomycetes). Mycotaxon 69:208–248

Roberts P (1998c) *Heteroacanthella*: a surprising addition to the British Mycota. Mycologist 12:146–147

Roberts P (1998d) Synonymy of *Tofispora* and *Thanatephorus*, with notes on a new collection from Puerto Rico. Mycotaxon 69:35–38

Roberts PJ, Spooner BM (1998) Heterobasidiomycetes from Brunei Darussalam. Kew Bull 53:631–650

Rockett TR, Kramer CL (1974) The biology of sporulation of selected Tremellales. Mycologia 66:926–941

Rogers DP (1932) A cytological study of *Tulasnella*. Bot Gaz 94:86–105

Rogers DP (1933) A taxonomic review of the Tulasnellaceae. Ann Mycol 31:181–203

Rogers DP (1935) Notes on the lower Basidiomycetes. Stud Nat Hist Iowa Univ 17:1–43

Rogers DP (1947) A new gymnocarpous heterobasidiomycete with gasteromycetous basidia. Mycologia 39:556–564

Sakagami Y, Yoshida M, Isogai A, Suzuki A (1981) Peptidal sex hormones inducing conjugation tube formation in compatible mating-type cells of *Tremella mesenterica*. Science 212:1525–1527

Seifert KA (1983) Decay of wood by the Dacrymycetales. Mycologia 75:1011–1018

Sneh B, Jabaji-Hare S, Neate S, Dijst G (eds) (1996) *Rhizoctonia* species: taxonomy, molecular biology, ecology, pathology and disease control. Kluwer Academic Publishers, Dordrecht

Swann EC, Taylor JW (1993) Higher taxa of basidiomycetes: an 18S rRNA gene perspective. Mycologia 85:923–936

Swann EC, Taylor JW (1995a) Phylogenetic diversity of yeast-producing basidiomycetes. Mycol Res 99:1205–1210

Swann EC, Taylor JW (1995b) Phylogenetic perspectives on basidiomycete systematics: evidence from the 18S rRNA gene. Can J Bot 73 (Suppl 1):S862–S868

Talbot PHB (1965) Studies of "*Pellicularia*" and associated genera of Hymenomycetes. Persoonia 3:371–406

Talbot PHB (1968) Fossilized pre-Patouillardian taxonomy? Taxon 17:620–628

Talbot PHB (1973) Towards uniformity in basidial terminology. Trans Br Mycol Soc 61:497–512

Taylor JW (1985) Mitosis in the basidiomycete fungus *Tulasnella araneosa*. Protoplasma 126:1–18

Tu CC, Kimbrough JW (1978) Systematics and phylogeny of fungi in the *Rhizoctonia* complex. Bot Gaz 139:454–466

Tu CC, Kimbrough JW, Aldrich HC (1977) Cytology and ultrastructure of *Thanatephorus cucumeris* and related taxa of the *Rhizoctonia* complex. Can J Bot 55:2419–2436

Tulasne LR (1853) Observations sur l'organisation des trémellinées. Ann Sci Nat Bot Sér 3 19:193–231

Tulasne LR, Tulasne C (1872) Nouvelles notes sur les fungi Tremellini et leur allies. Ann Sci Nat Bot V, 15:215–235

Tulasne LR, Tulasne C (1873) New notes upon the tremellineous fungi and their analogues. J Linn Soc Bot 13:31–42

Warcup JH, Talbot PHB (1967) Perfect states of rhizoctonias associated with orchids. New Phytol 66:631–641

Wells K (1964a) The basidia of *Exidia nucleata*. I. Ultrastructure. Mycologia 56:327–341

Wells K (1964b) The basidia of *Exidia nucleata*. II. Development. Am J Bot 51:360–370

Wells K (1977) Meiotic and mitotic divisions in the Basidiomycotina. In: Rost TL, Gifford EM (eds) Mechanisms and control of cell division. Dowden, Hutchinson and Ross, Stroudsburg, Pennsylvania, pp 337–374

Wells K (1987) Comparative morphology, intracompatibility, and interincompatibility of several species of *Exidiopsis* (Exidiaceae). Mycologia 79:274–288

Wells K (1994) Jelly fungi, then and now! Mycologia 86:18–48

Wells K, Oberwinkler F (1982) *Tremelloscypha gelatinosa*, a species of a new family, Sebacinaceae. Mycologia 74:325–331

Wells K, Raitviir A (1975) The species of *Bourdotia* and *Basidiodendron* (Tremellaceae) of the U.S.S.R. Mycologia 67:904–922

Wells K, Wong G (1985) Interfertility and comparative morphological studies of *Exidiopsis plumbescens* from the West Coast. Mycologia 77:285–299

Whelden RM (1934) Cytological studies in the Tremellaceae I. *Tremella*. Mycologia 26:415–435

Whelden RM (1935a) Cytological studies in the Tremellaceae II. *Exidia*. Mycologia 27:41–57

Whelden RM (1935b) Cytological studies of the Tremellaceae III. *Sebacina*. Mycologia 27:503–520

Whelden RM (1937) Cytological studies of the Tremellaceae. IV. *Protodontia* and *Tremellodendron*. Mycologia 29:100–115

Whitney HS, Parmeter JR (1963) Synthesis of heterokaryons in *Rhizoctonia solani* Kühn. Can J Bot 41:879–886

Wojewoda W (1977) Grzyby (Mycota), vol 8. Podstawczaki (Basidiomycetes), Trzęsakowe (Tremellales), Uszakowe (Auriculariales), Czerwcogrzybowe (Septobasidiales). Państwowe Wydawnictwo Naukowe, Warsaw

Wong GJ, Wells K (1985) Modified bifactorial incompatibility in *Tremella mesenterica*. Trans Br Mycol Soc 84:95–109

Wong GJ, Wells K, Bandoni RJ (1985) Interfertility and comparative morphological studies of *Tremella mesenterica*. Mycologia 77:36–49

Worrall JJ, Anagnost SE, Zabel RA (1997) Comparison of wood decay among diverse lignicolous fungi. Mycologia 89:199–219

Yen HC (1949) Contribution a l'étude de la sexualité et du mycelium des basidiomycètes saprophytes. Ann Univ Lyon 6:5–127

Yoshida M, Sakagami Y, Isogai A, Suzuki A (1981) Isolation of Tremerogen a-12, a peptidal sex hormone of *Tremella mesenterica*. Agric Biol Chem 45:1043–1044

Zugmaier W, Bauer R, Oberwinkler F (1994) Mycoparasitism of some *Tremella* species. Mycologia 86:49–56

5 Basidiomycota: Homobasidiomycetes

D.S. Hibbett[1] and R.G. Thorn[2]

CONTENTS

I.	Introduction	121
II.	Phylogeny	122
A.	Higher-Level Relationships of Homobasidiomycetes	122
B.	Overview of Homobasidiomycete Taxonomy	122
C.	Phylogeny Within Homobasidiomycetes	124
	1. Polyporoid Clade	124
	2. Euagarics Clade	129
	3. Bolete Clade	131
	4. Thelephoroid Clade	131
	5. Russuloid Clade	132
	6. Hymenochaetoid Clade	133
	7. Cantharelloid Clade	133
	8. Gomphoid-Phalloid Clade	134
III.	Characters	135
A.	Fruiting Body Macromorphology	135
	1. Agarics	135
	2. Corticioid Fungi	137
	3. Gasteroid Forms	138
B.	Anatomy, Cytology, and Ultrastructure	139
	1. Basidia	139
	2. Basidiospores	139
	3. Hyphal Systems of Fruiting Bodies	141
	4. Nuclear Behavior in Meiosis and Basidiosporogenesis	143
	5. Parenthesome Ultrastructure	144
C.	Asexual Reproduction and Somatic Morphology	145
D.	Mating Genetics	149
E.	Pigments and Bioluminescence	149
	1. Shikimate-Chorismate Pathway Derivatives	149
	2. Acetate-Malonate Pathway Derivatives	150
	3. Mevalonate Pathway Derivatives	150
	4. Nitrogen-Containing Compounds	151
	5. Bioluminescence	151
F.	Nutritional Modes	151
	1. Saprotrophs	151
	2. Mycorrhizae	153
	3. Plant Pathogens	154
	4. Mycoparasites, Bacteriovores, and Nematode-Trappers	155
	5. Lichens, Algal Parasites, and Bryophyte Associates	156
	6. Insect Symbionts	157
IV.	Conclusions	159
	References	160

I. Introduction

Homobasidiomycetes include the mushroom-forming fungi and related taxa. Over 13000 species of homobasidiomycetes have been described, which is equal to approximately 23% of all known species of eumycota (Hawksworth et al. 1995). Homobasidiomycetes occur in all terrestrial ecosystems, including deserts, and there are also a few aquatic species, in both marine and freshwater habitats (Kohlmeyer and Kohlmeyer 1979; Desjardin et al. 1995). The oldest unambiguous homobasidiomycete fossils are from the mid-Cretaceous, but indirect evidence, including molecular clock dating, suggests that the group may have been in existence by the late Triassic (ca. 200 ma; Berbee and Taylor 1993; Hibbett et al. 1997a). In contemporary ecosystems, homobasidiomycetes function as saprotrophs, plant pathogens, and partners in diverse symbioses, including ectomycorrhizae. Thus, homobasidiomycetes play a significant role in the carbon cycle, and they have a profound economic impact on agricultural industries, especially forestry. Finally, homobasidiomycetes are culturally significant, having served as food, drugs, and spiritual symbols in diverse human societies.

Homobasidiomycetes have been studied extensively, but there is still no comprehensive phylogenetic classification of the group. In this chapter, we synthesize results of recent phylogenetic studies in homobasidiomycetes and provide a preliminary phylogenetic outline for the group as a whole. We also discuss selected characters that appear to be phylogenetically informative or that have particular ecological or functional signifi-

---

[1] Department of Biology, Clark University, Worcester, Massachusetts 01610, USA
[2] Department of Botany, University of Wyoming, Laramie, Wyoming 82071, USA

cance. Our goal is to provide an overview of current knowledge regarding homobasidiomycete phylogeny as well as a framework for further studies.

II. Phylogeny

A. Higher-Level Relationships of Homobasidiomycetes

Molecular and ultrastructural evidence suggests that the homobasidiomycetes are nested in a clade of Basidiomycota that have biglobular spindle pole bodies, complex dolipore septa, and (usually) membrane-bound parenthesomes (Wells 1978; Moore 1980; McLaughlin 1981; Swann and Taylor 1993, 1995a,b). This group has been termed the hymenomycete lineage by Swann and Taylor (1993). In addition to homobasidiomycetes, the hymenomycete lineage includes the heterobasidiomycete orders Auriculariales sensu stricto, Dacrymycetales, Tremellales, Ceratobasidiales, and Tulasnellales (Oberwinkler 1972; Bandoni 1984; Tehler 1988; Wells 1994; Sjamsuridzal et al. 1997). Major characters that have been used to distinguish homobasidiomycetes from heterobasidiomycetes include basidial morphology, parenthesome ultrastructure, mode of spore germination, and gelatinization of fruiting bodies (Talbot 1973a; Tu and Kimbrough 1978; Jülich 1981; Ingold 1985, 1992; Oberwinkler 1985; Wells 1994; see Wells et al., Chap. 4, this Vol.). In this chapter, we generally follow the classification of heterobasidiomycetes proposed by Wells (1994).

The homobasidiomycetes have generally been accepted as monophyletic (e.g., Savile 1955; Schaffer 1975; Oberwinkler 1982). However, in a morphological cladistic analysis of fungal phylogeny, Tehler (1988) was unable to find any uncontradicted synapomorphies supporting the monophyly of the homobasidiomycetes. Higher-level relationships of homobasidiomycetes have been investigated in molecular studies using sequences of nuclear small-subunit ribosomal DNA (nuc-ssu rDNA; Swann and Taylor 1993, 1995a,b; Gargas et al. 1995; Fig. 3 in Sjamsuridzal et al. 1997), and nuclear large-subunit rDNA (nuc-lsu rDNA; Begerow et al. 1997). These studies support the monophyly of Auriculariales, Dacrymycetales, and Tremellales, but they give mixed results regarding the monophyly of the homobasidiomycetes (Fig. 1). In some analyses the Auriculariales and Ceratobasidiales appear to be nested within the homobasidiomycetes, whereas in others they are outside the homobasidiomycetes (Fig. 1). The nuc rDNA studies all support or are consistent with the view that the homobasidiomycetes plus Auriculariales and Ceratobasidiales form a monophyletic group, and that the Tremellales and Dacrymycetales are near the base of the hymenomycete lineage (their positions are interchangeable, however; Fig. 1). Tulasnellales have not been included in the studies just cited. Analyses of nuc-ssu rDNA and partial mitochondrial small-subunit (mt-ssu) rDNA sequences (Hibbett et al. 1997b, D.S. Hibbett, unpubl.; Lee and Jung 1997) suggest that the Tulasnellales and Ceratobasidiales are nested within the homobasidiomycetes, and that the Auriculariales is the sister group to this clade (Fig. 1). This is consistent with results of analyses of mt-lsu rDNA sequences (Bruns et al. 1998), which suggest that *Tulasnella* (Tulasnellales) and *Waitea* (Ceratobasidiales) are in the homobasidiomycetes.

B. Overview of Homobasidiomycete Taxonomy

Classification of homobasidiomycetes is in a period of major revision. One of the earliest broad classifications for homobasidiomycetes (and other macrofungi) was the Friesian system (Fries 1874), which grouped taxa based on gross morphology of the hymenophore, as follows (emphasizing Hymenomycetes, after Donk 1971):

Gasteromycetes	(enclosed hymenophore)
Hymenomycetes	(exposed hymenophore)
Agaricales	(lamellate hymenophore)
Agaricaceae	
Aphyllophorales	(nonlamellate hymenophore)
Cantharellaceae	(wrinkled hymenophore, erect, pileate)
Clavariaceae	(smooth hymenophore, erect, branched or unbranched)
Hydnaceae	(toothed hymenophore)
Meruliacae	(wrinkled hymenophore, resupinate or pileate)
Polyporaceae	(poroid hymenophore)
Thelephoraceae	(smooth hymenophore, resupinate or erect)

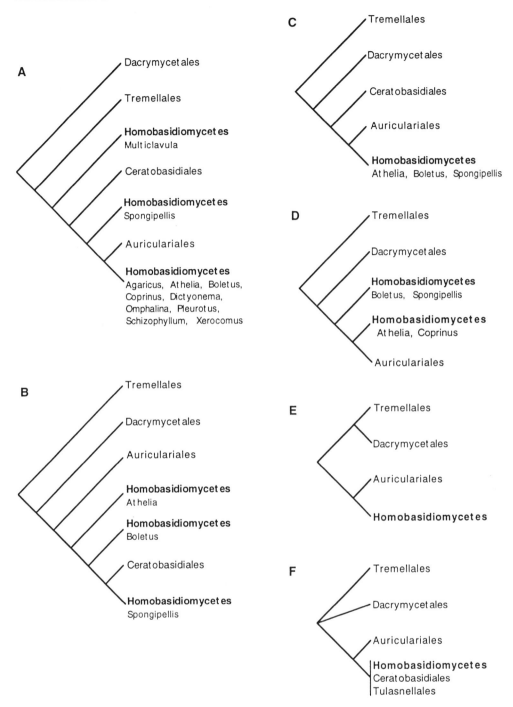

Fig. 1A–F. Higher-level phylogenetic relationships of homobasidiomycetes. **A–E** Simplified cladograms based on prior studies using nuc rDNA sequences (**A–D** nuc-ssu rDNA, **E** nuc-lsu rDNA). Note that studies disagree about the monophyly of homobasidiomycetes and the relative positions of heterobasidiomycete orders. **A** (Gargas et al. 1995, Fig. 1). **B** (Swann and Taylor 1993, Fig. 2; neighbor-joining analysis). **C** (Swann and Taylor 1993, Fig. 1; parsimony analysis). **D** (Swann and Taylor 1995a, Fig. 1). **E** (Begerow et al. 1997). **F** Summary cladogram incorporating results of Bruns et al. (1998), Hibbett et al. (1997b), and D.S. Hibbett (unpubl.; see text)

The Friesian system is intuitively accessible, but it is artificial (see Table 2). Anatomical studies since the turn of the century (Fayod 1889; Patouillard 1900) have resulted in major reclassifications of the Friesian higher taxa (for reviews, see Jülich 1981; Walker 1996). Among the most influential modern treatments are those of Donk (1964, 1971), who divided the 6 Friesian families of the Aphyllophorales into 23 families, Singer (1986), who divided the Agaricales into 17 families, and Dring (1973), who divided the Gasteromycetes into 9 orders with 23 families. In general, the taxonomic names used here are based on these works.

The classifications of Donk, Singer, and Dring each include a number of putatively monophyletic groups that have distinguishing morphological features (e.g., Ganodermataceae, Coprinaceae, Phallales), and therefore represent major advances toward a phylogenetic classification of homobasidiomycetes. Nevertheless, these classifications also include large residual taxa that lack synapomorphies and that are presumably polyphyletic (e.g., Clavariaceae, Corticiaceae, Polyporaceae, Tricholomataceae, Hymenogastrales). Moreover, they maintain the Friesian division of macrofungi into Aphyllophorales, Agaricales, and Gasteromycetes (with some modifications, such as the inclusion of boletes in Agaricales), in spite of strong evidence that these taxa are artificial.

The Friesian higher taxa must be integrated if a natural classification of the homobasidiomycetes is to be achieved. This was attempted by Jülich (1981), who based his phylogenetic hypothesis and classification on anatomy and morphology. Jülich's system was a bold attempt to unify homobasidiomycete classification, but it was widely criticized (e.g., Nuss 1983; Redhead and Ginns 1983) and has not been generally adopted. Alternative higher-level classifications of homobasidiomycetes were proposed by Kreisl (1969), Oberwinker (1977), Pegler (in Hawksworth et al. 1995), and Walker (1996), but there is still no consensus about the broad outlines of homobasidiomycete phylogeny.

C. Phylogeny Within Homobasidiomycetes

In recent years, there have been many phylogenetic studies centered on specific groups of homobasidiomycetes, but there has been no comprehensive analysis of the entire group. The most taxonomically inclusive phylogenetic study in homobasidiomycetes so far is that of Hibbett et al. (1997b), which was based on nuc-ssu and mt-ssu rDNA sequences. The analysis of Hibbett et al. (1997b) included 81 species of homobasidiomycetes, which represent 10 families of Agaricales sensu Singer (1986), 18 families of Aphyllophorales sensu Donk (1964), and 7 families of Gasteromycetes sensu Dring (1973). In this chapter, we have used the results of Hibbett et al. (1997b) as a framework for developing a preliminary phylogenetic outline of the homobasidiomycetes (Table 1).

The strict consensus of the shortest trees found by Hibbett et al. (1997b) is shown in Fig. 2. The data matrix used to infer the tree included about 2400 base pairs of sequence data per terminal taxon. Nevertheless, many nodes in the tree were weakly supported (as measured by bootstrapping), especially those near the base of the tree (Fig. 2). Consequently, the higher-order relationships and the position of the root of the homobasidiomycetes remain unclear. To resolve these issues, it will be necessary to perform phylogenetic analyses that include additional rDNA and protein-coding gene sequences, or morphological characters (see Sect. III).

We have tentatively divided the homobasidiomycetes into eight major clades (Fig. 2, Table 1). By referring to the results of other molecular studies (see Table 1 for references), we have increased the number of taxa that can be assigned to these clades to 246 genera (from a total of 79 genera sampled by Hibbett et al. 1997b). The genera in Table 1 represent most of the major groups of homobasidiomycetes recognized in current taxonomy. However, many small, taxonomically controversial groups have yet to be studied, such as *Astraeus*, *Cantharocybe*, *Horakia*, *Pachykytospora*, and cyphelloid forms. In addition, sampling is still limited in certain large polyphyletic groups, such as Hymenogastrales, Tricholomataceae, Polyporaceae, and especially Corticiaceae. With that caveat, we present crude estimates of the numbers of described species in each clade based on species counts for higher taxa of homobasidiomycetes in *The Dictionary of the Fungi*, 8th ed. (Hawksworth et al. 1995; Fig. 2). Exemplars of the eight major clades are shown in Figs. 3–14.

1. Polyporoid Clade

This group includes members of Corticiaceae, Ganodermataceae, Polyporaceae, and Sparassidaceae (Fig. 2, Table 1). The polyporoid clade is

Table 1. Preliminary phylogenetic outline of homobasidiomycetes based on molecular characters

Clade	Exemplar genera sampled in molecular studies, and references[1]
1. Polyporoid clade	
Hymenomycetes	
Aphyllophorales	
Corticiaceae	*Candelabrochaete*[26], *Cotylidia*[26], *Crustoderma*[26], *Dendrocorticium*[26], *Dentocorticium*[2], *Epithele*[26], *Galzinia*[26], *Hyphoderma*[26], *Lopharia*[26], *Mycoacia*[26], *Phanerochaete*[2,3,16,26], *Phlebia*[2,3,26], *Phlebiopsis*[25,26], *Porogramme*[26], *Pulcherricium*[3,25,26], *Punctularia*[26], *Scopuloides*[26], *Sistotrema*[2], *Vuilleminia*[26], symbionts of *Xyleborus* and *Dendroctonus* bark beetles[25]
Ganodermataceae	*Ganoderma*[2,3,4,6,26]
Hydnaceae ("residual")	*Climacodon*[26], *Steccherinum*[26]
Polyporaceae	*Abortiporus*[4], *Albatrellus*[2,11], *Antrodia*[2,3,4,25,26], *Antrodiella*[4,26], *Aurantioporus*[4], *Bjerkandera*[2,3,4,26], *Ceriporia*[2,3,26], *Ceriporiopsis*[4,26], *Climacocystis*[4], *Cryptoporus*[3], *Daedalea*[2,3], *Daedaleopsis*[3], *Datronia*[3], *Faerberia*[16], *Fomes*[2,3], *Fomitopsis*[2,3,26], *Gloeophyllum*[16,26], *Gloeoporus*[4], *Irpex*[16], *Juhnghunia*[26], *Laetiporus*[2,3], *Lentinus*[2,3,24], *Lenzites*[3,26], *Leptoporus*[4,26], *Meripilus*[2,25,26], *Oligoporus* (= *Postia*)[4,16,26], *Panus*[2,4,11,24], *Perenniporia*[2,26], *Phaeolus*[2,3], *Physisporinus*[26], *Piptoporus*[3], *Polyporoletus*[11], *Polyporus*[2,3,24], *Pycnoporus*[3], *Rigidoporus*[26], *Skeletocutis*[4], *Spongipellis*[4,9,26], *Trametes*[2,3,26], *Tyromyces*[4], *Wolfiporia*[2]
Sparassidaceae	*Sparassis*[2,3]
2. Euagarics clade	
Hymenomycetes	
Agaricales	
Agaricaceae	*Agaricus*[2,6,7,9,11,12,16,22], *Chlorophyllum*[6], *Cystoderma*[6,22], *Lepiota*[2,6,7,12,22], *Leucoagaricus*[6,7,12], *Leucocoprinus*[6,7,16,22], G1 and G3 attine ant symbionts[6,7]
Amanitaceae	*Amanita*[2,6,11,12,17], *Limacella*[6]
Bolbitiaceae	*Agrocybe*[6,8,16], *Bolbitius*[6,7,8,11,12,16,17], *Conocybe*[6]
Coprinaceae	*Annellaria*[6], *Coprinus*[2,6,8,9,16], *Lacrymaria*[6,16], *Paneolina*[6,7], *Paneolus*[6,8], *Psathyrella*[6,8]
Cortinariaceae	*Cortinarius*[2,6,7,11,12,17,22], *Dermocybe*[6], *Inocybe*[6,11], *Hebeloma*[6,7,11,12,22]
Entolomataceae	*Clitopilus*[6,16], *Entoloma*[6,12,17]
Hygrophoraceae	*Hygrocybe*[5,6,11,12,17,23], *Hygrophorus*[5,6,11,12,17]
Paxillaceae	*Lampteromyces*[6,16,22], *Omphalotus*[6,22], *Ripartites*[22]
Pluteaceae	*Pluteus*[2,6,17]
Strophariaceae	*Hypholoma*[6], *Kuhneromyces*[6], *Pholiota*[6], *Psilocybe*[6,16], *Stropharia*[2,6,16]
Tricholomataceae	*Armillaria*[6,11,12,16], *Arrhenia*[6,23], *Asterophora*[6,11], *Clitocybe*[6,16,22,23], *Collybia*[6,16], *Crinipellis*[6,7,12], *Flammulina*[6], *Gerronema*[5,6,23], *Hohenbuehelia*[6,16], *Laccaria*[6,11,13], *Lentinula*[2,3,6,24,25], *Lepista*[16], *Lyophyllum*[6,16], *Marasmiellus*[6], *Marasmius*[6,7,12], *Melanoleuca*[16], *Mycena*[6], *Omphalina* pro parte[6,9,23], *Panellus*[2,3], *Phaeotellus*[6,23], *Resupinatus*[6], *Termitomyces*[6], *Tricholoma*[6,11,12,17,22], *Xeromphalina*[6], G2 attine ant symbionts[7]
Aphyllophorales	
Clavariaceae	*Clavaria*[5], *Clavulinopsis*[5], *Macrotyphula*[5], *Pterula*[5], *Typhula*[2,5]
Corticiaceae	*Athelia*[9], *Piloderma*[11,17], *Gloeocystidium ipidophilum*[25]
Dictyonemataceae	*Dictyonema*[9]
Fistulinaceae	*Fistulina*[2,3]
Polyporaceae	*Phyllotopsis*[6,24], *Pleurotus*[2,3,6,9,12,16,22,24,25]
Schizophyllaceae	*Schizophyllum* (incl. *Auriculariopsis* pro parte)[2,9,18]
Gasteromycetes	
Hymenogastrales	
Coprinaceae	*Montagnea*[8]
Tricholomataceae	*Hydnangium*[13], *Podohydnangium*[13]
Strophariaceae	*Weraroa*[22]
unplaced	*Leratia*[22]
Lycoperdales	
Lycoperdaceae	*Calvatia*[2], *Lycoperdon*[2]
Nidulariales	
Nidulariaceae	*Crucibulum*[2,12,16,20], *Cyathus*[2,12,16,20]
Podaxales	
Podaxaceae	*Podaxis*[8]
Tulostomatales	
Tulostomataceae	*Tulostoma*[2]
Heterobasidiomycetes	
Ceratobasidiales	
Ceratobasidiaceae	*Ceratobasidium*[2], *Waitea*[11]

Table 1. *Continued*

Clade	Exemplar genera sampled in molecular studies, and references[1]
3. Bolete clade	
Hymenomycetes	
Agaricales	
Boletaceae	*Austroboletus*[11,17], *Boletus*[2,3,6,7,9,10,11,12,17,22], *Gyroporus*[11,17], *Paragyrodon*[10,11,17], *Phylloporus*[6,7,10,11,12], *Strobilomyces*[11,12], *Suillus*[3,6,10,11,12,26], *Tylopilus*[11], *Xerocomus*[9,10,11]
Gomphidiaceae	*Chroogomphus*[10,11,14], *Gomphidius*[10,11,12,14]
Paxillaceae	*Hygrophoropsis*[11,12,22], *Paxillus* (including *Tapinella*)[2,10,11,12,22]
Aphyllophorales	
Coniophoraceae	*Coniophora*[11,12,17], *Serpula*[11,12,16]
Gasteromycetes	
Hymenogastrales	
Cortinariaceae	*Chamonixia*[11]
Gomphidiaceae	*Brauniellula*[11]
Hymenogastraceae	*Hymenogaster*[11]
Rhizopogonaceae	*Gastrosuillus*[14,26], *Rhizopogon*[10,11,12,14,26], *Truncocolumella*[11,14]
Melanogastrales	
Melanogastraceae	*Alpova*[11,17], *Melanogaster*[11]
Sclerodermatales	
Sclerodermataceae	*Pisolithus*[11], *Scleroderma*[2]
Tulostomatales	
Calostomataceae	*Calostoma*[19]
4. Thelephoroid clade	
Hymenomycetes	
Aphyllophorales	
Thelephoraceae	*Boletopsis*[17], *Hydnellum*[2,12], *Pseudotomentella*[11], *Sarcodon*[11], *Thelephora*[2,11,12,17], *Tomentella*[11,17]
5. Russuloid clade	
Hymenomycetes	
Agaricales	
Russulaceae	*Lactarius*[6,11,12,17,21,22], *Russula*[2,3,6,11,12,17,21,22,25,26]
Aphyllophorales	
Auriscalpiaceae	*Auriscalpium*[2,3,21,26], *Gloeodontia*[21,26], *Gloiodon*[26], *Lentinellus*[2,3,11]
Bondarzewiaceae	*Bondarzewia*[2,3,6,11,17,21]
Corticiaceae	*Acanthophysium*[2,26], *Aleurodiscus*[2,26], *Boidinia*[21,26], *Byssoporia*[11], *Conferticium*[21], *Entomocorticium*[25], *Peniophora* (incl. *Dendrophora*, *Duportella*)[2,3,25,26], *Pseudoxenasma*[21], *Vesiculomyces*[21]
Echinodontiaceae	*Echinodontium*[2,3,21,26]
Hericiaceae	*Clavicorona*[2,5,21], *Creolophus*[21], *Dentipellis*[21,26], *Gloeocystidiellum*[2,21,26], *Hericium*[2,3,21], *Laxitextum*[2,21,26]
Lachnocladiaceae	*Dichostereum*[21,26], *Scytinostroma*[2,26], *Vararia*[21,26]
Polyporaceae	*Albatrellus*[11,17], *Heterobasidion*[2,3,11,16,21,25,26]
Stereaceae	*Amylostereum*[2,25,26], *Stereum* (incl. *Xylobolus*)[2,3,25,26]
Gasteromycetes	
Hymenogastrales	
Astrogastraceae	*Cystangium*[21], *Gymnomyces*[21], *Macowanites*[21], *Martellia*[21], *Zelleromyces*[21]
6. Hymenochaetoid clade	
Hymenomycetes	
Aphyllophorales	
Corticiaceae	*Basidioradulum*[2], *Hyphodontia*[2]
Hymenochaetaceae	*Coltricia*[2,3,15], *Hymenochaete*[26], *Inonotus*[2,3,15,26], *Phellinus*[2,3,26], *Phylloporia*[3]
Polyporaceae	*Oxyporus*[2,3], *Schizopora*[2], *Trichaptum*[2,3,15]
7. Cantharelloid clade	
Hymenomycetes	
Aphyllophorales	
Cantharellaceae	*Cantharellus*[2,5,11], *Craterellus*[5]
Clavariaceae	*Multiclavula*[2,5,9,23]
Clavulinaceae	*Clavulina*[2,5]
Corticiaceae	*Botryobasidium*[2,5]
Hydnaceae	*Hydnum*[2,5]
Heterobasidiomycetes	
Tulasnellales	
Tulasnellaceae	*Tulasnella*[2,11]

Table 1. *Continued*

Clade	Exemplar genera sampled in molecular studies, and references[1]
8. Gomphoid-phalloid clade	
Hymenomycetes	
Aphyllophorales	
Clavariaceae	*Clavariadelphus*[2,5,12]
Gomphaceae	*Gloeocantharellus*[5], *Gomphus*[2,5,11,12,17], *Kavinia*[11], *Lentaria*[5], *Ramaria*[2,5,11,12,16,17]
Gasteromycetes	
Gautieriales	
Gautieriaceae	*Gautieria*[11,12]
Hymenogastrales	
Cortinariaceae	*Kjeldsenia*[12]
Hymenogastraceae	*Chondrogaster*[12]
Lycoperdales	
Geastraceae	*Geastrum*[2,5]
Phallales	
Clathraceae	*Aseroe*[12,16], *Clathrus*[12], *Lysurus*[12], *Pseudocolus*[2,5]
Hysterangiaceae	*Hysterangium*[12], *Trappea*[12]
Phallaceae	*Phallus*[12]
Protophallaceae	*Protubera*[12]
Nidulariales	
Sphaerobolaceae	*Sphaerobolus*[2,5,16]

[1] List does not include all taxa sampled in studies of Gardes and Bruns (1996), Moncalvo et al. (2000, and unpubl.), Bruns et al. (1998), Colgan et al. (1997), J. Spatafora (unpubl.), R.E. Thorn (unpubl.), D.S. Hibbett (unpubl.), and Boidin et al. (1998).
[2] Hibbett et al. (1997 and unpubl.).
[3] Hibbett and Donoghue (1995).
[4] Y.-J. Yao and D.S. Hibbett (unpubl.).
[5] Pine et al. (1999).
[6] Moncalvo et al. (2000, and unpubl.).
[7] Chapela et al. (1994).
[8] Hopple and Vilgalys (1994).
[9] Gargas et al. (1995).
[10] Bruns and Szaro (1992).
[11] Bruns et al. (1998; see also Cullings et al. 1996).
[12] Colgan et al. (1997) and J. Spatafora (unpubl.).
[13] Mueller and Pine (1994).
[14] Kretzer and Bruns (1997).
[15] Ko et al. (1997).
[16] Thorn et al. (2000, and unpubl.).
[17] Gardes and Bruns (1996).
[18] Nakasone (1996).
[19] Hughey et al. (2000).
[20] A. Gargas (unpubl.).
[21] S. Miller and E. Larsson (unpubl.).
[22] Binder et al. (1997).
[23] Lutzoni (1997) and Lutzoni and Pagel (1997).
[24] Neda and Nakai (1995).
[25] Hsiau (1996).
[26] Boidin et al. (1998).

primarily composed of polypores and corticioid fungi, but also includes the gilled mushrooms *Lentinus*, *Panus*, and *Faerberia* (= *Geopetalum*), as well as the "cauliflower fungus" *Sparassis* (which appears to be closely related to the polypores *Laetiporus* and *Phaeolus*; Fig. 2). The tree in Fig. 2 suggests that the poroid habit is plesiomorphic in the polyporoid clade and has given rise to gilled, toothed, and corticioid forms. Monophyly of the polyporoid clade is only weakly supported by bootstrapping, and in certain analyses of mt-ssu rDNA (Hibbett and Donoghue 1995) or mt-ssu rDNA and nuc-ssu rDNA (Hibbett 1996) it appears to be polyphyletic (but see Ko et al. 1997).

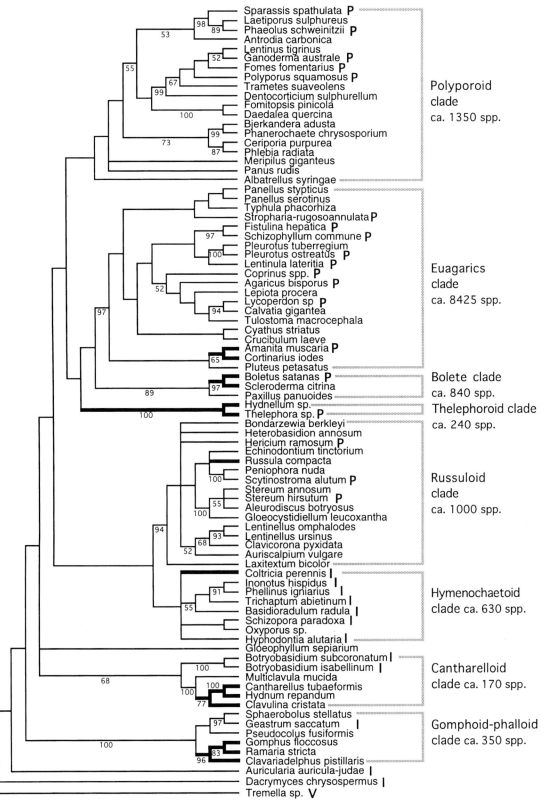

Fig. 2. Phylogenetic relationships of homobasidiomycetes. Strict consensus of 52 equally parsimonious trees found in analysis of nuc-ssu and mt-ssu rDNA sequences (Hibbett et al. 1997b). *Heavy lines* indicate ectomycorrhizal lineages. *Capital letters after taxon names* indicate parenthesome morphology, if known (*P* perforate; *I* imperforate; *V* vesiculate; see Table 4 for references). *Numbers by branches* are frequencies (%) of occurrence out of 100 bootstrap replicates (values <50% are not shown). Branch lengths do *not* correspond to numbers of nucleotide substitutions. *Bracketed groups to right of tree* are discussed in text. Estimates of diversity refer to described species only, based on Hawksworth et al. (1995)

An especially problematic result concerns *Gloeophyllum*; analyses of nuc-ssu and mt-ssu rDNA sequences have placed *Gloeophyllum* outside the polyporoid clade (Hibbett et al. 1997b; Fig. 2), but analyses of nuc-lsu rDNA sequences have placed it inside the polyporoid clade (Thorn et al. 2000). The latter placement is consistent with morphology, and it is accepted here. Despite the weak or conflicting results of molecular analyses, many members of the polyporoid clade have similar morphology and anatomy (i.e., the polypore habit, dimitic hyphal construction) and nutritional modes (wood decay). Thus, we tentatively conclude that the polyporoid clade is monophyletic, but this hypothesis should be tested in additional studies. Although the polyporoid clade as a whole is weakly supported, there are four groups within the polyporoid clade that are strongly supported and that have corroborating anatomical and physiological characters. Examples include the *Fomitopsis-Daedalea-Piptoporus* group (brown rot, bipolar mating system), and the *Polyporus-Lentinus-Ganoderma* group (white rot, tetrapolar mating systems, binding hyphae). For discussion of these and other lineages in the polyporoid clade, see Hibbett and Donoghue (1995).

The study of Hibbett et al. (1997b) included 19 genera in the polyporoid clade. An additional 27 genera of this group were sampled in studies by Bruns et al. (1998), Hibbett and Donoghue (1995), D.S. Hibbett (unpubl.), and Y.-J. Yao et al. (unpubl.; Table 1). In addition, Boidin et al. (1998) sampled approximately 37 genera in the polyporoid clade, including 19 that have not been examined elsewhere.[1] These additional genera represent the Corticiaceae, "residual" Hydnaceae (Donk 1964), and Polyporaceae. The latter includes *Faerberia* (Singer 1986; Thorn et al. 2000), and *Polyporoletus*, which is a terrestrial polypore that resembles *Albatrellus* (Singer et al. 1945; Gilbertson and Ryvarden 1986). Analyses of mt-lsu rDNA by Bruns et al. (1998; see also Gardes and Bruns 1996) suggest that *Polyporoletus* and some species of *Albatrellus* form a lineage in the polyporoid clade. However, these studies also suggests that *Albatrellus* is polyphyletic, with some species in the polyporoid clade and others in the russuloid clade (see below). *Albatrellus* has been interpreted as entirely mycorrhizal (Gilbertson and Ryvarden 1986), but Ginns (1997) suggested that *A. syringae* (polyporoid clade) is lignicolous.

The main families represented in the polyporoid clade, Polyporaceae and Corticiaceae, are polyphyletic, which makes it difficult to estimate the number of described species that can be assigned to this group. Ryvarden (1991) listed 11 groups of "related genera" in the Polyporaceae. Eight of these groups are represented in Table 1, and each of these has at least one member in the polyporoid clade (several of the groups appear to be polyphyletic). Parmasto (1986) divvided the Corticiaceae into 11 subfamilies. Eight of Parmasto's subfamilies are represented in Table 1, and three of these have species in the polyporoid clade (again, some subfamilies appear to be polyphyletic). Taking Ryvarden and Parmasto's divisions of the Polyporaceae and Corticiaceae into account, we estimate that the polyporoid clade contains roughly 1350 described species of homobasidiomycetes, including 90% of known Polyporaceae and 25% of Corticiaceae, as well as all Ganodermataceae and Sparassidaceae.

2. Euagarics Clade

This large clade is composed mostly of Agaricales (gilled mushrooms), but it also includes Aphyllophorales, Gasteromycetes, and possibly certain Ceratobasidiales (see below). In the study of Hibbett et al. (1997b), the euagarics clade was found to include exemplars of seven families of gilled Agaricales (Agaricaceae, Amanitaceae, Coprinaceae, Cortinariaceae, Pluteaceae, Strophariaceae, Tricholomataceae), four families of Aphyllophorales (Clavariaceae, Fistulinaceae, Polyporaceae, Schizophyllaceae), and three families of Gasteromycetes (Lycoperdaceae, Nidulariaceae, Tulostomataceae; Fig. 2, Table 1). It appears that the agaricoid habit is plesiomorphic in the euagarics clade and has given rise to multiple lineages of nongilled hymenomycetes and gasteroid forms (Hibbett et al. 1997b).

Members of the euagarics clade have been investigated in numerous phylogenetic studies, most notably the studies of Moncalvo et al. (2000, and unpubl.), who have sampled over 300 species in this group. The studies of Moncalvo et al., and others (see Table 1), have included many exem-

---

[1] Boidin et al. (1998) analyzed rDNA internal transcribed spacer (ITS) sequences from 360 species of basidiomycetes, mostly Aphyllophorales. The use of such a highly variable gene for broad phylogenetic analyses is questionable, and there were no bootstrap values reported. Taxonomic conclusions based solely on Boidin et al.'s study should be interpreted with caution.

plars of the families sampled by Hibbett et al. (1997b; Table 1), as well as five families of Agaricales (Bolbitiaceae, Crepidotaceae, Entolomataceae, Hygrophoraceae, Paxillaceae pro parte), two famillies of Aphyllophorales (Corticiaceae, Dictyonemataceae), and several secotioid Gasteromycetes that were not sampled by Hibbett et al.

The Paxillaceae are represented in the euagarics clade by *Lampteromyces*, *Omphalotus*, and *Ripartites* (Binder et al. 1997; Moncalvo et al. 2000, and unpubl.; Thorn et al. 2000). However, *Hygrophoropsis* and *Paxillus* (including *Tapinella*) occur in the bolete clade (see below), which indicates that the Paxillaceae is polyphyetic. There are no published reports on the relationships of Crepidotaceae, but preliminary analyses of nuc-lsu rDNA sequences suggest that *Crepidotus crocophyllus* is closely related to the euagarics clade or bolete clade (Thorn et al. 2000).

Monophyly of the euagarics clade was strongly supported in the analysis of Hibbett et al. (1997b), but other studies, with different samples of taxa, have found weak support for the group, or have suggested that it is polyphyletic. For example, analyses by Bruns et al. (1998) and Gardes and Bruns (1996) suggested that the Hygrophoraceae is outside the euagarics clade, contrary to the results of Pine et al. (1999) and Moncalvo et al. (2000). Even without the Hygrophoraceae, the euagarics clade would contain the majority of the gilled mushrooms in the Agaricales sensu Singer (1986). The largest groups of gilled mushrooms that are not in the euagarics clade are the Russulaceae (russuloid clade), and Gomphidiaceae and Paxillaceae pro parte (bolete clade, see below).

Aphyllophorales make up a relatively small part of the euagarics clade (Table 1). Several small, morphologically distinctive families are represented, including the Fistulinaceae, Schizophyllaceae, and Dictyonemataceae (Fig. 2, Table 1), as well as the large, polyphyletic families Corticiaceae and Clavariaceae (Table 1). In the analysis of Hibbett et al. (1997b), the euagarics clade contains a single species of the Clavariaceae, *Typhula phacorhiza* (Donk 1964, 1971). The analysis of Pine et al. (1999) suggested that the euagarics clade may contain four additional genera of Clavariaceae (Table 1), including *Clavaria* and *Clavulinopsis*, but bootstrap support was weak. The Corticiaceae is represented by *Athelia bombacina* (Gargas et al. 1995) and possibly *Piloderma fallax* (as *P. croceum*; Bruns et al. 1998; Gardes and Bruns 1996; see below). In addition, the corticioid bark beetle symbiont *Gloeocystidium ipidophilum* appears to be in the euagarics clade (Hsiau 1996).

Heterobasidiomycetes are represented in the euagarics clade by two members of Ceratobasidiales, *Ceratobasidium sp.* (D.S. Hibbett, unpubl.) and *Waitea circinata* (Bruns et al. 1998). *Ceratobasidium* has "ceratobasidioid" basidia (with large, fingerlike sterigmata) and parenthesomes with large perforations (Wells 1994). *Waitea* has typical homobasidia, but it also has parenthesomes with large perforations and has therefore been classified in the Ceratobasidiales (Tu and Kimbrough 1978) or Botryobasidiales (e.g., Müller et al. 1998). In the study of Bruns et al. (1998), *Waitea* and *Piloderma* (Corticiaceae) form a monophyletic group that is weakly supported as the sister group of the rest of the euagarics clade (excluding Hygrophoraceae). Analyses of mt-ssu and nuc-ssu rDNA sequences (D.S. Hibbett, unpubl.) suggest that *Ceratobasidium* is nested in the euagarics clade, although the exact placement is uncertain. Surprisingly, analyses including *Thanatephorus*, which is also classified in the Ceratobasidiales, have repeatedly placed this taxon outside of the euagarics clade (Swann and Taylor 1993; Gargas et al. 1995; Lee and Jung 1997; D.S. Hibbett, unpubl.). Taken at face value, these results suggest that the Ceratobasidiales is polyphyletic. However, Lee and Jung (1997) observed that two isolates identified as *Thanatephorus cucumeris* (= *T. praticola*) did not form a monophyletic group in analyses of nuc-ssu rDNA sequences, which suggests that some cultures or DNA preparations of *Thanatephorus* have been misidentified or mislableled. Clearly, additional studies are needed to resolve the relationships between Ceratobasidiales, corticioid members of the euagarics clade, and other homobasidiomycetes.

Gasteromycetes that have been sampled in the euagarics clade include bird's nest fungi, puffballs, false truffles, and secotioid fungi (Table 1). *Lycoperdon* and *Calvatia* (Lycoperdales) and *Tulostoma* (Tulostomatales pro parte) form a weakly supported monophyletic group that is nested in the Agaricaceae (Fig. 2; Hibbett et al. 1997b). The exact placement of the bird's nest fungi (Nidulariales) is not resolved with confidence. Secotioid fungi and false truffles in the euagarics clade that have been studied using molecular techniques include *Hydnangium*, which is closely related to *Laccaria* (Tricholomataceae),

Podaxis and *Montagnea*, which are closely related to Coprinaceae, *Weraroa*, which is thought to be related to Strophariaceae, and *Leratia*, which is of uncertain placement (Heim 1971; Hopple and Vilgalys 1994; Mueller and Pine 1994; Binder et al. 1997). Based on anatomical characters, at least five additional families of euagarics are thought to contain secotioid fungi and false truffles, including Agaricaceae (these secotioid forms may be related to Lycoperdales and Tulostomatales), Amanitaceae, Bolbitiaceae, Cortinariaceae, and Entolomataceae (Dring 1973; Smith 1973; Thiers 1984; Miller and Miller 1988; Castellano et al. 1989).

We estimate that the euagarics clade contains approximately 7400 species of Agaricales and 1025 species of Aphyllophorales and Gasteromycetes. This represents over half of all known homobasidiomycetes, including approximately 87% of all known gilled mushrooms (Hawksworth et al. 1995). The euagarics clade includes saprotrophs, pathogens, and symbionts of both plants and animals (see Sect. III F).

3. Bolete Clade

Hibbett et al. (1997b) sampled only three species in the bolete clade, representing the Boletaceae, Paxillaceae (which appears to be polyphyletic, see above), and Sclerodermataceae (Fig. 2, Table 1). However, the bolete clade has been sampled extensively by Bruns and colleagues, and others (see Table 1), who have demonstrated that the bolete clade also includes members of the Gomphidiaceae, Coniophoraceae, Hymenogastrales, Melanogastraceae, and Calostomataceae, as well as many additional exemplars of the groups studied by Hibbett et al. (1997; Table 1).

The bolete clade contains poroid forms, gilled mushrooms, resupinate fungi, false truffles, secotioid fungi, and puffballs, including the unusual stalked, gelationous puffball *Calostoma* (Hughey et al. 2000). Analyses by Bruns et al. (1998) show a basal trichotomy in the bolete clade, including one mostly agaricoid lineage that includes *Paxillus*, *Hygrophoropsis*, and *Serpula*. The bolete clade is weakly supported as the sister group of the euagarics clade in the analysis of Hibbett et al. (1997b), as well as the analysis of Begerow et al. (1997). Taken together, these results suggest that the ancestor of the bolete clade may have been a gilled mushroom.

Monophyly of the bolete clade has been moderately to strongly supported in analyses based on nuclear and mitochondrial rDNAs (Colgan et al. 1997; Hibbett et al. 1997b; Bruns et al. 1998). In many cases, close relationships have been suspected between the poroid boletes (e.g., *Boletus* and *Suillus*) and certain gilled (e.g., *Phylloporus*), resupinate (e.g., *Coniophora*), and gasteroid (e.g., *Rhizopogon* and *Melanogaster*) forms based on anatomy, spore morphology, wood decay chemistry, susceptibility to certain fungal pathogens, and pigments (e.g., Nilsson and Ginns 1979; Singer 1986; Gill and Steglich 1987; Besl et al. 1996). The Boletaceae, Paxillaceae, and Gomphidiaceae were placed in the suborder Boletineae of the Agaricales by Singer (1986).

The majority of the taxa in the bolete clade occur in two main ectomycorrhizal lineages that were termed the boletoid group and the suilloid group by Bruns et al. (1998). In both groups there are secotioid fungi and false truffles along with pileate poroid and gilled forms. Outside the boletoid and suilloid groups, there are several lineages that contain resupinate and pileate brown-rotting saprotrophs as well as ectomycorrhizal poroid forms and puffballs (Fig. 2 in Bruns et al. 1998). We estimate that the bolete clade includes about 840 described species.

4. Thelephoroid Clade

In the study of Hibbett et al. (1997b), the thelephoroid clade was represented by *Hydnellum* and *Thelephora*, which were strongly supported as monophyletic. Based on studies by Bruns et al. (1998) and Gardes and Bruns (1996), *Boletopsis*, *Sarcodon*, *Thelephora*, *Tomentella*, and *Pseudotomentella* are also in the thelephoroid clade. These taxa represent the majority of the Thelephoraceae sensu Donk (1964). Characters used by Donk to support the Thelephoraceae include dark, ornamented spores with an angular outline (in addition to the ornamentations), pigmentation of the fruiting body, and presence of thelephoric acid (see Sect. III.E).

The Thelephoraceae has been generally accepted as monophyletic, but its relationship to the Bankeraceae (which includes *Bankera* and *Phellodon*) is controversial. *Bankera* and *Phellodon* strongly resemble *Hydnellum*, but they have subglobose, spinose spores and often have light-colored fruiting bodies. Donk (1964, p. 247) suggested that the similarity of the Bankeraceae to certain Thelephoraceae is "an example of extreme convergence", but other authors have suggested that the Bankeraceae and Thelephoraceae are closely related (e.g., Jülich 1981; Stalpers 1993). As far as we are aware, *Bankera* and *Phellodon* have yet to be included in molecular studies.

With or without the Bankeraceae, the thelephoroid clade is a morphologically diverse group that includes corticioid fungi (*Tomentella*), clavarioid forms (*Thelephora*), and pileate forms with poroid (*Boletopsis*), toothed (*Hydnellum*, *Sarcodon*), smooth to wrinkled or tuberculate (*Thelephora*), or lamellate hymenophores (*Lenzitopsis*). In addition, Oberwinkler (1975) suggested that the agaricoid fungus *Horakia* (= *Verrucospora*) is related to Thelephoraceae based on spore morphology. Nevertheless, Singer (1986) placed *Horakia* in the Agaricaceae. Except for *Lenzitopsis*, which is apparently lignicolous (and has not been included in molecular studies), all members of this group are thought to be ectomycorrhizal or orchid symbionts (Cullings et al. 1996; Taylor and Bruns 1997; Bruns et al. 1998). Assuming that the Bankeraceae and Thelephoraceae form a monophyletic group, we estimate that the thelephoroid clade includes about 240 described species of homobasidiomycetes.

5. Russuloid Clade

In the analysis of Hibbett et al. (1997b), the russuloid clade is a strongly supported group that contains representatives of Russulaceae (Agaricales), and eight families of Aphyllophorales: Auriscalpiaceae, Bondarzewiaceae, Corticiaceae, Echinodontiaceace, Hericiaceae, Lachnocladiaceae, Polyporaceae, and Stereaceae (Fig. 2, Table 1). Analyses of mt-ssu rDNA sequences alone (Hibbett and Donoghue 1995) placed the Stereaceae outside of this group (with low bootstrap support), but otherwise supported monophyly of the russuloid clade (not all of the same taxa were sampled, however). The russuloid clade has been intensively sampled by Boidin et al. (1998), who examined ITS sequences, and S. Miller and E. Larsson (unpubl.), who examined nuc-lsu rDNA sequences. Miller and Larsson have investigated over 80 species of the russuloid clade, including every family that was studied by Hibbett et al. (197b), except Stereaceae sensu stricto. In addition, Miller and Larsson sampled five gasteroid genera in the Astrogastraceae (Hymenogastrales sensu Smith 1973), which their results show to be scattered throughout a clade that is otherwise composed of *Russula* and *Lactarius*

species. The russuloid clade has also been examined using mt-lsu rDNA sequences by Bruns et al. (1998) and Gardes and Bruns (1996), who have demonstrated relationships between Russulaceae, Bondarzewiaceae, and Polyporaceae, the latter being represented by *Albatrellus* pro parte, *Heterobasidion*, and *Byssoporia*. Taken together, the studies of Bruns et al. (1988), Gardes and Bruns (1996), and Hibbett et al. (1997b) suggest that *Albatrellus* consists of two separate lineages: one group of species in the polyporoid clade, along with *Polyporoletus*, and another in the russuloid clade, with *Byssoporia* (which is a resupinate, mycorrhizal polypore with broadly elliptic to subglobose spores, similar to those of *Albatrellus*; Ryvarden and Gilbertson 1993).

The russuloid clade has a remarkable diversity of fruiting body morphologies (Table 2). The hymenomycetous forms are resupinate, pileate, or coralloid, with smooth, toothed, lamellate, or poroid hymenophores; the gasteroid forms include false truffles and secotioid fungi. The russuloid clade is also ecologically variable, having ectomycorrhizal, pathogenic, saprotrophic, and possibly lichenized species (see Sect. III.F; Singer 1984). There is no obvious morphological synapomorphy for the russuloid clade. As discussed by Donk (1964), many members of the russuloid clade have spores with amyloid ornamentations and gloeoplerous cystidia or hyphae, but these characters are variable within the group (consequently, they might be useful for resolving phylogenetic relationships within the russuloid clade). Based on anatomical characters, close relationships have been proposed to exist between certain subsets of the russuloid clade, such as Bondarzewiaceae and Russulaceae (Singer 1984), Stereaceae, *Peniophora*, and Echinodontiaceae (Jülich 1981), *Lentinellus* and Auriscalpiaceae (Maas Geesteranus 1963), *Albatrellus* and Hericiaceae (Stalpers 1992), and various gasteroid taxa (e.g., *Arcangeliella*, *Macowanites*) and Russulaceae (Smith 1973; Pegler and Young 1979; Castellano et al. 1989).

Most of the families in the russuloid clade have distinctive characters and are probably monophyletic, or at least restricted to the russuloid clade (e.g., Russulaceae, Bondarzewiaceae, Hericiaceae). However, the Corticiaceae is certainly polyphyletic. We estimate that the russuloid clade contains approximately 1000 described species of homobasidiomycetes, including roughly 20% of Corticiaceae.

Table 2. Distribution of fruiting body morphotypes over the major clades of homobasidiomycetes

Major clades	Fruiting body types						
	Agaricoid	Poroid	Toothed	Club-shaped or coralloid	Corticioid	Epigeous gasteroid-secotioid	Hypogeous gasteroid
Polyporoid clade	✓	✓	✓	✓[a]	✓	✓[b]	
Euagarics clade	✓	✓		✓	✓	✓	✓
Bolete clade	✓	✓	✓[c]		✓	✓	✓
Thelephoroid clade	✓[d]	✓	✓	✓	✓		
Russuloid clade	✓	✓	✓	✓	✓	✓	✓
Hymenochaetoid clade	✓[e]	✓	✓	✓[f]	✓		
Cantharelloid clade	✓[g]		✓	✓	✓		
Gomphoid-phalloid clade	✓		✓	✓	✓[h]	✓	✓

[a] Environmentally induced coralloid forms of *Neolentinus lepideus*.
[b] Secotioid form of *Lentinus tigrinus*.
[c] *Gyrodontium*.
[d] *Horakia*? *Lenzitopsis*.
[e] *Cyclomyces*.
[f] *Clavariachaete*?
[g] *Cantharellus* spp.
[h] *Ramaricium*.

6. Hymenochaetoid Clade

The hymenochaetoid clade includes the Hymenochaetaceae, as well as certain members of the Corticiaceae (*Basidioradulum*, *Hyphodontia*) and Polyporaceae (*Trichaptum*, *Oxyporus*, *Schizopora*; Fig. 2, Table 1). Thus, the hymenochaetoid clade includes poroid, toothed, and corticoid forms (Tables 1, 2). In addition, the clavarioid *Clavariachaete* and the stipitate-sterioid *Stipitochaete* are probably in the hymenochaetoid clade, based on anatomical features (Corner 1991).

In analyses of mt-ssu rDNA sequences alone (Hibbett and Donoghue 1995), the hymenochaetoid clade was moderately supported (bootstrap = 75%), but with the inclusion of nuc-ssu rDNA sequences support dropped below 50% (Hibbett et al. 1997b; Fig. 2). An analysis of nuc-ssu rDNA sequences alone (Ko et al. 1997) supported monophyly of *Trichaptum* and *Inonotus*, but excluded *Coltricia* (no other members of the hymenochaetoid clade were sampled in the study).

The Hymenochaetaceae has been generally regarded as monophyletic based on the following suite of characters: clampless generative hyphae, fruiting bodies darkening in KOH, production of white rot, and presence of setae (in some species). The inclusion in the hymenochaetoid clade of Corticiaceae and Polyporaceae, which lack this combination of characters, conflicts with the generally accepted delimitation of the Hymenochaetaceae.

Nevertheless, members of the hymenochaetoid clade appear to be united by the possession of an imperforate parenthosome, which has been observed in *Basidioradulum*, *Coltricia* (W. Müller, pers. comm., contra Moore 1980), *Hyphodontia*, *Inonotus*, *Phellinus*, *Schizopora*, and *Trichaptum* (Traquair and McKeen 1978; Moore 1980, 1985; E. Langer 1994; Hibbett and Donoghue 1995; see Sect. III.B.5).

The hymenochaetoid clade is primarily composed of lignicolous species. However, *Coltricia* is ectomycorrhizal (Danielson 1984) and a number of lignicolous hymenochaetoid fungi have been shown to form orchid mycorrhizae (Umata 1995). It is likely that all the Hymenochaetaceae are in the hymenochaetoid clade, but it is difficult to estimate how many Corticiaceae and Polyporaceae are in this group. We conservatively estimate that the hymenochaetoid clade includes about 630 described species of homobasidiomycetes, which includes Hymenochaetaceae and the genera of the Corticiaceae and Polyporaceae discussed above.

7. Cantharelloid Clade

The analysis of Hibbett et al. (1997b) suggested that the cantharelloid clade contains Cantharellaceae (*Cantharellus*), Hydnaceae (*Hydnum*), Clavariaceae (*Multiclavula*), Clavulinaceae (*Clavulina*), and Corticiaceae (*Botryobasidium*).

These results are consistent with the findings of Pine et al. (1999), who performed analyses focused on the cantharelloid and clavarioid homobasidiomycetes. Pine et al's study also included *Craterellus* and additional species of *Cantharellus* and *Clavulina* that were not sampled in the study of Hibbett et al. (1997b).

Analyses of mt-rDNA sequences suggest that the heterobasidiomycete order Tulasnellales is in the cantharelloid clade (Bruns et al. 1998; D.S. Hibbett, unpubl.). Based on mt-ssu rDNA sequences (D.S. Hibbett unpubl.), *Tulasnella* is the sister group of *Botryobasidium*. *Tulasnella* and *Botryobasidium* are both corticioid fungi that have imperforate parenthesomes (G. Langer 1994; Wells 1994). The major difference between the two is in basidial morphology; *Tulasnella* has tulasnelloid basidia with inflated epibasidia Wells 1994), whereas *Botryobasidium* has rounded to subcylindrical homobasidia. In addition, analyses of mt-lsu rDNA sequences (Bruns et al. 1998) suggest that the sister group of *Cantharellus* is a clade composed of *Tulasnella* and an isolate that was tentatively identified as "*Sebacina* sp." (Auriculariales sensu Wells). Ignoring the questionable *Sebacina* isolate, the mt-rDNA studies collectively suggest that a lineage that contains *Tulasnella* and *Botryobasidium* is the sister group of the rest of the cantharelloid clade.

The analysis of Hibbett et al. (1997b) placed *Multiclavula* within the cantharelloid clade. However, the nuc-ssu rDNA analysis of Gargas et al. (1995) suggested that *Multiclavula* is the sister group of all other homobasidiomycetes plus Auriculariales and Ceratobasidiales. As discussed by Pine et al. (1999), the rate of sequence evolution of nuc-ssu rDNA appears to be greater in the cantharelloid clade than in most other homobasidiomycetes. Thus, it is possible that the phylogenetic artifact known as long branch attraction (Felsenstein 1978) is responsible for the placement of *Multiclavula* in the analysis of Gargas et al. (1995).

The cantharelloid clade includes cantharelloid to agaricoid, hydnoid, coralloid, clavarioid, and corticioid fungi. A distinctive feature of the cantharelloid clade is the possession of stichic basidia (see Sect. III.B.4), which have been observed in *Cantharellus*, *Clavulina*, *Craterellus*, *Hydnum*, and *Multiclavula* (Donk 1964; Restivo and Petersen 1976; Hubbard and Petersen 1979; Juel 1898; Pine et al. 1999). Most members of the cantharelloid clade are known or presumed to be mycorrhizal, but *Multiclavula* is a basidiolichen (see Sect. III.F.5). *Botryobasidium* produces fruiting bodies on wood or soil, but it is not known whether it is saprotrophic or mycorrhizal. We estimate that the cantharelloid clade contains about 170 described species of homobasidiomycetes.

8. Gomphoid-Phalloid Clade

In the study of Hibbett et al. (1997b; Fig. 2), the gomphoid-phalloid clade included club- and coral-shaped hymenomycetes in Clavariaceae (*Clavariadelphus*) and Gomphaceae (*Gomphus* and *Ramaria*), and morphologically diverse Gasteromycetes in Lycoperdales (*Geastrum*), Nidulariales (*Sphaerobolus*), and Phallales (*Pseudocolus*). The gomphoid-phalloid clade has also been studied by Bruns et al. (1998), Colgan et al. (1997), Pine et al. (1999), Spatafora (unpubl.), and Thorn et al. (2000; see Table 1). These studies suggest that the gomphoid-phalloid clade includes additional Gomphaceae (*Gloeocantharellus*, *Kavinia*, *Lentaria*), as well as epigeous and hypogeous Gasteromycetes in Gautieriales, Hymenogastrales, and Phallales (Table 1).

The internal topology of the gomphoid-phalloid clade varies from study to study. Nevertheless, the results of Hibbett et al. (1997b) and most of the studies cited above, suggest that there are two groups. One group contains the Gomphaceae sensu Donk (1964; *Gomphus*, *Gloeocantharellus*, *Kavinia*, *Lentaria*, *Ramaria*), *Clavariadelphus*, and the Gautieriales. This clade includes club-shaped fungi (*Gomphus*, *Clavariadelphus*), coralloid fungi (*Lentaria*, *Ramaria*), hydnoid resupinate fungi (*Kavinia*), gilled mushrooms (*Gloeocantharellus*), and false truffles (*Gautieria*). The corticioid fungus *Ramaricium* is also in this group, based on spore morphology and staining reactions to iron salts (Eriksson 1954; Eriksson and Ryvarden 1976; Ginns 1979). The Gomphaceae are united by cyanophilic, ornamented (occasionally smooth) spores and green staining reactions to iron salts (Donk 1964; Petersen 1967). *Clavariadelphus* has smooth, non-cyanophilic spores, and *Gautieria* has ridged, non-cyanophilic spores, but both stain green or red in response to iron salts (Donk 1964; Stewart 1974). The other group in the gomphoid-phalloid clade contains the earthstar *Geastrum*, the cannon-ball fungus *Sphaerobolus*, stinkhorns (Clathraceae and Phallaceae), and various hypogeous Phallales and Hymenogastrales (see Table 1). Other than the lack of ballistospory, and certain pigments (see Sect. III.E), there are no obvious characters that unite these taxa or that suggest a relationship to the Gomphaceae-*Clavariadelphus-Gautieria* lineage. However, Colgan et al. (1997) noted similarities in the ectomycorrhizal mats formed by *Gautieria* and Hysterangiaceae. We estimate that the gomphoid-phalloid clade includes about 350 described species.

III. Characters

Most recent phylogenetic studies in homobasidiomycetes have used molecular characters, especially those derived from rDNA (Bruns et al. 1991; Hibbett 1992; Hibbett and Donoghue 1998). In this section, we present an overview of selected "morphological" (= nonmolecular) characters that may be phylogenetically informative at high taxonomic levels in homobasidiomycetes, or that provide insight into the evolution of ecological strategies. For additional general discussions of taxonomic characters in homobasidiomycetes, including many characters that are not discussed here (e.g., cystidia, tramal anatomy, fruiting body development, etc.), see Clémençon (1997), Donk (1964), Singer (1986), Jülich (1981), Kühner (1980, 1984), Oberwinkler (1985), Petersen (1971), and Reijnders and Stalpers (1992).

A. Fruiting Body Macromorphology

The major fruiting body morphotypes of homobasidiomycetes include gilled mushrooms, polypores, toothed fungi, club and coral fungi, corticioid fungi, and epigeous and hypogeous gasteroid fungi (Figs. 3–14). For taxonomic overviews of these morphologically defined groups, see Kühner (1980), Singer (1986), Smith (1973), and Moser (1978) for agarics and boletes; Gilbertson and Ryvarden (1986), Ryvarden (1991), and Pegler (1973) for polypores (excluding boletes and poroid Agaricales); Harrison (1971, 1973) and Gilbertson (1971) for toothed fungi; Petersen (1971, 1973) and Corner (1950, 1966, 1970) for club and coral fungi; Parmasto (1986), Hjortstam et al. (1988, and subsequent volumes), Talbot (1973b), Christiansen (1960), Ginns and Lefebvre (1993), and Jülich and Stalpers (1980) for resupinate fungi; and Coker and Couch (1928), Smith (1973), Dring (1973), Castellano et al. (1989), and Miller and Miller (1988) for gasteroid fungi.

The phylogenetic distribution of fruiting body forms in homobasidiomycetes is summarized in Table 2 (see also Table 1 in Donk 1971). Each of the eight major clades of homobasidiomycetes that we recognize contains at least four of the basic fruiting body types. Notable gaps in the Table include the absence of toothed forms from the euagarics clade, and gasteroid forms from the thelephoroid and cantharelloid clades. Only the russuloid clade contains all the morphotypes.

Phylogenetic uncertainty and limited taxon sampling make it difficult to infer the history of morphological transformations in homobasidiomycetes in detail. Nevertheless, on a coarse scale it appears that the nongilled hymenomycetes make up a paraphyletic assemblage from which gilled mushrooms and Gasteromycetes have been repeatedly derived (Hibbett et al. 1997b). This underscores the need to integrate the classification of the Agaricales, Aphyllophorales, and Gasteromycetes. Based on comparison to heterobasidiomycetes, the ancestor of the homobasidiomycetes may have been a resupinate or sessile pileate form. In the following sections, we discuss general patterns of morphological evolution in homobasidiomycetes, highlighting agarics, corticioid fungi, and gasteroid forms.

1. Agarics

The Agaricales sensu Fries has long been recognized as an artificial taxon, but the number of independent lineages of gilled mushrooms, and their higher-order relationships, have remained controversial (Kühner 1980; Singer 1986). One of the most influential hypotheses regarding the origin of agarics is Corner's (1950) "*Clavaria* theory" (Jülich 1981; Miller and Watling 1987; Pine et al. 1999), which holds that simple club-shaped fruiting bodies are plesiomorphic in the homobasidiomycetes and were transformed in multiple lineages into cantharelloid and agaricoid forms (via hydnoid intermediates in some cases). Our present understanding of homobasidiomycete phylogeny is partially consistent with the *Clavaria* theory; agarics are polyphyletic, but there is no indication that they have all been derived from paraphyletic grades of club and coral fungi.

Hibbett et al. (1997b) suggested that agaricoid forms evolved at least six times. However, their analysis omitted a number of gilled taxa that could

represent additional independent origins of agaricoid fruiting bodies (e.g., *Phylloporus, Gomphidius, Faerberia, Gloeocantharellus*). As indicated in Table 2, taxa with some kind of lamellate hymenophore probably occur in all of the eight major clades that we recognize. The main concentrations of gilled mushrooms are in the euagarics, bolete, and russuloid clades.

Gilled mushrooms have probably been derived from morphologically diverse precursors. In a few cases, we have a good idea of what the ancestors of agarics may have looked like. For example, there is strong evidence that *Lentinus* sensu stricto was derived from stipitate polypores (Hibbett and Vilgalys 1993; Hibbett et al. 1997b). Developmental modifications involved in this pore-gill transformation were described by Hibbett et al. (1993b). Certain gilled taxa in the bolete clade are also probably derived from poroid forms. Examples include *Phylloporus*, which is nested in the boletoid group sensu Bruns, and *Gomphidius*, which is in the suilloid group (Bruns et al. 1998). Other lamellate taxa of the bolete clade are in lineages outside the boletoid and suilloid groups (e.g., *Paxillus*). As discussed previously (Sect. II.C.3), it is plausible that the plesiomorphic morphology of the bolete clade is lamellate. In the cantharelloid clade, corticioid, clavarioid, and coralloid taxa (*Botryobasidium, Tulasnella, Multiclavula, Clavulina*) form a paraphyletic assemblage from which agaricoid (*Cantharellus*) and other pileate forms (*Craterellus, Hydnum*) were derived (Fig. 2; Hibbett et al. 1997b; Pine et al. 1999). Thus, in the cantharelloid clade gills and teeth were probably derived by elaboration from smooth hymenophores. similar pattern may exist in the gomphoid-phalloid clade, where the agaricoid *Gloeocantharellus* is closely related to coralloid, clavarioid, and cantharelloid forms (*Ramaria, Clavariadelphus, Gomphus*; Pine et al. 1999). In both the cantharelloid and gomphoid-phalloid clade, the precise sequence of morphological transformations is unclear, however.

Finally, in the russuloid clade, the agaricoid *Lentinellus* is closely related to coralloid and hydnoid forms (*Clavicorona, Auriscalpium*), but again the order of transformations is unresolved. To better understand the pathways of morphological evolution that have led to agarics, a combination of phylogenetic and comparative developmental approaches will be necessary.

2. Corticioid Fungi

Corticioid forms occur in each of the eight major clades that we recognize (Table 2). The morphological simplicity of corticioid fungi has led to contradictory views that they are either (1) a plesiomorphic, paraphyletic group that has given rise to multiple lineages of complex forms (e.g., Parmasto 1995), or (2) a polyphyletic group that has been repeatedly derived by reduction (e.g., Corner 1991). Molecular studies suggest that in many cases corticioid fungi have been derived from pileate and erect forms. Apparently derived corticioid forms include *Dentocorticium, Phanerochaete*, and *Phlebia* in the polyporoid clade (Fig. 2), *Athelia* in the euagarics clade (Gargas et al. 1995), *Serpula* and *Coniophora* in the bolete clade (Bruns et al. 1998), and possibly *Tomentella* in the thelephoroid clade (Bruns et al. 1998). In addition, the minute, cupulate *Schizophyllum ampla* (= *Auriculariopsis ampla*) appears to be in the euagarics clade (Nakasone 1996), which supports the view that certain cyphelloid fungi (with minute, pendent cups lacking lamellae) are reduced agarics (Donk 1959, 1962, 1964; Reid 1963; Singer 1962, 1986). The contrasting view that the cyphelloid habit is primitive (Agerer 1986) seems unlikely.

Despite these examples, it would be premature to conclude that corticioid forms have never given rise to pileate or erect forms. There are three clades within the homobasidiomycetes in which the polarity of corticioid-erect transformations are particularly ambiguous: (1) The hymenochaetoid

Figs. 3–14. Examples of major fruiting body forms in the eight major clades of homobasidiomycetes. **Fig. 3.** *Daedaleopsis confragosa*. Poroid, polyporoid clade. **Fig. 4.** *Hyphodontia sambuci*. Corticioid, hymenochaetoid clade. **Fig. 5.** *Hydnellum* sp. Hydnoid, thelephoroid clade. **Fig. 6.** *Geastrum saccatum*. Epigeous gasteroid (puffball), gomphoid-phalloid clade. **Fig. 7.** *Rhizopogon* sp. Hypogeous gasteroid (false truffle), bolete clade. **Fig. 8.** *Heimiomyces* sp. Agaricoid, euagarics clade. **Fig. 9.** *Clavulinopsis fusiformis*. Coralloid, euagarics clade. **Fig. 10.** *Boletus* sp. Poroid, bolete clade. **Fig. 11.** *Auriscalpium vulgare*. Hydnoid, russuloid clade. **Fig. 12.** *Clavariadelphus truncatus*. Clavarioid, gomphoid-phalloid clade. **Fig. 13.** *Phallus* cf. *tenuis*. Epigeous gasteroid (stinkhorn), gomphoid-phalloid clade. **Fig. 14.** *Cantharellus cibarius*. Agaricoid, cantharelloid clade. Photo credits: **Figs. 4, 14** R.G. Thorn; **Figs. 7, 9** R.G. Thorn and G.L. Barron; others: D. Hibbett

clade contains a mix of corticioid and erect forms, but the topology of the clade is too poorly resolved (and the sampling of the corticioid taxa is too limited) to infer the pattern of morphological transformations (Fig. 2; Hibbett et al. 1997b). (2) In the russuloid clade, the work of S.L. Miller and E. Larsson (unpubl.) demonstrates that there is a complex pattern of relationships between corticioid and pileate-erect forms, which may include derivation of pileate-erect forms from corticioid forms. (3) In the cantharelloid clade, the basal split is between a corticioid lineage (containing *Botryobasidium* and *Tulasnella*) and a pileate-erect lineage, but the ancestral morphology of the cantharelloid clade is unresolved. Finally, because of uncertainty about higher-level relationships, and limited sampling of corticioid taxa, it is not clear whether the plesiomorphic morphology of the homobasidiomycetes as a whole is corticioid or pileate-erect.

3. Gasteroid Forms

Gasteromycetes produce spores internally and lack forcible spore discharge (see Sect. III.B.2). Some Gasteromycetes, including most false truffles and secotioid fungi, have strong anatomical similarities to certain taxa of Hymenomycetes (Smith 1973; Thiers 1984; Castellano et al. 1989), but many others, such as stinkhorns and other "true" Gasteromycetes, have no obvious morphological resemblance to any group of Hymenomycetes. Gasteromycetes have been interpreted as (1) a paraphyletic assemblage from which the Agaricales were derived (Singer 1986), or (2) a polyphyletic group that was derived from Hymenomycetes (e.g., Savile 1955; Schaffer 1975; Thiers 1984; Oberwinkler 1985; Miller and Watling 1987). Molecular studies indicate that there have been multiple origins of Gasteromycetes in the euagarics, bolete, russuloid, and gomphoid-phalloid clades (Fig. 2, Table 1). The genetic and developmental bases of transformations from Hymenomycetes and Gasteromycetes, and the ecological forces that may have selected for gasteroid fruiting bodies, have been discussed by Baura et al. (1992), Bruns et al. (1989), Hibbett et al. (1994b, 1997b), Miller et al. (1994), and Thiers (1984).

Diverse fruiting body morphologies and spore-dispersal mechanisms have evolved in Gasteromycetes (Ingold 1971). Some gasteroid forms have evolved multiple times, including puffballs, false truffles, and secotioid fungi. In puffballs, spores are dispersed into air through cracks or pores in the peridium. Many puffballs have a "capillitium" of sterile hyphae in the gleba, which is thought to aid dispersal by disrupting the spores or by supporting the structure of the fruiting body (Ingold 1971; Miller and Miller 1988; Pegler 1996). Most capillitial elements resemble skeletal or generative hyphae (see Sect. III.B.3), but others are highly modified, especially the elaters of *Battarrea*, which are sparingly branched cells that have spiral wall thickenings. Puffballs have evolved at least three times, in the euagarics clade (Lycoperdaceae, Tulostomataceae), bolete clade (Sclerodermataceae, Calostomataceae), and gomphoid-phalloid clade (Geastraceae, Fig. 2, Table 1). However, there are many puffballs that have yet to be included in molecular studies, such as *Astraeus*, which is an earthstar that has been classified in the Sclerodermatales (Dring 1973; see Sect. III.F.4).

In false truffles, spores are dispersed by rodents that consume the fruiting bodies, or are released directly into the soil as the fruiting body breaks down (Johnson 1996; Miller et al. 1994; Trappe and Maser 1977). The latter mode may facilitate the establishment of mycorrhizae (Miller et al. 1994). As far as we know, all false truffles are ectomycorrhizal (Trappe and Maser 1977). Thiers (1984) and others have suggested that false truffles were derived from agarics and boletes via secotioid intermediates. In addition, Oberwinkler and Horak (1979) suggested (on the basis of spore morphology) that the false truffle *Stephanospora* is closely related to the corticioid fungus *Lindtneria*. If so, this might represent a unique case of a corticioid-gasteroid transformation. Based on a combination of molecular and morphological evidence, false truffles and secotioid fungi have evolved repeatedly in the euagarics, bolete, russuloid, and gomphoid-phalloid clades (Table 1).

Puffballs, false truffles, and secotioid fungi are clearly polyphyletic, but certain other gasteroid forms have evolved only once. Uniquely derived gasteroid forms in the gomphoid-phalloid clade include the stinkhorns (epigeous Phallales), which disperse spores via flies and other insects that are attracted by the foul-smelling gleba and showy receptacle, and cannon ball fungi (*Sphaerobolus*), which forcibly eject a glebal mass by the sudden evagination of the endoperidium. The pathways of morphological evolution leading to these unusual forms are not well understood, but it seems likely that they were derived from Hymenomycetes via puffball, false truffle, or secotioid intermediates. As mentioned previously, the work of Colgan et al. (1997) and J. Spatafora (unpubl.) suggests that stinkhorns were derived from false truffles (Hysterangiaceae), which suggests that in the Phallales olfactory attractants evolved before showy epigeous receptacles. *Sphaerobolus* and the earthstar *Geastrum* appear to be closely related to the Phallales, but the precise relationships between these taxa are not known.

Another uniquely derived gasteroid form is found in the bird's nest fungi (Nidulariaceae, euagarics clade), which disperse packets of spores (peridioles) by a splash-cup mechanism (e.g., *Cyathus*, *Crucibulum*; Brodie 1975). The Nidulariaceae also includes forms that lack splash cups (Brodie 1975). For example, the fruiting bodies of *Nidularia* resemble minute puffballs, except that the spores are contained in peridioles, which are passively released as the exoperidium breaks down. Brodie (1975) suggested that forms that lack splash cups are primitive in the Nidulariaceae, but this hypothesis has yet to be evaluated through phylogenetic studies.

B. Anatomy, Cytology, and Ultrastructure

1. Basidia

The defining character of the homobasidiomycetes is the homobasidium. Typical homobasidia are nonseptate, clavate to cylindric, with four short, apically positioned sterigmata. However, there is considerable departure from this archetype in the homobasidiomycetes. For example, the stalked puffball *Tulostoma* has laterally attached basidiospores; basidia of *Pulcherricium* develop dendrohyphidial branches; *Tulasnella* has inflated epibasidia that develop adventitious septa at their bases; *Ceratobasidium* has subglobose basidia; and *Cantharellus*, *Sistotrema*, and others have up to eight sterigmata per basidium. Other variable characters of homobasidia include presence and form of a basal clamp, mode of attachment (i.e., whether terminal or *pleural*, as in *Xenasma*, Corticiaceae), and presence or absence of nested, or *repeating* basidia (as in *Repetobasidium*, Corticiaceae). For reviews of basidium morphology, see Oberwinkler (1982), Jülich (1981), and Talbot (1973a). Basidium morphology has been emphasized in the taxonomy of corticioid fungi and other Aphyllophorales (e.g., Hjortstam et al. 1988), but it has had relatively little impact on the taxonomy of Agaricales (Jülich 1981; Oberwinkler 1982). Basidial shape is a subtle, essentially quantitative character, and it is therefore difficult to code for phylogenetic analysis. As discussed by Oberwinkler (1982), there has been convergence in basidial form in unrelated taxa, such as *Sistotrema* and *Multiclavula* (which have urniform basidia), *Phlebia* and *Panellus* (meruliaceous basidia), and *Cantharellus* and *Hygrophorus* (elongate, slender basidia). In addition, there are several cases identified through molecular studies of closely related taxa that have strongly divergent basidial morphology. A prime example concerns *Tulasnella*; according to mt-ssu rDNA sequences (D.S. Hibbett, unpubl.), *Tulasnella* is closely related to *Botryobasidium*, which has nonseptate, urniform basidia, with up to eight sterigmata. These observations cast doubt on the utility of basidial morphology for resolving major clades of homobasidiomycetes, although it may be informative at lower taxonomic levels (e.g., Larsen and Gilbertson 1977).

2. Basidiospores

Basidiospores and an important source of characters for homobasidiomycete systematics. Variable aspects of basidiospore morphology include shape, size, color, staining reactions, ornamentation, wall structure, apical pores, hilar appendix morphology, presence and form of a suprahilar disk, and germination mode. For surveys of basidiospore morphology in homobasidiomycetes, see Singer (1986), Kühner (1972, 1980, 1984), Jülich (1981), Donk (1964), Castellano et al. (1989), Coker and Couch (1928), and Pegler and Young (1971). This section focuses on broad patterns of basidiospore evolution in homobasidiomycetes. The first part of this discussion considers only the basidiospores of Hymenomycetes.

Based on comparison to basidiospores of Auriculariales, Dacrymycetales, and Tremellales, it is likely that smooth, hyaline, inamyloid basidiospores are plesiomorphic in the homobasidiomycete. Spores with this combination of characters (with varying shapes) are typical of the polyporoid, cantharelloid, and hymenochaetoid clades, although other, presumably derived, morphologies also occur in these groups. For example, in the polyporoid clade there is strong evidence that *Ganoderma*, which has brown, ornamented spores with multilayered walls, is nested in a lineage that also includes *Polyporus*, *Lentinus*, and *Fomes* (and others), which have smooth, hyaline spores (Hibbett and Donoghue 1995; Hibbett et al. 1997b). Similarly, the strongly warted spores of *Botryobasidium isabellinum* are probably a derived feature in the cantharelloid clade.

The euagarics clade has a wide range of spore morphologies, which were discussed at length by Pegler and Young (1971), Singer (1986), and Kühner (1980, 1984). A few examples of putatively monophyletic groups in the euagarics clade that are distinguished by spore morphology are the Cortinariaceae (brown, generally warted spores, without an apical pore), Entolomataceae (angular, pink, cyanophilic spores), and Coprinaceae (dark, thick-walled spores with an apical pore). It is beyond the scope of this chapter to comment in detail on spore morphology and phylogenetic relationships within the euagarics clade. Nevertheless, we note that there is an apparent, if imperfect, correlation between the possession of an "open pore" type hilum, apical pore, and dark spore print in the Agaricaceae, Bolbitiaceae, Coprinaceae, Crepidotaceae, and Strophariaceae (Pegler and Young 1971), suggesting that these taxa may be closely related. In addition, Kühner (1984) emphasized double wall layers as an important character supporting his concept of the Pluteales, which encompasses the Entolomataceae and Pluteaceae sensu Singer [as well as *Macrocystidia*, which Singer (1986) placed in the Tricholomataceae]. Whether these basidiospore characters are actually synapomorphies of major lineages within the euagarics clade will only be determined when the phylogeny of the euagarics clade is better understood.

Basidiospores in the bolete clade are quite variable, but typically they are smooth, some shade of brown or yellow-brown, and fusoid (elongate with rounded, tapering ends), but elliptic to subglobose spores also occur. Corner (1972) suggested that there is a functional correlation between elongate basidiospores and poroid hymenophores in boletes. Based on measurements of spore shape and positioning, Corner concluded that fusoid spores use space more efficiently in narrow poroid hymenophores than spores that are more globose. By this reasoning, he suggested that elongate spores are derived in the boletes. Pegler (1983, p. 10) noted that elongate, cylindric spores are also common in polypores, and suggested that: "Spore elongation may therefore be seen as a phyletic trend occurring in both the Aphyllophorales and the Boletales". Based on the work of Bruns et al. (1998), it seems plausible that the plesiomorphic condition in the bolete clade is to have smooth, elliptic to ovoid spores, which would be consistent with Corner's hypothesis. Elliptic-ovoid spores are found in *Serpula*, *Paxillus*, *Hygrophoropsis*, *Coniophora*, *Gyroporus*, and *Gyrodon*, which form two lineages outside the main "boletoid" and "suilloid" groups, in which fusoid spores prevail (Pegler and Young 1971; Bruns et al. 1998). Derived ornamented spores in the boletoid group include those of *Boletellus*, *Austroboletus*, and *Strobilomyces*. Corner (1972) suggested that the subglobose spores of *Strobilomyces* are primitive in the boletes, but the analysis of Bruns et al. (1998) suggests that *Strobilomyces* is nested in the boletoid group (indicating that its subglobose spores are derived in the bolete clade).

Basidiospores of the thelephoroid, russuloid, and gomphoid-phalloid clades were mentioned previously (see Sects. II. C. 4, 5, 8). Each clade is characterized by derived spore morphologies, including (1) angular or globose-elliptic, pigmented, ornamented spores in the thelephoroid clade, (2) hyaline spores with amyloid ornamentations in the russuloid clade, and (3) warted, cyanophilic spores in the gomphoid-phalloid clade (Donk 1964). Spore morphologies are not uniform in these clades, however. For example, in the russuloid clade, there are taxa with smooth, amyloid or inamyloid spores (e.g., *Albatrellus*, *Peniophora*, *Stereum*), and ornamented inamyloid spores (*Heterobasidion*; Redhead and Norvell 1993, but see Stalpers 1979). In the gomphoid-phalloid clade, *Clavariadelphus* has smooth, non-cyanophilic spores (which may represent the plesiomorphic state in this clade).

So far, we have mentioned only the ballistosporic (forcibly discharged) basidiospores of Hymenomycetes. Structural features associated with ballistospory include short, curved sterigmata, bilateral spore symmetry, heterotropy (excentric positioning of spores on the sterigmata), and formation of a droplet of liquid at the time of discharge (Buller 1922; Ingold 1971; Webster and Chien 1990). Ballistospory has been repeatedly lost with the evolution of gasteroid fruiting bodies (Thiers 1984). Gasteroid basidiospores are statismosporic (not forcibly discharged) and, with the exception of some Hymenogastrales, radially symmetric and orthotropic (positioned directly over the apex of the sterigmata). Aside from these features, gasteroid basidiospores are morphologically variable. The spores of many false truffles and secotioid fungi resemble the spores of their putative Hymenomycete ancestors (e.g., fusoid spores in *Rhizopogon* and *Suillus*, angular, pink spores in *Richoniella* and *Entoloma*, and elliptic, ornamented spores in *Thaxterogaster* and *Cortinarius*; Castellano et al. 1989). In addition, there are certain recurring spore morphologies in Gasteromycetes that may be correlated with the mode of dispersal. For example, dark, globose to broadly elliptic, ornamented spores have evolved repeatedly in puffballs (e.g., *Lycoperdon*, *Scleroderma*, *Geastrum*), which have aerial spore dispersal, and smooth, elliptic, thick-walled spores have evolved independently in Nidulariaceae and Sphaerobolaceae, which have herbivore dispersal (Brodie 1975). For reviews of gasteroid basidiospore morphology, see Castellano et al. (1989), Coker and Couch (1928), Miller and Miller (1988), and Pegler et al. (1993, 1995). Besides terrestrial Gasteromycetes, the only statismosporic homobasidiomycetes are certain marine fungi (see below) and corticioid bark beetle symbionts (see Sect. III. F. 6).

Ballistospory has been lost in the lignicolous marine homobasidiomycetes *Nia vibrissa*, which is gasteroid, and *Digitatispora marina*, which is corticioid (Douget 1962, 1967; Kohlmeyer and Kohlmeyer 1979). In addition to being statismosporic, the basidiospores of *Nia* and *Digitatispora* have elongate processes (appendages) that probably function in anchoring the spores to substrates (Ingold 1971). Similarly appendaged conidia are found in aero-aquatic basidiomycetous hyphomycetes, such as *Ingoldiella* (see Sect. III. C), as well as the spores of many aquatic ascomycetes (Ingold 1971; Shaw 1972; Kohlmeyer and Kohlmeyer 1979). (see Seifert and Gams, Chap. 14, Vol. VII Part A). Ballistospory has also been lost in the gasteroid marine homobasidiomycete *Mycaureola dilseae*, which is a parasite of red algae that produces sigmoid, sessile basidiospores (Porter and Farnham 1986). The only other homobasidiomycetes known to produce basidiomata underwater are *Gloiocephala aquatica*, which is a minute agaric that grows in freshwater on *Scirpus* culms (Desjardin et al. 1995), and *Halocyphina villosa*, which is a cyphelloid fungus that grows on mangrove roots and other woody marine substrates (Ginns and Malloch 1977; Kohlmeyer and Kohlmeyer 1979). The clavate-cylindric basidia and smooth, asymmetric spores of *G. aquatica* and *H. villosa* are anatomically indistinguishable from those of ballistosporic homobasidiomycetes, which may indicate that they were recently derived from terrestrial ancestors (Ginns and Malloch 1977; Desjardin et al. 1995).

Basidiospores of homobasidiomycetes sensu stricto germinate to form hyphae directly. In contrast, basidiospores of many heterobasidiomycetes produce secondary ballistospores or conidia prior to the production of mycelium (Ingold 1985, 1992). "Germination by repetition" occurs in every order of heterobasidiomycetes, including Tulasnellales and Ceratobasidiales, but not all species of heterobasidiomycetes demonstrate this capability (Donk 1964; Tu and Kimbrough 1978; Jülich 1981; Oberwinkler 1982; Ingold 1985, 1992; Wells 1994). Diverse germination modes and secondary spore types occur in heterobasidiomycetes, and these may or may not be homologous. Nevertheless, it is plausible that germination by repetition is plesiomorphic for the hymenomycete lineage (it is also present in Uredinomycetes and Ustilaginomycetes *sensu* Swann and Taylor 1993) (see Swann and Frieders, Chap. 2, and Bauer et al., Chap. 3, this Vol.). If so, then it may have been lost in the lineage leading to the homobasidiomycetes (as well as certain heterobasidiomycetes), and secondarily derived in the Tulasnellales and Ceratobasidiales.

3. Hyphal Systems of Fruiting Bodies

Techniques for studying the micromorphology of hyphae that make up fruiting bodies were developed by Corner [1932, and later; see Pegler (1996) for a historical review of hyphal analysis], who described three main types of hyphae: skeletal hyphae (unbranched, thick-walled), binding hyphae (thick-walled, branched, and twining), and generative hyphae (thin-walled, branched, septate). Based on the presence or absence of these hyphal forms, context tissues were called monomitic (only generative hyphae), dimitic (generative hyphae and skeletal or binding hyphae), or trimitic (all three types present). Corner's early work centered on the polypores *Phellinus* (hymenochaetoid clade) and *Microporus* (unplaced, possibly polyporoid clade). As other homobasidiomycetes were examined, numerous intergrading hyphal types were described, which resulted in a proliferation of terms such as skeleto-ligative hyphae or sclerified generative hyphae. In 1966, Corner described sarcodimitic and sarcotrimitic hyphae, in which there are inflated and thick-walled cells, combined with intertwining generative hyphae. Pegler (1996) recognized seven principle forms of hyphal systems in homobasidiomycetes (monomitic, sarcomitic, intermediate dimitic, dimitic with mostly unbranched skeletal hyphae, dimitic with skeletoligative elements, sarcotrimitic, and trimitic), which he divided into 32 subcategories. For descriptions of the morphology and taxonomic distribution of hyphal forms, see Donk (1964), Gilbertson and Ryvarden (1986), Largent et al. (1977), Lentz (1971), Pegler (1996), and Singer (1986). The following discussion highlights patterns of evolution in hyphal systems and their potential phylogenetic utility.

Hyphal anatomy has been overemphasized in some classifications, which has led to taxonomic errors. Polyphyletic taxa that are strongly based on hyphal characters include *Lentinus* sensu Pegler (1983; dimitic), *Trogia* sensu Corner (1966; sarcodimitic), and Xerulaceae sensu Redhead (1986; sarcodimitic). Nevertheless, molecular studies suggest that variation in hyphal morphology can be phylogenetically informative at some levels see Hibbett and Donoghue 1995 for examples in polypores. For example, analyses of rDNA sequences support the division of *Lentinus* sensu Pegler into three groups that can be differentiated by their hyphal anatomy: *Lentinus* sensu stricto (dimitic with binding hyphae), *Panus* (dimitic with strongly developed skeletals), and *Neolentinus* (dimitic with weakly developed skeletals; Corner 1981; Hibbett and Vilgalys 1993; Pegler 1975, 1983; Redhead and Ginns 1985). Similarly, *Hydropus* and *Baeospora* (Xerulaceae sensu Redhead, with sarcodimitic stipe trama) appear to be closely related, based on molecular and morphological evidence (Bas et al. 1990; Moncalvo et al. 2000). Taken together, molecular and morphological studies indicate that hyphal characters are taxonomically informative, but that their use is limited by homoplasy and the difficulty of delimiting character states.

Based on comparison to outgroups (Auriculariales, Dacrymycetales, Tremellales), the plesiomorphic condition of the homobasidiomycetes is probably monomitic. Monomitic context construction occurs in each of the eight major clades that we recognize, and it is apparently the sole form of construction in the bolete, thelephoroid, cantharelloid, and gomphoid-phalloid clades. The majority of species with dimitic and trimitic construction are in the polyporoid, hymenochaetoid, and russuloid clades. Each of these clades includes dimitic and trimitic perennial forms, such as *Fomes*, *Fomitopsis*, and *Ganoderma* (trimitic) in the polyporoid clade, *Phellinus* (dimitic) in the hymenochaetoid clade, and *Echinodontium* and *Heterobasidion* (dimitic; Gilbertson and Ryvarden 1986) in the russuloid clade. The repeated evolution of perennial fruiting bodies in these groups appears to be correlated with decay of large wood substrates. The euagarics clade is mostly monomitic, but some species of *Pleurotus* (e.g., *P. tuberregium*) are dimitic, as is the clavarioid *Pterula* (Corner 1970; Pine et al. 1999). In addition, certain desert-adapted Gasteromycetes in the euagarics clade have tough, woody stipes, and these may also be interpreted as dimitic (e.g., Tulostomataceae, *Podaxis*, *Montagnea*; Miller and Miller 1988). Sarcodimitic construction occurs in diverse agarics (Xerulaceae sensu Redhead, euagarics clade), as well as in *Meripilus* (polyporoid clade; Corner 1984), and the stereoid and cantharelloid species of *Trogia* sensu Corner (1966), which are of uncertain taxonomic placement. Overall, dimitic, trimitic, and sarcodimitic construction appears to have evolved repeatedly as a mechanism for toughening fruiting bodies in diverse ecological circumstances.

Conducting elements are hyphae with refractive or oily-looking contents (often with characteristic staining reactions) that are presumed to transport or sequester substances of unknown function (Largent et al. 1977; Singer 1986; Pegler 1996). Some are of rather limited taxonomic distribution, such as laticiferous hyphae (e.g., *Lactarius*, *Bondarzewia*; Redhead and Norvell 1993), chryso-vessels (Strophariaceae), and oleiferous hyphae (*Russula*, *Amanita*; Largent et al. 1977; Singer 1986). In contrast, gloeoplerous hyphae (including gloeocystidia; Larsen and Burdsall 1976) are broadly distributed. Gloeoplerous hyphae are common in the russuloid clade (e.g., *Auriscalpium*, *Bondarzewia*, *Clavicorona*, *Gloeocystidiellum*, *Lentinellus*, *Russula*), but they are also reported in *Fistulina*, *Laetiporus*, *Albatrellus*, and others. The taxonomic distribution of conducting elements suggests that they are phylogenetically informative, although they also display homoplasy. Other specialized hyphae of limited taxonomic distribution include sphaerocysts of Russulales, dichohyphae and asterosetae of

Lachnocladiaceae (Reid 1965), and "pressure cells" of *Amanita* (Bas 1969).

4. Nuclear Behaviour in Meiosis and Basidiosporogenesis

Taxonomic characters for homobasidiomycetes can be drawn from the orientation of first-division meiotic spindles and patterns of postmeiotic mitosis and nuclear migration during basidiosporogenesis. Beginning with the work of Juel (1898, 1916) and Maire (1902), two main patterns of meiosis have been recognized, with two corresponding types of basidia: chiastic basidia and stichic basidia. In chiastic basidia, the first meiotic division takes place in the apex of the basidium, with the spindle transverse to the long axis of the basdium, whereas in stichic basidia the first division takes place in the middle of the basidium, with the spindle parallel to the long axis of the basidium. Intermediate forms are known, which may complicate efforts to code this character for phylogenetic analysis. For example, Boidin (Fig. 14 in 1958) observed both transverse and longitudinal spindles in *Sistotrema*, which he nevertheless interpreted as stichic. Boidin also described "hemichiastic" basidia that have variable orientation of the spindles, but Donk (1964, p. 221) concluded that hemichiastic basidia "may be subordinated as a subtype to the chiastic".

Among the heterobasidiomycetes, *Auricularia*, *Dacrymyces*, and *Dacryopinax* are reportedly stichic, and *Exidia*, *Sebacina*, and *Phlogiotis* are chiastic (Juel 1898; Maire 1902; Bodman 1938; Furtado 1968). Most of the homobasidiomycetes that have been examined have chiastic basidia, including all species of the euagarics, bolete, thelephoroid, russuloid, gomphoid-phalloid, and polyporoid clades (Juel 1898, 1916; Maire 1902; Ehrlich and McDonough 1949; Boidin 1958; Penancier 1961; Donk 1964; Restivo and Petersen 1976; Hubbard and Petersen 1979; Hibbett et al. 1994a), except for *Sistotrema*, which is stichic and is supported as a member of the polyporoid clade by mt-ssu rDNA sequences (D.S. Hibbett, unpubl.). Other stichic homobasidiomycetes include *Cantharellus*, *Clavulina*, *Craterellus*, *Hydnum*, and *Multiclavula*, which are all in the cantharelloid clade (Pine et al. 1999), and *Clavulicium* (Clavulinaceae; Parmasto 1986), which is a corticioid fungus that has yet to be included in phylogenetic studies. Possession of stichic basidia would appear to be a synapomorphy of the cantharelloid clade (Pine et al. 1999) except that *Tulasnella* has chiastic basidia (Figs. 12–15 in Rogers 1932). Based on the taxonomic distribution of stichic and chiastic basidia, and the tree in Fig. 2, we tentatively conclude that chiastic basidia are plesiomorphic in the homobasidiomycetes, including the cantharelloid clade, and were transformed into stichic basidia on the branches leading to *Multiclavula* and *Cantharellus*, and *Sistotrema*.

Between meiosis and basidiospore discharge, the haploid nuclei usually undergo a mitotic division, which may be completed in the basidia, sterigmata, or basidiospores. If postmeiotic mitosis is completed outside of the basidium, there is often a back-migration of one daughter nucleus into the basidium. Duncan and Galbraith (1972) described four kinds of postmeiotic nuclear behavior, which they termed patterns A, B, C, and D. Mueller et al. (1993) designated two additional patterns, E and F, based on observations by Arita (1979) and Tommerup et al. (1991). Briefly, the postmeiotic nuclear behavior patterns are defined as follows: Pattern A: postmeiotic mitosis occurs in the basidium, one nucleus enters each basidiospore; Pattern B: postmeiotic mitosis occurs in sterigmata, one daughter nucleus backmigrates to the basidium; Pattern C: postmeiotic mitosis occurs in basidiospores, one nucleus backmigrates to the basidium; Pattern D: postmeiotic mitosis occurs in basidiospores, backmigration does not occur; Pattern E: postmeiotic mitosis does not occur; Pattern F: postmeiotic mitosis occurs in the basidium, multiple nuclei enter each basidiospore. Typically, patterns A, B, C, and E result in uninucleate basidiospores, whereas patterns D and F result in binucleate basidiospores. The number of nuclei per spore is also affected by the number of spores per basidium (e.g., Petersen 1995a).

There are few reports of postmeiotic nuclear behavior in heterobasidiomycetes, Urediniomycetes, or Ustilaginomycetes. In *Exidia nucleata* (Auriculariales) postmeiotic mitosis is completed in the elongate sterigmata, with backmigration of one daughter nucleus into the basidium (Furtado 1968). In *Auricularia fuscosuccinea* postmeiotic mitosis is completed in the basidiospores, but it is not known whether backmigration occurs (McLaughlin 1981). In *Dacryopinax* (Dacrymycetales) there appears to be no postmeiotic mitosis, and spores are uninucleate

Table 3. Patterns of postmeiotic nuclear behavior in homobasidiomycetes. The number of genera reported to have each pattern is indicated

Major clades	Nuclear behavior patterns[a]					
	A	B	C	D	E	F
Polyporoid clade			4			
Euagarics clade	40+	1	4	23+	3	3
Bolete clade			4			
Thelephoroid clade		3?[b]	3?[b]			
Russuloid clade			3 (+1?[c])	1 (+1?[c])		
Hymenochaetoid clade						
Cantharelloid clade	2				3[d]	1[d]
Gomphoid-phalloid clade	1?[e]	1?[e]				

[a] A: postmeiotic mitosis occurs in the basidium; one nucleus enters each basidiospore. B: postmeiotic mitosis occurs in sterigmata; one daughter nucleus backmigrates to the basidium. C: postmeiotic mitosis occurs in basidiospores; one nucleus backmigrates to the basidium. D: postmeiotic mitosis occurs in basidiospores; backmigration does not occur. E: postmeiotic mitosis does not occur. F: postmeiotic mitosis occurs in the basidium; multiple nuclei enter each basidiospore.
[b] *Boletopsis*, *Hydnellum*, and *Sarcodon* are probably B or C, but not D, E, or F (Penancier 1961).
[c] *Clavicorona* could be C or D (Berbee and Wells 1989; Wilson et al. 1967).
[d] Including *Tulasnella* (Rogers 1932).
[e] *Ramaria* (Penancier 1961).

(Bodman 1938). In *Ustilago maydis* (Ustilaginomycetes sensu Swann and Taylor 1993) and *Eocronartium musicola* (Urediniomycetes sensu Swann and Taylor 1993), postmeiotic mitosis is completed in the basidiospores, followed by backmigration of one daughter nucleus into the basidium or sterigmata (O'Donnell and McLaughlin 1984; Boehm and McLaughlin 1989). Taken together, these observations suggest that postmeiotic mitosis with backmigration of daughter nuclei (pattern C, or possibly B) is plesiomorphic in the homobasidiomycetes, and possibly the entire hymenomycete lineage (if so, it may have been lost in the Dacrymycetales).

Mueller and Ammirati (1993) reviewed the literature on postmeiotic nuclear behavior in Agaricales. Additional information on Aphyllophorales, Gasteromycetes, and heterobasidiomycetes was presented by Berbee and Wells (1989), Brodie (1975), Ehrlich and McDonough (1949), Hibbett et al. (1994a), Hubbard and Petersen (1979), Penancier (1961), Restivo and Petersen (1976), Rogers (1932), and Wilson et al. (1967). A summary of the distribution of postmeiotic nuclear behavior patterns among the eight major clades of homobasidiomycetes that we recognize is shown in Table 3. Pattern C occurs in four or five of the eight clades, and it is the only pattern reported in the polyporoid and bolete clades (Table 3). As noted by Mueller and Ammirati (1993), there is considerable homoplasy in the evolution of postmeiotic nuclear behavior (Table 3). For example, there appears to have been a parallel loss of postmeiotic mitosis in the euagarics clade (e.g., *Pholiota*; Arita 1979) and cantharelloid clade (e.g., *Multiclavula*; Hubbard and Petersen 1979). Nevertheless, there is also some congruence between groupings based on nuclear behavior and those based on molecular characters, which suggests that this character could be phylogenetically informative. At present, it is difficult to assess broad patterns of evolution in postmeiotic nuclear behavior patterns, owing to inadequate sampling outside of the euagarics clade. In particular, there are no reports of which we are aware from the hymenochaetoid clade, and only a single genus (*Ramaria*) has been examined in the gomphoid-phalloid clade.

5. Parenthesome Ultrastructure

Parenthesomes are membrane-bound organelles that flank dolipore septa. Among the heterobasidiomycetes, Auriculariales and Dacrymycetales have imperforate (continuous) parenthesomes, Ceratobasidiales have parenthesomes with a few

Table 4. Parenthesome structure in representative homobasidiomycetes

Perforate	Imperforate
1. Polyporoid clade	
Fomes (Moore 1980)	*Phanerochaete* (Keller 1997)
Ganoderma (Mims and Seabury 1989)	
Phaeolus (Moore 1980)	
Polyporus (Moore 1980)	
Sparassis (Patrignani and Pellgrini 1986)	
2. Euagarics clade	
Agaricus (Thielke 1972)	
Amanita (Flegler et al. 1976)	
Ceratobasidium (Tu and Kimbrough 1978)	
Coprinus (Oberwinkler 1985)	
Dictyonema (Slocum 1980)	
Fistulina (Patrignani and Pellgrini 1986)	
Lentinula (Tsuneda 1983)	
Lycoperdon (Flegler et al. 1976)	
Pleurotus (Moore and Patton 1975)	
Schizophyllum (Moore and Patton 1975)	
Stropharia (Thielke 1972)	
Waitea (Tu and Kimbrough 1978)	
3. Bolete clade	
Boletus (Patrignani and Pellgrini 1986)	
Coniophora (Langvad 1971)	
Pulveroboletus (Keller 1997)	
Serpula (Keller 1997)	
4. Thelephoroid clade	
Bankera (Keller 1997)	
Hydnellum (Keller 1997)	
Thelephora (G. Langer 1994)	
Tomentella (Calonge 1969)	
5. Russuloid clade	
Auriscalpium (Keller 1997)	
Hericium (Flegler et al. 1976)	
Laxitextum (Keller 1997)	
Scytinostroma (Besson and Froment 1968)	
Stereum (Patrignani and Pellgrini 1986)	
Zelleromyces (Keller 1997)	
6. Hymenochaetoid clade	
Basidioradulum (Langer and Oberwinkler 1993)	
Coltricia (W. Müller, pers. comm.)	
Hymenochaete (Oberwinkler 1985)	
Hyphodontia (E. Langer 1994)	
Onnia (Moore 1980)	
Phellinus (Moore 1985)	
Schizopora (Langer and Oberwinkler 1993)	
Trichaptum (Traquair and Mckeen 1978)	
7. Cantharelloid clade	
Botryobasidium (G. Langer 1994)	
Cantharellus (Keller 1997)	
Clavulicium? (Oberwinkler 1985)	
Tulasnella (Khan and Talbot 1976)	
[*Stilbotulasnella* = missing (Bandoni and Oberwinkler 1982)]	
8. Gomphoid-phalloid clade	
Clathrus (Eyme and Parriaud 1970)	*Geastrum* (E. Langer, pers. comm.)
	Ramaria? (Patrignani and Pellegrini 1986)

large perforations, Tulasnellales have parenthesomes that are imperforate or absent, and Tremellales have parenthesomes that are vesiculate, absent, or reticulate (Adams et al. 1995; Bandoni and Oberwinkler 1982; Keller 1997; Moore 1985). Homobasidiomycetes have parenthesomes that have numerous small perforations or that are imperforate (Table 4).

Six of the eight major clades that we recognize are apparently monomorphic for parenthosome type: as far as we know, all members of the euagarics, bolete, thelephoroid, and russuloid clades have perforate parenthosomes, whereas the hymenochaetoid clade and the few members of the cantharelloid clade that have been examined have imperforate parenthosomes (Table 4). *Coltricia*, in the hymenochaetoid clade, has been reported to have a perforate parenthosome (Moore 1980). However, reanalysis of this taxon has shown that it has an imperforate parenthosome (W. Müller, pers. comm.), as do all other members of the hymenochaetoid clade that have been examined (Traquair and McKeen 1978; Moore 1980, 1985; E. Langer 1994). Only the polyporoid and gomphoid-phalloid clades are reported to have both perforate and imperforate parenthosomes (Table 4). In the polyporoid clade, all taxa that have been examined have perforate parenthesomes, except *Phanerochaete sordida*, which is reported to have an imperforate parenthesome (Keller 1997). In the gomphoid-phalloid clade, imperforate parenthesomes are reported from *Geastrum* (E. Langer unpubl.) and *Ramaria ignicolor* (as *Clavaria ignicolor*; Patrignani and Pellegrini 1986), but *Clathrus* is reported to have perforate parenthesomes (Eyme and Parriaud 1970). Unplaced homobasidiomycetes with imperforate parenthosomes include the corticioid fungi *Paullicorticium pearsonii*, *Radulomyces confluens*, and *Subulicystidium longisporum*, and the clavarioid fungus *Typhula uncialis* (which may be related to *Multiclavula*; Oberwinkler 1985; Patrignani and Pellegrini 1986; Keller 1997).

The tree in Fig. 2 suggests that imperforate parenthosomes are plesiomorphic in the homobasidiomycetes and are homologous with the parenthosomes of Auriculariales and Dacrymycetales. Nevertheless, the position of the hymenochaetoid clade (as sister taxon of the russuloid clade) and the apparent co-occurrence of perforate and imperforate parenthosome types in the polyporoid and gomphoid-phalloid clades suggest that there has been some homplasy in the evolution of this character. In addition, the probable placement of *Ceratobasidium* and *Waitea* (which have parenthosomes with a few large perforations) in the eugarics clade suggests that there can be considerable variation in the size and number of parenthosome perforations among closely related taxa. Finally, the apparent loss of parenthosomes in *Stilbotulasnella* of the Tulasnellales (Bandoni and Oberwinkler 1982) seems to parellel their loss in the Tremellales.

Parenthosome ultrastructure may contain clues to the deepest splits in the homobasidiomycetes. However, inferences about the evolution of parenthosome types are limited by the weak support for nodes deep in the homobasidiomycete tree (Fig. 2) and poor sampling in many taxa (Table 4). To assess the phylogenetic significance of parenthosome ultrastructure in homobasidiomycetes, it will be necessary to improve resolution of higher-order relationships, confirm certain reports (i.e., *Phanerochaete* and *Clathrus*), and increase the number of observations of parenthosome types, especially in the cantharelloid and gomphoid-phalloid clades.

C. Asexual Reproduction and Somatic Morphology

Asexual reproductive forms have received considerably less attention in homobasidiomycetes than in ascomycetes. Nevertheless, there is a diversity of anamorphic forms in homobasidiomycetes, including an abundance of simply arthroconidial forms in soil- and wood-inhabiting species (see Seifert and Gams, Chap. 14, Vol. VII, Part A). Much of the information on asexual reproductive structures in homobasidiomycetes comes from cultural studies (Nobles 1965, 1971; e.g., Miller 1971; Stalpers 1978; Nakasone 1990). Many saprotrophic homobasidiomyceetes produce abundant mitospores in culture, especially on nutrient-poor media or when confronted by other organisms (bacteria, nematodes, other fungi; e.g., Tsuneda et al. 1992). In contrast, it appears that ectomycorrhizal homobasidiomycetes (and ascomycetes) lack the ability to form mitrospores in culture (Hutchison 1989, 1991a). The functional significance (if any) of this apparent correlation between nutritional mode and asexual reproduction is unclear. As in ascomycetes, many basidiomycete anamorphs have no known teleomorph, but with the application of molecular techniques their relationships to sexual forms should be resolved. In the following sections, we discuss selected asexual reproductive structures of homobasidiomycetes, emphasizing their morphological variability and potential phylogenetic utility. An excellent review of this topic was provided by Stalpers (1987), which should be consulted for many taxa that could not be included here (also see Kendrick and Watling 1979).

As far as we know, asexual reproductive structures occur in each of the eight major clades of homobasidiomycetes that we recognize, with the possible exceptions of the thelephoroid and gomphoid-phalloid clades. Nevertheless, the possession of particular kinds of asexual reproductive structures may be phylogenetically informative within the homobasidiomycetes. For example, the formation of *Sporotrichum* anamorphs (Stalpers 1984) in *Laetiporus* and *Pycnoporellus* suggests that they are closely related, which is consistent with the morphology of basidiomata (Ryvarden 1991). However, *Phanerochaete* also has a *Sporotrichum* anamorph, which suggests that there has been homoplasy in the evolution of this form (Fig. 2). Similarly, the presence of *Spiniger* and *Spiniger*-like anamorphs in *Bondarzewia*, *Dichostereum*, *Heterobasidion*, and *Laurilia* is consistent with rDNA analyses and morphology, which suggest that these taxa are all in the russuloid clade (similar anamorphs are also formed by *Mutatoderma* (*Hyphoderma*), and *Resinicium*, which are of uncertain taxonomic placement). As a final example, cylindric to barrel-shaped arthroconidia with schizolytic dehiscence are frequent in *Hypholoma*, *Pholiota*, and *Psilocybe* (Jacobsson 1989; Klán et al. 1989; R.G. Thorn, unpubl.), which are all placed in Strophariaceae (euagarics clade), but similar arthroconidia have also been recorded in *Phlebia* (polyporoid clade; Sigler and Carmichael 1976), and in cultures isolated from human specimens and pupal chambers of the mountain pine beetle *Dendroctonus ponderosae* (Tsuneda et al. 1993). Taken together, these observations indicate that asexual reproductive structures may be phylogenetically informative, but that certain anamorphic form-taxa are polyphyletic.

The simplest anamorphic forms are sterile mycelia, which lack conidia or multicellular structures for reproduction or dispersal. Among these are *Rhizoctonia* sensu lato and *Ozonium*. *Rhizoctonia* has been subdivided on the basis of nuclear states and associated teleomorphs into *Rhizoctonia* (typonym *Thanatophyton*; multinucleate anamorphs of *Helicobasidium*, with simple septal pores, Urediniomycetes), *Moniliopsis* (multinucleate anamorphs of *Thanatephorus*, parenthesomes with few, regular pores, Ceratobasidiales), *Ceratorhiza* (binucleate anamorphs of *Ceratobasidium*, parenthesomes with few, regular pores, Ceratobasidiales), *Chrysorhiza* (multinucleate anamorphs of *Waitea*, parenthesomes with few, irregular pores, Botryobasidiales or Ceratobasidiales), *Epulorhiza* (binucleate anamorphs of *Tulasnella*, parenthesomes imperforate, Tulasnellales), and *Opadorhiza* (binucleate anamorphs of *Sebacina*, parenthesomes imperforate, Auriculariales, Exidiaceae: Moore 1987; Müller et al. 1998; Stalpers and Andersen 1996). As mentioned previously, molecular studies tentatively suggest that *Ceratobasidium*, *Thanatephorus*, and *Waitea* (all with variously perforated parenthesomes) are in the euagarics clade, and that *Botryobasidium* and *Tulasnella* (imperforate parenthesomes) are in the cantharelloid clade (Table 1). However, *Waitea* has also been classified in the Botryobasidiales (e.g., Müller et al. 1998). *Moniliopsis* includes the economically important plant pathogen *M. solani* (teleomorph *Thanatephorus cucumeris*). *Ozonium* is now restricted in use to anamorphs of the *Coprinus domesticus* group (Watling 1979, euagarics clade).

Complex condiomata are found in relatively few homobasidiomycetes. Examples include the acervular *Necator* anamorph of *Phanerochaete salmonicolor*, unnamed sporodochial anamorphs formed in combs of *Termitomyces*, and the coremioid or synnematous *Antromycopsis* (teleomorph *Pleurotus*) and *Tilachlidiopsis* (= *Sclerostilbum*; teleomorph *Collybia*). Some basidiomycetes with complex conidiomata have no known teleomorph, including the pycnidial *Ellula*, the sporodochial *Glutinoagger*, and the coremioid or synnematous *Gloeosynnema* and *Riessia* (Botha and Eicker 1991; Seifert and Okada 1988).

Splash-cup anamorphs are formed by a number of polyporoid and corticioid basidiomycetes (Brodie 1951a,b). In the *Michenera* anamorph of *Licrostroma*, the *Matula* anamorph of *Aleurocystis*, and unnamed anamorphs of *Trametes conchifer* and *Phaeotrametes decipiens* the splash cups disperse arthroconidia; in contrast, the splash-cup anamorphs of *Corticium minnsiae* contain single hyphal bodies (Brodie 1951a; Jackson 1950; Wright 1966; all polyporoid clade, or unplaced). In all of these, the splash-cup dispersal of mitotic diaspores occurs during the summer or fall and is followed by the development of the basidiomata and meiospores. The splash-cup dispersal of mitospores in these fungi is a remarkable parallel to the splash-cup dispersal of basidiospores in the birds' nest fungi *Cyathus* and *Crucibulum* (euagarics clade; Fig. 2).

Many species of homobasidiomycetes produce mitospores within or on the surface of basidiomata. Examples include *Nyctalis* (anamorph *Asterophora*), *Mycena* Sect. Sacchariferae, and *Pleurotus* (anamorphs unnamed, euagarics clade), *Inonotus* (anamorph *Ptychogaster*, hymenochaetoid clade), *Abortiporus*, *Antrodia*, *Fomitopsis*, *Oligoporus*, *Phlebia*, and *Punctularia* (*Ptychogaster* and similar anamorph forms, polyporoid clade or unplaced), *Botryobasidium* (anamorphs *Haplotrichum* and *Allescheriella*, cantharelloid clade), and *Arthrosporella* (*Nothoclavulina*, anamorph unnamed, unplaced; Desjardin 1995; G. Langer 1994; Ryvarden 1991; Stalpers 1987). In addition, the mycoparasite *Squamanita* produces chlamydospores on the remains of its host's basidiomata (Redhead et al. 1994).

Several taxa of aero-aquatic hyphomycetes have clamp connections or dolipore septa, which indicates that they are basidiomycetes, but not all are necessarily homobasidiomycetes (Nawawi 1985; Webster 1992). For example, *Tricladiomyces* and *Dendrosporomyces* have branched conidia and dolipore septa with perforate parenthesomes (Nawawi 1985) and *Fibulotaeniella* has sigmoid, one- or two-celled conidia with clamp connections (Marvanová and Bärlocher 1988). The teleomorphs of these taxa are unknown. Teleomorphs of *Taeniospora* and *Ingoldiella*, which have clamped, tetraradiate conidia, are found in the corticioid genera *Fibulomyces*, *Leptosporomyces*, and *Sistotrema* (polyporoid clade or unplaced; Nawawi et al. 1977; Nawawi and Webster 1982; Marvanova and Stalpers 1987). *Cyrenella*, isolated from the stipe base of the arenicolous-maritime agaric *Laccaria trullisata*, forms mycelia with clamp connections and simple, obovate conidia with four apical appendages and one basal appendage, which superficially resemble conidia of some aeroaquatic hyphomycetes (Gochenauer 1981). However, monokaryons of *Cyrenella* form a *Rhodotorula*-like yeast phase with budding balastospores, which suggests that *Cyrenella* is a member of the Tremellales, not homobasidiomycetes.

In addition to asexual spores, many homobasidiomycetes produce aggregates of somatic hyphae that function in dispersal, colonization, and persistence. Examples include sclerotia (compact hyphal aggregates with a differentiated rind of thick-walled cells), bulbils (usually small hyphal aggregates without a differentiated rind), and pseudosclerotia (hyphal masses including nonfungal tissues), as well as rhizomorphs and mycelial cords (not discussed here). Selected examples of these forms, each of which has evolved repeatedly, are discussed below.

Teleomorphs of the form-genus *Sclerotium* are found in the euagarics clade (*Agrocybe*, *Athelia*, *Ceratobasidium*, *Clitopilus*, *Collybia* (= *Microcollybia*), *Coprinus*, *Hypholoma*, *Leucocoprinus*, *Panaeolus*, *Stropharia*, and *Typhula*) and bolete clade (*Boletinellus*, *Hygrophoropsis*, *Leucogyrophana*, *Paxillus*, and *Pisolithus*), as well as in *Thanatephorus*, which is of uncertain placement (Cotter and Miller 1985; Grenville et al. 1985a, b; Redhead and Kroeger 1987; Stalpers 1987; Hutchison 1991b; Ginns and Lefebvre 1993; Lee and Jung 1997; R.G., Thorn, unpubl.). Bulbils are found in a number of anamorph genera with corticioid teleomorphs, including *Aegerita* (teleomorphs *Bulbilomyces* and *Subulicystidium*), *Burgoa* (*Sistotrema*, polyporoid clade), *Hyphelia* (*Corticium*), and *Myriococcum* (*Athelia*), as well as the corticioid homobasidiomycetes *Ceraceomyces*, *Crustoderma*, *Dendrothele*, and *Limonomyces* (Eriksson and Ryvarden 1976; Ginns and Lefebvre 1993). Large sclerotia and pseudosclerotia are formed in the polyporoid clade by *Polyporus*, such as *P. tuberaster* (with anamorphs *Mylitta* and *Pietraia*), *Wolfiporia* (*Pachyma* anamorph), and *Panus*, and in the euagarics clade by *Psilocybe* and *Pleurotus*, including *P. tuberregium* (Petch 1915; Corner 1981; Pegler 1983; Stamets and Chilton 1983; Hibbett et al. 1993a; Hibbett and Thorn 1994). Sclerotia and pseudosclerotia of some wood-decaying homobasidiomycetes are massive, and these have often been utilized in various ways by indigenous peoples. For example, the *Pachyma* anamorph of *Wolfiporia*, known as a tuckahoe, was eaten by native Americans (Weber 1929), and pseudosclerotia of the pantropical *Pleurotus tuberregium* have been used as food or medicine in Africa and elsewhere (Walleyn and Rammeloo 1994). In Papua New Guinea, sclerotia identified as *P. tuberregium* (but perhaps actually *Panus fulvus*?) have been used to make club heads (Price et al. 1978).

Finally, yeastlike forms are reported from only two homobasidiomycetes, although they are widespread in other groups of Basidiomycota (Swann and Taylor 1995b). Both homobasidiomycete yeasts are insect symbionts. Symbionts of the leafcutter ant *Cyphomyrmex* (euagarics clade) are dimorphic, existing in both a yeast and hyphal form (Weber 1979). Similarly, an unnamed symbiont of the bark beetle *Dendroctonus frontalis* (russuloid clade) exists as a yeast in the beetle mycangia, but becomes mycelial when free-living (Happ et al. 1976). Reports that the parasitic agaric *Asterophora lycoperdoides* has a yeast phase (Koller and Jahrmann 1985) have been shown to be in error (Laaser et al. 1989).

D. Mating Genetics

Sexual compatibility in heterothallic (outcrossing) homobasidiomycetes is regulated by multiallelic factors that occur at a single locus in bipolar (unifactorial) species, or at two loci (the A and B factors) in tetrapolar (bifactorial) species. Based on figures from Esser (1967), about 65% of homobasidiomycetes are tetrapolar, 25% are bipolar, and 10% are homothallic. Nobles (1971) and Ryvarden (1991) suggested that bipolarity is the primitive condition in the homobasidiomycetes, which implies that there have been multiple derivations of tetrapolar systems from bipolar systems. However, as noted by Raper and Flexer (1971), the functional similarity of tetrapolar mating systems in diverse homobasidiomycetes suggests that they are homologous. Bipolar mating systems could be derived from tetrapolar mating systems by two mechanisms: (1) self-compatible mutations in either the A or B factors could lead to effectively bipolar mating systems (or homothallism if both the A and B factors are affected), as has been demonstrated in *Coprinus* (Casselton and Kües 1994; Raper and Flexer 1971); (2) close linkage of A and B loci resulting from chromosome rearrangements could result in cosegregation of A and B loci, thus creating bipolarity (Bakkeren et al. 1992). Evidence for derivation of bipolar mating systems from tetrapolar systems is found in *Marasmius*, which contains both bipolar and tetrapolar species; molecular phylogenies suggest that the bipolar species form a clade that is derived from a paraphyletic assemblage of tetrapolar species (Owings and Desjardin 1997). Many other genera of homobasidiomycetes have been shown to contain both bipolar and tetrapolar mating species (as well as homothallic and heterothallic forms), such as *Sistotrema, Marasmius, Collybia*, and *Coprinus* (Lange 1952; Raper and Flexer 1971; Murphy and Miller 1993; Petersen 1995b). These observations indicate that the presence of tetrapolarity or bipolarity per se (ignoring the fact that bipolarity may arise by at least three different mechanisms) is not useful for delimiting major clades within homobasidiomycetes (it may be informative at low taxonomic levels, however, as in *Marasmius*). Nevertheless, the genetic architecture of mating systems may be informative for resolving higher-level relationships of homobasidiomycetes and heterobasidiomycetes. Among the heterobasidiomycetes, only the Auriculariales is known to have a bifactorial mating system with multiallelic loci such as that found in homobasidiomycetes, which supports the view that the Auriculariales is the sister group of the homobasidiomycetes. Tremellales (as well as Ustilaginales) have bipolar mating or "modified bifactorial compatibility" in which one locus is multiallelic and the other is biallelic (Wells 1994). In the Dacrymycetales, *Cerinomyces* has been shown to be tetrapolar (Maekawa 1987), but it is not known whether both the A and B factors are multiallelic.

E. Pigments and Bioluminescence

Pigments of macrofungi, including compounds responsible for bluing and other color reactions, have drawn the attention of organic chemists for more than a century (see reviews by Arpin and Fiasson 1971; Gill and Steglich 1987; Gill 1996; Johnson and Schroeder 1996; see also Tyler 1971 for a general review of chemotaxonomy in homobasidiomycetes). Nevertheless, only a small proportion of the species of homobasidiomycetes have been examined and, until recently, the findings were not widely used for taxonomic purposes (Kühner 1980, 1984; Singer 1986). The available data suggest that pigment compounds may be phylogenetically informative in some groups, but that they also display considerable homoplasy. The following discussion of selected pigments follows the outlines of Gill (1996) and Gill and Steglich (1987), who sort pigments into those derived from the shikimate-chorismate pathway, the acetate-malonate pathway, the mevalonate pathway, nitrogen-heterocycles (not discussed here), and other nitrogen-containing compounds.

1. Shikimate-Chorismate Pathway Derivatives

The shikimate-chorismate pathway produces a number of compounds that are characteristic of the bolete clade, including atrotomentin, pulvinic acid derivatives, cyclopentanoids, and polyprenylquinones. These compounds have been found in diverse poroid, lamellate, corticioid, and gasteroid-secotioid taxa, including *Boletus*, *Chamonixia*, *Chroogomphus*, *Coniophora*, *Gomphidius*, *Gyrodon*, *Hygrophoropsis*, *Leucogyrophana*, *Paxillus*, *Phylloporus*, *Pisolithus*, *Rhizopogon*, *Scleroderma*, *Serpula*, *Suillus*, and others (for details, see Gill and Steglich 1987).

Atrotomentin, pulvinic acid derivatives, and cyclopentanoids have also been found in the lignicolous agarics *Omphalotus* and *Lampteromyces*. Singer (1986) placed *Omphalotus* and *Lampteromyces* in the Paxillaceae (suborder Boletineae), largely on the basis of the presence of these pigments (see also Moser 1978; Kühner 1980). However, analyses of rDNA sequences suggest that *Omphalotus* and *Lampteromyces* are in the euagarics clade (Binder et al. 1997; Moncalvo et al. 2000; Thorn et al. 2000). This supports Jülich (1981), who placed *Omphalotus* and *Lampteromyces* in the Tricholomatales. In addition, atrotomentin and cyclopentanoids are found outside the bolete clade in *Albatrellus* (russuloid clade or polyporoid clade) and *Hydnellum* (thelephoroid clade). These observations imply that the production of atrotomentin, pulvinic acid derivatives, and cyclopentanoids has evolved repeatedly.

Thelephoric acid (which is a terphenylquinone, similar to atrotomentin) is found in *Bankera*, *Boletopsis*, *Hydnellum*, *Phellodon*, *Polyozellus*, *Pseudotomentella*, *Sarcodon*, and *Thelephora*. Thelephoric acid has been used as a defining character of the Thelephoraceae (including Bankeraceae; Bresinsky and Rennschmid 1971). This is consistent with molecular characters, which strongly support monophyly of the Thelephoraceae (Table 1). Nevertheless, thelephoric acid also occurs in *Suillus* and *Rhizopogon* (bolete clade), *Omphalotus* and *Lampteromyces* (euagarics clade), *Trametes* (polyporoid clade), and *Punctularia* (probably polyporoid clade).

Derivatives of cinnamic acids, the styrylpyrone pigments bisnoryangonin and hispidin, occur in *Hymenochaete*, *Inonotus*, *Onnia*, and *Phellinus* (Hymenochaetaceae, hymenochaetoid clade). These compounds also occur in *Phaeolus*, which has been classified in both the Hymenochaetaceae and Polyporaceae. According to molecular studies, *Phaeolus* is closely related to the polypore *Laetiporus* (polyporoid clade), which suggests that the presence of bisnoryangonin and hispidin in Hymenochaetaceae and *Phaeolus* is due to convergence. Furthermore, bisnoryangonin and hispidin also occur in *Gymnopilus*, *Hypholoma*, and *Pholiota* of the Strophariaceae (euagarics clade). In this context, it is worth noting that the xanthochroic reaction (tissues becoming dark on exposure to alkali), which has been an important character in delimiting the Hymenochaetaceae, is found in numerous Aphyllophorales (Parmasto and Parmasto 1979). A number of chemically dissimilar compounds darken in alkali, which may account for the occurrence of the xanthochroic reaction in diverse lineages of Aphyllophorales (Parmasto and Parmasto 1979).

Finally, the colorless-red-black reactions found in species of *Agaricus*, *Hygrocybe*, and *Rhodocybe* (euagarics clade), *Daedaleopsis* (polyporoid clade), *Russula* (russuloid clade), and *Strobilomyces* (bolete clade) are a result of oxidation by tyrosinase of L-DOPA, which is derived from tyrosine or phenylalanine via the shikimate-chorismate pathway (Gill and Steglich 1987). It is perhaps not surprising that color reactions based on such physiologically fundamental compounds are phylogenetically uninformative at high taxonomic levels.

2. Acetate-Malonate Pathway Derivatives

Acetate-malonate pathway derivatives include ketides and anthraquinones. Anthraquinones have been much used in the specific- and generic-level taxonomies of *Cortinarius* and its subgenus or segregate *Dermocybe* (Høiland 1980, 1986; Keller 1982; Liu et al. 1997). Chemically similar pigments occur in *Claviceps* and *Hypomyces* (Ascomycota, Hypocreales) and in certain species of *Leucopaxillus* and *Tricholoma* (euagarics clade; Gill and Steglich 1987).

Fatty acid or higher polyketide pigments include merulic acids, which are found in *Phlebia* (= *Merulius*), ceriporiones, from *Ceriporia*, and sarcodontic acids, from *Sarcodontia* (a resupinate, hydnoid fungus that has yet to be included in molecular studies). Based on rDNA sequences, *Phlebia* and *Ceriporia* are part of a monophyletic group within the polyporoid clade that also includes *Bjerkandera*, other polypores, and corticioid fungi (Fig. 2; Y.-J. Yao et al., unpubl.).

3. Mevalonate Pathway Derivatives

The mevalonate pathway gives rise to carotenoid and sesquiterpenoid pigments. Too little is known of sesquiterpenoid distribution among homobasidiomycetes for phylogenetic comparisons, although the search for antibiotics and other pharmaceuticals may gradually yield this information (Gill and Steglich 1987; Anke et al. 1995). Carotenoid pigments are widespread in the fungi, including operculate and inoperculate discomycetes (Ascomycota) and Dacrymycetales (Basidiomycota: Heterobasidiomycetes). Nevertheless, their distribution among the homobasidiomycetes suggests that they could be phylogenetically informative in some groups. Several groups of species that appear to be closely related based on rDNA sequences share similar carotenoids (ψ-ψ carotenoids, β carotenoids, and γ carotenoids), including: (1) Phallales and *Sphaerobolus* (gomphoid-phalloid clade); (2) *Cantharellus* and *Craterellus* (cantharelloid clade); (3) *Peniophora* and *Stereum* (russuloid clade); (4) *Clavulinopsis*, *Chrysomphalina* (as *Gerronema*), *Haasiella*, and *Phyllotopsis* (euagarics clade; see Gill and Steglich 1987 for details).

4. Nitrogen-Containing Compounds

Among the diverse nitrogen-based pigments, an assortment of chemically similar quinone imines, phenyldiazonins, arylazoxycyanides and rubroflavins are found in *Agaricus* and *Calvatia*, which supports the view that they are closely related, as suggested by rDNA sequences (Fig. 2, Table 1).

Pistillarin, the compound responsible for colorless to dark green reactions with ferric chloride in fruiting bodies of *Ramaria*, is also found in *Clavariadelphus* and *Gomphus*, which supports the view that *Clavariadelphus* is closely related to the Gomphaceae (Fig. 2; Steglich et al. 1984; Gill and Steglich 1987). Similar staining reactions to iron salts are found in the false truffle *Gautieria* (Stewart 1974), supporting its placement in the gomphoid-ramarioid lineage, in contrast to the bolete lineage, where it has also been placed (Stewart 1974; Castellano et al. 1989; Pegler et al. 1993).

5. Bioluminescence

Bioluminescence has been reported in mycelia, fruiting bodies, and spores of *Armillaria*, *Dictyopanus*, *Favolaschia*, *Lampteromyces*, *Mycena*, *Omphalotus*, *Panellus*, and *Pleurocybella* (as *Pleurotus noctilucens*; Kobayasi 1952; Burdsall and Miller 1975; Wassink 1978; O'Kane et al. 1990). It has never been suggested that these taxa form a monophyletic group. Furthermore, there is variation for the presence or absence of bioluminescence within these taxa, even within single species (Petersen and Bermudes 1992). Nevertheless, all the taxa from which bioluminescence has been reported are white-spored saprotrophs in the euagarics clade. Bioluminescence in these fungi may result from modifications of a homologous metabolic pathway that is associated with saprotrophy.

F. Nutritional Modes

Homobasidiomycetes have diverse mechanisms for obtaining nutrition. Although it is necessary to divide these mechanisms into discrete categories for the purpose of discussion, it has long been recognized that individual species may display more than one mode (Garrett 1981; Hudson 1972). For example, some wood-decaying fungi trap and consume animals, and certain ectomycorrhizal species can degrade components of plant tissues, or parasitize other fungi (e.g., Thorn and Barron 1984; Trojanowski et al. 1984; Zhao and Guo 1989; Durrall et al. 1994; Hutchison and Barron 1996). Clearly, simple definitions of homobasidiomycetes as mutualists, pathogens, or saprotrophs should be qualified by identifying the organisms involved in the interaction.

Among the heterobasidiomycetes, Tremellales may be exclusively mycoparasites, but the Auriculariales and Dacrymycetales are wood decayers (Bandoni 1987; Wells 1994). This supports the view that saprotrophy is the plesiomorphic nutritional mode of the homobasidiomycetes, and that pathogenic and mycorrhizal symbioses are derived (Hacskaylo 1971; Malloch 1987). The following discussion focuses on the evolution of different nutritional modes in homobasidiomycetes. For reviews of ecological and physiological aspects, see Ahmadjian (1993), Barron (1992), Cooke and Whipps (1993), Dix and Webster (1995), Jeffries and Young (1994), Rayner and Boddy (1988), Smith and Read (1997), and Thorn (1997).

1. Saprotrophs

Homobasidiomycetes include the major decomposers of plant tissues, which represent the vast majority of terrestrial biomass. (Decomposition of the remains of other organisms falls mostly to Eubacteria). Decomposition of wood is by the far the best-studied form of saprotrophy in homobasidiomycetes, owing in large part to its great economic importance in forestry and its potential applications in biopulping and livestock forage treatment (Rayner and Boddy 1988; Eriksson et al. 1990; Blanchette 1991; Eaton and Hale 1993; Simpson 1996). Decay of leaf litter and other fine plant debris appears to occur through the same enzymatic mechanisms as wood decay (Orth et al. 1993; Tanesaka et al. 1993). Two main forms of wood decay are recognized: white rot and brown rot (also called cubical brown rot, or dry rot). In white rot, cellulose (50–70% of wood by weight), hemicellulose (10–20%), and lignin (10–20%) are all decayed, leaving a soft, white, often stringy residue, whereas in brown rot, only hemicellulose and cellulose are appreciably degraded, leaving a reddish brown, crumbly residue that may be almost pure lignin (Rayner and Boddy 1988; Preston et al. 1990; Green and Highley 1997; Worrall et al. 1997). Selective delignification (a form of white rot) also occurs, as in *Ceriporiopsis*, *Crucibulum*, and others (Blanchette et al. 1988; Worrall et al. 1997).

The primary lignin-modifying enzymes of white rot fungi are lignin peroxidases (LiPs), manganese peroxidases (MnPs), and laccases (Rayner and Boddy 1988; Hatakka 1994; Reddy and D'Souza 1994; Thurston 1994). Presence or absence of tyrosinase has been used as a taxonomic character in studies of cultures of wood-decaying fungi (Boidin 1951; Stalpers 1978), but this enzyme is not known to be directly related to the degradation of wood polymers.

Determination of rot type has traditionally been based on observations of natural substrates, or culture studies involving spot tests with reagents intended to detect wood-decaying enzymes (Davidson et al. 1942; Nobles 1965; Rayner and Boddy 1988). However, many colorimetric substrates used in such tests (e.g., gum guaiac and pyrocatechol) can be oxidized by any LiPs, MnPs, laccase, or tyrosinase, or nonbiological oxidants such as malt extract in culture media (Maehly and Chance 1954; Kratochvil et al. 1971). Consequently, older literature reporting "polyphenol oxidases" as physiological indicators or taxonomic characters must be interpreted with caution. On the basis of spot tests, it has been widely assumed that brown rot fungi and certain white rot fungi, such as *Phanerochaete chrysosporium*, do not produce laccases (Nobles 1965, 1971; Hatakka 1994). However, recent studies have shown that laccase-specific gene sequences and laccase activity occur in both brown rot fungi (D'Souza et al. 1996) and *P. chrysosporium* (Srinivasan et al. 1995).

Most wood- and litter-decaying homobasidiomycetes, including soil fungi that decay fine litter in and on soil, cause white rots. Based on species counts in Hawksworth et al. (1995) and Gilbertson (1980, 1981), homobasidiomycetes probably include about 8500 described white rot species, and 200 brown rot species. In North America, about 6% of wood decaying fungi are brown rot species (Gilbertson 1980, 1981), and approximately 85% of these occur on conifer substrates (Gilbertson 1980). Under controlled conditions, brown rot fungi remove biomass from wood more rapidly than white rot fungi (Gilbertson 1980). On this basis, it has been argued that production of a brown rot has adaptive value in the conifer-dominated forests of high latitudes and elevations, which have short growing seasons (Gilbertson 1980).

Brown rot has been interpreted as either a plesiomorphic (Nobles 1965, 1971) or derived condition in the homobasidiomycetes (Gilbertson 1981; Ryvarden 1991; Worrall et al. 1997). Among the heterobasidiomycetes, the Dacrymycetales contains species causing brown rots (and possibly white rots; Seifert 1983; Worrall et al. 1997), whereas the Auriculariales sensu stricto (the putative sister group of the homobasidiomycetes) contains only white rot species. Within the homobasidiomycetes, species with white rot activity occur in each of the eight major clades that we recognize, except the bolete clade (Ginns and Lefebvre 1993; Rayner and Boddy 1988). In contrast, brown rot species occur only in the euagarics, bolete, and polyporoid clades (as far as we know). This distribution suggests that white rot is the plesiomorphic form in the homobasidiomycetes.

In the euagarics clade, brown rot species are found in *Fistulina*, *Hypsizygus*, and, possibly, *Ossicaulis* (previous reports suggesting that *Coprinus* contains brown rot species have been shown to be erroneous; Redhead and Ginns 1985). The agaricoid genera *Neolentinus* and *Heliocybe* also produce brown rots, but their higher-level relationships are unclear (Hibbett and Vilgalys 1993). *Neolentinus* and *Heliocybe* were segregated from *Lentinus* largely because species of *Lentinus sensu stricto* (as well as *Lentinula* and *Panus*) produce a white rot (Redhead and Ginns 1985). Molecular studies support the segregation of *Neolentinus* and *Heliocybe* from *Lentinus* and *Panus* (Hibbett and Vilgalys 1993).

In the bolete clade, brown rot species are found in *Hygrophoropsis*, *Paxillus* (including *Tapinella*), *Coniophora*, *Serpula*, and others. Brown rot fungi in the bolete clade are distinguished by their ability to degrade pure cellulose in culture (however, *Postia* and *Gloeophyllum* have been shown to degrade cellulose in soil-block tests; Nilsson 1974; Nilsson and Ginns 1979; Highley 1988; Green and Highley 1997). As discussed by Redhead and Ginns (1985), *Omphalotus* produces a white rot, which supports the view that it is not in the bolete clade (see Sect. II. C. 2). Phylogenetic studies suggest that production of a brown rot evolved once in the bolete clade, and several times in the euagarics clade, and that the production of brown rots in the euagarics and bolete clades are not homologous (Bruns et al. 1998; Moncalvo et al. 2000, and unpubl.).

The majority of brown rot species that have been studied using molecular characters occur in the polyporoid clade. However, brown rot fungi in this clade do not form a monophyletic group, suggesting that there is a complex pattern of transformations between decay types. Nevertheless, two strongly supported groups of genera are united by the production of a brown rot: (1) *Laetiporus-Phaeolus-Sparassis* (*Pycnoporellus* is probably in this group), and (2) *Fomitopsis-Daedalea-Piptoporus* (Hibbett and Donoghue 1995; Hibbett et al. 1997b). There are many other brown rot polypores that are of uncertain placement, however, such as *Antrodia*, *Postia* (= *Oligoporus*), *Wolfiporia*, and *Gloeophyllum* [as noted previously, the placement of *Gloeophyllum* outside of the polyporoid clade (Fig. 2) is contradicted by morphology and independent molecular studies]. In addition, there are several brown rot genera of Corticiaceae and Stereaceae that have yet to be investigated using molecular approaches, including *Chaetoderma* and *Crustoderma* (Corticiaceae), and *Columnocystis* (= *Veluticeps*, Stereaceae).

Taken together, the evidence discussed above suggests that the production of a brown rot has evolved multiple times in the homobasidiomycetes (Worrall et al. 1997), with most brown rot species occurring in the polyporoid clade. Although there is homoplasy in this character, it is nonetheless phylogenetically informative for some groups. As noted by Redhead and Ginns (1985), division of saproptrophic homobasidiomycetes into white rot and brown rot categories fails to represent the evolutionary diversity of decay systems. Detailed comparative studies of decay mechanisms (e.g., Vares and Hatakka 1997), especially analyses of the genes encoding decay enzymes, may provide many characters for homobasidiomycete systematics.

2. Mycorrhizae

Homobasidiomycetes include fungi that form ectomycorrhizae (including both arbutoid and monotropoid mycorrhizae), orchid mycorrhizae, and possibly ericoid mycorrhizae (Smith and Read 1997; Umata 1995). Historically, basic studies of mycorrhizae have been limited by the reliance of basidiomycete taxonomy on fruiting bodies, but molecular sequence databases are providing powerful new tools for identifying mycorrhizae in the absence of fruiting bodies, as well as for inferring evolution of mycorrhizae (Cullings et al. 1996; Gardes and Bruns 1996; Taylor and Bruns 1997; Bruns et al. 1998). The physiology, ecology, and morphology of mycorrhizae have been discussed at length elsewhere (e.g., Smith and Read 1997; Varma and Hock 1995). In this section, we consider the phylogenetic distribution of mycorrhizae in the homobasidiomycetes.

Ectomycorrhizae are the most widespread form of mycorrhizae in homobasidiomycetes. Ectomycorrhizae are also formed by certain ascomycetes, but they are absent from Auriculariales, Dacrymycetales, and Tremellales, which supports the view that the plesiomorphic condition of the homobasidiomycetes is saprotrophic (Hacskaylo 1971; Malloch 1987). Within the homobasidiomycetes, ectomycorrhizal species are concentrated in the bolete clade (Boletales, including Sclerodermataceae), thelephoroid clade (Thelephoraceae, Scutigeraceae), russuloid clade (*Albatrellus* pro parte, Russulaceae, including gasteroid forms), cantharelloid clade (Cantharellaceae, Hydnaceae sensu stricto), gomphoid-phalloid clade (Gomphaceae, *Gautieria*), and euagarics clade (*Amanita*, *Hygrophorus* sensu stricto, *Tricholoma*, *Inocybe*, *Cortinarius*, *Hebeloma*, and *Laccaria*; Trappe 1962). In addition, one ectomycorrhizal genus occurs in the predominantly saprotrophic hymenochaetoid clade (*Coltricia*). Within the euagarics clade, ectomycorrhizal taxa are absent from the Agaricaceae-Lycoperdaceae clade, Coprinaceae, Strophariaceae, Pleurotaceae, and indeed most major groups of euagarics (Fig. 2). Most lineages containing ectomycorrhizal taxa also contain (and may be derived from) saprotrophic taxa (Malloch 1987). Based on species counts in Hawksworth et al. (1995), we estimate that approximately 4500 described species of homobasidiomycetes form ectomycorrhizae.

Phylogenetic studies (Hibbett et al. 1997b; Bruns et al. 1998; Moncalvo et al. 2000) indicate that there have been repeated transformations between ectomycorrhizal and saprotrophic forms in homobasidiomycetes, but the precise number of gains and losses is not resolved. We inferred the evolution of ectomycorrhizae in the homobasidiomycetes by mapping the distribution of ectomycorrhizae onto the tree from Hibbett et al. (1997b, Fig. 2) using parsimony, under the assumption that gains and losses of ectomycorrhizae are equally likely. The results suggest that there have been seven independent gains of ectomycorrhizae in the homobasidiomycetes, one in each of the eight major clades that we recognize, except the polyporoid clade (Fig. 2). However, this inference is sensitive to rearrangements in weakly supported parts of the tree topology, and might be affected by the addition of more ectomycorrhizal taxa (noteworthy ectomycorrhizal taxa that are missing from Fig. 2 include *Hygrophorus*, *Laccaria*, and *Tricholoma* in the euagarics clade, and *Albatrellus* pro parte in the russuloid clade). Furthermore, the optimization of character states on the tree is sensitive to changes in the relative weight given to gains vs. losses. It is surprising to find that a complex character such as ectomycorrhizae, which involves specialized physiological as well as morphological attributes, could have arisen as many as seven times. In contrast, the fact that some ectomycorrhizal species have apparently retained lignolytic or cellulolytic enzyme systems (Trojanowski et al. 1984; Durrall et al. 1994) suggests that transformations from ectomycorrhizal to saprotrophic habits might be easily reversible. Thus, it may be more appropriate to infer the evolution of ectomycorrhizae using a model that favors losses over gains rather than the "flat-weighted" model used in Fig. 2. To rigorously infer the evolution of ectomycorrhizae, it will be necessary to add more ectomycorrhizal lineages to the tree, resolve weakly supported nodes, and map character states under a range of realistic models.

Orchid associates are found in the euagarics clade (*Armillaria*, *Ceratobasidium*, and *Lentinula*), cantharelloid clade (*Tulasnella*), polypore clade (*Ganoderma*, *Loweporus*, *Microporus*), and hymenochaetoid clade (*Erythromyces* and *Phellinus*) (Currah and Zelmer 1992; Umata 1995, 1998), as well as in the Auriculariales (Umata 1997). However, it is not known whether *Lentinula* and the polypores and hymenochaetoid fungi listed form fully functional orchid mycorrhizae in nature, or only promote orchid seed germination (Umata 1995, 1998).

Early reports of clavarioid fruiting bodies associated with pot cultures of *Vaccinium* (Ericaceae) and of clamped hyphae in ericoid mycorrhizal roots (Seviour et al. 1973) have not, to our knowledge, been substantiated with synthesis of mycorrhizae between Ericaceae and *Clavaria argillacea*, the putative symbiont. With the exception of the study by Seviour et al. (1973), ericoid mycorrhizae appear to be predominantly formed by ascomycetous fungi (Read 1974; Currah et al. 1990).

3. Plant Pathogens

The ability to attack living plants has evolved repeatedly in the homobasidiomycetes. This section provides a very brief overview of the phylogenetic distribution of plant pathogens in the homobasidiomycetes (see also Sect. III.F.5, below, on bryophyte parasites). For much more comprehensive lists of plant pathogenic homobasidiomycetes, see Farr et al. (1989).

Plant pathogens occur in at least five of the eight major clades of homobasidiomycetes that we recognize, but (as far as we know) they are absent from the thelephoroid, cantharelloid, and gomphoid-phalloid clades. From a purely economic standpoint, the most important plant pathogens include *Thanatephorus cucumeris*, *Ceratobasidium* spp., and *Athelia rolfsii*, and their anamorphs, which cause damping off and root rot diseases of diverse crops. Phylogenetic placement of these taxa is controversial; some or all may be in the euagarics clade (see Sect. II.C.2). Other economically important pathogens that occur in the euagarics clade include *Typhula* spp. (which cause snow molds of turf and cereal crops), *Marasmius* and *Marasmiellus* spp. (leaf blights of maize), *Crinipellis perniciosa* (witches' broom of cacao), and *Mycena citricolor* (leaf blight of coffee). Similar plant diseases are caused by several corticioid taxa of uncertain placement, including *Erythricium salmonicolor* (pink disease on various fruit trees), and *Laetisaria fuciformis*, *Limonomyces roseipellis*, and *Trechispora alnicola* (various turf diseases; Stalpers and Loerakker 1982; Ginns and Lefebvre 1993).

Ecologically similar timber pathogens have evolved repeatedly in the homobasidiomycetes. Forest pathogens that cause root and butt rots are primarily concentrated in the polyporoid clade (e.g., *Ganoderma*, *Phaeolus*, *Sparassis*), but also occur in the euagarics clade (*Armillaria*, *Coprinus*), russuloid clade (*Bondarzewia*, *Heterobasidion*), and hymenochaetoid clade (*Inonotus*, *Phellinus*). Similarly, pathogens that form cankers and heart rots of forest trees include members of the polyporoid clade (*Fomitopsis*, *Piptoporus*, *Polyporus*, *Postia*), euagarics clade (*Hypsizygus*, *Pholiota*, *Pleurotus*, *Volvariella*), russuloid clade (*Echinodontium*, *Hericium*), and hymenochaetoid clade (*Inonotus*, *Oxyporus*, *Phellinus*).

4. Mycoparasites, Bacteriovores, and Nematode-Trappers

Nitrogen is a limiting factor in many of the substrates occupied by homobasidiomycetes, especially wood (Cowling and Merrill 1966; Rayner and Boddy 1988). Consequently, microorganisms such as bacteria, nematodes, and other fungi are potentially important sources of nutrition for homobasidiomycetes. In this section, we discuss cases in which homobasidiomycetes appear to obtain nutrition by attacking and consuming fungi, bacteria, and nematodes.

Much of what is known about mycoparasitism in homobasidiomycetes is based on culture studies. Although it is difficult to extrapolate to natural systems, some interactions in culture seem to correspond well with those in nature. For example, in culture *Lenzites betulina* parasitizes and grows through mycelia of *Trametes versicolor*, just as in nature *L. betulina* frequently colonizes hardwood substrates that are already inhabited by *T. versicolor*, replacing *T. versicolor* in the process (Rayner et al. 1987). In addition, *L. betulina* is apparently not an effective parasite of other common wood-decaying fungi, suggesting that it employs selective parasitism of *T. versicolor* as a resource capture strategy (Rayner et al. 1987). Facultative mycoparasitism of filamentous fungi and yeasts appears to be widespread among primarily saprotrophic species of homobasidiomycetes in the polyporoid, euagarics, and gomphoid-phalloid clades, and is also reported in certain ectomycorrhizal species in the bolete and thelephoroid clades (Griffith and Barnett 1967; Rayner et al. 1987; Zhao and Guo 1989; Jeffries and Young 1994; Owens et al. 1994; Hutchison and Barron 1996; R.G. Thorn, unpubl.).

A few homobasidiomycetes produce fruiting bodies directly on the fruiting bodies of other homobasidiomycetes. Examples are known in the bolete clade (*Boletus parasiticus* fruits on *Scleroderma*; *B. astraeicola* on *Astraeus*), but most occur in the euagarics clade (e.g., *Nyctalis*/*Asterophora* spp. on Russulaceae; *Entoloma parasitica* on *Polyporus*, *Coltricia*, or *Cantharellus*; *Psathyrella epimyces* on *Coprinus* spp.; *Volvariella surrecta* on various Tricholomataceae; *Squamanita* spp. on various agarics; Rayner et al. 1985; Spurr et al. 1985; Jeffries and Young 1994; Redhead et al. 1994). Such forms have generally been interpreted as necrotrophic parasites, which kill and then digest living tissue (Rayner et al. 1985). However, Rayner et al. (1985, p. 10) noted that *B. parasiticus* and *B. astraeicola* do not appear to be parasitic on *Scleroderma* and *Astraeus*, "but simply require the presence of the latter to stimulate fruiting." Another enigmatic interaction involves *Entoloma abortivum*, which forms irregular masses of tissue termed *carpophoroids* when infected by the mycelium of *Armillaria mellea* (Watling 1974). Many host-parasite (or host-necrotroph) relationships involving homobasidiomycetes have a high degree of taxon specificity. For this reason, the association of *B. parasiticus* and *B. astraeicola* with *Scleroderma* and *Astraeus*, respectively, may indicate that *Astraeus*, which has yet to be included in molecular studies, is closely related to *Scleroderma*.

Many species of saprotrophic homobasidiomycetes are capable of lysing bacterial cells in vitro (Barron 1988; Barron and Thorn 1987; Thorn and Tsuneda 1992). In *Agaricus* and other filamentous fungi, this is accomplished by muramidases, which degrade bacterial cell walls (Fermor 1983; Grant et al. 1986). Bacteriolytic ability has been found extensively in the polyporoid and euagarics clades, as well as in the Auriculariales and certain corticioid fungi of uncertain placement (e.g., *Dendrothele*; Thorn and Tsuneda 1992). Because of limited sampling, it is difficult to infer the pattern of evolution of bacteriolytic activity. The available data suggest that it is widespread, as is mycoparasitism, and may therefore be of limited use for inferring higher-level phylogenetic relationships of homobasidiomycetes.

In contrast to mycoparasitism and bacteriolytic activity, the ability to attack and consume living nematodes has a limited taxonomic distribution among the homobasidiomycetes (nematode-trapping fungi also occur among ascomycetes; Barron 1977, 1988; Thorn and Barron 1984; Liou and Tzean 1992; Thorn and Tsuneda 1992). Species of *Pleurotus* and *Hohenbuehelia* capture and consume nematodes by immobilizing them with a toxin (*Pleurotus*; Barron and Thorn 1987) or adhering to them with adhesive knobs produced on hyphae or germinated spores (*Hohenbuehelia*; Barron and Dierkes 1977; Thorn and Barron 1986). This capability has been used as biological evidence for delimiting both *Pleurotus* and *Hohenbuehelia* (Hibbett and Thorn 1994; Redhead and Ginns 1985; Thorn and Barron 1986). Species of the dark-spored agarics *Conocybe* and *Panaeolina* produce droplets of nematotoxin of similar appearance to those in *Pleurotus*, but the nematodes are not colonized and consumed once immobilized (Hutchison et al. 1996). Toxin droplets in these taxa may serve a primarily defensive function. Droplets of similar appearance are also produced by species of *Psilocybe* (Heim and Wasson 1959), *Schizophyllum* (Parag 1965), *Resupinatus*, and *Stigmatolemma* (Thorn and Barron 1986), but their function in nature is as yet unknown. The taxa mentioned so far are all members of the euagarics clade, but there is no evidence that they form a monophyletic group. Thus, it seems likely that production of toxin droplets with nematode-trapping or antifeedant functions has evolved repeatedly within the euagarics clade. In addition, certain members of the corticioid genus *Hyphoderma* (which is unplaced) capture nematodes by production of specialized adhesive cells called stephanocysts, or by a toxin that apparently kills nematodes that have consumed hyphal cells (Tzean and Liou 1993).

5. Lichens, Algal Parasites, and Bryophyte Associates

A variety of fungi have independently formed symbioses with phototrophic green algae, cyanobacteria, and bryophytes (Hale 1983; Ahmadjian 1993; Gargas et al. 1995). Fungi in these associations form a continuum from necrotrophic parasites to mutualistic lichenized forms (Hawksworth 1988). This section discusses the diversity and evolution of symbioses involving homobasidiomycetes with algae and bryophytes.

More than 99% of the 20 000 or more known species of lichens are formed by ascomycetes, but a few are formed by homobasidiomycetes (Oberwinkler 1970, 1984; Redhead and Kuyper 1987; Hawksworth et al. 1995). Basidiolichens have arisen at least twice in the euagarics clade, in the agaricoid *Omphalina* and the stereoid *Dictyonema* (Parmasto 1978). The lichenized forms of *Omphalina* are nested in a clade of saprotrophic *Omphalina* species (and other genera), but the precise placement of *Dictyonema* is not resolved with confidence (Gargas et al. 1995; Lutzoni 1997; Lutzoni and Pagel 1997). *Semiomphalina leptoglossoides* is another basidiolichen that is probably in the euagarics clade (Redhead 1984; Redhead and Kuyper 1987), but its phylogenetic relationships have yet to be investigated. In the cantharelloid clade, basidiolichens are formed by the minute clavarioid *Multiclavula*, which appears to be the sister group of an exclusively ectomycorrhizal clade that includes *Cantharellus*, *Hydnum*, and *Clavulina* (Hibbett et al. 1997; Fig. 2). In addition to these, Singer (1973, 1984) has suggested that *Marasmiellus affixus* (presumably euagarics clade) and *Lactarius* (= *Pleurogala*) *igapoensis* (a pleurotoid member of the russuloid clade; Redhead and Norvell 1993) are basidiolichens. However, Singer's conclusions were based only on the cooccurrence of the fungi and algae, not a rigorous demonstration of hyphal connections. Although the lichenized status of *M. affixus* and *L. igapoensis* is questionable, it is clear that there have been multiple derivations of lichenized forms in the homobasidiomycetes. Anatomical differences among fungus-alga interactions of basidiolichens support the view that they are polyphyletic (Oberwinkler 1984).

Lichen symbioses involve formation of a stable dual-organism thallus, but a range of less stable or morphologically differentiated associations is also possible (Hawksworth 1988; Hutchison and Barron 1997). Microscopic observations of complex substrates such as well-rotted wood frequently reveal clamped hyphae in close contact with or coiling around cells of green algae (Thorn and Barron 1986). For example, Oberwinkler (1970) has described parasitism of unicellular green algae by the corticioid fungi *Athelia epiphylla*, *Resinicium bicolor*, and *Sistotrema brinkmanii* (the latter in culture). Recently, Hutchison and Barron (1997) found that 37 species of homobasidiomycetes (out of 74 species tested) were capable of attacking and parasitizing cells of the green alga *Protococcus* or the cyanobacterium *Synechococcus* in pure culture. Directional growth of hyphae toward algal or cyanobacterial colonies was followed by coralloid branching within the colonies, followed by cell death, lysis, and absorption of the cell contents by the invading fungi (Hutchison and Barron 1997). This pattern is similar to that seen in the attack of bacteria and yeasts by homobasidiomycetes (Barron 1988; Hutchison and Barron 1996). All of the species that attacked the algae or cyanobacteria were saprotrophs in the euagarics clade (e.g., *Crucibulum*, *Langermannia*, and *Typhula*), polyporoid clade (*Bjerkandera* and *Lenzites*), and russuloid clade (*Stereum*; the study did not include representatives of the thelephoroid, cantharelloid, or gomphoid-phalloid clades). Despite limited sampling, the observations cited above suggest that the ability to parasitize algae and cyanobacteria is widespread among saprotrophic homobasidiomycetes.

Homobasidiomycetes are often found growing among living bryophytes (Gulden et al. 1985; Gulden and Jenssen 1988; Redhead 1980, 1981, 1984). In some cases, necrotic zones have been described in the areas immediately adjacent to the fungi, but often the bryophytes have been described as healthy. Based on in vitro coculturing experiments with the moss *Sphagnum capillaceum*, Redhead (1981) demonstrated that *Lyophyllum* (*Tephrocybe*) *palustre* is capable of necrotrophic parasitism, which ultimately kills the host (Untiedt and Müller 1985), whereas *Galerina paludosa* forms a "balanced" parasitism, in which the moss is not killed. In addition to symbionts of mosses, homobasidiomycetes include symbionts of thallose liverworts, including *Rickenella pseudogrisella*, which is a symbiont of *Blasia*, and "*Gerronema*" *marchantiae*, which is a symbiont of *Marchantia* (Redhead 1980, 1981; Senn-Irlet et al. 1990). In cases of balanced parasitism of bryophytes, homobasidiomycetes form haustoria or appressoria on the rhizoids of the bryophytes (Redhead 1980, 1981). Redhead (1981) noted that the appressoria and haustoria formed by bryophyte parasites are similar to structures formed by homobasidiomycetous algal parasites and lichens, which suggests that they may be homologous. Based on the results of Moncalvo et al. (2000), *Rickenella*, *Lyophyllum*, and *Galerina* are in three separate lineages in the euagarics clade, which implies that bryophyte symbiosis has evolved repeatedly in the homobasidiomycetes.

6. Insect Symbionts

Homobasidiomycetes form ecological associations with diverse insects that feed on hyphae, spores, and fruiting bodies, or that colonize decayed wood (Batra 1979; Wheeler and Blackwell 1984; Wilding et al. 1989). In addition, there is a limited number of insect-homobasidiomycete symbioses, in which the life cycles of the fungi and insects appear to be tightly linked. The homobasidiomycetes in these symbioses make up a polyphyletic group of saprotrophs, each of which is closely related to (or perhaps conspecific with) free-living forms. Many have been successfully cultured, and some are probably able to complete their life cycles in the absence of the insect. As in the fungi, the insects involved are polyphyletic and are closely related to free-living forms. However, judged by their constant occurrence with fungi in nature, the insects appear to be obligate symbionts. In this section, we survey the phylogenetic distribution of homobasidiomycetes involved in symbioses with insects, which include (1) woodwasp symbionts, (2) bark and ambrosia beetle symbionts, (3) attine ant symbionts, and (4) termite symbionts.

Female woodwasps (Siricidae) have intersegmental pouches (mycetangia) that contain oidia or hyphal fragments of wood-rotting fungi, which are inoculated into trees at the time of oviposition (Stillwell 1966; Madden and Coutts 1979; Gilbertson 1984; Tabata and Abe 1995). Larvae feed on the mycelium that is produced by the oidia, and female larvae develop mycetangia, which house fungal inoculum for the next generation. Fungi involved in this symbiosis include species of *Amylostereum*, which are resupinate fungi that are associated with the wasp genera *Sirex* and *Urocerus*, and *Cerrena unicolor*, which is a polypore that is associated with the wasp *Tremex* (Gilbertson 1984; Tabata and Abe 1995). *Amylostereum* is in the russuloid clade (Table 1), but *Cerrena* is of uncertain placement. Ryvarden (1991) suggested that *Cerrena* is related to *Trametes*, which is in the polyporoid clade. If this is correct, then there must have been at least one switch of the fungal partner during the evolution of the siricid-homobasidiomycete symbiosis. According to Stillwell (1966) and King (1966), *Amylostereum* fruiting bodies are apparently absent in certain areas where the siricid-fungus association is common (i.e., the fungus is found only as mycelium in association with wasps). Citing these observations, Gilbertson (1984, p. 144) suggested the possibility that *Amylostereum* species associated with wood wasps have become completely dependent on the insect for dissemination".

Bark and ambrosia beetles (Scolytidae and Platypodidae) inhabit dead tree trunks, where they form galleries in either phloem (bark beetles) or xylem (ambrosia beetles; Beaver 1989). The different taxa of beetles vary in the degree to which they feed on fungal tissue vs. wood, and in the anatomical locations of the mycangia by which they transmit fungi (Beaver 1989). Many homobasidiomycetes are found growing on trees colonized by bark and ambrosia beetles, and the work of Castello et al. (1976), among others, indicates that bark beetles can be effective vectors of wood-rotting homobasidiomycetes, such as *Cryptoporus volvatus* and *Fomitopsis pinicola* (polyporoid clade). However, only a few homobasidiomycetes have been directly isolated from beetle mycangia or galleries (most bark beetle symbionts are ascomycetes). One of these is the unusual corticioid fungus, *Entomocorticium dendroctoni*, which is apparently limited to the galleries of the bark beetle *Dendroctonus ponderosae* (Whitney et al. 1987). Whitney et al. (1987) suggested that the fungus is distributed exclusively by the insect, which seems plausible considering that its fruiting bodies are produced inside the galleries. Furthermore, the spores of *Entomocorticium* are symmetrically positioned over broad sterigmata, with no evidence of a hilar appendix, indicating that *Entomocorticium* is no longer ballistosporic.

Other unnamed bark-beetle symbionts have been recovered from mycangia of *Dendroctonus frontalis* (known only as SJB 122; Happ et al. 1976) and *D. brevicomis* (Whitney and Cobb 1972). Although no fruiting body is known, the presence of perforate parenthesomes strongly suggests that SJB 122 is a homobasidiomycete (Happ et al. 1976). Based on cultural studies, Whitney et al. (1987) concluded that SJB 122 and the fungus from *D. brevicomis* are similar, but that *E. dendroctoni* is a separate taxon.

Recent molecular and cultural studies of the fungal symbionts of Scolytidae suggest that there have been at least four independent origins of bark beetle symbionts in the homobasidiomycetes (Hsiau 1996). *Entomocorticium dendroctoni* and eight other unnamed taxa (including symbionts of *D. brevicomis*, *D. frontalis*, and other beetle species) form a monophyletic group that is nested in *Peniophora*, which is in the russuloid clade. The close relationship between these taxa is also supported by the presence of similar incrusted cystidia in *Peniophora* and *Entomocorticium*. Two other unnamed *Dendroctonus* symbionts are very closely related to the corticioid fungus *Phlebiopsis gigantea* (mt-ssu rDNA sequences of the beetle symbionts differ from that of *Phlebiopsis* by only three single base indels). In Hsiau's analysis, the sister group of the beetle symbiont-*Phlebiopsis* clade is *Pulcherricium*, suggesting that this group is in the polyporoid clade (also see Boidin et al. 1998). The fungal symbiont of the beetle *Xyleborus dispar* appears to be closely related to the brown rot polypores *Antrodia carbonica* and *Meripilus giganteus*, which are also in the polyporoid clade. The *X. dispar* associate is the only bark beetle symbiont that has a negative reaction on gallic and tannic acid agar, which supports the molecular phylogeny. Finally, *Gloeocystidium ipidophilum*, which is associated with the beetle *Ips typographus*, appears to be in the euagarics clade, which was represented in Hsiau's analysis by *Lentinula* and *Pleurotus*.

Attine leafcutter ants (Formicidae, Attini) of the Neotropics cultivate homobasidiomycetes on plant material inside large subterranean nests (Weber 1979; Cherrett et al. 1989). Occasional fruiting bodies produced on ant-fungus nests (e.g., Fisher et al. 1994) have indicated that some of the fungi involved in the symbiosis are members of the Agaricaceae sensu Singer (1986; euagarics clade). Recent molecular studies by Chapela et al. (1994) have shown that the attine ant fungi are composed of three groups: (1) The G1 group is nested in the Agaricaceae and is characterized by the production of modified hyphal tips termed gongylidia that are consumed by the ants. (2) The G2 group is a paraphyletic assemblage that is also nested in the Agaricaceae, and from which the G1 group was derived. Some ant symbionts in G2 are more closely related to free-living Agaricaceae (*Leucoagaricus*, *Leucocoprinus*) than they are to other ant symbionts. (3) The G3 group is unique among attine symbionts in being closely related to white-spored agarics in the Tricholomataceae (represented in their study by *Marasmius* and *Crinipellis*). The results of Chapela et al. (1994) indicate that there has been repeated acquisition of symbionts from free-living fungi in the Agaricaceae and Tricholomataceae.

Finally, termites (Macrotermitinae) of the Paleotropics exist in symbioses with the homobasidiomycete *Termitomyces* that in many ways parallel the association of attine ants and their fungi in the Neotropics (Batra and Batra 1979; Wood and Thomas 1989). Termite nests are large subterranean structures or mounds that include combs made up primarily of termite feces, on which *Termitomyces* produces "spherules" of conidiogenous hyphae that are eaten by the termites (Batra and Batra 1979; Wood and Thomas 1989). *Termitomyces* fruiting bodies are commonly produced on termite nests, and the elongate pseudorhizas can be traced to the combs. According to Singer (1986), there are about 13 species of *Termitomyces*, all of which are termite symbionts. Singer placed *Termitomyces* in the Termitomyceteae of the Tricholomataceae, along with *Podabrella*, which is found on dead organic matter, or occasionally on termite nests. Singer (1986) suggested that the Termitomyceteae is closely related to the Lyophylleae (*Lyophyllum*, etc.) based on their shared possession of siderophilous granules (which turn violet-black in acetocarmine) in the mature basidia. Recent molecular studies by Moncalvo et al. (2000) suggest that *Podabrella* is the sister group of *Termitomyces*, and that they are nested within *Lyophyllum*.

Summarizing the preceding discussion, insect symbionts have evolved repeatedly in the homobasidiomycetes, and occur in the euagarics clade (attine symbionts in Agaricaceae and Tricholomataceae, *Termitomyces*, *Gloeocystidium ipidophilum*), russuloid clade (*Amylostereum*, *Entomocorticium*), and polyporoid clade (*Cerrena*, *Xyleborus* symbionts, *Dendroctonus* symbionts). Each of these symbioses appears to be uniquely derived. Nevertheless, there are striking similarities between some systems: Symbioses with ants and termites occur in the tropics and involve cultivation of saprotrophic agarics inside massive communal nests, with fungal cells specialized for feeding insects. Symbioses with woodwasps and bark/ambrosia beetles occur primarily in temperate forests and involve insect-vectored transmission of wood-decaying corticioid fungi and polypores as oidia or mycelial fragments in mycangia, perhaps coupled with a loss of spore dispersal ability by the fungus (loss of ballistospory or fruiting body production).

IV. Conclusions

In recent years there has been rapid progress in homobasidiomycete phylogenetics, which has been brought about largely through the analysis of molecular characters (Table 1). Molecular studies have supported many morphology-based taxonomic hypotheses, such as the placement of *Rhizopogon* in the Boletales (Bruns et al. 1989), or *Lentinus* sensu stricto among the polypores (Hibbett and Vilgalys 1993). Other hypotheses have been refuted, however, such as the inclusion of *Omphalotus* and *Lampteromyces* in the Boletales (Binder et al. 1997; Moncalvo et al. 2000). In addition to resolving preexisting controversies, molecular studies have also provided novel insights, such as the realization that the Gomphaceae and Phallales are closely related (Colgan et al. 1997; Hibbett et al. 1997b), or that bark beetle associates have evolved four times in the homobasidiomycetes (Hsiau 1996). Most importantly, molecular characters derived from universal genes such as rDNA provide comparative data that will eventually form the basis of a phylogenetic classification for all homobasidiomycetes, including asexual forms.

Summarizing current research, we have divided the homobasidiomycetes into eight mutually exclusive clades. Not all clades are equally well supported, however, and there is considerable danger in extrapolating from exemplar-based phylogenetic studies to a classification for over 13000 described species. Much more work will be needed to attain a comprehensive phylogenetic classification of the homobasidiomycetes. Specific challenges that remain include evaluating the monophyly, higher-order relationships, and internal structure of the eight major clades that we recognize, identifying the root and sister group of the homobasidiomycetes, and resolving relationships of certain controversial taxa, such as *Thanatephorus*. Molecular data will be essential for resolving homobasidiomycete phylogeny, but there are also numerous potentially informative morphological characters, many of which have never been included in phylogenetic analyses. We have highlighted selected characters that appear to be phylogenetically informative or ecologically significant.

However, many of our conclusions about character evolution are speculative, and we have often been limited by the number of published observations. We hope that the work presented here will provide a phylogenetic framework for studies of many non-molecular attributes of homobasidiomycetes, such as cytology and ultrastructure, developmental morphology, and mechanisms of symbiosis.

Acknowledgments. We thank David McLaughlin for inviting us to contribute this chapter, Dennis Desjardin, Jim Ginns, Leif Ryvarden, and Scott Redhead for helpful comments, George Barron and the Friends of Algonguin Park for permission to reproduce the images in Figs. 7, 9, and 14, and the following colleagues for generously sharing unpublished phylogenetic trees and other valuable information: Gerry Adams, Tom Bruns, Andrea Gargas, Tom Harrington, Portia Hsiau, Brandi Hughey, Ewald Langer, Gitta Langar, Ellen Larsson, Steve Miller, Jean-Marc Moncalvo, Wally Müller, Liz Pine, Joey Spatafora, Rytas Vilgalys, Yi-Jian Yao. DSH was supported by the United States National Science Foundation (DEB-9629427).

References

Adams GC, Klomparens KL, Hennon PE (1995) Unusual reticulated parenthesomes surround the dolipore of a hyphomycete with clamp connections, *Ditangifibulae dikaryotae* gen. et sp. nov. Mycologia 87:909–921

Agerer R (1986) "Cyphellaceae" versus Tricholomataceae, or what is a family? In: La Famiglia delle Tricholomataceae, Atti del Convegno Internazionale del 10-15 settembre 1984. Centro Studi per la Flora Mediterranea, Borgo Val di Taro, Italy, pp 9–27

Ahmadjian V (1993) The lichen symbiosis. John Wiley, New York

Anke H, Stadler M, Mayer A, Sterner O (1995) Secondary metabolites with nematicidal and antimicrobial activity from nematophagous fungi and Ascomycetes Can J Bot 73 (Suppl 1):S932–S939

Arita I (1979) Cytological studies on *Pholiota*. Rep Tottori Mycol Inst 17:1–118

Arpin N, Fiasson J-L (1971) The pigments of basidiomycetes: their chemotaxonomic interest. In: Petersen RH (ed) Evolution in the higher basisidiomycetes. University of Tennessee Press, Knoxville, pp 63–98

Bakkeren G, Gibbard B, Yee A, Froelinger E, Leong S, Kronstad J (1992) The a and b loci of *Ustilago maydis* hybridize to DNAs from other smut fungi. Mol Plant Microbe Interact 5:347–355

Bandoni RJ (1984) The Tremellales and Auriculariales: an alternative classification. Trans Mycol Soc Jpn 25:489–530

Bandoni RJ (1987) Taxonomic overview of the Tremellales. Stud Mycol 30:87–110

Bandoni RJ, Oberwinkler F (1982) *Stilbotulasnella*: a new genus in the Tulasnellaceae. Can J Bot 60:1875–1879

Barron GL (1977) The nematode-destroying fungi. Canadian Biological Publications, Guelph

Barron GL (1988) Microcolonies of bacteria as a nutrient source for lignicolous and other fungi. Can J Bot 66:2505–2510

Barron GL (1992) Lignolytic and cellulolytic fungi as predators and parasites. In: Carroll GC, Wicklow DT (eds) The fungal community: its organization and role in the ecosystem. Marcel Dekker, New York, pp 311–326

Barron GL, Dierkes Y (1977) Nematophagous fungi: *Hohenbuehelia*, the perfect state of *Nematoctonus*. Can J Bot 55:3054–3062

Barron GL, Thorn RG (1987) Destruction of nematodes by species of *Pleurotus*. Can J Bot 65:774–778

Bas C (1969) Morphology and subdivision of *Amanita* and a monograph of its section *Lepidella*. Persoonia 5:285–579

Bas C, Noordeloos ME, Vellinga EC (1990) Flora agaricina Neerlandica, vol 2. AA Balkema, Rotterdam

Batra LR (1979) Insect-fungus symbiosis: nutrition, mutualism, commensalism. John Wiley, New York

Batra LR, Batra SWT (1979) Termite-fungus mutualism. In: Batra LR (ed) Insect-fungus symbiosis: nutrition, mutualism, commensalism. John Wiley, New York, pp 117–163

Baura G, Szaro TM, Bruns TD (1992) *Gastrosuillus laricinus* is a recent derivative of *Suillus grevillei*: molecular evidence. Mycologia 84:592–597

Beaver RA (1989) Insect-fungus realtionships in the bark and ambrosia beetles. In: Wilding N, Collins NM, Hammond PM, Webber JF (eds) Insect-fungus interactions. Academic Press, London, pp 121–144

Begerow D, Bauer R, Oberwinkler F (1997) Phylogenetic studies on nuclear large subunit ribosomal DNA sequences of smut fungi and related taxa. Can J Bot 75:2045–2056

Berbee ML, Taylor JW (1993) Dating the evolutionary radiations of the true fungi. Can J Bot 71:1114–1127

Berbee ML, Wells K (1989) Light and electron microscopic studies of meiosis and basidium ontogeny in *Clavicorona pyxidata*. Mycologia 81:20–41

Besl H, Dorsch R, Fischer M (1996) Zur verwandtschaftlichen Stellung der Gattung *Melanogaster* (Melanogastraceae, Basidiomycetes). Z Mykol 62:195–199

Besson M, Froment A (1968) Observation d'un capuchon septal de type polypore hors des polyporacées. Bull Soc Mycol Fr 84:485–488

Binder M, Besl H, Bresinsky A (1997) *Agaricales* oder *Boletales*? Molekularbiologische Befunde zur Zuordnung einiger umstrittener Taxa. Z Mykol 63:189–196

Blanchette RA (1991) Delignification by wood-decay fungi. Annu Rev Phytopathol 29:381–398

Blanchette RA, Burnes TA, Leatham GF, Effland MJ (1988) Selection of white-rot fungi for biopulping. Biomass 15:93–101

Bodman MC (1938) Morphology and cytology of *Guepinia spathularia*. Mycologia 30:635–653

Boehm EWA, McLaughlin DJ (1989) Phylogeny and ultrastructure in *Eocronartium musicola*: meiosis and basidial development. Mycologia 81:98–114

Boidin J (1951) Recherche de la tyrosinase et alccase chez les Basidiomycetes en culture pure. Rev Mycol 16:173–197

Boidin J (1958) Essai biotaxonomique sur les hydnés et les corticiés. Rev Mycol Mem 6:1–388

Boidin J, Mugnier J, Canales R (1998) Taxonomie moleculaire des Aphyllophorales. Mycotaxon 66:445–491

Botha WJ, Eicker A (1991) Cultural studies on the genus *Termitomyces* in South Africa. I. Macro- and microscopic characters of basidiome context cultures. Mycol Res 95:435–443

Bresinsky A, Rennschmid A (1971) Pigmentmerkmale, Organisationsstufen und systematische Gruppen bei höheren Pilzen. Ber Dtsch Bot Ges 84:313–329

Brodie HJ (1951a) The function of the cups of *Polyporus conchifer*. Science 114:636

Brodie HJ (1951b) The splash-cup dispersal mechanism in plants. Can J Bot 29:224–234

Brodie HJ (1975) The bird's nest fungi. University of Toronto Press, Toronto

Bruns TD, Szaro TM (1992) Rate and mode differences between nuclear and mitochondrial small-subunit rRNA genes in mushrooms. Mol Biol Evol 9:836–855

Bruns TD, Fogel R, White TJ, Palmer JD (1989) Accelerated evolution of a false truffle from a mushroom ancestor. Nature 339:140–142

Bruns TD, White TJ, Taylor JW (1991) Fungal molecular systematics. Annu Rev Ecol Syst 22:525–564

Bruns TD, Szaro TM, Gardes M, Cullings KW, Pan JJ, Taylor DL, Horton TR, Kretzer A, Garbelotto M, Li Y (1998) A sequence database for the identification of ectomycorrhizal basidiomycetes by phylogenetic analysis. Mol Ecol 7:257–272

Buller AHR (1922) Researches on fungi, vol 2. Longmans, Green, and Co, London

Burdsall HH, Miller OK (1975) A reevaluation of *Panellus* and *Dictyopanus* (Agaricales). Beih Nova Hedwigia 51:79–91

Calonge FD (1969) Electron microscope studies on *Tomentella*. I. Ultrastructure of the vegetative hyphae. Arch Mikrobiol 65:136–145

Casselton LA, Kües U (1994) Mating-type genes in homobasidiomycetes. The Mycota I. Growth, differentiation and sexuality. Springer, Berlin Heidelberg New York, pp 307–321

Castellano MA, Trappe JM, Maser Z, Maser C (1989) Key to spores of the genera of hypogeous fungi of north temperate forests. Mad River Press, Eureka, California

Castello JD, Shaw CG, Furniss MM (1976) Isolation of *Cryptoporus volvatus* and *Fomes pinicola* from *Dendroctonus pseudotsugae*. Phytopathology 66:1431–1434

Chapela IH, Rehner SA, Schultz TR, Mueller UG (1994) Evolutionary history of the symbiosis between fungus-growing ants and their fungi. Science 266:1691–1694

Cherrett JM, Powell RJ, Stradling DJ (1989) The mutualism between leaf-cutting ants and their fungus. In: Wicding N, Collins NM, Hammond PM, Webber JF (eds) Insect-fungus interactions. Academic Press, London, pp 93–120

Christiansen MP (1960) Danish resupinate fungi, part II. Homobasidiomycetes. Dan Bot Ark 19:63–388

Clémençon H (1997) Anatomie der Hymenomyceten. F Flück-Wirth, Teufen

Coker WC, Couch JN (1928) The Gasteromycetes of the eastern United States and Canada. University of North Carolina Press, Chapel Hill

Colgan W, Castellano MA, Spatafora JW (1997) Systematics of the Hysterangiaceae Inoculum 48(3):7 (Abstr)

Cooke RC, Whipps JM (1993) Ecophysiology of the fungi. Blackwell, Oxford

Corner EJH (1932) The fruit-body of *Polystictus xanthopus* Fr. Ann Bot 46:71–111

Corner EJH (1950) A monograph of *Clavaria* and allied genera. Ann Bot Mem 2:1–740

Corner EJH (1966) A monograph of cantharelloid fungi. Ann Bot Mem 2:1–255

Corner EJH (1970) Supplement to "A monograph of *Clavaria* and allied genera." Beih Nova Hedwigia 33:1–299

Corner EJH (1972) Studies in the basidium-spore spacing and the *Boletus* spore. Gard Bull 26:159–194

Corner EJH (1981) The agaric genera *Lentinus*, *Panus*, and *Pleurotus*. Beih Nova Hedwigia 69:1–169

Corner EJH (1984) Ad Polyporaceas III: *Piptoporus*, *Buglossoporus*, *Laetiporus*, *Meripilus*, and *Bondarzewia*. Beih Nova Hedwigia 78:136–222

Corner EJH (1991) Ad Polyporaceas VII: the xanthochroic polypores. Beih Nova Hedwigia 101:1–175

Cotter HVT, Miller OK (1985) Sclerotia of *Boletinellus merulioides* in nature. Mycologia 77:927–931

Cowling EB, Merrill W (1966) Nitrogen in wood and its role in wood deterioration. Can J Bot 44:1539–1554

Cullings KW, Szaro TM, Bruns TD (1996) Evolution of extreme specialization within a lineage of ectomycorrhizal epiparasites. Nature 379:63–66

Currah RS, Zalmer C (1992) A key and notes for the genera of fungi mycorrhizal with orchids and a new species in the genus *Epulorhiza*. Rep Tottori Mycol Inst 30:43–59

Currah RS, Smreciu EA, Hambleton S (1990) Mycorrhizae and mycorrhizal fungi of boreal species of *Platanthera* and *Coeloglossum* (Orchidaceae). Can J Bot 68:1171–1181

Danielson RM (1984) Ectomycorrhizal associations in jack pine stands in northeastern Alberta. Can J Bot 62:932–939

Davidson RW, Campbell WA, Vaughn DB (1942) Fungi causing decay of living oaks in the eastern United States and their identification. US Dept Agric Tech Bull 785, Washington, DC

Desjardin DE (1995) A preliminary accounting of the worldwide members of *Mycena* sect. *Sacchariferae*. Bibl Mycol 159:1–89

Desjardin DE, Martínez-Peck L, Rajchenberg M (1995) An unusual psychrophilic aquatic agaric from Argentina. Mycologia 87:547–550

Dix NJ, Webster J (1995) Fungal ecology. Chapman and Hall, London

Donk MA (1959) Notes on Cyphellaceae I. Persoonia 1:25–110

Donk MA (1962) Notes on Cyphellaceae II. Persoonia 3:331–348

Donk MA (1964) A conspectus of the families of the Aphyllophorales. Persoonia 3:199–324

Donk MA (1971) Progress in the study of the classification of the higher basidiomycetes. In: Petersen RH (ed) Evolution in the higher basidiomycetes. University of Tennessee Press, Knoxville, pp 3–25

Douget G (1962) *Digitatispora marina* n.g., n.sp., Basidiomycète marin. CR Hebd Séances Acad Sci Paris 254:4336–4338

Douget G (1967) *Nia vibrissa* Moore et Meyers, remarquable basidiomycète marin. CR Hebd Séances Acad Sci Paris, Sér D, 265:1780–1783

Dring DM (1973) Gasteromycetes. In: Ainsworth GC, Sparrow FK, Sussman AS (eds) The Fungi, an advanced treatise, vol IV B. Academic Press, New York, pp 451–478

D'Souza TM, Boominathan K, Reddy CA (1996) Isolation of laccase gene-specific sequences from white rot and brown rot fungi by PCR. Appl Environ Microbiol 62:3739–3744

Duncan EG, Galbraith MH (1972) Post-meiotic events in the Homobasidiomycetidae. Trans Br Mycol Soc 58:387–392

Durrall DM, Todd AW, Trappe JM (1994) Decomposition of ^{14}C-labelled substrates by ectomycorrhizal fungi in association with Douglas fir. New Phytol 127:725–729

Eaton RA, Hale MDC (1993) Wood: decay, pests and protection. Chapman and Hall, London

Ehrlich HG, McDonough ES (1949) The nuclear history in the basidia and basidiospores of *Schizophyllum commune* Fries. Am J Bot 36:360–363

Eriksson J (1954) *Ramaricium* n. gen., a corticioid member of the *Ramaria* group. Sven Bot Tids Sci 48:188–198

Eriksson J, Ryvarden L (1976) The Corticiaceae of North Europe, vol 4, *Hyphodermella – Mycoacia*. Fungiflora, Oslo

Eriksson KE, Blanchette RA, Ander P (1990) Microbial and enzymatic degradation of wood and wood components. Springer, Berlin Heidelberg New York

Esser K (1967) Die Verbreitung der Incompatibilität bei Thallophyten. In: Ruhland W (ed) Handb Pflanzenphysiol 18. Springer, Berlin Heidelberg New York, pp 321–343

Eyme J, Parriaud H (1970) Au sujet de l'infrastructure des hyphes de *Clathrus cancellatus* Tournefort, champignon gasteromycète. CR Hebd Séances Acad Sci Paris, Ser D, 270:1890–1892

Farr DF, Bills GF, Chamuris GP, Rossman AY (1989) Fungi on plants and plant products in the United States. American Phytopathological Society, St Paul

Fayod V (1889) Prodrome d'une histoire naturelle des agaracinées. Ann Sci Bot, Sér 7,9:181–411

Felsenstein J (1978) Cases in which parsimony or compatibility methods will be positively misleading. Syst Zool 27:401–440

Fermor TR (1983) Fungal enzymes produced during degradation of bacteria. Trans Br Mycol Soc 80:357–360

Fisher PJ, Stradling DJ, Pegler DN (1994) *Leucoagaricus* basidiomata from a live nest of the leaf-cutting ant *Atta cephalotes*. Mycol Res 98:884–888

Flegler SL, Hooper GR, Fields WG (1976) Ultrastructural and cytochemical changes in the basidiomycete dolipore septum associated with fruiting, Can J Bot 54:2243–2253

Fries EM (1874) Hymenomycetes Europaei. Upsaliae. E Berling, Uppsala

Furtado JS (1968) Basidial cytology of *Exidia nucleata*. Mycologia 60:9–15

Gardes M, Bruns TD (1996) Community structure of ectomycorrhizal fungi in a *Pinus muricata* forest: above- and below-ground views. Can J Bot 74:1572–1583

Gargas A, DePriest PT, Grube M, Tehler A (1995) Multiple origins of lichen symbioses in fungi suggested by ssu rDNA phylogeny. Science 268:1492–1495

Garrett SD (1981) Soil fungi and soil fertility, 2nd edn. Pergamon Press, Oxford

Gilbertson RL (1971) Phylogenetic relationships of hymenomycetes with resupinate, hydnaceous basidiocarps. In: Petersen RH (ed) Evolution in the higher basidiomycetes. University of Tennessee Press, Knoxville, pp 275–307

Gilbertson RL (1980) Wood-rotting fungi of North America. Mycologia 72:1–49

Gilbertson RL (1981) North American wood-rotting fungi that cause brown rots. Mycotaxon 12:372–416

Gilbertson RL (1984) Relationships between insects and wood-rotting basidiomycetes. In: Wheeler Q, Blackwell M (eds) Fungus-insect relationships. Columbia University Press, New York, pp 130–165

Gilbertson RL, Ryvarden L (1986) North American polypores vol 1. Fungiflora, Oslo

Gill M (1996) Pigments of fungi (macromycetes). Nat Prod Rep 1996:513–528

Gill M, Steglich W (1987) Pigments of fungi (macromycetes). Prog Chem Nat Prod 51:1–317

Ginns J (1979) The genus *Ramaricium* (Gomphaceae). Bot Not 132:93–102

Ginns J (1997) The taxonomy and distribution of rare or uncommon species of *Albatrellus* in western North America. Can J Bot 75:261–273

Ginns J, Lefebvre MNL (1993) Lignicolous corticioid fungi (Basidiomycota) of North America: systematics, distribution, and ecology. Mycol Mem 19:1–247

Ginns JH, Malloch D (1977) *Halocyphina*, a marine Basidiomycete (Aphyllophorales). Mycologia 69:53–58

Gochenauer SE (1981) *Cyrenella elegans* gen. et sp. nov., a dikaryotic anamorph. Mycotaxon 13:267–277

Grant WD, Rhodes LL, Prosser BA, Asher RA (1986) Production of bacteriolytic enzymes and degradation of bacteria by filamentous fungi. J Gen Microbiol 132:2353–2358

Green F, Highley TL (1997) Mechanism of brown-rot decay: paradigm or paradox. Int Biodet Biodegrad 39:113–124

Grenville DJ, Peterson RL, Riche Y (1985a) The development, structure, and histochemistry of sclerotia of ectomycorrhizal fungi. I. *Pisolithus tinctorius*. Can J Bot 63:1402–1411

Grenville DJ, Peterson RL, Piche Y (1985b) The development, structure, and histochemistry of sclerotia of ectomycorrhizal fungi. II. *Paxillus involutus*. Can J Bot 63:1402–1411

Griffith NT, Barnett HL (1967) Mycoparasitism by basidiomycetes in culture. Mycologia 59:149–154

Gulden G, Jenssen KM (1988) Arctic and alpine fungi-2. Soppkonsulenten, Oslo

Gulden G, Jenssen KM, Stordal J (1985) Arctic and alpine fungi-1. Soppkonsulenten, Oslo

Hacskaylo E (1971) The role of mycorrhizal associations in the evolution of the higher basidiomycetes. In: Petersen RH (ed) Evolution in the higher basidiomycetes. University of Tennessee Press, Knoxville, pp 217–240

Hale ME (1983) The biology of lichens, 3d edn. Edward Arnold, Baltimore

Happ GM, Happ CM, Barras SJ (1976) Bark beetle-fungus symbiosis. II. Fine structure of a basidiomycetous ectosymbiont of the southern pine beetle. Can J Bot 54:1049–1062

Harrison KA (1971) The evolutionary lines in the fungi with spines supporting the hymenium. In: Petersen RH (ed) Evolution in the higher basidiomycetes. University of Tennessee Press, Knoxville, pp 375–392

Harrison KA (1973) Aphyllophorales III: Hydnaceae and Echinodontiaceae. In: Ainsworth GC, Sparrow FK, Sussman AS (eds) The Fungi, an advanced treatise, vol IV B. Academic Press, New York, pp 369–396

Hatakka A (1994) Lignin-modifying enzymes from selected white-rot fungi: production and role in lignin degredation. FEMS Microbiol Rev 13:125–135

Hawksworth DL (1988) Coevolution of fungi with algae and cyanobacteria in lichen symbioses. In: Pirozynski KA, Hawksworth DL (eds) Coevolution of fungi with plants and animals. Academic Press, London, pp 125–148

Hawksworth DL, Kirk PM, Sutton BC, Pegler DN (1995) Dictionary of the fungi, 8th edn. CAB International, Wallingford, UK

Heim R (1971) The interrelationships between the Agaricales and Gasteromycetes. In: Petersen RH (ed) Evolution in the higher basidiomycetes. University of Tennessee Press, Knoxville, pp 505–534

Heim R, Wasson RG (1959) Les champignons hallucigènes du Mexique. Muséum National d'Histoire Naturelle, Paris

Hibbett DS (1992) Ribosomal RNA and fungal systematics. Trans Mycol Soc Jpn 33:533–556

Hibbett DS (1996) Phylogenetic evidence for horizontal transmission of group I introns in the nuclear ribosomal DNA of mushroom-forming fungi. Mol Biol Evol 13:903–917

Hibbett DS, Donoghue MJ (1995) Progress toward a phylogenetic classification of the Polyporaceae through parsimony analyses of mitochondrial ribosomal DNA sequences. Can J Bot 73 (Suppl 1):s853–s861

Hibbett DS, Donoghue MJ (1998) Integrating phylogenetic analysis and classification in fungi. Mycologia 90:347–356

Hibbett DS, Thorn RG (1994) Nematode-trapping by *Pleurotus tuber-regium* Mycologia 86:696–699

Hibbett DS, Vilgalys R (1993) Phylogenetic relationships of *Lentinus* (Basidiomycotina) inferred from molecular and morphological characters. Syst Bot 18:409–433

Hibbett DS, Murakami S, Tsuneda A (1993a) Sporocarp ontogeny in *Panus*: evolution and classification. Am J Bot 80:1336–1348

Hibbett DS, Murakami S, Tsuneda A (1993b) Hymenophore development and evolution in *Lentinus*. Mycologia 85:428–443

Hibbett DS, Murakami S, Tsuneda A (1994a) Postmeiotic nuclear behavior in *Lentinus*, *Panus*, and *Neolentinus*. Mycologia 86:725–732

Hibbett DS, Tsuneda A, Murakami S (1994b) The secotioid form of *Lentinus tigrinus*: genetics and development of a fungal morphological innovation. Am J Bot 81:466–478

Hibbett DS, Grimaldi D, Donoghue MJ (1997a) Fossil mushrooms from Miocene and Cretaceous ambers and the evolution of homobasidiomycetes. Am J Bot 84:981–991

Hibbett DS, Pine EM, Langer E, Langer G, Donoghue MJ (1997b) Evolution of gilled mushrooms and puffballs inferred from ribosomal DNA sequences. Proc Natc Acad Sci USA 94:12002–12006

Highley TL (1988) Celluloytic activity of brown-rot and white-rot fungi on solid media. Holzforschung 42:211–216

Hjortstam K, Larsson K-H, Ryvarden L, Eriksson J (1988) The Corticiaceae of North Europe, vol 1. Introduction and keys. Fungiflora, Oslo

Høiland K (1980) *Cortinarius* subgenus *Leprocybe* in Norway. Norw J Bot 27:101–126

Høiland K (1986) Contribution to the nomenclature of *Cortinarius* subgenus *Dermocybe*. Nord J Bot 5:625–627

Hopple JS, Vilgalys R (1994) Phylogenetic relationships among coprinoid taxa and allies based on data from restriction site mapping of nuclear rDNA. Mycologia 86:96–107

Hsiau P (1996) The taxonomy and phylogeny of the mycangial fungi from *Dendroctonus brevicomis* and *D. frontalis* (Coleoptera: Scolytidae). PhD Thesis, Iowa State University, Ames, Iowa

Hubbard M, Petersen RH (1979) Studies in basidial nuclear behavior of selected species of clavarioid and cantharelloid fungi. Beih Sydowia 8:209–223

Hudson HJ (1972) Fungal saprophytism. Studies in Biology no 32. Edward Arnold, Oxford

Hughey BD, Adams GC, Bruns TD, Hibbett DS (2000) Phylogeny of *Calostoma*, the gelatinous-stalked puffball, based on nuclear and mitochondrial ribosomal DNA sequences. Mycologia 92:94–104

Hutchison LJ (1989) Absence of conidia as a morphological character in ectomycorrhizal fungi. Mycologia 81:587–594

Hutchison LJ (1991a) Description and identification of cultures of ectomycorrhizal fungi found in North America. Mycotaxon 42:387–504

Hutchison LJ (1991b) Formation of sclerotia by *Hygrophoropsis aurantiaca* in nature. Trans Mycol Soc Jpn 32:235–245

Hutchison LJ, Barron GL (1996) Parasitism of yeasts by lignicolous Basidiomycota and other fungi. Can J Bot 74:735–742

Hutchison LJ, Barron GL (1997) Parasitism of algae by lignicolous Basidiomycota and other fungi. Can J Bot 75:1006–1011

Hutchison LJ, Madzia SE, Barron GL (1996) The presence and antifeedant function of toxin-producing secretory cells on hyphae of the lawn-inhabiting agaric *Conocybe lactea*. Can J Bot 74:431–434

Ingold CT (1971) Fungal spores, their liberation and dispersal. Clarendon Press, Oxford

Ingold CT (1985) Observations on spores and their germination in certain heterobasidiomycetes. Trans Br Mycol Soc 85:417–423

Ingold CT (1992) The conidial stage in *Exidiopsis effusa* and *E. longispora*. Mycol Res 96:932–934

Jackson HS (1950) Studies of Canadian Thelephoraceae. V. Two new species of *Aleurodiscus*. Can J Res C, 28:63–77

Jacobsson S (1989) Studies on *Pholiota* in culture. Mycotaxon 36:95–145

Jeffries P, Young TWK (1994) Interfungal parasitic relationships. CAB International, Wallingford, UK

Johnson CN (1996) Interactions between mammals and ectomycorrhizal fungi. TREE 11:503–507

Johnson EA, Schroeder WA (1996) Microbial carotenoids. Adv Biochem Eng Biotechnol 53:119–178

Juel HO (1898) Die Kerntheilungen in den Basidien und die Phylogenie der Basidiomyceten. Jahrb Wiss Bot 32:361–388

Juel HO (1916) Cytologische Pilzstudien I. Die Basidien der Gattungen *Cantharellus*, *Craterellus* und *Clavaria*. Nova Acta Regiae Soc Sci Ups, Ser IV, 4:1–40

Jülich W (1981) Higher taxa of basidiomycetes. J Cramer, Vaduz

Jülich W, Stalpers J (1980) The resupinate non-poroid Aphyllophorales of the temperate northern hemisphere. North Holland, Amsterdam

Keller G (1982) Pigmentationsuntersuchungen bei europäischen Arten aus der gattung *Dermocybe* (Fr) Wånsche. Sydowia 35:110–126

Keller J (1997) Atlas des Basidiomycetes. Union des Societes Suisses de Mycologie, Neuchâtel

Kendrick B, Watling R (1979) Mitospores in Basidiomycetes. In: Kendrick B (ed) The whole fungus, vol 2. National Museum of Natural Sciences, Ottawa, Canada, pp 473–546

Khan SR, Talbot PHB (1976) Ultrastructure of septa in hyphae and basidia of *Tulasnella*. Mycologia 68:1027–1036

King JM (1966) Some aspects of the biology of the fungal symbiont of *Sirex noctilio*. Aust J Bot 14:25–30

Klán J, Baudisová D, Rulfová I (1989) Cultural, enzymatic and cytological studies in the genus *Pholiota*. Mycotaxon 36:249–271

Ko KS, Hong SG, Jung HS (1997) Phylogenetic analysis of *Trichaptum* based on nuclear 18S, 5.8S and ITS ribosomal DNA sequences. Mycologia 89:727–734

Kobayasi Y (1952) On the genus *Favolaschia* and *Campanella* from Japan. J Hattori Bot Lab 8:1–4

Kohlmeyer J, Kohlmeyer E (1979) Marine mycology. Academic Press, New York

Koller B, Jahrmann HJ (1985) Life-cycle and physiological description of the yeast-form of the homobasidiomycete *Asterophora lycoperdoides*. Antonie van Leeuwenhoek 51:255–261

Kratochvil JF, Burris RH, Seikel MK, Harkin JM (1971) Isolation and characterization of α-guaiaconic acid and the nature of guaiacum blue. Phytochemistry 10:2529–2531

Kreisel H (1969) Grunndzüge eines natürlichen Systems der Pilze. J Cramer, Lehre

Kretzer A, Bruns TD (1997) Molecular revisitation of the genus *Gastrosuillus*. Mycologia 89:586–589

Kühner R (1972) Architecture de la paroi sporique des hyménomycètes et de ses différenciations. Persoonia 7:217–248

Kühner R (1980) Les Hyménomycètes agaricoïdes (Agaricales, Tricholomatales, Pluteales, Russulales). Bull Soc Linn Lyon, Num Spec, pp 1–1027

Kühner R (1984) Some mainlines of classification in the gill fungi. Mycologia 76:1059–1074

Laaser G, Möller E, Jahnke K-D, Bahnweg G, Prillinger H, Prell HH (1989) Ribosomal DNA restriction fragment analysis as a taxonomic tool in separating physiologically similar basidiomycetous yeasts. Syst Appl Microbiol 11:170–175

Lange M (1952) Species concept in the genus *Coprinus*. Dan Bot Ark 14:1–164

Langer E (1994) Die Gattung *Hyphodontia* John Eriksson. Bibl Mycol 154:1–298

Langer E, Oberwinkler F (1993) Corticioid basidiomycetes. I. Morphology and ultrastructure. Windahlia 20:1–28

Langer G (1994) Die Gattung *Botryobasidium* Donk (Corticiaceae, Basidiomycetes). Bibl Mycol 158:1–459

Langvad F (1971) New structures in the basidiomycete *Coniophora cerebella*. J Bacteriol 106:679–682

Largent D, Johnson D, Watling R (1977) How to identify mushrooms to genus III: microscopic features. Mad River Press, Eureka, California

Larsen M, Burdsall HH (1976) A consideration of the term gloeocystidium. Mem New York Bot Gard 28:123–130

Larsen MJ, Gilbertson RL (1977) Studies in *Laeticorticium* (Aphyllophorales, Corticiaceae) and related genera. Norw J Bot 24:99–121

Lee S-S, Jung HS (1997) Phylogenetic analysis of the Corticiaceae based on gene sequences of nuclear 18S ribosomal DNAs. J Microbiol 35:253–258

Lentz PL (1971) Analysis of modified hyphae as a tool in taxonomic research in the higher basidiomycetes. In: Petersen RH (ed) Evolution in the higher basidiomycetes. University of Tennessee Press, Knoxville, pp 99–128

Liou JY, Tzean SS (1992) Stephanocysts as nematode-trapping and infecting propagules. Mycologia 84:786–790

Liu YJ, Rogers SO, Ammirati JF (1997) Phylogenetic relationships in *Dermocybe* and related *Cortinarius* taxa based on nuclear ribosomal DNA internal transcribed spacers. Can J Bot 75:519–532

Lutzoni FM (1997) Phylogeny of lichen- and non-lichen-forming omphalinoid mushrooms and the utility of testing for combinability among multiple data sets. Syst Biol 46:373–406

Lutzoni FM, Pagel M (1997) Accelerated evolution as a consequence of transitions to mutualism. Proc Nate Acad Sci USA 94:11422–11427

Maas Geesteranus RA (1963) Hyphal structures in Hydnums. II. Proc K Ned Akad Wet, Ser C, 66:426–436

Madden JL, Coutts MP (1979) The role of fungi in the biology and ecology of woodwasps (Hymenoptera: Siricidae). In: Batra LR (ed) Insect-fungus symbiosis. John Wiley, New York, pp 165–174

Maehly AC, Chance B (1954) The assay of catalases and peroxidases. Meth Biochem Anal 1:357–408

Maekawa N (1987) A new species of the genus *Cerinomyces*. Can J Bot 65:583–588

Maire R (1902) Recherchs cytologiques and taxonomiques sur les basidiomycètes. Bull Soc Mycol Fr 18:(Suppl):1–192

Malloch DW (1987) The evolution of mycorrhizae. Can J Plant Pathol 9:398–402

Marvanová L, Bärlocher F (1988) Hyphomycetes from Canadian streams. I. Basidiomyceteous anamorphs. Mycotaxon 32:339–351

Marvanová L, Stalpers JA (1987) The genus *Taeniospora* and its teleomorphs. Trans Br Mycol Soc 89:489–498

McLaughlin DJ (1981) The spindle pole body and postmeiotic meiosis in *Auricularia fuscosuccinea*. Can J Bot 59:1196–1206

Miller OK (1971) The relationship of cultural characters to the taxonomy of the agarics. In: Petersen RH (ed) Evolution in the higher basidiomycetes. University of Tennessee Press, Knoxville, pp 197–216

Miller OK, Miller H (1988) Gasteromycetes: morphological and developmental features. Mad River Press, Eureka, California

Miller OK, Watling R (1987) Whence cometh the agarics? A reappraisal. In: Rayner ADM, Brasier CM, Moore D (eds) Evolutionary biology of the fungi. Cambridge University Press, Cambridge, pp 435–448

Miller SL, Torres P, McClean TM (1994) Persistence of basidiospores and sclerotia of ectomycorrhizal fungi and *Morchella* in soil. Mycologia 86:89–95

Mims CW, Seabury F (1989) Ultrastructure of tube formation and basidiospore development in *Ganoderma lucidum*. Mycologia 81:754–764

Moncalvo JM, Lutzoni FM, Rehner SA, Johnson J, Vilgalys R (2000) Phylogenetic relationships of agaric fungi based on nuclear large subunit ribosomal DNA sequences. Syst Biol 49 (in press)

Moore RT (1980) Taxonomic significance of septal ultrastructure in the genus *Onnia* Karsten (Polyporineae/Hymenochaetaceae). Bot Not 133:169–175

Moore RT (1985) The challenge of the dolipore septum. In: Moore D, Casselton LA, Wood DA, Frankland JC (eds) Developmental biology of higher fungi. Cambridge University Press, Cambridge, pp 175–212

Moore RT (1987) The genera of *Rhizoctonia*-like fungi: *Ascorhizoctonia*, *Ceratorhiza* gen. nov., *Epulorhiza* gen. nov., *Moniliopsis*, and *Rhizoctonia*. Mycotaxon 29:91–99

Moore RT, Patton AM (1975) Parenthesome fine structure in *Pleurotus cystidiosus* and *Schizophyllum commune*. Mycologia 67:1200–1205

Moser M (1978) Keys to agarics and boleti. Phillips, London

Mueller GJ, Mueller GM, Shih L-H, Ammirati JF (1993) Cytological studies in *Laccaria* (Agaricales). I. Meiosis and postmeiotic mitosis. Am J Bot 80:316–321

Mueller GM, Ammirati JF (1993) Cytological studies in *Laccaria* (Agaricales). II. Assessing phylogenetic relationships among *Laccaria*, *Hydnangium*, and other Agaricales. Am J Bot 80:322–329

Mueller GM, Pine EM (1994) DNA data provide evidence on the evolutionary relationships between mushrooms and false truffles. McIlvainea 11:61–74

Müller WH, Stalpers JA, van Aelst AC, van der Krift TP, Boekhout T (1998) Field emission gun-scanning electron microscopy of septal pore caps of selected species in the *Rhizoctonia* s.l. complex. Mycologia 90:170–179

Murphy JF, Miller OK (1993) Diversity and local distribution of mating alleles in *Marasmiellus praeacutus* and *Collybia subnuda* (Basidiomycetes, Agaricales). Can J Bot 75:8–17

Nakasone KK (1990) Cultural studies and identification of wood-inhabiting Corticiaceae and selected Hymenomycetes from North America. Mycol Mem 15:1–412

Nakasone KK (1996) Morphological and molecular studies on *Auriculariopsis albomellea* and *Phlebia albida* and a reassessment of *A. ampla*. Mycologia 88:762–775

Nawawi A (1985) Basidiomycetes with branched, waterborne conidia. Bot J Linn Soc (Lond) 91:51–60

Nawawi A, Webster J (1982) *Sistotrema hamatum* sp. nov., the teleomorph of *Ingoldiela hamata*. Trans Br Mycol Soc 78:287–291

Nawawi A, Descals E, Webster J (1977) *Leptosporomyces galzinii*, the basidial state of a clamped branched conidium from fresh water. Trans Br Mycol Soc 68:31–36

Neda H, Nakai T (1995) Phylogenetic analysis of *Pleurotus* based on data from partial sequences of 18SrDNA and ITS-1 regions. In: Elliot T (ed) Science and cultivation of edible fungi. Balkema, Rotterdam, pp 161–168

Nilsson T (1974) Comparative study on the cellulolytic activity of white-rot and brown-rot fungi. Mater Org 9:173–198

Nilsson T, Ginns JH (1979) Cellulolytic activity and the taxonomic position of selected brown-rot fungi. Mycologia 71:170–177

Nobles MK (1965) Identification of cultures of wood-inhabiting Hymenomycetes. Can J Bot 43:1097–1139

Nobles MK (1971) Cultural characters as a guide to the taxonomy of the Polyporaceae. In: Petersen RH (ed) Evolution in the higher basidiomycetes. University of Tennessee Press, Knoxville, pp 169–196

Nuss I (1983) Bemerkungen zu dem Buch von Jülich (1982) Higher taxa of basidiomycetes. Westf Pilzbr 10–11:260–271

Oberwinkler F (1970) Die Gattungen der Basidiolichenen. Dtsch Bot Ges Neue Folge 4:139–169

Oberwinkler F (1972) The relationships between the Tremellales and the Aphyllophorales. Persoonia 7:1–16

Oberwinkler F (1975) Eine agaricoide Gattung Thelephorales. Sydowia 28:359

Oberwinkler F (1977) Das neue System der Basidiomyceten. In: Frey H, Hurka H, Oberwinkler F (eds) Beiträge zur Biologie der niederen Pflanzen. G Fischer, Stuttgart, pp 59–105

Oberwinkler F (1982) The significance of the morphology of the basidium in the phylogeny of basidiomycetes. In: Wells K, Wells EK (eds) Basidium and basidocarp, evolution, cytology, function, and development. Springer, Berlin Heidelberg, New York, pp 9–35

Oberwinkler F (1984) Fungus-alga interactions in basidiolichens. Beih Nova Hedwigia 79:739–774

Oberwinkler F (1985) Anmerkungen zur Evolution and Systematik der Basidiomyceten. Bot Jahrb Syst 107:541–580

Oberwinkler F, Horak E (1979) Stephanosporaceae–eine neue Familie der Basdiomycetes mit aphyllophoralean und gastroiden Fruchtkörpern. Plant Syst Evol 131:157–164

O'Donnell KL, McLaughlin DJ (1984) Postmeiotic mitosis, basidiospore development, and septation in *Ustilago maydis*. Mycologia 76:486–502

O'Kane DJ, Lingle WL, Porter D, Wampler JE (1990) Localization of bioluminescent tissues during basidiocarp development in *Panellus stypticus*. Mycologia 82:595–606

Orth AB, Royse DJ, Tien M (1993) Uniquity of lignin-degrading peroxidases among various wood-degrading fungi. Appl Environ Microbiol 59:4017–4023

Owens EM, Reddy CA, Grethlein HE (1994) Outcome of interspecific interactions among brown-rot and white-rot wood decay fungi. FEMS Microbiol Ecol 14:19–24

Owings P, Desjardin DE (1997) A molecular phylogeny of *Marasmius* and selected genera. Inoculum 48(3):29

Parag Y (1965) Papillae secreting water droplets on aerial mycelia of *Schizophyllum commune*. Isr J Bot 14:192–195

Parmasto E (1978) The genus *Dictyonema* ("Thelephorolichenes"). Nova Hedwigia 29:99–144

Parmasto E (1986) On the origin of hymenomycetes (what are corticioid fungi?). Windahlia 16:3–19

Parmasto E (1995) Corticioid fungi: a cladistic study of a paraphyletic group. Can J Bot 73 (Suppl 1):s843–s852

Parmasto E, Parmasto I (1979) The xanthochroic reaction in Aphyllophorales. Mycotaxon 8:201–232

Patouillard N (1900) Essai taxonomique sur les familles et les genres des hyménomycètes. Lucien Declume, Lons-le-Saunier

Patrignani G, Pellegrini S (1986) Fine structures of the fungal septa on varieties of basidiomycetes. Caryologia 39:239–250

Pegler DN (1973) Aphyllophorales IV: Poroid families. In: Ainsworth GC, Sparrow FK, Sussman AS (eds) The

Fungi, an advanced treatise, vol IVB. Academic Press, New York, pp 397–420
Pegler DN (1975) The classification of the genus *Lentinus* Fr. (Basidiomycota). Kavaka 3:11–20
Pegler DN (1983) The genus *Lentinus*, a world monograph. Kew Bull Add it Ser 10:1–281
Pegler DN (1996) Hyphal analysis of basidiomata. Mycol Res 100:129–142
Pegler DN, Young TWK (1971) Basidiospore morphology in the Agaricales. Beih Nova Hedwigia 35:1–210
Pegler DN, Young TWK (1979) The gasteroid Russulales. Trans Br Mycol Soc 72:353–388
Pegler DN, Spooner BM, Young TWK (1993) British truffles. Royal Botanic Gardens, Kew
Pegler DN, Læssøe T, Spooner BM (1995) British puffballs, earthstars and stinkhorns. Royal Botanic Gardens, Kew
Penancier N (1961) Recherches sur l'orientation des fuseaux mitotiques dans la baside des Aphyllophorales. Trav Lab La Jaysinia 2:57–71
Petch T (1915) The pseudo-sclerotia of *Lentinus similis* and *Lentinus infundibuliformis*. Ann R Bot Gard Peradeniya 6:1–18
Petersen RH (1967) Evidence on the interrelationships of the families of clavarioid fungi. Trans Br Mycol Soc 50:641–648
Petersen RH (1971) Evolution in the higher basidiomycetes. University of Tennessee Press, Knoxville
Petersen RH (1973) Aphyllophorales II: The clavarioid and cantharelloid basidiomycetes. In: Ainsworth GC, Sparrow FK, Sussman AS (eds) The Fungi, an advanced treatise, vol IVB. Academic Press, New York, pp 351–368
Petersen RH (1995a) There's more to a mushroom than meets the eye: mating studies in the Agaricales. Mycologia 87:1–17
Petersen RH (1995b) Contributions of mating studies to mushroom systematics. Can J Bot 73 (Suppl 1):s831–842
Petersen RH, Bermudes D (1992) Intercontinental compatibility in *Panellus stypticus* with a note on bioluminescence. Persoonia 14:457–463
Pine EM, Hibbett DS, Donoghue MJ (1999) Phylogenetic relationships of cantharelloid and clavarioid homobasidiomycetes based on mitochondrial and nuclear rDNA sequences. Mycologia 91:944–963
Porter D, Farnham WF (1986) *Mycaureola dilseae*, a marine basidiomycete parasite of the redalga, *Dilsea carnosa*. Trans Br Mycol Soc 87:575–582
Price TV, Baldwin JA, Simpson JA (1978) Fungal club-heads in Papua New Guinea. Nature 273:374–375
Raper JR, Flexer AS (1971) Mating systems and evolution of the basidiomycetes. In: Petersen RH (ed) Evolution in the higher basidiomycetes. University of Tennessee Press, Knoxville, pp 149–168
Preston CM, Sollin P, Sayer BG (1990) Changes in organic components for fallen logs in old-growth Douglas-fir forests monitored by ^{13}C nuclear resonance spectroscopy. Can J For Res 20:1382–1391
Rayner ADM, Boddy L (1988) Fungal decomposition of wood: its biology and ecology. John Wiley, Chichester
Rayner ADM, Watling R, Frankland JC (1985) Resource relations—an overview. In: Moore D, Casselton LA, Wood DA, Frankland JC (eds) Developmental biology of higher fungi. Cambridge University Press, Cambridge, pp 1–40
Rayner ADM, Boddy L, Dowson CG (1987) Temporary parasitism of *Coriolus* spp. by *Lenzites betulina*: a strategy for domain capture in wood decay fungi. FEMS Microbiol Ecol 45:53–58
Read DJ (1974) *Pezizella ericae* sp. nov. the perfect state of a typical mycorrhizal endophyte of Ericaceae. Trans Br Mycol Soc 63:381–383
Reddy CA, D'Souza TM (1994) Physiology and molecular biology of the lignin peroxidases of *Phanerochaete chrysosporium*. FEMS Microbiol Rev 13:137–152
Redhead SA (1980) *Gerronema pseudogrisella*, No. 170. Fungi Canadenses. National Mycological Herbarium, Ottawa
Redhead SA (1981) Parasitism of bryophytes by agarics. Can J Bot 59:63–67
Redhead SA (1984) *Arrhenia* and *Rimbachia*, expanded generic concepts, and a reevaluation of *Leptoglossum* with emphasis on muscicolous North American taxa. Can J Bot 62:865–892
Redhead SA (1986) The Xerulaceae (Basidiomycetes), a family with sarcodimitic tissues. Can J Bot 65:1551–1562
Redhead SA, Ginns JH (1983) Book review of "higher Taxa of Basidiomycetes" by W. Jülich. Mycologia 75:567–578
Redhead SA, Ginns JH (1985) A reappraisal of agaric genera associated with brown rots of wood. Trans Mycol Soc Jpn 26:349–381
Redhead SA, Kroeger P (1987) A sclerotium-producing *Hypholoma* from British Columbia. Mycotaxon 29:457–465
Redhead SA, Kuyper TW (1987) Lichenized agarics: taxonomic and nomenclatural riddles. In: Laursen GA, Ammirati JF, Redhead SA (eds) Arctic and alpine mycology II. Plenum Press, New York, pp 319–349
Redhead SA, Norvell L (1993) Notes on *Bondarzewia*, *Heterobasidion* and *Pleurogala*. Mycotaxon 48:371–380
Redhead SA, Ammirati JF, Walker GR, Norvell LL, Puccio MB (1994) *Squamanita contortipes*, the Rosetta Stone of a mycoparasitic agaric genus. Can J Bot 72:1812–1824
Reid DA (1963) Notes on some fungi of Michigan–I. Cyphellaceae. Persoonia 3:97–154
Reid DA (1965) A monograph of the stipitate stereoid fungi. Beih Nova Hedwigia 18:1–382
Reijnders AFM, Stalpers JS (1992) The development of the hymenophoral trama in the Aphyllophorales and the Agaricales. Stud Mycol 34:1–109
Restivo JH, Petersen RH (1976) Studies on nuclear division and behavior within basidia I. *Hydnum umbilicatum*. Mycologia 68:666–672
Rogers DP (1932) A cytological study of *Tulasnella*. Bot Gaz 94:86–105
Ryvarden L (1991) Genera of polypores: nomenclature and taxonomy. Synopsis Fungorum 5:1–363
Ryvarden L, Gilbertson RL (1993) European polypores, part 1. Fungiflora, Oslo
Savile DBO (1955) A phylogeny of the basidiomycetes. Can J Bot 33:60–104
Schaffer RL (1975) The major groups of basidiomycetes. Mycologia 67:1–12
Seifert KA (1983) Decay of wood by the Dacrymycetales. Mycologia 75:1011–1018
Seifert KA, Okada G (1988) *Gloeosynnema*, a new genus of synnematous Hyphomycetes with basidiomycetous affinities. Mycotaxon 32:471–476

Senn-Irlet B, Jenssen KM, Gulden G (1990) Arctic and alpine fungi-3. Soppkonsulenten, Oslo

Seviour RJ, Willing RR, Chilvers GA (1973) Basidiocarps associated with ericoid mycorrhizas. New Phytol 72:381–385

Shaw DE (1972) *Ingoldiella hamata* gen. et sp. nov., a fungus with clamp connexions from a stream in North Queensland. Trans Br Mycol Soc 59:255–259

Sigler L, Carmichael JW (1976) Taxonomy of *Malbranchea* and some other Hyphomycetes with arthroconidia. Mycotaxon 4:349–488

Simpson JA (1996) Wood decay fungi. In: Orchard AE, Mallett K, Grgurinovic C (eds) Fungi of Australia, vol 1B. Australian Biological Resources Study, Canberra, pp 95–136

Singer R (1962) The Agaricales (mushrooms) in modern taxonomy, 2nd edn. Cramer, Weinheim

Singer R (1973) The genera *Marasmiellus*, *Crepidotus* and *Simocybe* in the Neotropics. Beih Nova Hedwigia 44:1–517

Singer R (1984) Tropical Russulaceae II. *Lactarius* sect. *Panuoidei*. Nova Hedwigia 40:435–447

Singer R (1986) The Agaricales in modern taxonomy, 4th edn. Koeltz Scientific Books, Koenigstein

Singer R, Snell WH, White WL (1945) The taxonomic position of *Polyporoletus sublividus*. Mycologia 37:124–128

Sjamsuridzal W, Tajiri Y, Nishida H, Thuan TB, Kawasaki H, Hirata A, Yokota A, Sugiyama J (1997) Evolutionary relationships of members of the genera *Taphrina*, *Protomyces*, *Schizosaccharomyces*, and related taxa within the archiascomycetes: integrated analysis of genotypic and phenotypic characters. Mycoscience 38:267–280

Slocum RD (1980) Light and electron microscopic investigations in the Dictyonemataceae (basidiolichens). *Dictyonema irpicinum*. Can J Bot 58:1005–1015

Smith AH (1973) Agaricales and related secotioid Gasteromycetes. In: Ainsworth GC, Sparrow FK, Sussman AS (eds) The Fungi, an advanced treatise, vol IVB. Academic Press, New York, pp 397–420

Smith SE, Read DJ (1997) Mycorrhizal symbiosis. Academic Press, San Diego

Spurr R, Spurr J, Ammirati J (1985) A parasitic mushroom on the white chantarelle. McIlvainea 7:29–31

Srinivasan C, D'Souza TM, Boominathan K, Reddy CA (1995) Demonstration of laccase in the white rot basidiomycete *Phanerochaete chrysosporium* BKM-F1767. Appl Environ Microbiol 61:4274–4277

Stalpers JA (1978) Identification of wood-inhabiting Aphyllophorales in pure culture. Stud Mycol 16:1–248

Stalpers JA (1979) *Heterobasidion* (*Fomes*) *annosum* and the Bondarzewiaceae. Taxon 28:414–417

Stalpers JA (1984) A revision of the genus *Sporotrichum*. Stud Mycol 24:1–105

Stalpers JA (1987) Pleoanamorphy in Holobasidiomycetes. In: Sugiyama J (ed) Pleomorphic fungi: the diversity and its taxonomic implications. Kodansha, Tokyo, pp 201–220

Stalpers JA (1992) *Albatrellus* and the Hericiaceae. Persoonia 14:537–541

Stalpers JA (1993) The aphyllophoraceous fungi I. Keys to the species of the Thelephorales. Stud Mycol 35:1–168

Stalpers JA, Andersen TF (1996) A synopsis of the taxonomy of teleomorphs connected with *Rhizoctonia* s.l. In: Sneh B, Jabaji-Hare S, Neate S, Dijst G (eds) *Rhizoctonia* species: taxonomy, molecular biology, ecology, pathology and control. Kluwer, Dordrecht, pp 49–63

Stalpers JA, Loerakker WM (1982) *Laetisaria* and *Limonmyces* species (Corticiaceae) causing pink diseases in turf grasses. Can J Bot 60:529–537

Stamets P, Chilton JS (1983) The mushroom cultivator: a practical guide to growing mushrooms at home. Agarikon Press, Olympia, Washington

Steglich W, Steffan B, Stroech K, Wolf M (1984) Pistillarin, a characteristic metabolite of *Clavariadelphus pistillaris* and several *Ramaria* species. [Pistillarin, ein charakteristischer Inhaltddtoff der Herkuleskeule (*Clavariadelphus pistillaris*) und einiger *Ramaria*-arten (Basidiomycetes).] Z Naturforsch 39c:10–12

Stewart EL (1974) The genus *Gautieria* Vitt. PhD Thesis, Oregon State University, Corvallis, Oregon

Stillwell MA (1966) Woodwasps (Siricidae) in conifers and the associated fungus, *Stereum chailletii*, in eastern Canada. For Sci 12:121–128

Swann EC, Taylor JW (1993) Higher taxa of basidiomycetes: an 18S rRNA gene perspective. Mycologia 85:923–936

Swann EC, Taylor JW (1995a) Phylogenetic perspectives on basidiomycete systematics: evidence from the 18S rRNA gene. Can J Bot 73 (Suppl):s862–s868

Swann EC, Taylor JW (1995b) Phylogenetic diversity of yeast-producing basidiomycetes. Mycol Res 99:1205–1210

Tabata M, Abe Y (1995) *Cerrena unicolor* isolated from the mycangia of a horntail, *Tremex longicollis*, in Kochi Prefecture, Japan. Mycoscience 36:447–450

Talbot PHB (1973a) Towards uniformity in basidial terminology. Trans Br Mycol Soc 61:497–512

Talbot PHB (1973b) Aphyllophorales I: General characteristics; thelphoroid and cupuloid families. In: Ainsworth GC, Sparrow FK, Sussman AS (eds) The Fungi, an advanced treatise, vol IV B. Academic Press, New York, pp 327–350

Tanesaka E, Masuda H, Kinugawa K (1993) Wood degrading ability of basidiomycetes that are wood decomposers, litter decomposers, or mycorrhizal symbionts. Mycologia 85:347–354

Taylor DL, Bruns TD (1997) Independent, specialized invasions of ectomycorrhizal mutualism by two non-photosynthetic orchids. Proc Natl Acad Sci USA 94:4510–4515

Tehler A (1988) A cladistic outline of the Eumycota. Cladistics 4:227–277

Thielke C (1972) Die Dolipore der Basidiomyceten. Arch Mikrobiol 82:31–37

Thiers HD (1984) The secotioid syndrome. Mycologia 76:1–8

Thorn RG (1997) The fungi in soil. In: van Elsas JD, Trevors JT, Wellington EMH (eds) Modern soil microbiology. Marcel Dekker, New York, pp 63–127

Thorn RG, Barron GL (1984) Carnivorous mushrooms. Science 224:76–78

Thorn RG, Barron GL (1986) *Nematoctonus* and the tribe Resupinateae in Ontario, Canada. Mycotaxon 25:321–453

Thorn RG, Tsuneda A (1992) Interactions between various wood-decay fungi and bacteria: antibiosis, attack, lysis, or inhibition. Rep Tottori Mycol Inst 30:13–20

Thorn RG, Moncalvo JM, Reddy CA, Vilgalys R (2000) Phylogenetic analyses and the distribution of nematophagy support a monophyletic Pleurotaceae within the polyphyletic pleurotoid-lentinoid fungi. Mycologia 92 (in press)

Thurston CF (1994) The structure and function of fungal laccases. Microbiology 140:19–26

Tommerup IC, Bougher NL, Malajczuk N (1991) *Laccaria fraterna*, a common ectomycorrhizal fungus with mono- and bi-sporic basidia and multinucleate spores: comparison with the quadristerigmate, binucleate spored *L. laccata* and the hypogeous relative *Hydnangium carneum*. Mycol Res 95:689–698

Trappe JM (1962) Fungus associates of ectotrophic mycorrhizae. Bot Rev 28:538–606

Trappe JM, Maser C (1977) Ectomycorrhizal fungi: Interactions of mushrooms and truffles with beasts and trees. In: Walters T (ed) Mushrooms and man, an interdisciplinary approach to mycology. Linn-Benton Community College, Albany, pp 165–179

Traquair JA, McKeen WE (1978) Ultrastructure of the dolipore septum in *Hirschiporus paragamenus* (Polyporaceae). Can J Microbiol 24:767–771

Trojanowski J, Haider K, Hütterman A (1984) Decomposition of ^{14}C-labelled lignin, holocellulose and lignocellulose by mycorrhizal fungi. Arch Microbiol 139:202–206

Tsuneda A (1983) Fungal morphology and ecology. Tottori Mycological Institute, Tottori

Tsuneda A, Thorn RG, Hibbett DS (1992) *Lentinus tigrinus*: chlamydospores and interaction with *Pseudomonas fluorescens*. Rep Tottori Mycol Inst 30:1–12

Tsuneda A, Murakami S, Sigler S, Hiratsuka Y (1993) Schizolysis of dolipore-parenthesome septa in an arthroconidial fungus associated with *Dendroctonus ponderosae* and in similar anamorphic fungi. Can J Bot 71:1032–1038

Tu CC, Kimbrough JW (1978) Systematics and phylogeny of fungi in the *Rhizoctonia* complex. Bot Gaz 139:454–466

Tyler VE (1971) Chemotaxonomy in the basidiomycetes. In: Petersen RH (ed) Evolution in the higher basidiomycetes. University of Tennessee Press, Knoxville, pp 29–62

Tzean SS, Liou JY (1993) Nematophagous resupinate basidiomycetous fungi. Phytopathology 83:1015–1020

Umata H (1995) Seed germination of *Galeola altissima*, an achlorophyloous orchid, with aphyllophorales fungi. Mycoscience 36:369–372

Umata H (1997) Formation of endomycorrhizas by an achlorophyllous orchid, *Erythrorchis ochobiensis*, and *Auricularia polytricha*. Mycoscience 38:335–339

Umata H (1998) A new biological function of Shiitake mushroom, *Lentinula edodes*, in a myco-heterotrophic orchid, *Erythrorchis ochobiensis*. Mycoscience 39:85–88

Untiedt E, Müller K (1985) Colonization of *Sphagnum* cells by *Lyophyllum palustre*. Can J Bot 63:757–761

Vares T, Hatakka A (1997) Lignin-degrading activity and ligninolytic enzymes of different white-rot fungi: effects of manganese and malonate. Can J Bot 75:61–71

Varma A, Hock B (1995) Mycorrhiza: structure, function, molecular biology and biotechnology. Springer, Berlin Heidelberg New York

Walker J (1996) The classification of the fungi: history, current status and usage in the *Fungi of Australia*. In: Orchard AE, Grgurinovic C, Mallett K (eds) Fungi of Australia, vol 1A. Australian Biological Resources Study, Canberra, pp 1–66

Walleyn R, Rammeloo J (1994) The poisonous and useful fungi of Africa south of the Sahara. National Botanic Garden of Belgium, Meise

Wassink EC (1978) Luminescence in fungi. In: Herring PJ (ed) Bioluminescence in action. Academic Press, London, pp 171–197

Watling R (1974) Dimorphism in *Entoloma abortivum*. Bull Soc Linn Lyon, Num Spec 43:449–470

Watling R (1979) The morphology, variation and ecological significance of anamorphs in the Agaricales. In: Kendrick B (ed) The whole fungus, vol 2. National Museum of Natural Sciences, Ottawa, Canada, pp 453–472

Weber GF (1929) The occurrence of tuckahoes and *Poria cocos* in Forida. Mycologia 21:113–130

Weber NA (1979) Fungus-culturing by ants. In: Batra LR (ed) Insect-fungus symbiosis. John Wiley, New York, pp 77–116

Webster J (1992) Anamorph-teleomorph relationships. In: Bärlocher F (ed) The ecology of aquatic Hyphomycetes. Springer, Berlin Heidelberg New York, pp 99–117

Webster J, Chien C-Y (1990) Ballistopore discharge. Trans Mycol Soc Jpn 31:301–315

Wells K (1978) The fine structure of septal pore apparatus in the lamellae of *Pholiota terrestris*. Can J Bot 56:2915–2924

Wells K (1994) Jelly fungi, then and now! Mycologia 86:18–48

Wheeler Q, Blackwell M (1984) Fungus-insect relationships. Columbia University Press, New York

Whitney HS, Cobb FW (1972) Non-staining fungi associated with the bark beetle *Dendroctonus brevicomis* (Coleoptera: Scolytidae) on *Pinus Ponderosa*. Can J Bot 50:1943–1945

Whitney HS, Bandoni RJ, Oberwinkler F (1987) *Entomocorticium dendroctoni* gen. et sp. nov. (Basidiomycotina), a possible nutritional symbiote of the mountain pine beetle in lodgepole pine in British Columbia. Can J Bot 65:95–102

Wilding N, Collins NM, Hammond PM, Webber JF (1989) Insect-fungus interactions. Academic Press, London

Wilson CL, Miller JC, Griffin BR (1967) Nuclear behavior in the basidium of *Fomes annosus*. Am J Bot 54:1186–1188

Wood TG, Thomas RJ (1989) The mutualistic association between Macrotermitinae and *Termitomyces*. In: Wilding N, Collins NM, Hammond PM, Webber JF (eds) Insect-fungus interactions. Academic Press, London, pp 69–92

Worrall JW, Anagnost SE, Zabel RA (1997) Comparison of wood decay among diverse lignicolous fungi. Mycologia 89:199–219

Wright JE (1966) The genus *Phaeotrametes*. Mycologia 58:529–540

Zhao Z, Guo X (1989) Study on hyperparasitic relationships between *Rhizoctonia solani* and ectomycorrhizal fungi. Acta Microbiol Sin 29:170–173

Nomenclature and Documentation

6 The Naming of Fungi

David L. Hawksworth

CONTENTS

I. Introduction 171
II. Why Fungal Names Change 173
III. The International Code
    of Botanical Nomenclature 174
    A. Ranks 174
    B. Author Citations 176
    C. Publication (Effective Publication) 177
    D. Establishment (Valid Publication) 177
    E. Spelling and Meaning 178
    F. Typification 179
    G. Acceptability (Legitimacy) 180
    H. Precedence (Priority) 181
    I. Autonyms 182
    J. Synonyms 182
    K. Pleomorphic Fungi 182
    L. Fossil Fungi 183
    M. Special Provisions 183
    N. Conservation, Rejection and Suppression ... 184
    O. Revising the Code 185
IV. Other Codes and the BioCode 185
V. Indices of Fungal Names 187
VI. Responsibility and the Stability of Names ... 187
    References 189

I. Introduction

Names applied to organisms are the key to all accumulated knowledge on them, their properties, and uses. They are the basis of communication in all aspects of biology from the molecular to the perspectives of global ecology. A consideration of names and their application is, consequently, not an esoteric exercise but fundamental to an understanding of access to information and communication throughout biology and its dependent disciplines.

From the earliest days of human evolution, people found a need to communicate the difference between the edible and poisonous, the safe and the dangerous, etc. Today, isolated indigenous peoples, even with no written language, employ oral names to differentiate organisms important to them; such folk taxonomies often coincide with the species concepts of professionals (Mayr 1969). Macrofungi and macrolichens, in particular, which are edible, poisonous, hallucinogenic, used in folk medicine or dyeing, or have religous connections, are given different names by different peoples; examples amongst mushrooms are included in Benjamin (1995).

Greek and Roman writers employed names for several different kinds of fungi, some of which can be equated with known species (Ainsworth 1976). The selection of names started to become more complex as the exploration of the globe accelerated in the 16th century. The use of descriptive phrase names (*polynomials*) developed, which in reality had a dual function, to state the differentiating features and to provide a convenient label. For example, the phrase name *Lichenoides pulmoneum reticulatum vulgare, marginibus peltiseris* Dillenius 1742 (i.e., the lung lichen with a coarse network and petal-like edges). Polynomials and associated texts were in Latin because that was the common language of educated Europeans at the time.

Polynomials became increasingly cumbersome as more and more similar species had to be separated, and individual authors adopted their own and did not necessarily follow either their predecessors or peers. A concept of *genus* arose in which a particular name was emphasized first, and species groups with this same starting name were treated together, sometimes subdivided by shared features. This practice was adopted by Micheli (1729), who used such familiar names as *Aspergillus*, *Clavaria*, *Phallus* and *Puccinia*. The Swede Carl Linnaeus (1707–78; von Linné after enoblement in 1757) followed a similar practice, but then started to use a second word in index entries in his *Flora lapponica* of 1737, and more consistently in the *Öländska och gothländska resa* of 1745. The two-word system (*binomial*), with a first generic name and a second

---

MycoNova, 114 Finchley Lane, Hendon, London, NW4 IDG, UK

specific one (*specific epithet*), started to be used outside indices in theses from 1751. Linnaeus (1753), in the monumental *Species plantarum*, proposed marginal second names for the some 5900 species he accepted in the world, including about 200 fungi, 106 of which were lichenized (Jørgensen et al. 1994); for example, he used the marginal name *pulmonarius* in the case of the species represented by Dillenius' polynomial cited above, including it in the genus *Lichen* (Fig. 1).

The binomial system proved eminently suited to verbal as well as written communication, and also for labelling specimens. Although Linnaeus did not initially intend these names to be more than shorthand for the preferred polynomial, they rapidly supplanted them throughout the world. The *Species plantarum* is now recognized as the formal start of botanical, including mycological, nomenclature. The origin and adoption of binomial nomenclature is a fascinating story (Blunt 1971; Ramsbottom 1938; Stafleu 1971; Stearn 1957, 1959).

In order to prevent chaos in the usage and application of names, internationally agreed rules and procedures are essential. In some fields of human endeavour, for example the names of chemicals, physical measurements, enzymes, proteins, the numbering of serial publications and books, and geographical names, systems to be adopted internationally have been agreed by bodies such as the International Standards Organization. However, biosystematics is a science continually refining and reconsidering the objects of its study. New research repeatedly leads to an improved understanding of relationships which necessitates successive reclassifications. International agreements relating to scientific names have consequently focused on the principles and procedures to be followed to determine the name to be used for an entity that research indicates as meriting recognition in a particular position and rank.

Taxonomy is the scientific act of classification, while *nomenclature* is the labelling procedure for the units it wishes to recognize. Both activities are amongst the components of *biosystematics* (biological systematics; systematics) which also involves the elucidation of evolutionary and biological relationships. The process of nomenclature is, in effect, an administrative device for the delivery and communication of research results. Nomenclature is not a scientific activity *per se*, but on occasion can involve considerable bibliographic and historical research. As nomenclature in biology is not restricted to the scientific names of whole organisms, the term *bionomenclature* is increasingly being employed.

Nomenclature often appears daunting at first. While some aspects can be complex, an understanding of the principles and commonly used procedures is required to both practice taxonomy and use names effectively for communication and information retrieval. All aspects of this subject cannot be covered in this chapter, but further information can be sought in Hawksworth (1974), Jeffrey (1989), Jong and Birmingham (1991), Naqshi (1993), Ride and Younés (1986), Sivarajan (1984), and authoritatively in the International Code of Botanical Nomenclature itself (Greuter et al. 1994a). A now somewhat dated glossary of nomenclatural terms used in the botanical Code is provided by McVaugh et al. (1968), and a compilation of 1175 "unofficial" as well as "official" terms used in organismal nomenclature across biology is also available (Hawksworth 1994). Most nomenclatural terms likely to be encountered by mycologists are defined in the "nomenclature" entry of the current edition of *Ainsworth & Bisby's Dictionary of the Fungi*

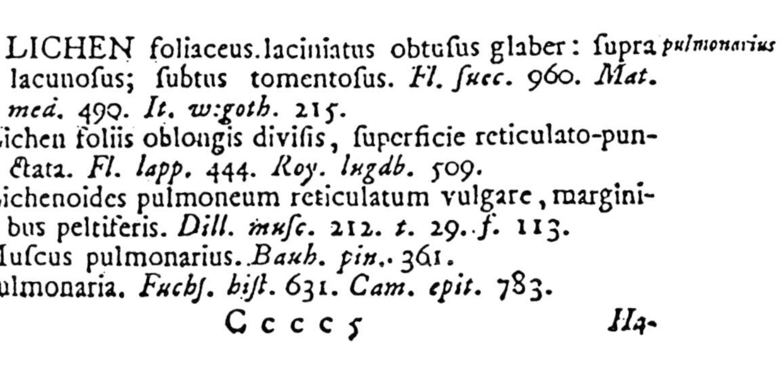
32. LICHEN foliaceus.laciniatus obtuſus glaber: ſupra *pulmonarius*
lacunoſus; ſubtus tomentoſus. *Fl. ſuec.* 960. *Mat. med.* 490. *It. w:goth.* 215.
Lichen foliis oblongis diviſis, ſuperficie reticulato-punctata. *Fl. lapp.* 444. *Roy. lugdb.* 509.
Lichenoides pulmoneum reticulatum vulgare, marginibus peltiferis. *Dill. muſc.* 212. *t.* 29. *f.* 113.
Muſcus pulmonarius. *Bauh. pin.* 361.
Pulmonaria. *Fuchſ. hiſt.* 631. *Cam. epit.* 783.
Cccc 5 *H4.*

Fig. 1. Sample species entry from the *Species plantarum* (Linnaeus 1753) showing the use of trivial names in the margin; these later became accepted as species epithets

Table 1. The principal nomenclatural terms adopted in the International Code of Botanical Nomenclature (Greuter et al. 1994a), together with the harmonized terms recommended for use in the future for all groups of organisms (Greuter et al. 1996; with minor subsequent ammendments made by the International Committee on Bionomenclature in 1997)

BioCode	Botanical Code
Publications and dates of names	
Published	Effectively published
Registerable	–
Date	Date (or priority)
Priority	Priority
Precedence	Priority
Earlier	Earlier
Later	Later
Nomenclatural status	
Established	Validly published
Registration	Registration
Acceptable	Legitimate
Taxonomic status	
Accepted	Correct
Types of names	
Name-bearing type	Nomenclatural type
Nominal taxon	Name and type
Synonymy	
Homotypic	Nomenclatural
Heterotypic	Taxonomic
Replacement name	Avowed substitute
Setting aside the rules	
Conserved	Conserved
Rejected	Rejected
Suppressed	Explicitly rejected

(Hawksworth et al. 1995). Proposals to recommend harmonized nomenclatural terms across all biological groups have been made by the International Committee on Bionomenclature (see below; Greuter et al. 1996); these harmonized terms are already starting to be adopted, and as they will become more familiar during the life of this Volume, they have been used where appropriate in this chapter with the more familiar "botanical" equivalents also indicated in parentheses (Table 1).

II. Why Fungal Names Change

Name changes hamper communication and are a cause of constant frustration to all who deal with fungi. Seemingly endless changes in the names of even familiar species contribute to the poor standing of taxonomists amongst their peers and consequently to the underresourcing of this vital subject (Bisby and Hawksworth 1991). Names are changed for three distinct reasons:

1. New research shows that a species has been placed in a genus to which it proves not to be related, or that what was recognized as a single species actually embraces more than one. Name changes for such *taxonomic* reasons should be welcomed by users as correct generic (and thus family and higher classifications) placements facilitate the prediction of other properties of the fungus, for example, the kinds of secondary metabolites that are produced, the form of any asexual stage in the case of pleomorphic fungi, or particular fungicides that could eradicate a disease. In the case of species which are divided into more than one, the segregates may have diverse biologies or ecologies and even cause different diseases or produce disparate mycotoxins – matters also of major concern to name users.

2. Applications of the Code result in changes from two basic causes: first, the discovery that a name is incorrect because, under the rules, a different one should have been adopted; for example, the discovery that a name that was assumed to have been available for use was not, or that another name exists for the same species that has precedence under the rules; second, because the rules in the Code change, requiring cases to be reassessed by criteria that did not exist when a particular name was first used. These categories of changes do not arise from any new scientific data, are *nomenclatural* and not taxonomic, and so are especially frustrating. However, recognizing this problem, the provisions of the Code are gradually being modified in ways that will increasingly reduce the numbers of changes in familiar names for nomenclatural reasons.

3. Misapplications of names arise, for example, when a name has been applied to a species which is not the same as that represented by its name-bearing type. This can result from insufficient attention being given to comparisons with type material, i.e., its misidentification, or the discovery that what had been considered the type was not and represented a different species. As in the case of changes for nomenclatural reasons, the Code now has mechanisms which can allow well-

known names that have been misapplied to continue to be used.

III. The International Code of Botanical Nomenclature

The need for internationally accepted nomenclatural procedures was addressed at the International Botanical Congress in Paris in 1867. This was formally adopted, after review, as A.P. de Candolle's *Lois de Nomenclature Botanique*. This document of 64 pages proved not to be exhaustive, and at each International Botanical Congress since that of 1905 in Vienna revisions have been considered. This Congress now meets at intervals of 6 years, and a new Code is published after each Congress. Accounts of the history and evolution of the Code are given by McNeill and Greuter (1986), Perry (1991), and Nicolson (1991). Codes adopted by particular Congresses are often referred to by linking them with the name of the venue. The current Code, adopted by the 15th International Botanical Congress in Tokyo in 1993 (Greuter et al. 1994a), is therefore termed the Tokyo Code.

The Tokyo Code has six *Principles* and 62 *Articles*. The Articles are mandatory, and the paragraphs within them are numbered; Art. 37.4 denotes the fourth paragraph in Article 37. In addition, some Articles include explanatory *Notes* and most worked Examples of their application. *Recommendations*, adherence to which is considered good practice but remains optional, follow certain articles; cited in the form Rec. 37A.1, this identifies the first paragraph in Recommendation A under Art. 37. Provisions for modification of the Code are detailed, and Appendices deal with the names of hybrids and list conserved and rejected names. The Tokyo Code is entirely in English; but "unofficial" translations in other languages can be approved by the General Committee. Translations already exist in Czech (Marhold 1995), French (Burdet 1995), German (Greuter and Hiepko 1995), Japanese (Ohashi 1997), and Russian (Egorova and Gladkova 1996), and others are expected.

In endeavouring to apply the Code in determining the name which should be applied to a taxon, it is convenient to view it as a Nomenclatural Filter (Fig. 2). Different Articles apply at successive stages, each a filter of finer gauge excluding further names until the one correct under the Code is reached.

A. Ranks

The different levels in the taxonomic hierarchy are ranks. The principle ranks (Art. 3.1) appear in capitals in Table 2, and the number can be increased if required by use of the prefix sub- (e.g., subfamily, subsection, subspecies, subform: Art. 4.2). The rank of phylum, a term familiar to zoologists, is an alternative to division, but is increasingly employed instead of division for consistency in general textbooks and interdisciplinary works.

Unless a special exception is made in the Code, names above the rank of genus and up to and including that of order are based on acceptable (see below) generic names and formed by adding the appropriate termination (Table 2), e.g., the family *Amanitaceae* from the generic name *Amanita*. Names above order can also be made in this way but they can also be based on some distinctive character (Art. 17.1), e.g., *Loculoascomycetes*.

There is no obstacle under the Code to prevent the special suffixes used for taxa above order in

Table 2. The principal taxonomic ranks used in mycology, and the terminations to be used in those above genus

Rank	Termination
KINGDOM	Fungi
Subkingdom	-bionta
PHYLUM [DIVISION][a]	-mycota
Subdivision	-mycotina
CLASS	-mycetes
Subclass	-mycetidae
ORDER	-ales
Suborder	-neae
FAMILY	-aceae
Subfamily	-oideae
Tribe	-eae
Subtribe	-inae
GENUS	
Subgenus	
Section	
Series	
SPECIES [sp., *pl.* ssp.][a]	
Subspecies [subsp./ssp., *pl.* subspp./sspp.]	
Variety [var., *pl.* vars.]	
Form [f., *pl.* ff.]	
[Special form] [f. sp.]	

[a] Commonly used abbreviations for names of ranks are included in squared brackets; *pl.* = plural form of abbreviation.

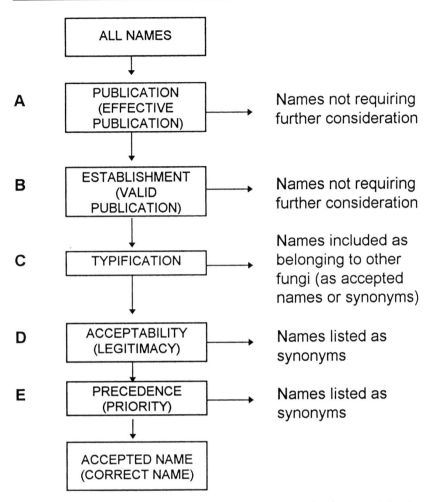

Fig. 2. The Nomenclatural Filter, indicating the steps to be taken in ascertaining the correct name for a taxon, adopting the harmonized BioCode (Table 1) with the Botanical Code equivalents in *parentheses*

fungi, such as *mycota* (phylum) and *-mycetes* (class), for all fungi traditionally studied by mycologists regardless of the kingdoms in which they are now classified. For example, *Oomycota* can continue to be used even though the phylum is referred to the kingdom *Chromista* and not the kingdom *Fungi* (see Chap. 1, this Vol.). This system was adopted in the most recent edition of the *Dictionary of the Fungi* (Hawksworth et al. 1995).

In the case of fossils, genera based on separated parts are termed *form genera* and cannot be grouped into families, although they may be referable to higher ranks. These are not to be confused with the alternative names permitted for asexual morphs of fungi with pleomorphic life cycles.

In mycology, the rank of *special form* (f.sp.) has been used for apparently indistinguishable fungi characterized by their ability to attack different host plants; for example, *Fusarium oxysporum* f.sp. *lycopersici* attacking tomatoes and f.sp. *elaedis* destroying oil palms. I expect the need for this rank eventually to decline as knowledge at the genetic and molecular level expands. The names of special forms are not regulated by the Code (Art. 4. Note 3).

In order for names to be validly published, the ranks accorded to them have to be clearly specified. This was not always the case, and some different terms may be encountered in the early literature, for example, *cohors* for order, and *ordo naturalis* or *ordo* for family. Other authors have used symbols to denote ranks, for example an asterisk (*) for subspecies as in *Alectoria ochroleuca* * *A. thulensis* Th.Fr.; these latter names were indexed incorrectly as if they were full species (i.e., "*Alectoria thulensis*") by some compilers exacerbating the problem. If symbols are encountered, it is necessary to carefully study any introduction to the work and the interpretations of later authors; in some instances, symbols may merely indicate localities, rarity, or seasonality and be nothing to do with nomenclature.

Scientific names of genera, species, and infraspecific entities are always italicized, but the practice at the ranks of family and above has varied through time and with different traditions and journal house-styles. Family names are sometimes italicized where orders and higher ranks are not, on the basis that they differ in also being subject to the rules relating to priority, as are generic and species names. The practice in the Code itself has varied between editions, but is not ruled upon. In the Tokyo Code all scientific names are italicized regardless of rank, e.g., *Fungi* (kingdom), *Chytridiomycota* (phylum), *Diatrypales* (order), *Cookellaceae* (family). This practice is now being followed by an increasing number of journals, and has the advantage of making it easy to recognize when a scientific or an informal name is being used, e.g., fungi vs. *Fungi*, pyrenomycetes vs. *Pyrenomycetes*.

B. Author Citations

The object of appending to a scientific name that of the person or persons responsible for it is to increase precision by providing a shortcut to the original publication, and also to avoid confusion with any similar names (Art. 46.1). *Epicoccum cocos* F. Stevens can be immediately seen to have been introduced in a publication by F. Stevens. If the generic position or rank is changed, that of the original author of the name-bringing epithet (*basionym*) is placed in brackets; *Pseudoepicoccum cocos* (F. Stevens) M.B. Ellis indicates that Ellis transferred into *Pseudoepicoccum* as a species of that genus an epithet *cocos* used by Stevens but under a different generic name (or at a different rank).

Author citations are not correctly used after the species names where infraspecific taxa other than that repeating the species name (i.e., autonyms) are being cited, for example *Xanthoria fallax* var. *subsorediosa* (Räs.) Awasthi not *X. fallax* (Hepp) DuRietz var. *subsorediora* (Räs.) Awasthi.

When the person coining the name did not, for any reason, establish (validly publish) it and it was taken up formally by another person, the names are connected by *ex* (Art. 46.4). *Sordaria decipiens* G. Winter ex Fuckel indicates that Winter first used this binomial but did not meet the requirements for valid publication under the Code, and that it was later adopted by Fuckel, who did meet those requirements; the critical part of this citation is that of the validating author and reference to Winter is optional, i.e., the form *Sordaria decipiens* Fuckel is acceptable.

When one author provides material validly publishing a name in the work of another, their names are sometimes linked by *in*, but in that case all names appearing after the first author can be omitted as only constituting further bibliographic information and not being part of the author citation *per se* (Art. 46.2 Note 1). *Sphaeropsis conica* Lév., in Demidoff indicates that Léveille contributed an entry including this name to a work published under Demidoff's name. The author citation is strictly only Lév., and when *in* is used to provide fuller information, it should always be preceded by a comma, as above, although that was not always done prior to the Tokyo Code. *Apud* was formerly used occasionally as an alternative to in.

The names of authors are regularly abbreviated to save space. Abbreviations need to be unambiguous, not only within the fungi but throughout botanical nomenclature, as many authors (especially from the last century) did not confine their interests to only one field. The International Working Group on Taxonomic Databases, a body recognized by the International Union of Biological Sciences (IUBS), has adopted a list of recommended standard forms of authors' names including abbreviations made according to common principles (Brummitt and Powell 1992). The 9288 fungal author entries are also available separately (Kirk and Ansell 1992). In these listings, names are not usually abbreviated unless more than two letters are eliminated, names of eight letters or fewer are not abbreviated where no strong tradition exists, and while two- or three-syllable names are generally abbreviated to one, those of four or more are most often abbreviated to two. Examples are: Nyl. for W. Nylander (1822–1899), Kremp. for A. von Krempelhuber (1813–1882), and Lév. for J.H. Léveille (1796–1870). Initials of forenames are included where necessary to distinguish between authors with the same surname, although for some well established abbreviations convention is maintained. For example, J.B. Ellis (1829–1905) appears as Ellis, while other, later, Ellis's have initials included, such as D. Ellis (1874–1937), J.W. Ellis (1857–1916), and M.B. Ellis (1911–1996).

An increasingly common and helpful practice in works such as checklists is to cite the date of the publication after the author citation, either inside or without brackets; for instance *Mollisia humidicola* Graddon 1997 and *M. melaleuca* (Fr.) Sacc. (1888).

Where names have been misapplied or employed in different senses, some other elements can be encountered within or in place of author citations, notably *auct.* (in the sense of some authors but not of the type). Sometimes, individual workers or those in a country misapplying a name are identified, e.g., *auct. amer.* (of American authors), *non* (not), and *sensu* (in the sense of, followed by an author citation, or to indicate a broad, *lato*, or narrow, *stricto*, usage). A colon (:) is sometimes used in author citations in sanctioned fungal names.

It has been the common practice of many journals and authors to insist that author citations are included routinely in all kinds of publications. However, these serve little purpose except as short-hand bibliographic clues in taxonomic works, and are rarely verified by those preparing nontaxonomic papers. Journals are therefore increasingly dropping this requirement, including *Mycologia* and *Mycological Research*; this is a sensible practice to be encouraged.

The case for dropping author citations in nontaxonomic papers is summarized by Garnock-Jones and Webb (1996), and Korf (1996) provides practical hints on the use and simplication of author citations of fungal names.

C. Publication (Effective Publication)

The first step in the Nomenclatural Filter (Fig. 2A) is *publication*, aimed at ensuring that information is actually available for consultation by scientists. This is achieved by the distribution of printed matter by sale, exchange, or gift, to at least publicly accessible botanical libraries (Art. 29.1). Using names in meetings or on specimen labels does not meet this criterion. From 1953, handwritten material (Art. 30.1), printed trade catalogues or nonscientific newspapers (Art. 30.3), and printed labels distributed only with sets of dried specimens (*exsiccates*; Art. 30.4) are also unacceptable. Nicolson (1980) provides a key to the identification of effectively published material.

Computers, and especially desktop publishing, are increasingly complicating this subject to the extent that it is not always clear whether distribution of a particular work has been effective in the sense intended by Art. 29. This problem is being addressed by a Special Committee appointed by the Tokyo Congress in 1993 to report to the next International Congress in 1999.

The date of publication is critical in considerations of priority. The date is that on which the material became available in the sense described above (Art. 31.1); normally this is the date material passed to a carrier (e.g., postal service) for distribution (Rec. 31A.1). While printed dates are generally correct (at least to the year), it is good practice to always question this. Was the work issued in parts? (Look for bound-in wrappers, changes in paper colour or quality, printer's codes on different signatures, and changes in typefaces.) What is the library date stamp? Were separates issued in advance (preprints)? What is the date of the Preface? Is there a date by the printer's name at the end? What are the most recent dates of papers or of collections cited? When did reviews or notices appear in other journals? Detective work is often needed, but botanists are fortunate in that this has already been done for most major botanical (including mycological) publications by Stafleu and Cowan (1976–88).

D. Establishment (Valid Publication)

The next sieve in the Nomenclatural Filter (Fig. 2B), is *establishment*, and this comprises a suite of criteria to be met before a name can be accepted for further consideration.

1. Names must be in the form appropriate for their rank (Arts. 16–24, 26–27). For example, a generic name is a substantive in the singular (or a word treated as such) (Art. 20.1) and, if introduced after 1911, may not be the same as a technical term; a specific epithet is an adjective, noun in the genetive, word in apposition, or several linked words (Art. 23.1). Further, it must not be identical to the generic name (a *tautonym*; Art. 23.4; e.g., *Crucibulum crucibulum* is unacceptable).
2. A description or *diagnosis* (a statement of what the author regards as the distinguishing features) or a reference to one already effectively published

is needed (Art. 32.1); if after 1 Jan. 1935 this must be in Latin (Art. 36.1), except for fossils. Examples of descriptions and diagnoses in Latin in a wide range of groups are included in Stearn (1992). Illustrations can, in some cases, satisfy the requirement for a description or diagnosis prior to 1908 (Arts. 42.3–4, 44.1–2). In the case of a transfer in rank or position (*new combination*) or the introduction of a replacement name (*new name*) for an illegitimate one, from the start of 1953 the literature reference to the basionym or replaced name must be to the actual page or plate concerned and the date (Art. 33.2).

3. The name-bearing type (i.e., a species name for a genus, a single specimen for a species) has had to be cited since 1 Jan. 1958 (Art. 37.1). From 1 Jan. 1990 the designation is required to include the word *typus* or *holotypus* or its equivalent (Art. 37.4) and the collection or institution in which it is preserved must be stated (Art. 37.5). Dried reference collections (including herbaria) are normally referred to by unique acronyms (Holmgren et al. 1990); for example, B (Botanical Garden and Museum, Free University of Berlin), K (Royal Botanic Gardens, Kew), RNG (University of Reading), and US (Smithsonian Institution, Washington, DC). Some are based on former names of the institutions but are nevertheless retained for continuity, for example BPI (US National Fungus Collection, UDSA Agricultural Research Service, Beltsville; formerly the Bureau of Plant Industry) and DAOM (Eastern Cereal and Oilseed Research Centre, Ottawa; originally Department of Agriculture, Ottawa, Mycology). Microbial genetic resources collections may need to be cited in some cases and their abbreviations follow Sugawara et al. (1993).

4. A name must also be accepted by the author proposing it (Art. 34.1). This may at first seem suprising, but sometimes names are listed only as synonyms, mentioned in discussions about placements, or anticipate future classifications. In rare cases where an author suggests more than one name for the same taxon (*alternative names*) in the same publication, since 1 Jan. 1953 none is accepted as established (Art. 34.2).

5. Names in which ranks are used out of sequence are not normally acceptable as valid (Art. 33.5); since 1 Jan. 1953, neither are ones where the rank is not clearly stated (Art. 35.1).

6. Subject to the approval of the detailed mechanism by the 16th International Botanical Congress in St Louis in 1999, names proposed on or after 1 Jan. 2000 will have to be registered. At the time of writing, the proposals being developed by the International Association for Plant Taxonomy (IAPT) to implement this requirement had not been released. However, it is anticipated that these will include a system of registered journals and both national and group-based registration offices.

7. All criteria which apply must be fulfilled in order for a name to be established. If these are not achieved simultaneously, names are valid only from the date the last requirement is met (Art. 45.1).

Additional further provisions include: acceptability of names published with a question mark, indicating taxonomic uncertainty (Art. 34.1); combined generic and specific descriptions for a new genus and species when both are simultaneously introduced (Art. 42.1); names below generic rank are valid only if the generic name to which they are referred is valid (Art. 43.1); and various rules applying to particular groups.

E. Spelling and Meaning

Generic and specific names can be derived from any source, and are then treated as Latin regardless of their origin. They can relate to the characters or the fungus, its ecology or host, geographic distribution or place of origin, or be based on the name of a person (usually the discoverer or as a tribute to another mycologist). Even acronyms (e.g., *Isia* after International Scientific Instruments Ltd) and anagrams (e.g., *Milopsium* from *Spilomium*) are permissible. For generic names without a classical or obvious gender, that used by the proposing author is adopted (Art. 62.1).

The Code does not advise on pronunciation. While it might be thought that classical practice should be followed (e.g., Allen 1978), this often differs from current usage in anglophone countries. Differences in pronunciation frequently cause problems in verbal exchanges at international scientific meetings, and in an international context it is preferable to adhere to classical usage.

Should an incorrect Latin termination be used in a newly proposed name, this is to be corrected

(Arts. 32.6, 60.11). Original spellings are otherwise retained, apart from the correction of typographic or orthographic errors (Art. 11.2). Hyphens are now deleted unless epithets are formed of words that usually stand independently (Art. 60.9; e.g., *pseudoplatanus* not *pseudo-platanus*; *novae-angliae* accepted), and letters with accents or other signs transcribed (Art. 60.6; e.g., ö to oe). The diaresis, indicating that a vowel is to be pronounced separately from the preceding vowel is, however, permissible (e.g., *Elsinoë*).

Botanical Latin has developed as a subject in its own right during the past 250 years. Many words have been adapted to have more precise meanings than in classical Latin, and others continually have to be invented for use in descriptions and names. The standard reference work which all practicising botanical taxonomists require is Stearn (1992). A specialist Latin glossary for mycologists is provided by Cash (1965).

F. Typification

A name-bearing type is "... a nomenclatural device, whose chief purpose is to provide for stability of the nomenclatural vessel in the often turbulent seas of taxonomy" (Rickett 1959). Every name in the rank of family and below has a nomenclatural type (*typus*) to which it is permanently attached (Art. 7.2), and is the final arbiter which determines its application (Art. 7.1). The name-bearing type of a family name is a generic name; that of a generic name a species name; that of a species name or variety a specimen or permamently preserved culture (see below); etc. It cannot be stressed too strongly that it is only the names of taxa which can have name-bearing types and not the taxa themselves, i.e., the types need not be representative of the taxon in which they are placed, but clearly must come within the range of variation that a taxonomist circumscribes and wishes to recognize.

It is now a requirement for establishment that the publishing author designates a single element to serve as the name-bearing type (Art. 37.1; see above). A specimen thus designated, or for earlier names a single element used, is the *holotype*. A duplicate or part of the collection from which the holotype was taken is an *isotype* (it is good practice to deposit such isotypes in different major collections for additional security and improved access).

Specimens other than the holotype detailed when a new taxon is introduced are *paratypes*.

Many cases are found where no holotype was designated or where this has been subsequently destroyed. The first step is then to ascertain if there is other original material studied by or cited by the author in existence that can be used. An element from the original material selected by a later author to serve as the nomenclatural type in the absence of a holotype is a *lectotype*. A lectotype may be chosen from one of the several specimens listed (*syntypes*) where none was distinguised as a holotype; from amongst any paratypes if the holotype has been destroyed; from amongst dated and(or) annotated specimens in the author's herbarium; from any published or unpublished illustrations; etc.

Where no original material that can be used as a lectotype remains, a new type (*neotype*) can be selected to serve as the name-bearing type. Should candidate material for lectotypification be discovered later, the lectotype has precedence (Art. 9.9), just as a lectotype would be supplanted should a "lost" holotype be rediscovered (Art. 10.5).

In some cases, a name-bearing type collection may be extant, but the characters needed to ascertain the species to which it belongs may be missing, for example, perithecia may be present but have no mature asci or ascospores inside. It is then possible to designate an interpretive type, an *epitype*, to ensure the precise application of the name (Art. 9.7).

A large number of terms related to types exist outside the framework of the Code and may be encountered from time to time. The most frequent in mycology are: *kleptotype* (a fragment removed from a type without authorization and kept in a different collection), *topotype* (material from the original locality from which a taxon was described but collected at a later date), and *typotype* (the specimen on which an illustration is based, where the illustration is the type).

The type of a species name or lower taxon is a single permanently conserved specimen or illustration (Art. 8.1); more than a single individual in a single preparation is acceptable for fungi. Living cultures per se are not accepted as nomenclatural types (Art. 8.2), although dried cultures have always been, but a major change was made in the Tokyo Code to permit metabolically inactive prepartions to serve as name-bearing types (Art.

8.2. Ex. 1). The Code recommends that living cultures of fungi prepared from the holotype should be preserved in at least two genetic resource or culture collections wherever possible, whether the type is a preserved culture or not (Rec. 8B.1). Living cultures derived from either the name-bearing type metabolically inactive culture or the same strain as a dried culture or specimen which serves as holotype, are commonly referred to as *ex-type* (i.e., from the type), ex-holotype, ex-isotype, etc. (Rec. 8B.2), but, pedantically, are not themselves name-bearing types.

In determining which species names or specimens should be recognized as nomenclatural types, considerable care is often needed. Attention must be paid to the *protologue* (everything associated with the name when it was introduced, including the description and illustrations) and to later usage. Lectotypifications and neotypifications must conform to the requirements related to publication (Art. 9.11; see above), and as in the case of holotype designation, from 1 Jan. 1990 the collections where the specimens are retained is required (Art. 9.14).

Type material is an irreplaceable resource. Great care is therefore necessary when examining such collections; portions for DNA extraction, metabolite analyses, scanning electron microscropy, or microscopic anatomical studies, etc., should only be removed when absolutely essential and, where possible, any such material should be kept with the specimen. Many insititutions apply strict regulations to loans of type material. The use of type specimens in the course of routine identifications, or in early stages of taxonomic revisions, should consequently be avoided. After taxonomic concepts are clear, types can then be checked to see to which of the entities they belong.

G. Acceptability (Legitimacy)

Established names must also fulfill some additional criteria in the Code to be acceptable (Fig. 2D); those which do are *acceptable* (*legitimate*) and have to be considered further, whereas any which do not are *not acceptable* or *illegitimate*, and can only be listed as synonyms. Fungal names can be unacceptable for a variety of reasons:

1. If they have been changed because they were considered inappropriate (Art. 51.1). For example, *Toninia aromatica* is acceptable even though the species has no smell (the type was received in a scented envelope by the describing author), as is *Phacidium musae* although it occurs not on *Musa* but on *Clusia*.
2. If the name was not necessary as the taxon as circumscribed by its author included the name-bearing type of a name that should have been adopted (Art. 52.1); such names are *superfluous*. Inclusion of a type can be by mentioning a name as a synonym or citing a type specimen (or species for generic names). When the generic name *Torrubia* was validly published in 1865, one of the species included was *Clavaria militaris*; however, that species name was the type of the generic name *Cordyceps* validly published earlier in 1833. *Torrubia* was therefore superfluous, as *Cordyceps* could have been used, and *Torrubia* is thus unacceptable (illegitimate).
3. If a name is spelled the same as another based on a different type, each is a *homonym*. The most recently published (later homonym) is illegitimate (Art. 53.1). The lichen-forming genus *Ancistrospora* Thor 1990 is unacceptable as a later homonym of *Ancistrospora* Menendez & Azcuy 1972 established for fossil spores; the *replacement name* (*new name*) *Ancistroporella* Thor 1994 was therefore introduced, based on the type species of *Ancistrospora* Thor. The principle of homonymy extends to names which are spelled so similiarly they are likely to be confused, but does not now apply outside groups treated under the Botanical Code (Art. 54.1). Under the current Code, the mushroom generic name *Drosophila* Quélet 1866 is therefore not regarded as unacceptable, even though there is an earlier fly genus *Drosophila* Fállen 1823.
4. If a name has come to be widely used in a sense other than its type, it is termed *misapplied* such hams can be rejected with the approval of appropriate Committees and placed on a special list of *rejected names* (Art. 56.1). The type of *Lycoperdon aurantium* L. described in 1753 was found not represent the *Scleroderma* species to which it had generally been understood to belong up to that time. Linnaeus' name was therefore rejected to prevent its being used in another sense, and *S. citrinum* taken up as the accepted name for the fungus.

Prior to the Leningrad Code adopted in 1975, names could be regarded as unacceptable because they were based on mixtures of more than one species (*discordant elements*) or monstrosities (gross deformations, from whatever cause), for example, a "fungus" based on a combination of characters from a lichenicolous fungus and the host lichen, such as *Arthonia versicolor* Ach. 1803 based on a thallus of *Opegrapha atra* infected by the hyphomycete *Milospium graphideorum*. In such instances, one of the two elements must now be selected as lectotype, and in this instance it was the *Opegrapha* in order to preserve current usage.

Where no acceptable (legitimate) alternative name exists, a *new name* must be chosen to replace the one which is unacceptable (Art. 58.1). In some cases, an epithet unacceptable in one combination or rank can be retained in another; in those instances it is attributed to the author using it in the acceptable sense (Art. 58.3). For example, *Cenomyce ecmocyna* Ach. 1810 was unacceptable (illegitimate) as *Lichen gracilis* L. 1753 was cited as a synonym and could have been used; however, when Leighton used the name *Cladonia ecomocyna*, he treated *L. gracilis* as a separate species, *C. gracilis*, and is thus ruled to have introduced the new name *C. ecmocyna* Leighton 1866 [not (Ach.) Leighton].

H. Precedence (Priority)

The accepted (*correct*) name for a taxon of the rank of family and below is determined by priority of publication (Fig. 2E) amongst the acceptable names in the same rank whose nomenclatural types fall within its range of variation (Art. 11.3). In the case of ranks below genus, that name is combined with the correct generic name (Art. 11.4), if necessary in a new combination (see above).

For example, when *Heterosporium echinulatum* Berk. 1870 and *H. dianthi* Sacc. & Roum. 1881 are treated as belonging to the same species and placed in the genus *Cladosporium*, the correct name is *C. echinulatum* (Berk.) G.A. de Vries 1952. However, when *Bryopogon divergens* var. *rigidum* Haszl. 1884 was found to belong to the species known as *Alectoria smithii* Du Rietz 1926, Du Rietz's epithet is retained, as Haszlin's name was used only at varietal rank and does not have priority at species rank.

Publication, and hence the principle of priority, is agreed to date from 1 May 1753, the date ascribed to the publication of Linnaeus' *Species plantarum*. Since the Sydney Code was adopted in 1981, this date now applies to the nomenclature of all fungi [including lichen-forming fungi and slime moulds; Art. 13.1(d)]. Prior to 1981, the nonlichen-forming fungi had a rather complex system of later starting point dates based on two different works: 31 Dec. 1801 for rusts, smuts, and gasteromycetes (Persoon 1801), 1 Jan. 1821 for all other fungi (Fries 1821–32). Names adopted in these works were protected from earlier homonyms and competing synonyms. Names published prior to those dates were not established, and usages meeting the necessary requirements (see above) after those dates had to be found to ascertain when and if they had to be taken into account, the names taking authorships and dates from the establishing works (linked by "ex", e.g., *Stilbospora angustata* Pers. ex Link 1824, not *S. angustata* Pers. 1801. This proved increasingly a very messy situation, especially as no comprehensive catalogues of usages of pre-1821 fungal names existed (although started by Petersen 1975), as explained by Demoulin et al. (1981), and Korf and Kohn (1980).

In order to avoid unfortunate name changes arising from this change, names adopted in the former starting point books are treated as *sanctioned* for use over earlier homonyms and competing synonyms (Art. 15.1). The symbol ":" with the abbreviation of the sanctioning author (i.e., Fries or Persoon) can be used after the author citations to indicate names are sanctioned (Rec. 50E.2), for example, *Boletus piperatus* Bull.: Fr. The date of establishment of sanctioned names is that of the establishing and not the sanctioning author; *B. piperatus* dates from Buillard 1791 and not Fries 1821. Fries becomes the key author to cite only where he specifically excluded the type of the original person who introduced the epithet or name; for instance, *Cantharellus tubaeformis* (Schaeff.) Bull. is attributed to Fr.: Fr. as Fries explicitly excluded Schaeffer's type by including it within a different species in the sanctioning work. The sanctioning provision includes lichen-forming fungi; this was confirmed for the Berlin Code agreed in 1987 following an analysis of the changes this entailed (Hawksworth 1986).

A detailed account of the sanctioning procedures is provided by Korf (1983), and the issues are discussed further by Gams and Kuyper (1995). Gams (1984) catalogued most sanctioned names (but not all those in Fries' Introduction, which included some lichen-forming genera). Korf

(1996) recommends that mycologists drop the colon and sanctioning author "unless preparing a taxonomic paper in which [the] reason for choosing a particular name may be tied to the sanctioned use of that name". I concur; this is the practice already adopted in the *Dictionary of the Fungi* and other publications from the International Mycological Institute.

Names above the rank of family are exempt from the precedence rule, but it is recommended that priority also be applied to them (Rec. 16B.1).

The object of the precedence provision was the conviction that stability would arise by usage of the earliest name. Regrettably, this has not proved to be the case, as overlooked names repeatedly come to light in rarely opened obscure volumes, and some authors go to considerable efforts to lectotypify and resurrect long-forgotten names. In recognition of this situation, the eligibility of names for conservation and rejection was broadened in the Tokyo Code (see below).

I. Autonyms

An *autonym* is an automatically established name (Arts 19.4, 22.3, 26.3). When a subdivision of a family, genus, species, or infraspecific taxon is introduced, a corresponding name of the same rank, including the type of the higher taxon and repeating its name, is established. Publication of the name *Hymenoscyphus fucatus* var. *badensis* Hengstm. in 1996 automatically established the autonym *fucatus* (W. Phillips) Baral & Hengstm. var. *fucatus* which includes the type of Phillip's epithet; autonyms do not bear any author citation. An autonym has priority over the name or names that created it (Art. 11.6).

J. Synonyms

A *synonym* is a name other than the correct one which has been applied to a single taxon. These are of two distinct types: (1) *homotypic* (nomenclatural, obligate) synonyms based on the same nomenclatural type (i.e., typonyms), for example, combinations into different genera which have the same basionym (see above); and (2) *heterotypic* (taxonomic, facultative) synonyms based on different name-bearing types, a taxonomist having decided that those types come within the range of one species. In some nomenclatural presentations, nomenclatural synonyms are kept together and clearly separated from taxonomic ones (Fig. 3); the symbols ≡ and = are sometimes used to indicate homotypic and heterotypic synonyms, respectively.

K. Pleomorphic Fungi

A major special provision applies to ascomycetous and basidiomycetous fungi, other than those which form lichens, but not to chytrids, oomycetes, slime moulds, or zygomycetes. This enables separate names to be adopted for sexual (meiotic; *teleomorph*) and asexual (mitotic; *anamorph*) morphs of pleomorphic fungi (Art. 59). This Article was substantially revised at the Sydney Congress in 1981 and only current practice is reviewed here. The correct name covering such a fungus in all its stages (*holomorph*) is that of the teleomorph. In order to qualify as a teleomorph name, the original description must include (or not certainly exclude) a description of the sexual state, and that must be represented in the type (Art. 59.2). Even if an anamorph name was published earlier than that for the teleomorph, the latter retains priority as the holomorph name (Art. 59.4).

For example, *Penicillium brefeldianum* B.O. Dodge 1933 was described inclusive of the sexual stage, which is also represented on its type. Although described under a generic name typified by an anamorphic fungus, Dodge's epithet was correctly transferred to a teleomorphic genus as *Eupenicllium brefeldianum* (B.O. Dodge) Stolk & D.B. Scott 1967. As Dodge's name is considered as typified by the sexual part of the type, it was correct for the name *P. dodgei* Pitt 1980 to be introduced to apply to the anamorph alone. A holomorph may have more than one anamorph and these can be separately named (*synanamorphs*). In addition, when describing new species of pleomorphic fungi there is no obstacle to the same epithet being used for the teleomorph and anamorph(s), provided they have different types and no earlier homonyms exist; indeed this is a not infrequent practice.

Some mycologists prefer always to use the teleomorph name, even when just working with the anamorph. For example, Ellis (1971) adopts "*Stemphylium* state of *Pleospora hebarum*" not *Stemphylium botryosum* Wallr. With the advent of molecular methods, anamorphs can now be placed with the appropriate meiotic group, often down to the genus level. This has led to the practice of maintaining a dual nomenclature becoming somewhat superfluous, and deletion of the entire

A   **Neottiospora caricina** (Desm.) Höhnel, Ber Dtsch Bot Ges 37: 158 (1919).
Sphaeria caricina Desm., Ann Sci Nat 6: 246 (1836).
Neottiospora caricum Desm., Ann Sci Nat. 19: 346 (1843).
Darluca caricum (Desm.) Fuckel, Jb Nass Ver Naturk 23/24: 380 (1870).
Hendersonia atramentarius Schröter, Hedwigia 17: 173 (1878).
Spilomyces atramentarius (Schröter) Petrak & Sydow, Beih Feddes Repert 42: 293 (1927).

Samukuta berkeleyi Subram. & Ramakr., J Ind Bot Soc 36: 75 (1957).

B   **Neottiospora caricina** (Desm.) Höhnel
Ber Dtsch Bot Ges 37: 158 (1919). -- *Sphaeria caricina* Desm., Ann Sci Nat 6: 246 (1836). -- *Neottiospora caricum* Desm., Ann Sci Nat. 19: 346 (1843). -- *Darluca caricum* (Desm.) Fuckel, Jb Nass Ver Naturk 23/24: 380 (1870).

Hendersonia atramentarius Schröter, Hedwigia 17: 173 (1878). -- *Spilomyces atramentarius* (Schröter) Petrak & Sydow, Beih Feddes Repert 42: 293 (1927).

Samukuta berkeleyi Subram. & Ramakr., J Ind Bot Soc 36: 75 (1957).

C   **Neottiospora caricina** (Desm.) Höhnel, Ber Dtsch Bot Ges 37: 158 (1919).
      ≡ *Sphaeria caricina* Desm., Ann Sci Nat 6: 246 (1836).
      ≡ *Neottiospora caricum* Desm., Ann Sci Nat. 19: 346 (1843).
      ≡ *Darluca caricum* (Desm.) Fuckel, Jb Nass Ver Naturk 23/24: 380 (1870).
= *Hendersonia atramentarius* Schröter, Hedwigia 17: 173 (1878).
      ≡ *Spilomyces atramentarius* (Schröter) Petrak & Sydow, Beih Feddes Repert 42: 293 (1927).
= *Samukuta berkeleyi* Subram. & Ramakr., J Ind Bot Soc 36: 75 (1957).

Fig. 3A–C. Three examples of the many possible layouts of synonyms: **A** by date of publication. **B** heterotrophic synonyms grouped together in single paragraphs and a homotypic one in different paragraphs. **C** Using symbols and indentation to separate heterotypic (=) and homotypic (≡) synonyms. The names *Sphaeria caricina*, *Neottiospora caricum*, and *Darluca caricum* are homotypic (nomenclatural) synonyms as they are based on the same type, as are *Hendersonia atramentosus* and *Spilomyces atramentarius* based on a second type. The names *Hendersonia atramentarius* and *Samukuta berkeleyi* are heterotypic (taxonomic) synonyms, as each epithet is based on a different type

Article has been advocated (Reynolds and Taylor 1991). While this is a commendable long-term aim, it is currently uncertain how this can be done without unfortunate changes in the names of many well-known agriculturally, industrially, and medically important fungi.

The application of Art. 59 is considered further by Korf (1982), Gams (1993), Reynolds (1993), and Hennebert (1993).

L. Fossil Fungi

Fossil representatives of all groups covered by the Botanical Code are subject to a variety of special provisions. A starting point of 31 Dec. 1820 (Sternberg 1820) applies (Art. 13.1f); an English or a Latin description or diagnosis is necessary for establishment (Art. 36.1), but there is a requirement from 1 Jan. 1912 for an illustration to be provided (Art. 38.1). The type specimen is the material on which the validating illustration was based (Art. 8.4).

Generic names based on particular fossil organs or *form genera* are not assignable to families (Art. 3.3).

Should fossil and nonfossil names compete for the same taxon, the nonfossil name automatically has priority (except for algae) regardless of precedence (Art. 11.7).

M. Special Provisions

The names of lichen-forming fungi strictly refer to the fungal partner (Art. 13.1d) so that the algal or cyanobacterial partners can bear separate names. "Naming" or "identifying" a lichen is strictly a contraction of naming or identifying a lichen-forming fungus. This provision does not prohibit the use of thallus characters in taxonomy even though they may not be expressed when the fungus is isolated and grown in pure culture. Problems arise when some lichenologists with to separate morphologically different thalli (*photomorphs*) which are produced by the same fungus interacting with a cyanobacterium as opposed to a green alga. In some cases, these different forms are placed in the same genus, for example, *Nephroma skottsbergii* with green algae and *N. papillosum* with cyanobacterium (White and James 1988). However, in others, the products from the same fungus are placed in different genera, as in *Sticta felix* (a leafy apotheciate lichen with green algae) and *Dendriscocaulon dendroides* (a shrubby sterile lichen with cyanobacteria). The name to be used for the fungus in such cases follows the regular provisions of the Code, with the first established species epithet having precedence. Heidmarsson et al. (1997) proposed that lichenologists wishing to indicate one of the photomorphs add "/chlor" (green algae only), "/cyan" (cyanobacteria only) or "/chlor + cyan" (green algae and cyanobacteria present).

As the rules relating to pleomorphic fungi do not apply to those forming lichens (see above), specimens which have only conidiomata have equal nomenclatural status to those with ascomata. The name-bearing type for a lichen-forming ascomycete may not have ascomata present on it, and can have conidiomata or be sterile. Many lichens with ascomata, in which the fungal partners are not known, are placed in ascomycete genera on the basis of thallus characters or secondary chemistry; there are no obstacles to this under the Code. A summary of current topics in naming lichens is provided by Jørgensen (1991).

The Tokyo Code makes it clear that slime moulds (*Myxomycota*; zoologically *Mycetozoa*), although now recognized as most appropriately classified in the *Protozoa*, are covered by the Botanical and not the Zoological Code as they have been traditionally studied by botanists (Pre. 7). Similarly, the *Oomycota* (*Oomycetes*) have traditionally been regarded as fungi, and, although now accepted as a phylum of *Chromista* (or *Protoctista*), they continue to be treated as fungi and not algae for the purposes of nomenclature. Weresub (1979) provided a comparison of the names of higher-ranked taxa of *Myxomycota* under the botanical and zoological systems.

Although not of immediate relevance to most mycologists, Appendix 1 of the Code is devoted to the nomenclature of hybrids. This Appendix is not limited in the groups of botanical organisms to which it applies and is available to mycologists who wish to employ it. That a name refers to a hydrid is indicated by either the use of a multiplication sign, ×, or adding the prefix *notho* to the rank (e.g., nothospecies, nothovar., nvar.) (Art. H.1.1); the symbol × can be used either between the names of the two parents in a hybrid formula (Art. H.2.1), in which the female parent is generally cited first (Rec. H.2A), or, if the hybrid is sufficiently frequent or important to merit an independent scientific name, before the start of the name itself (Art. H.3.1). The extent of sexually formed hybrids in fungi is unknown, and they appear to be very rare, but some have been reported amongst rusts and some rulers fungi. The existing rules for plants can be used if it is considered desirable to provide these with separate names.

N. Conservation, Rejection and Supression

The conservation procedure exists "... to avoid disadvantageous changes in the nomenclature of families, genera, and species entailed by the strict application of the rules, and especially the principle of priority [precedence]...." (Art. 14.1). It can be used to protect a name over an earlier one (e.g., *Phaeocollybia* R. Heim 1931 over *Quercella* Velen. 1921), to retain a later homonym (e.g., *Nidularia* Fr. 1817 over *Nidularia* Bull. 1791) or a particular spelling (*conserved orthography*; *orth. cons.*; e.g., *Lactarius* Pers. not *Lactaria*), or to fix a nomenclatural type (*conserved type*; *typ. cons.*; e.g., *Sordaria fimiseda* Ces. & De Not. 1883 with a specimen in Stockholm as the conserved type). Generic names can also be conserved with a generic type specimen rather than a species name (Art. 10.4; e.g., *Physconia* Poelt 1965 on a specimen preserved in Munich).

The conservation and rejection facility was for many years available only for family and generic names, but has been extended considerably in the last two editions of the Code. At the Sydney Congress in 1981 it was broadened to encompass species names, principally those of major economic importance, and at the Tokyo Congress in 1993, to any change seen as disadvantageous. The protected name is said to be *conserved* (*nomen conservandum*), and is to be retained over ones listed in Appendices to the Code and any names based on the same nomenclatural type whether listed or not (Art. 14.4). Conserved names are not protected against earlier names based on different types (Art. 14.5), and listed rejected names can be taken up in alternative taxonomies provided that they do not compete nomenclaturally with the conserved name (Art. 14.6; e.g., *Schizothecium* Corda 1838 is listed as rejected against *Podospora* Ces. 1856, but Corda's name can still be used when both genera are accepted as distinct).

Where a name has been misapplied widely and persistently to a species other than that represented by its name-bearing type, the Code states that the name is not to be used in a different sense until the conservation and rejection procedures have been tried (Art. 57.1).

Names in specified ranks, or in some cases all names, included in particular publications can be suppressed (Art. 32.8). Names in the suppressed works (*opera utique oppressa*) listed in an Appendix of the Code are treated as not published and so are not to be considered further in nomenclature. The Appendix in the Tokyo Code includes several works which include fungal names (e.g., Ehrhart 1780–85; generic names) as well as some which are entirely mycological (e.g., Secretan 1833; species and infraspecific taxa).

It is responsible taxonomic practice to always explore whether conservation or rejection can be achieved, especially before changing familiar generic names. The steps in this procedure (Table 3) are lengthy, rarely being completed in less than 3 years; in some instances where debates have been complex, the procedure has taken as much as 12 years. The steps required to add publications to the list of suppressed works are the same as for conservation or rejection, and guidelines for how to proceed have been published by Nicolson and Greuter (1994).

O. Revising the Code

Any botanist wishing to modify the Code must publish a proposal in *Taxon*, the journal of the International Association for Plant Taxonomy (IAPT). About 6 months before a Congress all proposals are collected together in a Synopsis in *Taxon* arranged by Article and with comments on their implications by those elected to serve as Rapporteurs at that Congress (Greuter and McNeill 1993). All members of the IAPT then participate in a Mail Vote, the results of which are made available to the Nomenclature Session of the Congress. The proposals are debated openly at that week-long Session, voted on by all registered for the Session (with additional votes allocated in advance to major institutions), and thereby accepted (often with modifications), rejected, or referred to an Editorial Committee appointed by the Session; a two-thirds majority is normally necessary for acceptance of any change. Special Committees can be established to consider particular problems and are instructed to report to the next Congress. Standing Committees are also appointed to comment on conservation proposals or other matters relating to particular groups, the Committee for Fungi (formerly the Committee for Fungi and Lichens) considers all organisms traditionally studied by mycologists, including lichen-forming fungi, oomycetes, and slime moulds.

Changes come into force immediately they are approved by the Congress, i.e., before the new Code is printed, unless an operative date is specified. Results of votes taken at the Session are, however, published in the next issue of *Taxon*. Of 321 formal proposals made to the Tokyo Congress, 147 were accepted or referred to a Committee. The Session is now taped and an edited transcript or Report published (e.g., Greuter et al. 1994b); these often entertaining documents provide important background to the intent of changes.

VI. Other Codes and the BioCode

Separate Codes of Nomenclature govern the names of bacteria and groups traditionally studied by zoologists. Only the major differences in those

Table 3. The steps in the conservation and rejection of fungal names or the suppression of names in particular publications

1. Proposal published in *Taxon*
2. Considered by appropriate Standing Committee (i.e., Committee for Fungi), who exchange comments
3. Voted on by Standing Committee (two-thirds majority usually required; an abstention taken as "no")
4. Report of Standing Committee published in *Taxon* including results of their voting and reasons for decisions
5. General Committee considers report of the Standing Committee and votes to accept or reject each of its decisions
6. General Committee report published in *Taxon*. Rejection or retention is then authorized, subject to the next Congress
7. International Botanical Congress accepts or rejects General Committee decisions
8. Conserved and rejected names and suppressed works listed in Appendixes to the next edition of the Code

Codes are highlighted here. For a general overview see Jeffrey (1989), and for more detailed comparisons Hawksworth et al. (1994), Ride (1988), and Ride and Younès (1986). The Code to be applied to any particular name is determined by the taxonomic position of its type, and not by where it was placed by its original author. *Pneumocystis* P. Delanoë & Delanoë 1912 was described as a protozoan, but is now known to be a fungus and so has to meet the requirements of the Botanical Code for establishment; as no Latin diagnosis was then required under the Botanical Code, it can continue in use unimpeded.

Bacteria were historically treated under the Botanical Code, with the first independent Code being issued in 1948 under the authority of the 1947 International Congress of Microbiology. Virus names were covered in the 1958 Code but excluded from recent editions. The most recent edition of the Bacteriological Code (Sneath 1992) includes a review of its history, which is also analysed by Sneath (in Ride and Younès 1986). Practical short guides to its use are provided by Austin and Priest (1986) and Bradbury (1983).

Under the Bacteriological Code from 1 Jan. 1980, effective publication is only possible in the *International Journal of Systematic Bacteriology*, does not require diagnoses in Latin, requires that a living culture be designated as the nomenclatural type, does not accept names if they are later homonyms of other microorganisms (algae, fungi, protozoa, etc.), recognizes only the single rank of subspecies below species, and has a unique starting point date of 1 Jan. 1980. This Code is accompanied by the *Approved Lists of Bacterial Names* (Skerman et al. 1980, 1989); names not in these lists published prior to 1980 are declared as not established (invalid). A rank of pathovar. (pv.) is approved by the International Society for Plant Pathology for use in a manner comparable to the special forms of fungi (Young et al. 1996).

The principle differences of the *International Code of Zoological Nomenclature* (Ride et al. 1985) from the Botanical Code are: the starting point date of 1 Jan. 1758; no requirement for Latin diagnoses; tautonyms are permitted (e.g., *Troglodytes troglodytes*); only one infraspecific rank (subspecies) is used and in a trinomial form (e.g., *Mus musculus domesticus*); a type relates to a *nominal taxon* rather than a name, so that misapplied names can be more easily retained; and precedence is not confined to names in the same rank, but to groups of names said to have *coordinate status* (e.g., an earlier subspecies name has precedence over a later species name). Some of the terminology used also varies from the Botanical Code, and this can be confusing. Major changes are expected in the next edition of this Code, particularly in relation to reducing name changes for nomenclatural reasons; that edition is due out in late 1999. A historical account of the evolution of the Zoological Code is available (Melville 1995).

A Code independent of the Botanical Code, the *International Code of Nomenclature for Cultivated Plants* (Trehane et al. 1995), governs the names of cultivated varieties of plants (cultivars) and graft-chimaeras. A cultivar (cv.) is a cultivated plant "... selected for a particular atribute or combination of attributes, and that is clearly distinct, uniform and stable in its characteristics and that, when propagated by appropriate means, retains those characteristics" (Art. 2.1[1]). Cultivar names are not italicized and follow the Latinized scientific names. The naming of graft-chimaeras, plants "composed of tissues from two or more different plants..." (Art. 5.1), mirrors that of hybrids under the Botanical Code (see above), but employs the symbol + (Art. 20.2); the epithet names used must be different from any hybrid one that might exist. Mycologists working with domesticated strains or selected variants of cultivated mushrooms may wish to mirror the systems for plants, and mechanical hybrids in lichens, thalli formed by fungal partners from more than one species, could be referred to by the graft notation were that ever felt to be desirable (Hawksworth 1988b).

Since 1985, the International Union of Biological Sciences (IUBS) has been striving to harmonize the different Codes of biological nomenclature, and progess up to 1995 is summarized by Hawksworth (1995). A unified series of terms has been proposed (Table 1), appropriate parallel provisions are being proposed for introduction into the different Codes; an International Committee on Bionomenclature has been established jointly with IUMS, and a *Draft BioCode: the Prospective International Rules for the Scientific*

---

[1] The Article numbers in this paragraph are those in the International Code of Nomenclature of Cultivated Plants (Trehane et al. 1995).

Names of Organisms has been prepared for wider discussion amongst the scientific community (Greuter et al. 1996). The key features of the *BioCode* from the botanical standpoint are summarized by Greuter and Nicolson (1996), and its concepts and provisions are currently under debate (Greuter 1996; Hawksworth 1997; Orchard et al. 1996; Reveal 1996). If in due course the BioCode is adopted by the relevant mandated bodies, it is anticipated that its provisions will be introduced gradually as lists of names in use are compiled, vetted, and endorsed. The next steps in this initiative were considered by IUBS at its General Assembly in Taipei in Nov. 1997.

V. Indices of Fungal Names

Around 300000 scientific names have been employed for species of fungi to date, although the number of accepted fungi is only about 72000 species (Hawksworth et al. 1995). This discrepancy is mainly allowed for by synonymy, but many species described long ago have never been reassessed by later mycologists. Before embarking on the Nomenclatural Filter (Fig. 2), it is necessary to compile a list of names from those which have been published which are known to or could conceivably apply to the taxon for which the correct name is being sought. The major catalogues of fungal names are listed in Table 4. The biannual *Index of Fungi* provides an ongoing catalogue produced by scanning world literature, and includes both names which have been recently published and ones omitted from earlier compilations which have been rediscovered.

A major initiative is currently underway to catalogue all groups of organisms through a series of interlinked global master species databases. Named SPECIES 2000 and sponsored by IUBS and CODATA (Committee on Data for Science and Technology), this ambitious programme includes a name-finder tool, which will be a master list of all generic and species names (Bisby 1994; Fig. 4). The name-finder tool already includes all the species names included in the fungal catalogues listed in Table 4, and a pilot verson can be accessed through the worldwide web. The names and where they are catalogued are included, but not the bibliographic and author citations, for which the original catalogue still needs to be consulted. With the prospects of this system and the probable requirement for registration of newly proposed names from 1 Jan. 2000 (see above), access to the existing body of names is becoming easier than ever before.

In coining names for newly recognized genera and species, and to confirm whether a name it is desired to take up has not previously been used, it is necessary to check candidate names to avoid the creation or adoption of later homonyms that would subsequently have to be rejected. For generic names, it is necessary to check lists of names in all botanical groups covered by the Code and not only those concerned with fungi. In addition, it is now good practice to endeavour to avoid names used in zoology because of the problems this creates in electronic information retrieval systems.

VI. Responsibility and the Stability of Names

A major feature of the Tokyo Code was the introduction of a series of proposals which together greatly reduce the need to change names for nontaxonomic reasons. This marked a major change in direction by opening up the possibilities for conservation and rejection of names. It is true to say that, provided a name is well-established, there are mechanisms that can be used to avoid the taking up of names only on the basis of precedence or other nomenclatural points. Further, the Nomenclature Section of the Congress also passed a resolution urging taxonomists "to avoid displacing well-established names for purely nomenclatural reasons, whether by change in their application or by resurrection of long-forgotten names" while procedures to improve stability further are under discussion (Greuter et al. 1994a). The international taxonomic community has effected a major change in direction towards improved stability (Hawksworth 1993; Greuter and Nicolson 1993), although some disputed this (Brummit 1994).

As stressed above, increased stability by reducing taxonomic changes arising from new data is not in the interests of science, provided that those proposed are soundly based. Regrettably, some taxonomists are too ready to shuffle names in the absence of conclusive new data, or use names without checking the identity of the types; the discovery that names have been misapplied is

Table 4. Indexes of fungal names, including selected works which cover all groups with which the International Code of Botanical Nomenclature is concerned

Group	Indexes

All groups
Farr E, Leussink JA, Stafleu FA, eds (1979) Index nominum genericorum (Plantarum), 3 vols. (Regnum Vegetabile Nos. 100–102) Bohn, Scheltema & Holkema, Utrecht
Farr E, Leussink JA, Zijlstra G, eds (1980) Index nominum genericorum (Plantarum). Supplementum 1. (Regnum vegetabile No. 113) Bohn, Scheltema & Holkema, Utrecht
Greuter W, Brummitt RK, Farr E, Kilian N, Kirk PM, Silva PC (1993) NCU-3: names in current use for extant plant genera. (Regnum Vegetabile No. 129) Koeltz Scientific Books, Königstein
Pfeiffer LKG (1871–74) Nomenclator Botanicus. 2 vols. Fischer, Kassel.
Steudel E (1821–24) Nomenclator Botanicus, 2 vols. Cotta, Stuttgart (Also 2nd edn, 1840–41)

Fungi
Deighton FC (1969) A Supplement to Petrak's Lists 1920–1939 (Index of Fungi Supplement) Commonwealth Mycological Institute, Kew
Hawksworth DL (1972) Lichens 1961–1969. (Index of Fungi Supplement) Commonwealth Mycological Institute, Kew
Hawksworth DL, David JC (1989) Family names. (Index of Fungi Supplement) CAB International Mycological Institute, Kew
Hawksworth DL, Kirk PM, Sutton BC, Pegler DM (1995) Ainsworth & Bisby's Dictionary of the Fungi, 8th edn. CAB International, Wallingford
Index of Fungi (1940 on). (Biannual) CAB International, Wallingford
Kirk PM (1985) Saccardo's Omissions. (Index of Fungi Supplement) Commonwealth Mycological Institute, Kew
Lamb IM (1963) Index nominum lichenum inter annos 1932 et 1960 divulgatorum. Ronald Press, New York
Petrak F (1930–44) Verzeichnis der neuen Arten, Varietäten, Formen, Namen und wichtigsten Synonyme. Just's Bot Jarhb 48(3), 49(2), 56(2), 57(2), 58(1), 60(1), 63(2)
Petrak F (1950) Index of Fungi 1936–1939. Commonwealth Mycological Institute, Kew
Saccardo PA (1882–1931, 1972) Sylloge fungorum, 26 vols. Saccardo, Padua and Johnson, New York
Streinz WM (1892) Nomenclator fungorum. Gorischek, Vienna
Zahlbruckner A (1921–40) Catalogus lichenum universalis, 10 vols. Borntraeger, Leipzig and Johnson, New York

Fossils (of all "botanical" groups)
Andrews HN (1970) Index of Generic Names of Fossil Plants, 1820–1965. (United States Geological Survey Bulletin No 1300) US Geological Survey, Washington, DC
Blazer AM (1975) Index of Generic Names of Fossil Plants, 1966–1973. (United States Geological Survey Bulletin No. 1396) US Geological Survey, Washington, DC
Holmes PL, Boulter MC, Woolliams PD, eds (1991) The Plant Fossil Record Database. International Organization of Palaeobotany, London
Jansonius J, Hills LV (1976) Genera file of fossil spores. Departmet of Geology, University of Calgary, Calgary
Jongmans W, ed (1913–1988) Fossilium Catalogus, Plantae, 93 vols. Junk, The Hague
Meyen SV (1990) Catalogue of fossil plants. Internat Org Palaeobot Circ 9:1–56
Watt AD (1982) Index of generic names of fossil plants 1974–1978. (United States Geological Survey Bulletin No. 1517) US Geological Survey, Washington, DC

a major cause of name instability. Improvements in the standard of taxonomic practice, and increased responsibility amongst taxonomists, constitute the best long-term prospects for reducing instability due to taxonomic changes. To this end, a code of practice for systematic mycologists has been devised by the International Commission on the Taxonomy of Fungi (ICTF; Sigler and Hawksworth 1987).

This problem of the scale of names and improved stability was solved for bacteria through the *Approved Lists*, and in 1988 the decision was taken to commence compilation of *Lists of Names in Current Use* (NCUs) in all groups covered by the Botanical Code, including fungi, and for which protected status is to be sought from Botanical Congresses (Hawksworth 1988a, 1991a; Hawksworth and Greuter 1989). NCU lists are a nomenclatural device and drawn up to allow for current alternative taxonomies, but it is envisaged that listed names will eventually be protected against earlier names and honomyms, but that unlisted names will not be disestablished, but remain available for use provided that they do not compete with a listed name. Name-bearing types are also listed, thus enabling applications as well as names to be fixed.

Lists of generic names of extant (not fossil) botanical genera, including fungi, in current use (Greuter et al. 1993), family names of vascular

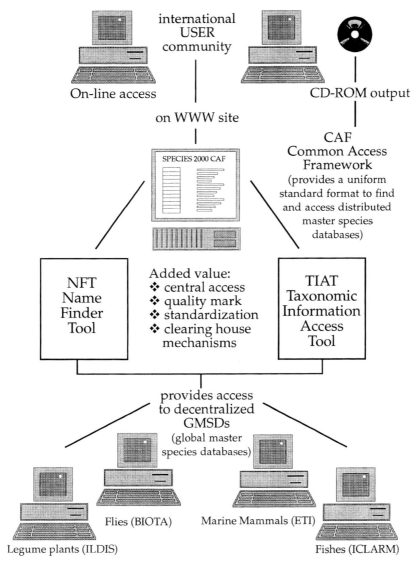

Fig. 4. Structure being developed for the SPECIES 2000 programme to link global taxonomic databases

plants, bryophytes, and fungi (Hoogland et al. 1993), and selected families including *Cladoniaceae* and *Trichocomaceae* (Pitt et al. 1993) were prepared for the Tokyo Congress in 1993, but the proposals were defeated by a small margin. However, a special resolution was passed in relation to the list of names in *Trichocomaceae*, developed by the International Commission on *Penicillium* and *Aspergillus*, which urged taxonomists not to adopt names that would compete with or change the application of any names on the list (Greuter et al. 1994a, b). This was encouraging for such a novel idea, and the matter will be debated further at the International Botanical Congress in St Louis in 1999.

A wide range of approaches to the improvement of stability in names, including detailed information on the NCU and Registration proposals, is to be found in the papers included in Hawksworth (1991b).

References

Ainsworth GC (1976) Introduction to the history of mycology. Cambridge University Press, Cambridge

Allen WS (1978) Vox latina. The pronunciation of classical Latin. Cambridge University Press, Cambridge

Austin B, Priest F (1986) Modern bacterial taxonomy. van Nostrand Reinhold, Wokingham

Benjamin DR (1995) Mushrooms: poisons and panaceas. WH Freeman, New York

Bisby FA (1994) Global master species databases and biodiversity. Biol Int 29:33–40

Bisby FA, Hawksworth DL (1991) What must be done to save systematics? In: Hawksworth DL (ed) Improving the stability of names: needs and options. (Regum Vegetabile No 123) Koeltz Scientific Books, Königstein, pp 323–336

Blunt W (1971) The compleat naturalist. A life of Linnaeus. Collins, London

Bradbury JF (1983) The new bacterial nomenclature – what to do. Phytopathology 73:1349–1350

Brummit RK (1994) What did we think we were voting for at Tokyo? Linnean 10(2):13–15

Brummit RK, Powell CE (1992) Authors of plant names. Royal Botanic Gardens, Kew

Burdet HM (1995) Code international de la nomenclature botanique (Code de Tokyo). Boissiera 49:1–185

Cash EK (1965) A mycological English-Latin glossary. Hafner, New York

Demoulin V, Hawksworth DL, Korf RP, Pouzar Z (1981) A solution to the starting point problem in the nomenclature of fungi. Taxon 30:52–63

Ehrhart JF (1780–85) Phytophylacium ehrhartianum. 10 decades. Hannover, privately printed

Ellis MB (1971) Dematiaceous Hyphomycetes. CAB International, Wallingford, UK

Ergorova TV, Gladkova VK (1996) International code of botanical nomenclature (Tokyo code). Russian translation. Komarov Botanical Institute, St Petersburg

Fries EM (1821–32) Systema mycologicum. 3 vols. Berling, Lund

Gams W (1984) An index to fungal names and epithets sanctioned by Persoon and Fries. Mycotaxon 19: 219–270

Gams W (1993) Anamorphic species and nomenclature. In: Reynolds DR, Taylor JW (eds) The fungal holomorph: mitotic, meiotic and pleomorphic speciation in fungal systematics. CAB International, Wallingford, pp 293–304

Gams W, Kuyper TW (1995) Elias Fries as sanctioning authority for fungal names. Symb Bot Ups 30(3):25–31

Garnock-Jones OJ, Webb CJ (1996) The requirement to cite authors of plant names in botanical journals. Taxon 45:285–286

Greuter W (1996) On a new BioCode, harmony, and expediency. Taxon 45:291–294

Greuter W, Hiepko P (1995) Internationaler Code der Botanischen Nomenclatur ins Deutsche übertragen. Englera 15:i–xxi, 1–150

Greuter W, McNeill J (1993) Synopsis of proposals on botanical nomenclature – Tokyo 1993. A review of the proposals concerning the International Code of botanical nomenclature submitted to the XV International Botanical Congress. Taxon 42:191–271

Greuter W, Nicolson DH (1993) On the threshold to a new nomenclature? Taxon 42:925–927

Greuter W, Nicolson DH (1996) Introductory comments on the draft Biocode, from a botanical point of view. Taxon 45:343–348

Greuter W, Brummitt RK, Farr E, Kilian N, Kirk PM, Silva PC (1993) NCU-3: names in current use for extant plant genera. (Regnum Vegetabile No 129) Koeltz Scientific Books, Königstein

Greuter W, Barrie FR, Burdet HM, Chaloner WG, Demoulin V, Hawksworth D, Jørgensen PM, Nicolson DH, Silva PC, Trehane P (eds) (1994a) International code of botanical nomenclature (Tokyo code). (Regnum Vegetabile No 131) Koeltz Scientific Books, Königstein

Greuter W, McNeill J, Barrie FR (1994b) Report on botanical nomenclature – Yokohama 1993. Englera 14:1–265

Greuter W, Hawksworth DL, McNeill J, Mayo MA, Minelli A, Sneath PHA, Tindall BJ, Trehane P, Tubbs P (1996) Draft BioCode: the prospective international rules for the scientific names of organisms. Taxon 45:349–372; Bull Zool Nomencl 53:148–166; International Union of Biological Sciences, Paris

Hawksworth DL (1974) Mycologist's handbook. An introduction to the principles of taxonomy and nomenclature in the fungi and lichens. Commonwealth Mycological Institute, Kew

Hawksworth DL (1986) Changes which would arise from the extension of sanctioning to the names of lichen-forming fungi. Taxon 35:787–793

Hawksworth DL (1988a) Improved stability for biological nomenclature. Nature 334:301

Hawksworth DL (1988b) Naming mechanical and sexual hybrids in lichen-forming fungi. Int Lichen Newsl 21:59–61

Hawksworth DL (1991a) Lists of names in current use: a new initiative to address a continuing problem. Mycotaxon 40:445–458

Hawksworth DL (ed) (1991b) Improving the stability of names: needs and options. (Regnum Vegetabile No 123) Koeltz Scientific Books, Königstein

Hawksworth DL (1993) Name changes for purely nomenclatural reasons are now avoidable. Syst Ascom 12:1–6

Hawksworth DL (1994) A draft glossary of terms used in bionomenclature. (IUBS Monograph No 9) International Union of Biological Sciences, Paris

Hawksworth DL (1995) Steps along the road to a harmonized bionomenclature. Taxon 44:447–456

Hawksworth DL (ed) (1997) The new bionomenclature: the BioCode debate. Biol Int Spec Issue 34:1–103

Hawksworth DL, Greuter W (1989) Improvement of stability in biological nomenclature. Biol Int 19:5–11

Hawksworth DL, McNeill J, Sneath PHA, Trehane P, Tubbs PK (1994) Towards a harmonized bionomenclature for life on earth. Biol Int Spec Issue 30:1–44

Hawksworth DL, Kirk PM, Sutton BC, Pegler ND (1995) Ainsworth and Bisby's dictionary of the Fungi, 8th edn. CAB International, Wallingford

Heidmarsson S, Mattsson J-E, Moberg R, Nordin A, Santesson R, Tibell L (1997) Classification of lichen photomorphs. Taxon 46:519–520

Hennebert GL (1993) Towards a natural classification of the fungi. In: Reynolds DR, Taylor JW (eds) The fungal holomorph: mitotic, meiotic and pleomorphic speciation in fungal systematics. CAB International, Wallingford, UK, pp 283–294

Holmgren PK, Holmgren NH, Barnett LC (1990) Index herbariorum. Part 1. The herbaria of the world, 8th edn. (Regnum Vegetabile No 120) New York Botanical Garden, New York

Hoogland RD, Reveal JL, Crosby MJ, Grolle R, Zijlstra G, David JC (1993) NCU-1: family names in current use for vascular plants, bryophytes and fungi. (Regnum Vegetabile No 126) Koeltz Scientific Books, Königstein

Jeffrey C (1989) Biological nomenclature, 3rd edn. Edward Arnold, London

Jong S-C, Birmingham JM (1991) Problems in standardization of fungal nomenclature. Adv Cult Coll 1:39–61

Jørgensen PM (1991) Difficulties in lichen nomenclature. Mycotaxon 40:497–501

Jørgensen PM, James PW, Jarvis CE (1994) Linnean lichen names and their typification. Bot J Linn Soc 115:261–405

Kirk PM, Ansell AE (1992) Authors of fungal names. (Index of Fungi Supplement) CAB International, Wallingford, UK

Korf RP (1982) Mycological and lichenological implications of changes in the code of nomenclature enacted in 1981. Mycotaxon 14:476–490

Korf RP (1983) Sanctioned epithets, sanctioned names, and cardinal principles in ": Pers." and ": Fr." citations. Mycotaxon 16:341–352

Korf RP (1996) Simplified author citations for fungi and some old traps and new complications. Mycologia 88:146–150

Korf RP, Kohn LM (1980) Later starting point blues. II. *Dumontinia tuberosa*: the thorny thickets of synonymy and some examples of nomenclatorialism. Mycotaxon 11:381–395

Linnaeus C (1753) Species plantarum. 2 vols. Laurentius Salvi, Stockholm

Marhold K (1995) Medzinarodny kód botanickej nomenklatúry (Tokijskjy kód). Zpravy Ceska Bot Spol. Bull Slovensk Bot Spol 1995(1):i–xix, 1–110

Mayr E (1969) The biological meaning of species. Biol J Linn Soc 1:311–320

McNeill J, Greuter W (1986) Botanical nomenclature. In: Ride WDL, Younès T (eds) Biological nomenclature today (IUBS Monograph series No 2) IRL Press, Eynsham, Oxon, pp 3–25

McVaugh R, Ross R, Stafleu FA (1968) An annotated glossary of botanical nomenclature. (Regnum Vegetabile No 56) International Bureau for Plant Taxonomy and Nomenclature, Utrecht)

Melville RV (1995) Towards stability in the names of animals: a history of the International Commission on Zoological Nomenclature 1895–1995. International Trust for Zoological Nomenclature, London

Micheli PA (1729) Nova plantarum genera iuxta Tournefortii methodum disposita. Florence. (Reprinted 1976, Richmond Publishing, Richmond)

Naqshi AR (1993) An introduction to botanical nomenclature. Scientific Publishers, Jodhpur

Nicolson DH (1980) Key to identification of effectively/ineffectively published material. Taxon 29:485–488

Nicolson DH (1991) A history of botanical nomenclature. Ann Mo Bot Gard 78:33–56

Nicolson DH, Greuter W (1994) Guidelines for proposals to conserve or reject. Taxon 43:109–112

Ohashi H (1997) International code of botanical nomenclature (Tokyo code). Japanese edn. Tsumura Laboratory, Ibaragi-ken

Orchard AE, Anderson WR, Gilbert MG, Sebsebe D, Stearn WT, Voss EG (1996) Harmonized bionomenclature – a recipe for disharmony. Taxon 45:287–290

Perry G (1991) Nomenclatural stability and the botanical code: a historical review. In: Hawksworth DL (ed) Improving the stability of names: needs and options. (Regnum Vegetabile No 123) Koeltz Scientific Books, pp 79–93

Persoon CH (1801) Synposis methodica fungorum. H Dieterich, Göttingen

Petersen RH (1975) Specific and infraspecific names for fungi used in 1821. Part I. Introduction, A & B. Mycotaxon 1:149–188

Pitt JI, Samson RA, Ahti T, Farjon A, Landolt E (1993) NCU-2: Names in current use in the families Trichocomaceae, Cladoniaceae, Pinaceae, and Lemnaceae. (Regnum Vegetabile No 128) Koeltz Scientific Books, Köuigstein

Ramsbottom J (1938) Linnaeus and the species concept. Proc Linn Soc Lond 150:192–219

Reveal J (ed) (1996) Biological nomenclature in the 21st century. www.life.umd.edu/bees/96sym.html

Reynolds DR (1993) The fungal holomorph: an overview. In: Reynolds DR, Taylor JW (eds) The fungal holomorph: mitotic, meiotic and pleomorphic speciation in fungal systematics. CAB International, Wallingford, UK, pp 15–25

Reynolds DR, Taylor JW (1991) Nucelic acids and nomenclature: name stability under Article 59. In: Hawksworth DL (ed) Improving the stability of names: needs and options. (Regnum Vegetabile No 123) Koeltz Scientific Books, Königstein, pp 171–177

Rickett HW (1959) The status of botanical nomenclature. Syst Zool 8:22–26

Ride WDL (1988) Towards a unified system of biological nomenclature. In: Hawksworth DL (ed) Prospects in systematics (Systematics Association Spec Vol 36) Clarendon Press, Oxford, pp 332–353

Ride WDL, Younès T (eds) (1986) Biological nomenclature today. (IUBS Monograph series No 2) IRL Press, Eynsham, Oxon

Ride WDL, Sabrosky CW, Bernardi G, Melville RV (1985) International code of zoological nomenclature adopted by the XX general assemby of the International Union of Biological Sciences. 3rd edn. International Trust for Zoological Nomenclature, London

Secretan L (1833) Mycographie Suisse. 3 vols. PA Bonnant, Geneva

Sigler L, Hawksworth DL (1987) International Commission on the Taxonomy of Fungi. Code of practice for systematic mycologists. Microbiol Sci 4:83–86; Mycopath 99:83–86; Mycologist 21:101–105; Acta Mycol Sin 8 (1989):154–159

Sivarajan VV (1984) Introduction to the principles of plant taxonomy. Oxford and IBH Publishing, New Delhi

Skerman VBD, McGowan V, Sneath PHA (1980) Approved lists of bacterial names. Int J Syst Bacteriol 30:225–420

Skerman VBD, McGowan V, Sneath PHA (1989) Approved lists of bacterial names. Amended edn. American Society for Microbiology, Washington, DC

Sneath PHA (ed) (1992) International code of nomenclature of bacteria, 1990 Revision. American Society for Microbiology, Washington DC

Stafleu FA (1971) Linnaeus and the Linneans. The spreading of their ideas in systematic botany, 1735–1789. (Regnum Vegetabile No 79) A. Oosthoek's Uitgeversmaatschappi, Utrecht

Stafleu FA, Cowan RS (1976–88) Taxonomic literature. A guide to botanical publications and collections with dates, commentaries and types. 7 vols. (Regnum Vegetabile Nos. 94, 98, 105, 110, 112, 115, 116) Bohn, Scheltema & Holkema, Utrecht

Stearn WT (1957) An introduction to the *Species plantarum* and cognate botanical works of Carl Linnaeus. In:

Linneaus C, Species plantarum (reprint), vol 1. Ray Society, London, pp 1–176

Stearn WT (1959) The background of Linnaeus's contributions to the nomenclature and methods of systematic biology. Syst Zool 8:4–22

Stearn WT (1992) Botanical Latin, 4th edn. David & Charles, Newton Abbot

Sternberg KM (1820–38) Versuch einer geognostisch-botanischen Darstellung der Flora der Vorwelt, 8 vols. CE Brenck, Regensburg

Sugawara H, Ma J, Miyazaki S, Shimura J, Takishima Y (1993) World directory of collections of cultures of microorganisms: bacteria, fungi and yeasts, 4th edn. World Federation for Culture Collection's World Data Center on Microorganisms, Saitama

Trehane P, Brickell CD, Baum BR, Hetterscheid WLA, Leslie AC, McNeill J, Spongberg SA, Vrugtman F (eds) (1995) International code of nomenclature for cultivated plants – 1995. (Regnum Vegetabile No 133) Quarterjack Publishing, Wimborne

Weresub LK (1979) Mycological nomenclature: reflections on its future in the light of its past. Sydowia Beih 8:16–431

White FJ, James PW (1988) Studies on the genus *Nephroma* II. The southern temperate species. Lichenologist 20:103–166

Young JM, Saddler GS, Takikawa Y, De Boer SH, Vauterin L, Gardin L, Gvozdyak RI, Stead DE (1996) Names of plant pathogenic bacteria 1864–1995. Rev Plant Pathol 75:721–763

Postscript: The original text of this chapter was completed in August 1997, and updated in March 1998. The XVIth International Botanical Congress met in St Louis in August 1999 and the St Louis Code is due to be published in June 2000. The St Louis Congress rejected the concept of registration of newly proposed names, reversing the decision of the Tokyo Congress, and despite a successful trial; the Congress also failed to adopt proposals for the protection of lists of names in current use. Amongst other changes made in the new Code are clarifications of typification procedures, including permanently preserved cultures, encouragement not to introduce names for anamorphs of pleomorphic fungi unnecessarily, a recommendation to avoid introducing generic names which already exist in bacteriology or zoology. The new edition of the Code should be consulted when describing or working on the nomenclature of fungi.

7 Cultivation and Preservation of Fungi in Culture

Shung-Chang Jong and Jeannette M. Birmingham

CONTENTS

I.	Introduction	193
II.	Cultivation	193
A.	Culture Media	194
B.	Culture Conditions	195
C.	Control of Mite Infestations	195
III.	Preservation	196
A.	Periodic Transfer	196
B.	Sterile Mineral Oil	196
C.	Water Storage	197
D.	Sterile Soil or Sand	197
E.	Silica Gel	197
F.	Drying on Organic Substrates or Paper	198
G.	L-Drying	198
H.	Mechanical Freezers	198
I.	Freeze-Drying	199
J.	Cryogenic Storage	200
	References	201

I. Introduction

Pure culture methodology allows an investigator to detect, isolate, identify, and quantify numbers and kinds of fungi from a wide array of environments and to define the nutritional, chemical, and environmental requirements for their growth and metabolism. Studies with pure cultures not only enhance understanding of any natural ecosystem within infected hosts or in the environment, but also help in the determination of molecular architecture and screening for novel activity/compounds. Pure culture methodology is the foundation of fundamental and applied research, as well as the commercial exploitation of fungi.

II. Cultivation

Pure cultures of fungi do not occur in nature. Many techniques have been devised to isolate fungi of interest from bacteria, contaminating fungi, or mites (Jong 1981; Stevens 1981; Jong and Atkins 1985). When the fungus produces abundant spores, single-spore isolation and dilution plate methods are commonly used.

Bacteria-free cultures can be obtained by growing fungi on ultraviolet-irradiated media, which inhibits the growth of bacteria with little or no effect on the fungi. The ability of many fungi to penetrate an agar medium may also be utilized to separate fungi from bacteria. A piece of agar with contaminated mycelial growth is removed with a sterile scalpel, inverted, and placed in a sterile petri dish. After a period of incubation, mycelia growing through the agar from the undersurface can be cut away to obtain a pure culture. Another technique places a small glass cylinder (van Tieghem cell) with three glass beads (0.3–0.5 mm in diameter) fused to one end in a petri dish so that it rests on the beads. Melted agar is poured into the dish until it fills the cylinder. The contaminated culture is inoculated inside the cylinder. As growth proceeds, the target fungus will grow underneath the free edges of the cylinder, while the bacteria remain confined within the cell. The hyphae outside the cylinder provide the inoculum for a pure culture. Glass beads may be replaced with small pellets of modeling clay. The clay secures the cylinder to the bottom of the plate so that the plate can be inverted during incubation. In another approach, a mixed culture is inoculated at the center of an agar plate and a sterile coverslip then pressed down over that region. Many fungi will pass through the agar to reach the air.

Selective media are especially useful to isolate fungi from mixed populations. They permit the growth of the desired fungus while suppressing the other microorganisms. Selective media are generally more complex than those used for culture after isolation. In many cases, the compounds included as the sole sources of carbon or nitrogen allow only a few types of microorganisms to grow. Toxic compounds or antimicrobics are frequently incorporated into media to suppress the growth of

American Type Culture Collection, 10801 University Boulevard, Manassas, Virginia 20110-2209, USA

contaminants while permitting the cultivation of the target fungus. Bile salts, selenite, tetrathionate, tellurite, azide, phenylethanol, sodium lauryl sulfate, high sodium chloride concentrations, and various dyes, such as eosin, crystal violet, and methylene blue, can be used as selective toxic chemicals. Antimicrobial agents include ampicillin, chloramphenicol, colistin, cycloheximide, gentamicin, kanamycin, nalidixic acid, sulfadiazine, and vancomycin. Various combinations of antimicrobics are effective in suppressing whole classes of microorganisms, such as enteric bacteria.

A. Culture Media

Because of their diversity and numerous metabolic pathways, there is no standard medium for all fungi. An understanding of the basic concepts of medium design and the physical conditions that limit fungal growth provides a rational approach to devising media and culture conditions. Useful references on fungal physiology include books by Hawker (1950), Lilly and Barnett (1951), Cochrane (1958), and Griffin (1994).

Culture media can range from simple formulations of commercially available media to those that require very meticulous and skillful assembly to prepare. The essential growth substances needed by fungi include simple sugars, such as glucose and sucrose, mineral salts, air, and water. Most fungi prefer ammonium to nitrate as a nitrogen source, while others require organic nitrogen in forms such as asparagine or amino acids. Many fungi are able to grow and produce spores on plain agar media made with tap water; others require specific growth factors, such as vitamins (thiamin, biotin, inositol, pyridoxine, nicotinic acid, and pantothenic acid), amino acids, fatty acids, trace metals, and blood components. Most often mixtures of growth factors are used. Acid hydrolysates of casein are common sources of amino acids. Extracts of yeast cells are also sources of amino acids and vitamins. Many media, particularly those employed in the clinical laboratory, contain blood or blood components for more fastidious microorganisms. Heme contained in hemoglobin and nicotinamide adenine dinucleotide are often supplied. A medium conducive to growth may not be suitable for sporulation. Low-nutrient media are generally recommended for sporulation.

On the basis of composition, culture media can be classified as natural, semisynthetic, or synthetic. Natural media are based on natural materials, such as cornmeal, V-8 vegetable juice, lima beans, carrots, potatoes, onions, prunes, dung, or soil. These materials are usually used in the form of an extract, infusion, or decoction. Semisynthetic media contain both natural ingredients and defined components. Potato-glucose agar, yeast extract-glucose agar, and peptone-glucose agar are examples. Synthetic media are of known composition, and each component and its concentration are controlled. They are primarily used for microbiological assays and enzymatic or other biochemical studies.

Culture media may be solid or liquid. Solid media are either natural substances, such as pieces of roots, plugs of potato or carrot, or string beans, or nutrient solutions solidified by the addition of agar or gelatin. Agar is a complex polysaccharide extracted from various red marine algae. It is almost exclusively used at a concentration of 1.5%. For distinctly acidic media, 2% or higher is required. Agar media do not melt until the temperature exceeds 95°C and do not resolidify until the temperature falls below 40°C. Gelatin melts easily at temperatures above 30°C. Agar media are commonly used in preliminary experiments; isolation, identification, and maintenance of cultures; and in sporulation studies. Liquid media are preferred for biochemical work, particularly in studies on metabolic byproducts, metabolic deficiencies, and microbial assays.

Ingredients are usually dissolved in a medium and sterilized by autoclaving. When agar is the solidifying agent, the medium must be heated gently, usually to boiling, to melt the agar. Where interaction of components, such as metals, would cause precipitates, the components are prepared and sterilized separately before mixing. The pH is usually adjusted prior to sterilization, but in some cases sterile acid or base must be added after sterilization to readjust the pH. A different sterilization procedure, namely membrane filtration, is employed when heat-labile compounds are included in media formulations.

A medium is usually designated by the name of the investigator who first used it, for example, Czapek's medium, Leonian's medium, and Sabouraud's agar, or by its principal ingredients, such as glucose-asparagine medium, sucrose-nitrate medium, or cornmeal agar. Formulations of some of the most commonly used media and the methods employed in their preparation can be found in the *Mycology Guidebook* (Stevens 1981),

Methods in Microbiology, Vol. 4 (Booth 1971b), and other laboratory manuals (Riker and Riker 1936; Conn 1957; Johnson et al. 1959; Raper and Fennell 1965; Atlas 1993). The American Type Culture Collection (ATCC) includes the formulations of all culture media for growth and maintenance of its fungal strains in its catalogs (Jong and Edwards 1995; Jong et al. 1996).

B. Culture Conditions

Temperature, pH, and light affect cultivation. The majority of fungi grow well at room temperature. However, the optimal temperature may vary among species of a genus and even among strains of a species. Some fungi, such as those that infect plants under snow or cause spoilage of refrigerated foods, are tolerant of low temperatures and can grow at or below 0 °C. Dermatophytes grow best in culture at 25–35 °C, a temperature below the host's body temperature, while fungi causing systemic mycoses frequently require incubation at 37 °C. Thermophilic fungi grow satisfactorily at elevated temperatures of 45–55 °C. Agar media on which thermophiles are grown tend to dry out rapidly, and extra care must be taken to ensure high humidity during incubation. The range of temperatures for spore formation is usually narrower than for mycelial development.

Fungi flourish in slightly acid media. Although most common fungi will grow within a pH range of 3–7, a much more restricted range is necessary for sporulation in those strains that produce spores. The pH of a medium can change according to the composition and buffering capacity of the medium, the temperature of incubation, and the type of metabolites produced by the growing culture. Since fungi grow optimally only within a certain range, pH is generally controlled by means of phosphate buffers. Control can be maintained within a few tenths of a unit by using varying amounts of equimolar concentrations of Na_2HPO_4 and NaH_2PO_4. Many fungi are differentiated by the acids produced from carbohydrate metabolism or by the decarboxylation of amino acids in their media. Indicators often included in media detect pH changes.

Light influences both growth and reproduction. Most common fungi seem to grow equally well in light or darkness. Some species are reported to do better in light than in dark, while mycelial growth of others is retarded by continuous exposure to light. Light can also have a stimulatory or an inhibitory effect on sporulation. Many fungi in culture produce alternate zones of sporulation in response to diurnal changes in light intensity. Light increases the production of conidia, resulting in zones of intense sporulation alternating with those of sparse or no sporulation. Near-ultraviolet light stimulates sporulation in certain fungi, particularly those in the dematiaceous group. A practical guide to the effects of light has been provided by Leach (1971).

C. Control of Mite Infestations

Mite infestation of fungus cultures is a common problem (Jong 1987). In the laboratory, mites not only eat pure cultures but also contaminate them with other fungi or bacteria. In nature, free-living mites inhabit leaf litter, humus, and topsoil, feeding on vegetation or decaying organic matter. They are frequently found on the plant and animal materials and soil particles brought into laboratories for examination. Mites can also enter the laboratory in new cultures or on clothing, shoes, laboratory supplies, and equipment.

Adult mites are about 0.02 mm in length, usually too small to be seen without magnification. Although larger, the eggs are transparent and even more difficult to detect. Since mites are able to pass through cotton plugs into tubes and into petri dishes, they can freely move from culture to culture and contaminate an entire culture collection before being discovered.

The first line of defense against mite infestation is exclusion. Living plants or cut flowers should never be brought into the laboratory. Plant materials such as leaves, hay, grains, or soil to be used in laboratory work should be confined to one area, preferably well away from the culture work area. If not immediately used or discarded, these materials should be kept on racks placed in oil-filled pans to prevent the migration of any mites present. In addition, areas in the laboratory can be periodically sprayed with a miticide solution.

Incoming cultures should be examined carefully under a stereoscope, kept isolated over oil when not being used, and discarded as soon as possible. Packing materials and wrappers should be disposed of immediately. Storage of cultures at 10 °C or below will slow migration and prevent the hatching of eggs. Once a tube is returned to room temperature, any mites present will become active and eggs will hatch.

Upon discovery of a mite infestation, immediate action is necessary. Every culture in the area must be examined. Incubators, storage areas, and all culture tubes must be sprayed with 70% alcohol. Any infested cultures which are duplicates or are ready for discard should be autoclaved. Infested cultures that must be saved can be frozen for 24h at −20°C. The mites and eggs will be killed, and the fungus can probably be revived and subcultured.

III. Preservation

There is no universal method for storing fungal cultures (Jong and Birmingham 1992; Smith and Onions 1994; Dhingra and Sinclair 1995). The primary goal in conservation is to preserve viability without contamination, genetic variation, or deterioration. In general, the technique aims either to minimize the risk of changes and eliminate frequent transfer by extending the periods between subculture or tries to bring cellular activity to a halt. Selection of a method is based on its advantages or disadvantages, the amenability of the culture to preservation and its future use, and the equipment, personnel, and financial resources available. Culture collections usually employ two different types of preservation and store backup stock off site as an added protection (Jong 1981; Jong and Atkins 1985).

A. Periodic Transfer

Pure cultures can be maintained in the laboratory for short periods by serial transfer. Fungi are usually grown on agar slants that support maximum growth. The interval between transfer from old slants to fresh slants depends on the requirements of the organism, as well as the storage conditions. At room temperature the periods will vary from days to months. Transfer every 3 months is sufficient for most fungi, although certain species may remain viable for years. For fungi that fail to produce specialized propagative cells, large pieces of mycelia from young, actively growing marginal areas can be transferred.

To avoid rapid drying of the agar medium and to reduce metabolism, fresh mature cultures, which are not sensitive to cold, can be stored in a refrigerator at about 5°C or in a refrigerator freezer at −10 to −20°C (see below). At low temperatures the viability of most fungi is increased, and the interval between transfers can be lengthened, even doubled.

Periodic subculturing is simple, inexpensive, and does not require specialized equipment, but it is labor-intensive. Properties of fungi in culture may be unstable through loss of plasmids, spontaneous mutations, or genetic recombination due to the presence of heterokaryons, the parasexual cycle, or normal sexual events. While periodic transfer maintains viability, it is frequently ineffective in conserving genetic integrity. With every transfer there is also a danger of introducing contamination by airborne spores or mites, of selecting cells with certain characteristics that may eliminate the "normal" character of the strain, and of the gradual weakening and of eventual death of the strain through continued growth under artificial conditions.

B. Sterile Mineral Oil

The storage life of an agar slant can be extended with an overlay of sterile mineral or paraffin oil. This technique decreases dehydration of the medium, retards metabolic activity of the culture by reducing oxygen tension (though some fungi continue to grow under oil), and lessens the chance of mite infestations. Oil overlay is particularly useful for mycelial or nonsporulating forms, which cannot be freeze-dried or frozen successfully, and in small collections, where freeze-drying or cryopreservation is not economical.

Cultures should be grown on suitable agar slants until acceptable growth or sporulation has occurred, then covered with sterile oil to 1cm above the edge of the agar under aseptic conditions. Slants can be stored in an upright position for years at either room or refrigeration temperature, depending on the sensitivity of the culture. At appropriate intervals cultures must be transferred to fresh agar slants and the process repeated. More than one subculture may be necessary after retrieval from storage, because the growth rate can be reduced due to adhering oil. Oil can be removed by washing the culture mass in sterile water or by making a transfer to a slant and incubating it upright, allowing the oil to drain away. Besides being messy, other disadvantages of this method include risk of contamination during transfer and the presence of adhering oil in microscope preparations. Fungi that produce acid or liquefy media are not suitable for storage under oil.

C. Water Storage

Compared to maintaining cultures on agar slants as described above, water storage has numerous advantages: culture viability or growth rate is not significantly influenced, isolates can be stored longer, genetic stability is greater, and pleomorphic changes are suppressed. Water storage is clean, quick, easy, inexpensive, and requires minimal storage space. It has been used successfully for all classes of fungi, including yeasts, ectomycorrhizal fungi, and plant and human pathogens.

The most important factors influencing survival in sterile water over a long period are the selection of good sporulating cultures and sufficient inoculum. Agar blocks permeated with hyphae cut from colony margins can also be used. Growth may sometimes occur if too much medium is transferred into the storage tube, but the chance is reduced if the spores or hyphae are removed from the surface of the agar medium before transfer. The advantages of this method are that bacterial contaminants can be eliminated with an antibiotic bath, and mite infections can be eradicated or prevented from spreading during storage in the refrigerator. This method is useful in preventing the development of pleomorphism, but it does not eliminate it once it has developed.

Sources of failure with water collections are due to variation of water from laboratory to laboratory and the fact that water may contain trace amounts of toxic substances even after distillation. Storage vessels may leach ions and other substances into the suspensions with possible adverse effects. Until 1991 glass containers were thought preferable to plastic, since all reports of the water method until that time involved glass. The possible toxic effects of plasticizers on aqueous fungal suspensions are still being determined.

Storage in sterile water is certainly simpler and less messy than the oil overlay method; however, cultures generally will survive longer under oil, reportedly, 1 to 32 years, than in water storage, 2 to 7 years.

D. Sterile Soil or Sand

While fungi stored at lowered temperatures, under oil, or in distilled water are in a state of reduced metabolism, those stored under dry conditions become dormant. Drying methods are best for cultures that produce spores or other resting structures. Most fungal spores have a lower water content than vegetative hyphae and are able to withstand desiccation. Only when water is restored will the fungus revive and grow. Drying methods are technically simple and do not require expensive equipment. Spores have been stored in soil, sand, and silica gel, on filter papers, and on porcelain or glass beads.

Since 1918 scientists have stored fungi in sterilized soil for months or even years with little or no variability when compared with the same cultures maintained by serial transfer. Sterile soil can be used as an absorbent medium for a small amount of inoculum, which is immediately dried and stored in the refrigerator, or it can be used as a growth medium, in which case the mycelium and spores of the second generation are stored. This method is used in many industrial collections to preserve viability and the characteristics of commercially important fungi, such as *Aspergillus*, *Penicillium*, and *Fusarium*. *Fusarium avenaceum* remains viable in sterile loam soil for 3 years at room temperature and up to 8 years under refrigeration at 5 °C.

The advantages of the soil culture method include increased longevity of the culture, reduction or elimination of morphological changes, and the availability of a uniform inoculum for many years. The most serious objection is the considerable time lag before dryness brings about dormancy. This period can allow vegetative strains to overgrow the wild-type or saprophytic mutants to overgrow a pathogenic strain (Booth 1971a).

E. Silica Gel

Yeasts and fungi that produce abundant spores, e.g., *Aspergillus nidulans* and *Neurospora crassa*, will survive storage in anhydrous silica gel. However, fungi with thin-walled spores, spores with a high water content, or spores with appendages are not preserved well with this method.

For yeasts, chromatographic grade silica gel (60–200 mesh) is dry-sterilized in uncapped 0.5-dram vials in a covered metal box at 365 °F for 3 h; caps are autoclaved separately and dried. After cooling, vials are capped and stored until needed. Prior to use, the vials are cooled by placing them into holes of a metal block, which has been refrigerated overnight. The block is put into a pan of ice. Yeast cells that have been grown up on agar plates are removed by scraping the surface with a sterile applicator stick, which is then inserted into the vial and mixed well with silica gel. The caps are tightened and the vials are stored in the refrigerator. Viability can be considerably increased by suspending the yeast in sterile milk and transferring a few drops of the suspension to the silica gel.

As an alternative, cotton-plugged test tubes filled 65 mm deep with anhydrous silica gel can be dry sterilized at 180 °C for 2 hr and stored until ready for use. The cultures to be preserved are grown on agar slants. About 0.5 ml of sterile water is added to the culture and the tube contents mixed with a vortex-type mixer, followed by the addition of 0.5 ml sterile skim milk. The entire suspension is pipetted drop by drop over the silica gel, and the tube is vibrated to distribute the inoculum. After an initial 15 min in an ice bath, the contents are dried at room temperature. The tubes are sealed with Parafilm and stored at 5 °C or −20 °C in moisture-proof boxes.

Cultures can be revived by transferring small portions of gel with a spatula onto a slant or plate and incubating for the recommended time. The same storage container can be used for successive samplings. Many cultures have been preserved for more than 8 years by this method, and they probably can remain viable for even longer periods (Perkins 1977; Raper 1984).

F. Drying on Organic Substrates or Paper

Fungi grown on various organic substrates, such as wood chips, cereal grains, straw, insect and plant tissues, or even agar media can be stored for at least a year by drying. For example, when colonies on agar media are fully developed, agar strips are removed with a sterile scalpel, placed between sterile blotter paper, and dried in a desiccator at 0 to 20 °C. After drying, samples are placed in sealed containers and stored at room temperature or in a refrigerator at low humidity. Some plant pathogens survive when infected host tissue is dried in a plant press and stored in a refrigerator. Rust-infected material can be placed between blotters in a tube with $CaCl_2$ for a month then stored in envelopes at 2–3 °C.

Yeasts have been successfully dried on filter paper and stored above desiccant silica gel. Sterile squares or disks of paper are inoculated by immersing them into drops of yeast suspension prepared in 5% skim milk. The inoculated paper is dried in a desiccator and stored in an airtight container at 4 °C. Cultures are revived by placing the filter paper on an agar plate or in a broth medium and incubating at a suitable temperature. This method is particularly good for the preservation of *Saccharomyces cerevisiae* with genetic markers.

G. L-Drying

L-drying, liquid drying in *vacuo*, is a method by which fungi are dried under vacuum without freezing. The operation takes place in ampules attached to a manifold and immersed in water at 20 °C. The vacuum level is adjusted to allow evaporation. The ampules are sealed to retain low pressure. Organisms are recovered by opening the ampule and adding growth medium.

L-drying has an advantage over freeze-drying (see below) in that the cells dry faster and nonabsorbent cotton wool plugs can be used to prevent contamination. This procedure has been successfully used in preservation of various cultures at the Institute for Fermentation in Osaka, Japan (Kaneko et al. 1985).

H. Mechanical Freezers

If fungal cultures can survive temperatures as low as −20 °C and rapid thawing, they can be successfully frozen at −20 to −80 °C in mechanical freezers. Storage in a deepfreeze allows most fungi to survive 4–5 years between transfers, though some freezing damage may occur. Storage at 0 to −20 °C is more destructive than below −20 °C. There is no danger of mite contamination at these lower temperatures.

While cultures grown on agar slants in test tubes with screw caps can be placed directly in the freezer, the addition of a cryoprotectant is recommended. For example, the anaerobic rumen fungus *Piromyces communis* has been frozen in 4% ethylene glycol at −84 °C with 80% viability after 1 year (Sakurada et al. 1995). Sporidial cultures of *Tilletia tritici* and *Tilletia controversa* have been preserved for at least 1 year by suspending a mixture of mycelia and sporidia in 15% glycerol and storing at −70 °C (Loomis and Leung 1995). In general, vigorously growing cultures survive the process better than less vigorous strains, and repeated freezing and thawing is detrimental. The cost of the freezer, maintenance, and the possibilities of power failure are limiting factors.

I. Freeze-Drying

Drying a culture reduces metabolic activity to a minimum by removal of water necessary for metabolism; freezing a culture accomplishes the same thing. Since both procedures have disadvantages, freeze-drying was developed. It extends shelf-life, and isolates generally retain their morphological, biochemical, and physiological characteristics.

Freeze-drying is a multistep dehydration process in which spore suspensions are prefrozen to form a mixture of ice and a solute-rich phase and dried under vacuum to sublime the ice from the frozen mass. It is a relatively low-cost form of permanent preservation, but not universally applicable. The protectant used, rate of cooling, final temperature, rate of heat input during drying, residual moisture, and storage conditions all affect the viability and stability of fungi. Mycelial cultures or cultures with fragile or complex spores do not usually survive the process.

At the ATCC cultures are grown under conditions to induce maximum sporulation and spore maturation. Optimum media and growth conditions are listed in the individual strain descriptions in ATCC reference catalogs. Cited literature may give further guidance on appropriate cultivation procedures (Jong et al. 1996).

Living cells will not survive freeze-drying without some sort of cryoprotectant. Many suspending media have been tested for their protective qualities to prevent total drying and to buffer the effects of electrolyte concentration during dehydration. Bovine serum was the earliest additive used. Others are sodium glutamate, peptone, various sugars or mixtures of sugars, and skim milk. At the ATCC a 20% solution of skim milk is prepared and autoclaved at 116 °C for 20 min in 10-ml tubes. One tube is usually more than enough for ten freeze-dried vials unless the culture is very mycelial. The milk is stored at 2–8 °C until needed so that it will be cold when used (Simione and Brown 1991).

A spore suspension is prepared by slowly introducing about 2 ml of skim milk into a culture tube or plate while gently scraping the surface of the culture with a pipette. Caution must be taken to avoid raising a cloud of spores, especially with *Aspergillus*, *Penicillium*, and *Neurospora*. Such cultures are usually grown on slanted agar in 250-ml Erlenmeyer flasks. The suspension is returned to the tube containing the remainder of the milk and mixed thoroughly. If more than one slant or plate is used, the procedure must be repeated for each and the suspensions pooled into one tube. A concentration of at least 10^6 spores ml^{-1} is needed to insure the survival of adequate numbers of cells from freezing injury and to minimize the selection of mutants. An aliquot of 0.2 ml is dispensed into each glass vial. Many spores will begin to germinate when suspended in liquid, so timing is critical. Filled vials must be refrigerated and processed within 2 h.

Although the method is labor-intensive, freeze-drying spore-forming fungi greatly facilitates their storage and distribution. After vials are prepared, they are processed in one of the following types of freeze-drying systems: (1) component freeze-dryer with benchtop vacuum pump and condenser, where samples are freeze-dried in cotton-plugged glass inner vials, which are then sealed inside glass outer ampules under vacuum; (2) commercial freeze-dryer, where samples are processed in glass vials in ampules as above with the freeze-drying being done in a commercial freeze-dryer; (3) ATCC Preceptrol, where samples are processed in glass serum vials sealed with butyl rubber stoppers and metal caps using a commercial freeze-dryer; and (4) manifold, where samples are processed in bulb-shaped or tubular glass ampules attached by latex tubing to a manifold. This last method is not used at the ATCC, but it is relatively inexpensive and utilizes equipment that the laboratory may already own.

As a note of caution, when freeze-drying fungi in vials or ampules without cotton plugs or other bacteriological filters, the fungi can contaminate the outside of the vial or ampule or parts of the freeze-drying system, such as the condenser.

A freeze-drying system should be designed to monitor the contamination level, and decontamination procedures should be implemented if necessary.

If a culture survives the freezing shock, it will probably retain viability for years. Storage of the freeze-dried material must exclude oxygen and water vapor; both cause rapid deterioration. This is achieved by filling the ampules with a dry inert gas before sealing. Freeze-dried material can be stored at room temperature without loss of significant viability, but storage at low temperature (4 °C) is recommended. Fresh isolates are more likely to have longer survival times than cultures maintained for years in subculture before being freeze-dried.

The rehydration procedure has been found to be an important factor in post-preservation viability. Freeze-dried cells must be allowed sufficient time to reabsorb water before transfer to a solid medium. Sterile distilled water and a sterile Pasteur pipette are used to transfer the contents of the freeze-dried preparation to approximately 5 ml of sterile distilled water in a test tube. The contents should be allowed to rehydrate for at least 1 hr before transfer of a few drops to broth or agar; 2h is better, and overnight is not too long. To ensure optimal recovery, the strain descriptions in the ATCC catalogs can be consulted for special media, growth conditions, and tips on maintenance. This is especially important for transformation hosts, genetic mutants, producers of secondary metabolites, and quality control strains. The rehydrated culture should be incubated at the appropriate temperature. The remainder of the suspension may be stored for a few days in the refrigerator, allowing for another recovery attempt if the first should fail. Given proper treatment and conditions, most cultures will grow in a few days. Some strains may exhibit a prolonged lag period and should be given twice the normal incubation time before being discarded as nonviable.

J. Cryogenic Storage

Since the rates of mitotic recombination and mutation are likely to correspond to those of cell division and metabolic activity, the ideal system for preservation is to store living cultures in such a manner as to achieve a complete cessation of cell division and total arrest of metabolism. This may be accomplished at temperatures below −139 °C, where there is no growth of ice crystals and the rates of other biophysical processes are too slow to affect cell survival. Fungi that survive cooling, freezing, and subsequent thawing can be stored indefinitely in liquid nitrogen (Jong 1989).

The cryogenic storage temperatures now commonly used are those of liquid nitrogen (−196 °C) and liquid nitrogen vapor (−150 °C and below). Storage in liquid nitrogen itself is not always convenient or safe; storage in liquid nitrogen vapor is more practical. Plastic screw-capped vials can present a hazard if stored directly in liquid nitrogen. Vials with an inadequate seal at the interface between the cap and the vial can fill with liquid nitrogen. Upon retrieval to warmer temperatures or upon opening the cap, the liquid can spray from the vial, potentially disseminating the contents. (A similar problem can occur with pinhole leaks in glass vials.) Material in plastic vials is therefore stored in liquid nitrogen vapor at the ATCC. The vapor will have a gradient which is near −196 °C at the level of the liquid and gradually becomes warmer near the top of the freezer. Frozen materials must be kept below −130 °C; storage at warmer temperatures can compromise the stability of many strains.

To reduce injury upon freezing and thawing, one feature common to most successful protocols is the application of chemical cryoprotectants. These are of two types: penetrating agents, such as glycerol and dimethyl sulfoxide (DMSO), which readily pass through the cell membrane and exercise their protective effect within intracellular and extracellular conditions; and nonpenetrating agents, such as sucrose, lactose, glucose, mannitol, sorbitol, dextran, polyvinylpyrrolidone, and hydroxyethyl starch, which exert their protective effect external to the cell membrane. Glycerol and DMSO have been proven by far most effective for fungi.

The ATCC uses 10% (v/v) glycerol and 5% (v/v) dimethyl sulfoxide (DMSO). Glycerol solutions are sterilized by autoclaving in the manner of ordinary culture media, while DMSO is sterilized by filtration. The ATCC has used this cryogenic technique for the conservation of a wide variety of living fungi since 1960 (Jong et al. 1996).

Both sporulating and nonsporulating fungi can be cryopreserved. Sporulating strains are grown on solid media as for freeze-drying; nonsporulating strains are grown on either solid or liquid media. If the mycelium is easily broken, the culture is grown in test tubes. Spores or mycelial fragments are harvested by flooding the slants or plates with 10% glycerol or 5% DMSO. Approximately a 0.5-ml suspension is pipetted into 2-ml screw-top polypropylene vials.

If the mycelium is sticky, will not break up, or grows embedded in the agar, the culture is grown on agar plates. Three agar plugs containing hyphal tips are cut out with a sterile cork borer or a sterile scalpel and placed in each plastic vial with approximately 0.4 ml of 10% glycerol.

If the mycelium cannot be cut from the agar, the culture is grown in liquid medium in flasks. Cultures in broth are fragmented for a few seconds in a sterile miniblender and suspended in equal parts of 20% glycerol and growth medium, or equal parts of 10% DMSO and growth medium, to give a final concentration of 10% glycerol or 5% DMSO. Aliquots of 0.5 ml are dispensed into freezing vials. Slow-growing strains are incubated for several days before freezing to allow for healing after blending. Fast-growing strains can be frozen immediately. Some strains must be concentrated by centrifugation to obtain sufficient material for freezing. Pathogens should be handled under a hood and not blended because of the hazard of aerosol dispersion. *Histoplasma*, *Paracoccidioides*, and *Blastomyces* are frozen in the yeast phase and *Coccidioides* in the young mycelial stage to avoid contamination from airborne spores and to minimize exposure of laboratory personnel.

The filled vials are put in labeled aluminum cans and then in boxes, which are then placed in the freezing chamber of a programmed freezer. The chamber is cooled to 4°C. Prior to the start of freezing, the material is allowed to chill to within 2°C of the chamber temperature. The best results in the survival and recovery of most fungal cultures have been obtained using a slow cooling rate and rapid thawing. The material is frozen at a rate of $1\,°C\,min^{-1}$ to $-40\,°C$, then $10\,°C\,min^{-1}$ to $-90\,°C$. The vials are immediately stored in the vapor phase of a liquid nitrogen unit.

Because of its extremely cold temperature, liquid nitrogen can be hazardous if improperly used. When handling liquid nitrogen, precautions must be taken to protect face and skin from exposure to the liquid. Protective clothing, including a laboratory coat, gloves designed for handling material at cryogenic temperatures, and a face shield must be worn.

To reduce exposure to cryogenic temperatures, inventory systems that allow for easy retrieval and minimize the time required to look for frozen specimens must be designed. When liquid nitrogen is used in confined and inadequately ventilated areas, the nitrogen can quickly displace the room air. Liquid nitrogen freezers should be located in well-ventilated areas, and special precautions should be taken during fill operations. In facilities with several liquid nitrogen freezers, an oxygen monitor should be installed to warn personnel of a deterioration in the air quality due to nitrogen gas.

To recover cultures, frozen plastic vials are thawed rapidly in a 55°C water bath until the last trace of ice has disappeared. This usually takes about 90s with moderately vigorous agitation. Glass vials thaw in less than a minute. Culture samples may then be aseptically transferred to an appropriate growth medium. Most will grow in a few days.

Cryopreservation causes little or no genetic change in the cells. Cost of the purchase of liquid nitrogen is the main reason it is not employed exclusively in some collections. For shipping purposes one disadvantage for a service collection is that strains preserved in liquid nitrogen must first be grown up on agar or in liquid medium to avoid the expense of shipping frozen materials.

References

Atlas RM (1993) Handbook of microbiological media. CRC Press, Boca Raton
Booth C (1971a) The genus *Fusarium*. Commonwealth Mycological Institute, Kew, Surrey, UK
Booth C (1971b) Fungal culture media. In: Booth C (ed) Methods in microbiology, vol 4. Academic Press, New York, pp 49–94
Cochrane VW (1958) Physiology of fungi. Wiley, New York
Conn HJ (ed) (1957) Manual of microbiological methods (by the Society of American Bacteriologists). McGraw-Hill, New York
Dhingra OD, Sinclair JB (1995) Basic plant pathology methods, 2nd edn. CRC Press, Boca Raton
Griffin DH (1994) Fungal physiology, 2nd edn. Wiley, New York
Hawker LE (1950) Physiology of the fungi. University of London Press, London
Johnson LF, Curl EA, Bond JH, Fribourg HA (1959) Methods for studying soil microflora plant-disease relationships. Burgess, Minneapolis
Jong SC (1981) Isolation, cultivation and maintenance of conidial fungi. In: Cole GT, Kendrick B (eds) Biology of conidial fungi, vol 2. Academic Press, New York, pp 551–575
Jong SC (1987) Prevention and control of mite infestations in fungus cultures. ATCC Q Newsl 7(1):127

Jong SC (1989) Germplasm preservation of edible fungi for mushroom cultivation. Mushroom Sci 12(Part 1):241–251

Jong SC, Atkins WF (1985) Conservation, collection and distribution of cultures. In: Howard DH (ed) Fungi pathogenic for humans and animals. Marcel Dekker, New York, pp 153–194

Jong SC, Birmingham JM (1992) Current status of fungal culture collections and their role in biotechnology. In: Arora DK, Elander RP, Mukerji KG (eds) Handbook of applied mycology. Marcel Dekker, New York, pp 993–1024

Jong SC, Edwards MJ (eds) (1995) ATCC yeasts, 19th edn. American Type Culture Collection, Rockville, MD

Jong SC, Dugan F, Edwards MJ (eds) (1996) ATCC filamentous fungi, 19th edn. American Type Culture Collection, Rockville, MD

Kaneko Y, Mikata K, Banno I (1985) Maintenance of recombinant plasmids in *Saccharomyces cerevisiae* after L-drying. IFO Res Commun 12:78–82

Leach CM (1971) A practical guide to the effects of visible and ultraviolet light on fungi. In: Booth C (ed) Methods in microbiology, vol 4. Academic Press, New York, pp 609–664

Lilly VG, Barnett HL (1951) Physiology of the fungi. McGraw Hill, New York

Loomis P, Leung H (1995) Low-temperature storage of monosporidial cultures of the wheat bunt fungi. Can J Bot 73:758–760

Perkins DD (1977) Details for preparing silica gel stocks. Neurospora Newslett 24:16–17

Raper KB (1984) The dictyostelids. Princeton University Press, Princeton

Raper KB, Fennell DI (1965) The genus *Aspergillus*. Williams and Wilkins, Baltimore

Riker AJ, Riker RS (1936) Introduction to research on plant diseases. John Swift St Louis

Sakurada M, Tsuzuki Y, Morgavi DP, Tomita Y, Onodera R (1995) Simple method for cryopreservation of an anaerobic rumen fungus using ethylene glycol and rumen fluid. FEMS Microbiol Lett 127:171–174

Simione FP, Brown EM (1991) ATCC preservation methods: freezing and freeze-drying, 2nd edn. American Type Culture Collection, Rockville, MD

Smith D, Onions AHS (1994) The preservation and maintenance of living fungi, 2nd edn. CAB International, Wallingford, UK

Stevens RB (ed) (1981) Mycology guidebook (by Mycology Guidebook Committee, Mycol Soc Am). University of Washington Press, Seattle

8 Computer-Assisted Taxonomy and Documentation

O. Petrini[1] and T.N. Sieber[2]

CONTENTS

I.	Introduction	203
II.	Numerical Taxonomy	204
A.	Introduction	204
B.	The Approach to the Problems	204
C.	Collecting Data	205
D.	Exploratory Data Analysis	205
E.	Tests and Numerical Analysis	205
F.	Classification and Ordination Analysis	207
	1. First Step: Detection of Groups	207
	a) Cluster Analysis	207
	b) Clustering Algorithms	208
	c) Principle Components Analysis, Factor Analysis, and Nonparametric Multidimensional Scaling	208
	d) Correspondence Analysis	208
	2. Second Step: Testing the Goodness of Fit	209
	a) Discriminant Analysis	209
III.	Cladistics	210
IV.	Databases and Identification Keys	210
V.	Computer Keys	211
VI.	The Future of Computers in Mycology	212
VII.	Computer Software for the Mycologist	212
	References	215

I. Introduction

In the past few years, the development of affordable, powerful personal computers has encouraged taxonomists to use methods that were formerly reserved for a selected group of experts whose knowledge of computing was broad, but whose skills in handling the complexity of taxonomic problems were often rudimentary. The development of inexpensive and user-friendly software enabled researchers to concentrate on the biological aspects of taxonomy, at the same time, however, allowing them to make use of the power offered by modern computers. Phenetic analysis of complex datasets, including both morphological and biochemical characters, as well as the use of cladistics for large DNA and protein sequences, have thus allowed us a better and more objective judgement of the relationships within and among taxa.

The main thrust of this chapter is the use of numerical methods in fungal taxonomy. It has been written for nonmathematicians and for computer users, not for experts in both fields. Although some reference will be made to relevant mathematical assumptions underlying the use of different methods, the chapter will focus on the practical aspects linked with the use of numerical techniques in fungal taxonomy. While we have tried to write accurately from a mathematical point of view, we have avoided any reference to complex mathematical formulas or computer jargon. Detailed explanation of each numerical taxonomy method can be found in textbooks on statistics and in the general references we have given here. The chapter is only a guide through the often complicated process of choosing an appropriate method of data analysis and management. We make no claims for completeness and we recommend that the help of a professional statistician or data manager should always be sought before starting any new taxonomic analysis.

We also briefly discuss the latest developments of databases and the potential use of new techniques, such as fuzzy logics and genetic algorithms, to develop new tools for cataloguing and identifying fungi. The almost universal presence of personal computers not only in research laboratories but also at home has fostered interest in the preparation of more or less sophisticated computer-based identification keys (Pankhurst 1991). Some attempts to include knowledge-based engines in the keys have also been described (Petrini et al. 1990). The latest developments in intelligent databases (Parsaye et al. 1989) and the possibility of using neural networks and algorithms that simulate evolutive processes (genetic

---

[1] Pharmaton SA, P.O. Box, 6903 Lugano, Switzerland
[2] Forest Protection and Dendrology, Swiss Technical Institute of Technology, ETH-Zentrum, 8092 Zurich, Switzerland

algorithms: Hedberg 1993) in personal computer-based software will soon produce powerful and reliable identification software shells to support identification and detection of new taxa.

II. Numerical Taxonomy

A. Introduction

The use of numerical methods in taxonomy is not new. Nearly 40 years ago in 1963 Sokal and Sneath published a comprehensive and useful book on the subject, replaced in 1973 by a monumental treatise by the same authors (Sneath and Sokal 1973), which is still the standard reference in numerical taxonomy. For a long time, however, the use of numerical taxonomy was restricted to research groups with direct access to mainframes. In recent years, first in plant taxonomy and later in mycology, taxonomists have increasingly used numerical methods to study taxonomic relationships within and among taxa (e.g., Bridge 1988; Thrane 1990; Jun et al. 1991; Trigiano et al. 1995; Leuchtmann et al. 1996) and new, easily understandable treatises now exist that help taxonomists who have little knowledge of numerical methods in the choice of adequate numerical tests (Dunn and Everitt 1982; Systat 1996).

Numerical taxonomy has been widely used to study relationships among and within plant species, yet mycologists have been relatively reluctant in applying it to taxonomic problems, perhaps because they considered it not adequate to handle the rather complex datasets that originate from morphological data, but probably also because of an unjustified feeling of mistrust in the methods underlying it. Physiological, biochemical, and molecular techniques such as isozyme, RFLP, and RAPD analysis have generated datasets that lend themselves to analysis by numerical taxonomy. A number of publications have thus started to include numerical analyses of isozyme and RFLP data by cluster or factor analysis (Petrunak and Christ 1992; Brunner and Petrini 1992; Assigbetse et al. 1994; Leuchtmann et al. 1996). At first, matrices used only data from a single type of analysis. O. Petrini et al. (1989), however, analyzed morphological, biochemical, and physiological data to express relationships among *Gremmeniella* species, and Sieber-Canavesi et al. (1991) used datasets that combined information originating from biochemical and physiological studies. Morphologists have for a long time been reluctant to use numerical methods in fungal taxonomy, mostly because of the fear of the problems linked with the coding of the mixed types (continuous and categorical) of data originating from morphological studies. In 1977 Whalley and Whalley applied cluster analysis to data derived from pigment analysis of *Hypoxylon* species. Chemical data have been used to carry out numerical analysis within species of *Fusarium* (Svendsen and Frisvad 1994). Discriminant analysis of spore sizes was used to distinguish intraspecific taxa within *Hypoxylon fuscum* (Petrini et al. 1988) and to separate species of *Rosellinia* within the *R. mammaeformis* complex (L. Petrini et al. 1989). More recently, both exploratory and confirmatory numerical analyses of mixed datasets were used as a base for the preparation of a taxonomic monograph of *Rosellinia* (L. Petrini 1992).

Exploratory and confirmatory statistics are of value in detecting relationships among fungal taxa. Cladistic methods are also becoming increasingly important in fungal taxonomy. In the past few years cladistics have been widely used and have become the standard tool to analyze molecular data (Swofford and Olsen 1990; Bruns et al. 1991). Phylogeny inference by cladistic techniques is not considered by many taxonomists to belong strictly to numerical taxonomy, because the assumptions underlying numerical methods are of strictly phenetic character, in contrast to the phylogenetic nature of the cladistic analysis. We consider numerical and cladistic methods, however, to be tightly linked to each other. Even if methodologically and philosophically they are distinct, nevertheless both techniques try to solve with a numerical approach the problem of defining relationships among organisms.

B. The Approach to the Problems

Although not all taxonomic problems can be approached using numerical methods, taxonomic studies often generate large amounts of data that can be evaluated statistically. The use of mathematical procedures with data that have been collected inadequately can generate problems that often lead to incorrect or at least incomplete results. Therefore, data should always be collected with statistical analysis in mind. The sample size should be set with standard procedures, described in most statistical textbooks (e.g., Cox 1958; Fleiss 1981; Krebs 1989), on data collected during preliminary experiments. When this is not possible, at least previous experience should be used to estimate a reasonable sample size. In general, before any analysis is planned, the data structure should be studied carefully.

C. Collecting Data

While data collection for a standard taxonomic work may appear a straightforward issue, inadequate planning often leads to collecting enormous amounts of data which, however, cannot be adequately evaluated by standard statistical tests. Such situations have been repeatedly addressed in standard textbooks of ecology (e.g., Green 1979; Ludwig and Reynolds 1988). The ten principles recommended to plan ecological experiments (Green 1979) can be applied, with slight modifications, to fungal taxonomy as well (Table 1).

D. Exploratory Data Analysis (EDA)

EDA (Tukey 1977) is probably the most effective way of using statistics in taxonomy and ecology. It mostly does not rely on any assumption of distribution, its application is straightforward, and the most important procedures are now found in most integrated statistical packages. EDA often saves valuable time and work and makes the use of additional numerical methods and tests superfluous. In general, if no trends can be seen in the data using EDA, any additional sophisticated data analysis is not likely to yield additional information.

There are a number of effective EDA methods which can easily be used to analyze taxonomic data. They include the use of frequency tables, the display of data with scatterplots, bar charts, boxplots, confidence ellipses, or SPLOM (Scatterplot matrices). Some examples are presented in Fig. 1 and a clear, exhaustive description of these and additional methods is given in Cleveland (1985).

E. Tests and Numerical Analysis

Parametric tests should be applied only to normally distributed data. For other kinds of data, a wide selection of nonparametric or distribution-free tests exist (Table 2). In taxonomy, as a rule, preference should be given to nonparametric tests, because taxonomic data are often not normally distributed. Moreover, nonparametric tests can be applied with virtually no loss of power to both nor-

Table 1. Planning of taxonomic experiments. (Adapted from Green 1979)

1. The questions to be asked have to be formulated clearly at the beginning of the study, together with the primary endpoints to be analyzed after completion of the research work, to check that the approach is correct and the problem understandable
2. Differences among groups can be best demonstrated by comparisons with differences within groups. Take as many samples of one group as possible to demonstrate that a given group is indeed different from a second one. In general, biometrics based only on one sample tend to be very weak
3. Take random samples. Do not measure only the largest or the smallest individuals!
4. Whenever possible, use a control (or an outgroup). This is particularly important in isozyme analysis and other molecular work if any kind of ordination or cladistic analysis of the data is envisaged
5. Carry out some preliminary sampling to determine the adequate sampling size. Although the determination of sample size is described in standard textbooks of statistics and can now be computed using appropriate software (e.g., nQuery Advisor), it is likely that formulas reported for sample size determination will yield sample sizes too large to be managed in taxonomic research. In such cases, there is no easy recipe for determining sample size requirements and, at the end, the decision will be taken based on estimates. Our experience has shown that at least 30 individuals per population are required to provide good estimates of the population parameters. The sample size, however, must be increased as
   a) the allowable error becomes smaller,
   b) the degree of confidence increases, and
   c) the variability of the population from which the sample is chosen increases. Be aware of the Type I and Type II errors
6. Verify that the population sampled is adequate to study all parameters needed to characterize it
7. If the sample is very large, break it up into subsamples
8. Verify that the sample unit size is appropriate
9. Check data for implausible values, outliers, and other mistakes
10. Test the data for deviations from normality assumptions and if necessary use "first aid" transformations
11. Having chosen the best statistical method to test a hypothesis, stick with the result. An unexpected or undesired result is NOT a valid reason for rejecting the method and hunting for a "better" one

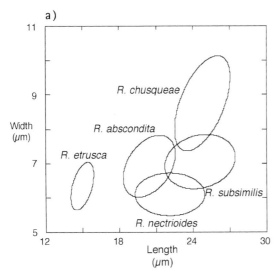
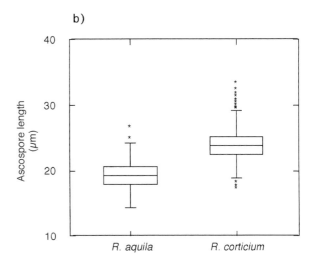

Fig. 1a,b. Examples of exploratory data analysis. **a** 50% confidence ellipses of the ascospore size of *Rosellinia abscondita, R. chusqueae, R. etrusca, R. nectrioides,* and *R. subsimilis.* (Petrini 1992). The graph shows, e.g., that *R. etrusca* has smaller and *R. chusqueae* larger spores than all other taxa. **b** Box-and-whisker plot depicting selected percentiles of spore length measurements of two *Rosellinia* species. (Courtesy L. E. Petrini, unpubl.). The *horizontal line in the center* of the box indicates the sample median. The *lines at the lower and upper end* of the box indicate the 25th and the 75th percentiles (lower and upper quartiles or hinges), respectively. The absolute value of the distance between the hinges is called the hspread. The whiskers extend to the last point occurring between each hinge and its inner fence, a distance 1.5 hspreads from the hinge. Two kinds of outliers can be distinguished. Points occurring between 1.5 hspreads and 3 hspreads (the outer fence) are indicated by an *asterisk*. Points occurring beyond the outer fence are indicated by an *open circle* (not present for these two samples)

Table 2. Statistical tests useful in fungal taxonomy

Test	Assumptions	Test for
t-test	Normal distribution, variance unknown	• Difference between two means • Difference between a single specimen and a sample • Paired comparisons
Parametric z-test	Normal distribution, variance known	• Difference between two means • Difference between a single specimen and a sample • Paired comparisons
F-test[a]	Normal distribution, variance unknown	• Differences between variances
Wilcoxon's signed ranks test	Any distribution which is symmetric with respect to the mean, variance unknown	• Difference between a single specimen and a sample • Paired comparisons
Wilcoxon's two-sample test (or Mann-Whitney U-test)	Any distribution – the same for all ubservations in both samples, variance unkown	• Difference between two samples
Nonparametric Kruskal-Wallis test	Any distribution – the same for all observations in both samples, variance unknown	• Difference between more than two samples
Sign test	Any distribution, variance unknown	• Difference between two medians • Difference between an individual specimen and a sample • Paired comparisons

[a] Small departures from normal distribution may lead to erroneous results.

mally and nonnormally distributed data, although they are slightly less powerful than parametric tests if the data are normally distributed. Unfortunately, we are not aware of easy-to-use distribution-free alternatives to the classical analysis of variance and regression analyses. An exception is Friedman's method for randomized blocks, a nonparametric alternative to analysis of variance of randomized block designs. A way to avoid the use of distribution-free tests is to normalize the data using adequate transformations. As "first aid" transformations, the logarithm (ln or log) can be used for concentrations, measurements, or quantities, the square root for counts, and the arcsine for proportions or percentages. On the other hand, some parametric methods are particularly robust against departure from normality, provided the residuals are normally distributed. Thus, an analysis of the residuals has to be performed after each statistical analysis. Particular care, however, must be used when dealing with percentages and frequency data expressed as ratios. Such data are not normally distributed (Atchley et al. 1976) and parametric tests are, therefore, not applicable. In particular, the use of analysis of variance procedures with ratios and percentages is fraught with mathematical problems. Percentage data and ratios are better dealt with using nonparametric procedures (Fleiss 1981).

F. Classification and Ordination Analysis

The main problem for a mycologist is obviously to classify collections into taxonomic entities, be they subspecies, species, or higher-rank taxa.

An appropriate way of dealing with the task is to take a two-step approach to the problem. In the first part of the analysis groups must be detected whilst the second part has to test the goodness of fit of all data in the groups found. Strictly speaking, the first step deals with the *identification* of operational taxonomic units (OPU), whilst the second refers to the *statistical testing* of the consistency of these units. Many examples of the detection of groups in the mycological literature use cluster analysis. By contrast, only a few researchers have undertaken the task of verifying the groupings resulting from the ordination analysis (Ludwig and Reynolds 1988).

1. First Step: Detection of Groups

a) Cluster Analysis

Cluster analysis is a multivariate procedure routinely used in numerical taxonomy to detect groupings in data. The method is thoroughly described in a number of textbooks (e.g., Sneath and Sokal 1973), and different variations have been discussed in specialized papers (e.g., Hartigan 1975). Numerical analyses of isozyme data in mycological papers most often rely on cluster analysis. The method is well established and the results are usually reliable if the correct measure of proximity is used. The most frequent errors stem from the difficulty of choosing an adequate measure of proximity among the many available (for some frequently used measures, see Table 3). A discussion of the most frequently used similarity indices can be found in Gower (1985). Covariance and correlation coefficients are very useful in treating mixed-character data. Along with the percentage of comparisons of values resulting in disagreements in two profiles (PCT; Table 3), they have been used successfully for isozyme and other biochemical analysis. The use of genetic similarity coefficients (Nei 1972, 1978; Nei and Li 1979) should be used only with truly genetic data. In all other cases, binary coefficients should be used instead. We refer to the review by Swofford and Olsen (1990) and to Sneath (1989) for additional details of the use of similarity coefficients in phylogeny reconstruction.

Table 3. Proximity measures most commonly used and their applications

Proximity measure	Use for:
Euclidean	Continuous or discrete variables or ratios
Manhattan or City-Block	Continuous or discrete variables or ratios
Gamma coefficient	Ranked or ordinal variables
Person product-moment correlation coefficient	Continuous or discrete variables or interval scales
Percentage or number of (dis-)agreements (PCT)	Categorical, nominal or binary variables
Dichotomy coefficients (e.g., Jaccard, Tanimoto, Anderberg)	Binary variables

There are two sorts of cluster: overlapping and exclusive. In the overlapping cluster the same object may appear in more than one cluster; in the exclusive, only in one.

Exclusive clusters can be either hierarchical or partitioned. Hierarchical clusters consist of clusters that completely contain other subclusters in a hierarchical way. In partitioned clustering (e.g., K-means; see Hartigan 1975; Hartigan and Wong 1979) the number of clusters is preset by the user and the program assigns each case (collection, isolate) to one of the clusters. Ideally, the number of clusters is set to two or three. The quality of the clusters thus formed can then be examined by discriminant analysis or logistic regression (see below). The mycological literature abounds with classification analysis based on hierarchical clustering, whereas K-means clustering has not yet found its way into fungal taxonomy. An additional type of clustering, the so-called fuzzy clustering, has already been used with success in epidemiology (Xu et al. 1991) and is likely to find application in taxonomy as well.

b) Clustering Algorithms

A number of clustering algorithms can be used. Sneath and Sokal (1973), Hartigan (1975) and Ludwig and Reynolds (1988) describe in detail the advantages and disadvantages of each method.

Single-linkage amalgamates the two closest objects or the two clusters in which the two nearest members are closest (nearest-neighbors).

Complete linkage amalgamates the two clusters in which the farthest members are closest (farthest-neighbors). This method should only be used if the clusters are pea-shaped (more or less spherical).

Centroid linkage joins the centroids of the two nearest clusters (the centroid is the point represented by the mean vector of all objects in a cluster). This method works only if the clusters are more or less elliptical or spherical.

Average linkage averages all distances between pairs of objects in different clusters to decide how far apart they are. A special form of average linkage clustering, the UPGMA (Unweighted Pair Grouping using Mathematical Averages) method described in detail by Sneath and Sokal (1973) has proven to be one of the most reliable methods in taxonomy.

Median linkage is a variant of the centroid method in which the new centroid is calculated as the unweighted mean of the two subcluster centroids [in contrast, the new centroid is calculated as the weighted mean (weighted by the number of objects in each cluster) of the two subcluster centroids in the centroid method].

Ward's method resembles centroid linkage but adjusts for covariances.

The computation of the cophenetic coefficient is offered as an option by many statistical packages to compute the goodness of fit of the grouping to the data.

c) Principal Components Analysis, Factor Analysis and Nonparametric Multidimensional Scaling

Principal components analysis (PCA) has appeal as an ecological and taxonomic method because it is a multivariate eigenanalysis method that produces an ordination by extracting axes of maximum variation from a covariance or correlation (similarity coefficients) matrix. It is also readily available, as it is included in several statistical packages. The method is quite robust to deviation from normality, but if there are strong nonlinearities in the data, PCA will poorly represent the true relationships (Sneath and Sokal 1973; Ludwig and Reynolds 1988). In such cases, alternative methods are offered by nonmetric multidimensional scaling (MDS; Kruskal 1964a,b; Guttman 1968; Ludwig and Reynolds 1988; Systat 1996) or multiple or joint correspondence analysis (Greenacre 1993).

d) Correspondence Analysis

Correspondence analysis (CA) is a widely used ordination technique. One of the attractions of CA is that corresponding variables and taxon ordination are obtained simultaneously, thus allowing the examination of the interrelationships between taxa and variables in a single analysis (Fig. 2). CA can be derived through the use of an eigenanalysis approach (similar to PCA) or through a series of weighted-average operations. Technical details can be found in Greenacre (1981, 1984, 1993) and Greenacre and Hastie (1987). On the other hand, CA has not yet often been applied to taxonomic problems, although it offers a number of advantages over other methods (Sieber et al. 1997) and is a powerful exploratory technique that can be used on both large and small

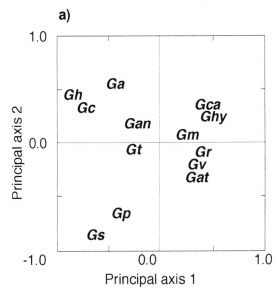
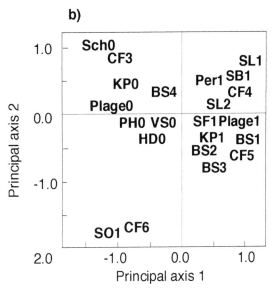

Fig. 2a,b. Graphical displays (symmetric maps) of the result of correspondence analysis for a set of morphological data of some *Galerina* species. The maps illustrate the relative positions of the *Galerina* species (**a**) and the studied characteristics (**b**) in the plane formed by the first two principal axes. Two taxa shown to be very close on the graph are morphologically highly similar. Comparison of maps **a** and **b** allows one to see which characters are typical for a given species or group of species. For example, *G. stagnina* and *G. pseudotundrae* are characterized by smooth spores (SO1) and polymorphic cheilocystidia (CF6), while *G. tundrae* has no hilar depression (HD0). In order not to clutter the display, only selected characters have been displayed on the graph **b**

Abbreviations used: Ga, *G. arctica*; Gan, *G. antheliae*; Gat, *G. atkinsoniana*; Gc, *G. clavata*; Gca, *G. calyptrata*; Gh, *G. heterocystis*; Ghy, *G. hypnorum*; Gm, *G. muricellospora*; Gp, *G. pseudotundrae*; Gr, *G. rubiginosa*; Gs, *G. stagnina*; Gt, *G. tundrae*; Gv, *G. vittaeformis*. Sch0, clamp connections absent; Plage0, plage absent; Plage1, plage present; KP0, germ pore absent; KP1, germ pore present; HD0, hilar depression absent; Per1, perispore present; VS0, velum on stipe absent; PH0, pleurocystidia absent; BS1, basidia with one sterigma; BS2, basidia with two sterigmata; BS3, basidia with three sterigmata; BS4, basidia with four sterigmata; SF1, spore shape mucronate; SL1, spores length 5.0–8.5 µm; SL2, spore length 8.6–12.0 µm; SB1, spore width 2.0–5.0 µm; CF3, shape of cheilocystidia tibiiform; CF4, cheilocystidia clavate; CF5, cheilocystidia fusoid; CF6, cheilocystidia polymorphic; SO1, spore surface smooth

datasets. CA can also be applied without any concern to mixed-type data. The result of the analysis, however, depends on how the different data types are weighted. Correspondence analysis has also been used in the identification of bacteria and can be used to perform identification of unknown samples (Greenacre 1992). For a detailed review of the possibilities offered by CA in fungal taxonomy, we refer to Sieber et al. (1997). Recent examples of the use of CA in fungal texonomy are provided by Samuels et al. (1998) and Lieckfeldt et al. (1999).

2. Second Step: Testing the Goodness of Fit

a) Discriminant Analysis

Once the grouping has been prepared, it often helps to check whether the groups obtained can be confirmed. This is best done by using discriminant analysis. This multivariate method is powerful and highly flexible in that it generates a function that can be used:

- to determine whether or not some of the variables (characteristics) examined are redundant. (If all the variables are redundant the groups are not significantly different);
- to determine the multivariate standard distances between the groups (Flury and Riedwyl 1988);
- to classify a new individual (spore, collection, etc.). (This is often referred to as identification analysis).

Additional methods such as loglinear models (Fienberg 1980) are comparatively complex and their use requires the help of experienced statisticians.

III. Cladistics

The beginning of modern cladistics can be set with Hennig's (1966) publication on phylogenetic systematics. The availability of powerful, specialized, and user-friendly software on personal computers has fostered the use of cladistic methods, first in zoology and botany, and more recently in fungal systematics. In contrast to phenetic analysis, cladistic analysis is based on some fundamental assumptions (synapomorphy, parsimony, descent from a common ancestor) and aims at reconstructing the phylogeny of organisms. It allows one to detect systematic relationships based on homology and natural, monophyletic groups. In addition, cladistics (in particular neighbor-joining and UPGMA methods) can be used to test the outcome of phenetic analysis. Phylogenetic analysis has been used most successfully in fungal taxonomy with molecular data (e.g., LoBuglio et al. 1993; Kuhls et al. 1996; Sreenivasaprasad et al. 1996), and cladistics is becoming the tool of choice for comparative biology. Caution is recommended, however, when dealing with data originating from nonspecific analysis methods such as random amplified polymorphic DNA (RAPD; Backeljau et al. 1995). In addition, phylogenetic inference methods often are inadequate in dealing with morphological data and the available methods do not yet handle mixed-type data satisfactorily (Lutzoni and Vilgalys 1995). Forey et al. (1992) provide a detailed and useful insight into the theoretical and practical aspects of cladistics, and Swofford and Olsen (1990) discuss all aspects related to cladistics in molecular systematics.

IV. Databases and Identification Keys

Databases have always been used to store information and, as such, have been known for a long time to taxonomists. Until a few years ago, however, the use of databases was reserved to only a limited number of people who had access to mainframe computers and who usually had some programming skills. The availability of fast personal computers with large storage facilities, as well as of user-friendly software, has now made large databases accessible to almost everybody.

In the past few years, software development has evolved and now relational, intelligent databases are available on most operating systems. Intelligent databases can be defined as software systems that can manage information in a natural way, making the storage, access, and use of the information easy (Parsaye et al. 1989). The "intelligent" database, as defined by Parsaye et al. (1989), is the result of a blend of high-level tools (which must manage knowledge discovery, data integrity and quality control, hypermedia management, data presentation and display, decision support and scenario analysis, data format and system design), good and intuitive user-interface systems, and inherent database management capabilities. Databases with these attributes are able to manage large volumes of information in a transparent way and make the access to such information easy. They will also help the user by providing mechanisms (e.g., search and learning engines) that actually allow one a more meaningful interpretation of data. In fungal taxonomy such tools and mechanisms can be used to construct either classical identification keys or expert systems (Petrini et al. 1990).

The choice of an adequate database software is often difficult and should be the result of a detailed analysis. Compatibility with coworkers and colleagues should receive high priority and during the database preparation a common data format should be agreed. Cross-platform compatibility (i.e., the availability of the same database sharing the same file format on different operating systems such as UNIX and DOS, Windows, and MacIntosh Computers) should also be an important issue. Most database vendors claim their products allow more or less seamless transfer of data across platforms and are able to read many foreign formats, but 100% warranty of cross-platform compatibility or interchange is usually granted in only rare cases.

Whenever possible, relational databases should be given preference. Relational databases link different files using a common pointer and have a number of advantages over so-called flat-file databases, in which links can be established either through lookups or with rather complicated techniques (Parsaye et al. 1989). The package considered should also have some SQL (Structured Query Language) capability, in particular if data have to be downloaded from a mainframe or queries on large, external databases need to be performed.

Dealing with large files (more than 5000 records per file) requires speed. It is thus advisable to obtain detailed information on the technical specifications of the database and on the search engines used by the software to avoid sluggish performance after a substantial increase in the size of the files to be used. Similarly, the possibility to customize layout, queries, and reports should be considered in detail.

Increasingly, databases are used to construct keys. This can be the case, e.g., for askSam (DOS, Windows) or HyperCard (Mac). For example, HyperCard can be used as a hypertext-based database to construct dichotomous or synoptic keys or even to prepare presentations. Thus, the possibility of interfacing the database with a hypertext-based knowledge engine should be considered and is discussed later in detail.

Basically, all databases can be used to prepare bibliographies. This is certainly the cheapest way of handling the issue, but not always the easiest. It is now possible to buy sophisticated, yet inexpensive and user-friendly database management software, also called reference managers, that are specific applications for storage, management, and search for bibliographic references. They can also be used to insert citations into word processing documents and later compile lists of cited works automatically in predefined formats. The applications are usually compatible with the major word processors, have comparatively fast search engines and intuitive interfaces.

V. Computer Keys

The identification of taxa is the ultimate aim of all taxonomic work, but it may be difficult and frustrating because of the often fuzzy definition of many characters. Dichotomous and synoptic keys have mostly been used by taxonomists to identify taxa. A dichotomous key is a form of hierarchical decision-making process leading through branches of a binary tree to reach the most suitable taxon, whereas the synoptic key checks several characters simultaneously and eliminates taxa not presenting given features until only one taxon is left.

Both types of keys present inherent problems. The morphological and physiological characters used are often ill-defined or not accurate. In a dichotomous key, information is most often presented in the form of disjunctive (OR), conjunctive (AND), or mixed (AND/OR) logic functions or by subjectively defined frequencies of occurrence (rarely, often, etc.). In addition, the definition of a hierarchy in the choice of characters to be used in the identification steps is strongly dependent on the judgment of the author of the key. Wrong decisions can lead to failure of a correct identification, depending on the skills of the user. On the other hand, the synoptic key relies heavily on the knowledge of characters and their reliability by the user.

In an attempt to overcome the difficulties described above, taxonomists have tried to develop tools and techniques that involve the use of computers. A few attempts have been made in mycology to introduce the use of computer keys (e.g., Korf and Zhuang 1985; Wolfaardt et al. 1992). In general, however, the computer applications are merely versions of synoptic keys, based on the use of the search engines of commercially available or proprietary databases and the only advantage over their paper counterparts is the presence of more or less appealing graphic interfaces, glossaries and on-line help, as well as fast search capability. The user, however, is not offered any additional help in choosing the best available characters or in judging the reliability of an identification.

Petrini et al. (1990) have described the possibility of using expert systems to cope with the problems linked with the identification of biological objects. Expert systems are computer programs that should reach approximately the level of performance of a human specialist and are based not only on formal textbook knowledge but also on the heuristic knowledge derived from the experience of an expert. They may also incorporate some mechanisms for reasoning with uncertainty (Petrini et al. 1990).

The development of expert system shells should ideally result from the collaboration between a taxonomist and a computer expert. According to Petrini et al. (1990), an expert system shell should be used by experts to prepare identification keys and by novices to identify fungi without guidance. The system would lead the users through the crucial questions, explaining any unknown terms, and shielding them from mistakes due to lack of expertise.

The ideal system should help the experts in preparing the knowledge base, warning them when inappropriate taxonomic characters are used. In addition, such a system should not be too sensitive to mistakes by the users, informing them of taxa closely related to the one identified, and should be able to explain and justify its reasoning and add new knowledge or alter existing rules.

We are not aware of any expert system shell currently available for use in taxonomy. A prototype, based on the ideas put forward by Petrini et al. (1990) is currently under development (Petrini et al., unpubl.).

VI. The Future of Computers in Mycology

Mycology has rapidly evolved in the last decade from a purely descriptive into an experimental science. Fungal taxonomists now combine morphology with ecological, physiological, biochemical, and molecular biology methods to describe and characterize taxa. The wealth of data deriving from this polyphasic approach (Petrini and Petrini 1996) can only be evaluated effectively using computer databases and numerical methods, and morphometrics and exploratory analysis of both qualitative and quantitative data now have become standard tools in mycology.

Expanding research in biodiversity requires the fast identification of organisms by ecologists, who often are not satisfactorily trained in taxonomy. As only comparatively few taxonomists are available to perform reliable identification of organisms, computer-aided identification based on the use of expert systems integrating fuzzy logic (Barron 1993; Cox 1994), hypertext, and neural networks (Edwards and Morse 1995) will become increasingly important.

While only a few years ago mycologists were exchanging data through conventional mail or fax systems and only a small number of scientists dared to send data on computer disks, the Internet, and in particular the World Wide Web, has now become the most popular means to exchange information and messages, and the Internet is now common-place. The Web is a convention to link together text, images, and interactive programs over the internet using hypertext languages. There are now a large number of Web sites dedicated to mycology, and *Inoculum*, the official newsletter of the Mycological Society of America, has dedicated a section (Mycology Online) to report new sites. The Web has also become a means of electronic publishing, and several journals can now be accessed online. For example, the *Rhizoctonia* newsletter or AnaNet (the Anamorph Information Network) can be consulted directly on the net and the most recent advances in phylogenetic research can be almost instantly seen in Tree of Life. Several disciplines have now started to publish directly on the Net and in the future we can expect to see peer-reviewed mycological journals in electronic form.

Until recently, Web browsers accessed static multimedia files and there was no means to interact with the contents of the files. The development of Java, a software originally designed for programming consumer electronics and linking domestic appliances, by Sun Microsystems, has now delivered the possibility of interacting with applications by downloading small programs, called applets, and running them embedded in a Web page (Oliver et al. 1996). Java applets are distributed as byte code and must then be interpreted by a Java virtual engine, a software that translates the code into instructions for the computer's operating system. The virtual engine is small and is incorporated into Web browsers such as Netscape or Microsoft Internet Explorer. Sun Microsystems describes Java as "a simple, object-oriented, distributed, interpreted, robust, secure, architecture neutral, portable, high-performance, multi-threaded, and dynamic language". In lay terms, Java allows programmers to write small applications that can be distributed over the Internet and used directly on all desktops. Java applets have an almost limitless potential which can be effectively used also by taxonomists. As Java is not platform-dependent, applets created by a programmer can run similarly on any platform (Unix, Windows, or Mac), provided a Java-compatible browser such as Netscape (Oliver et al. 1996) or Microsoft Internet Explorer is used. Thus, computer keys or expert systems for the identification of fungi could easily be put on the Web and used by taxonomists independently of the operating system and software they run on their home machines. First examples of Web-based keys are already available.

VII. Computer Software for the Mycologist

A large number of software products are available and they cannot all be discussed here. This chapter is restricted to programs running primarily on personal computers, although for some of them mainframe versions exist. We mention only software we have used; our list is by no means exhaustive and in no way reflects on the quality of the programmes not mentioned. We have worked with all those we mention more or less extensively, depending on the package, and, in a few cases, our opinion may be biased by the personal feeling derived from the tests we have performed. In addition, new software releases may have additional features and/or cross-platform compatibilities than those we have summarized here. We strongly encourage testing a software before buying it, because personal feeling cannot be replaced by any recommendation.

Minitab (DOS, Windows 3.1, Windows 95 and NT, PowerMac versions available; Minitab Ltd.) offers extensive testing and exploratory data analysis capabilites. It is a powerful, easy-to-use program, with good import/export capabilities and a good user interface. Its macrolanguage easily permits the creation of custom operations and the graphics capability of the package is very good.

MVSP (Windows 95 & NT; Kovacs Computing Services; http://www.kovcomp.com/) performs ordination and cluster analyses [principal components, principal coordinates, correspondence analysis (simple, canonical, detrended), as well as cluster analysis using 20 different distance or similarity measures and 7 clustering strategies]. The software is excellent, very user-friendly and the manual clear and understandable. Kovacs provides also XLSTAT, an add-in for Excel that performs a nice array of multivariate statistics. We found both programs to rank among the best packages dedicated to multivariate analyses that we were able to evaluate.

NCSS 6.0 (Windows only; Statistical Solutions, Ltd.) for Windows is a comprehensive, accurate package, with an outstanding array of options in the exploratory and confirmatory analysis. It is also user-friendly and the Windows interface is self-explanatory. Import and export of data from and to commercial packages such as spreadsheets and word processors is comparatively easy and straightforward.

nQuery Advisor (Windows only; Statistical Solutions Ltd.) is the only package we know that provides investigators and statisticians with accurate sample size calculations. nQuery Advisor can be used to compute sample size for a number of different study designs. Although it has been developed explicitly to aid in the planning of clinical trials, its use is by no means restricted to this field. The package provides a structured index to assist in specifying the research goal as well as an array of aids to the novice and experienced statistician. It also gives sample size justification statements in plain English which can be used in writing up grant proposals or research papers. Its use is straightforward and the user interface very intuitive.

NTSYS-pc (DOS; no Mac version available; Exeter Software) is the standard package used by most taxonomists. The software is intended for use in numerical taxonomy and integrates all standard methods needed in the field, including correspondence analysis. Its documentation is excellent and the user interface self-explanatory.

S-PLUS [Windows (version 3.1), Unix (version 3.3), no Mac version] is a tool for data analysis as well as a programming language. New functions and preferences can be added by the user, thus making the program very flexible. The package, however, is not very user-friendly, although communication between user and program is interactive. On the other hand, it is an excellent tool for exploratory data analysis especially because of its outstanding graphic modules. The software provides extensive possibilities for formulating and testing statistical models such as regression, analysis of variance, time series analysis or generalized linear models and it offers a flexibility equal to SAS.

SAS (Windows 3.1, Windows 95 and NT, OS/2, Mac, Unix; SAS Institute, Inc.) is an impressive package that features almost all statistical procedures that can be routinely used. The strong programming language allows complete customization of tests and interface. The software, on the other hand, is not user-friendly and requires good knowledge of statistics.

SimCA (DOS only; Greenacre 1990, 1993) is a dedicated statistical package that performs simple correspondence analysis, although multiple correspondence analysis can be carried out with the program as well. It is user-friendly, intuitive, and an excellent tool to perform correspondence analysis for numerical taxonomy purposes.

Statistica (DOS, Windows and Mac; StatSoft, Inc.) is very similar to Systat in its look and user-friendliness, but features a larger array of options, including correspondence analysis and useful AutoLayout wizards. The manuals are somewhat less well organized and exhaustive than those offered by Systat, but equally useful and clear.

Systat (DOS, Windows and Mac; now marketed by SPSS, Inc.) is a statistical package with strong exploratory data analysis capabilities. It also offers good models for multivariate analysis (nonlinear regression, loglinear models, principal components analysis, discriminant and factor analysis), smoothing, special designs, time series analysis, and a large range of clustering algorithms. A module for correspondence analysis, is present The software is very user-friendly: the manuals are exhaustive and give an excellent introduction to statistics as well.

Unistat (Windows only; Unistat Ltd.) is an excellent package that offers a large array of statistical procedures, including exploratory data analysis and multivariate models. The software integrates very well with Excel and, in general, with the Microsoft Office suite, yet the package cannot be defined exactly as the statistical software everybody can use. It is a statistics tool for statisticians, or at least researchers with good knowledge in the theory and practice of statistical analysis.

The major computer software packages available for *phylogenetic analysis* have been described in detail by Swofford and Olsen (1990). The new releases of PAUP (PAUP 3.0s + 4; Swofford 1993) and MacClade (Sinauer Associates, Sunderland, MA) are easy-to-use, yet very powerful and fast packages. They both run on MacIntosh computers and share the same file format, thus allowing easy data transfer from one application into the other. PHYLIP (J. Felsenstein, Seattle, WA) is a collection of several programmes and is available on both PC and Mac. Swofford et al. (1996) provide a more critical overview of phylogenetic software programs available on this subject.

Among the database software packages, *Microsoft Access* (Microsoft Inc.) is probably the best known. This relational database offers a wealth of options, including the possibility of interfacing the database with applications written in Visual Basic that could be used to construct computerized keys or expert systems. Unfortunately, the software is available only for Windows and no version for the Mac exists. Cross-platform compatibility is offered by Microsoft through *FoxPro*, a relational package with a very good performance.

FileMaker Pro (Mac and Windows; Claris Corp.) is a slick, relational database that allows the preparation of relatively sophisticated applications, even if the options offered cannot be compared to those of MS Access. On the other hand, the software is very user-friendly, with a simple and intuitive interface and a powerful Script-Maker programming tool. It also shares the same file format across the two platforms.

AskSam 3.0 (DOS and Windows; Seaside Software, Inc.) is a software that handles freeform text (such as E-mail, memos, etc.) as well as structured information. The application is easy to use and the search engine surprisingly fast. AskSam offers hypertext features that allows creation of HTML documents for use on the Net. The software has been used by some authors to prepare computer keys for the identification of fungal taxa (e.g., Wolfaardt et al. 1992). In the Windows version, graphics and pictures can be inserted in the documents using either OLE or the usual Clipboard exchange format.

HyperCard 2.4.1 (Mac only; Claris Corp.) is a software construction tool that allows the production of standalone applications. It integrates good database capabilities with excellent multimedia tools. The software can be used to create databases as well as to produce front-end tools to expert systems or other applications. HyperCard supports both Hypertalk, an easy-to-learn, yet powerful programming language as well as AppleScript. The software provides ideal conditions to prepare interactive identification keys. A key to *Rosellinia* species from the temperate zones is available (Petrini and Petrini, unpubl.). *SuperCard* (Incwell, Ltd; http://www.incwell.com) has the same structure of HyperCard, but offers a more comprehensive set of features, more programming power and an excellent web support.

Among the bibliography packages, *EndNote Plus* (Mac and Windows; Niles and Assoc., Inc.) is an excellent package, which can be used as a standalone or together with word processing software. It is easy to use and set up, and direct import of data from searches are good. It also runs smoothly across platforms.

Reference Manager (Mac and Windows; RIS) is a good package with essentially the same functionality of EndNote Plus. While it is very effective when dealing with large databases, we have been disappointed by its performance and user-friendliness when compared with EndNote Plus. The cross-platform compatibility is also not seamless.

ProCite (Mac and Windows; RIS) is a full-featured reference manager with all capabilities present in the previous packages. In addition, it offers excellent import and export facilities and manages information directly from the Web. The company claims that in the newest release a Web page can be launched directly from within ProCite, but we have not been able to test it.

References

Assigbetse KB, Fernandez D, Dubois MP, Geiger JP (1994) Differentiation of *Fusarium oxysporum* f.sp. *vasinfectum* races on cotton by random amplified polymorphic DNA (RAPD) analysis. Phytopathology 84:622–626

Atchley WR, Gaskins CT, Anderson D (1976) Statistical properties of ratios. I. Empirical results. Syst Zool 25:137–148

Backeljau T, De Bruyn L, De Wolf H, Jordaens K, Van Dongen S, Verhage R, Winnepenninckx B (1995) Random amplified polymorphic DNA (RAPD) and parsimony methods. Cladistics 11:119–130

Barron JJ (1993) Putting fuzzy logic into focus. Byte (April 1993):111–118

Bridge P (1988) Computer assisted taxonomy of filamentous microfungi. In: Houghton DR, Smith RN, Eggins HOW (eds) Biodeterioration 7. Selected papers presented at the 7th International Biodeterioration Symposium, Cambridge, UK, September 1987, Elsevier, Barking, UK, pp 73–77

Brunner F, Petrini O (1992) Taxonomy of some *Xylaria* species and xylariaceous endophytes by isozyme electrophoresis. Mycol Res 96:723–733

Bruns TD, White TJ, Taylor JW (1991) Fungal molecular systematics. Annu Rev Ecol Syst 22:525–564

Cleveland WS (1985) The elements of graphing data. Wadsworth Advanced Books and Software, Monterey, California

Cox DR (1958) Planning of experiments. Wiley, New York

Cox E (1994) The fuzzy systems handbook: a practitioner's guide to building, using, and maintaining fuzzy systems. AP Professional, Boston

Dunn G, Everitt BS (1982) An introduction to mathematical taxonomy. Cambridge University Press, Cambridge, UK

Edwards M, Morse DR (1995) The potential for computer-aided identification in biodiversity research. TREE 10:153–158

Fienberg SE (1980) The analysis of cross-classified categorical data. MIT Press, Cambridge, MA

Fleiss JL (1981) Statistical methods for rates and proportions. Wiley, New York

Flury B, Riedwyl H (1988) Multivariate statistics – a practical approach. Chapman and Hall, London

Forey PL, Humphries CJ, Kitching IJ, Scotland RW, Siebert DJ, Williams DM (1992) Cladistics – a practical course in systematics. Clarendon Press, Oxford

Gower JC (1985) Measures of similarity, dissimilarity, and distance. In: DeGroot MH, Ferber R, Frankel MR, Seneta E, Watson GS (eds) Encyclopedia of statistical sciences. Vol 5. Lindberg condition to multitrait-multimethod matrices. Wiley, New York, pp 397–405

Green RH (1979) Sampling design and statistical methods for environmental biologists. Wiley, New York

Greenacre MJ (1981) Practical correspondence analysis. In: Barnett V (ed) Interpreting multivariate data. Wiley, Chichester, pp 119–146

Greenacre MJ (1984) Theory and applications of correspondence analysis. Academic Press, London

Greenacre MJ (1990) SimCA Version 2 User's manual. Greenacre Research, Irene, South Africa

Greenacre MJ (1992) Correspondence analysis in medical research. Stat Meth Med Res 1:97–117

Greenacre MJ (1993) Correspondence analysis in practice. Academic Press, London

Greenacre MJ, Hastie T (1987) The geometric interpretation of correspondence analysis. J Am Stat Assoc 82:437–447

Guttman LA (1968) A general nonmetric technique for finding the smallest coordinate space for a configuration of points. Psychometrika 33:469–506

Hartigan JA (1975) Clustering algorithms. Wiley, New York

Hartigan JA, Wong MA (1979) A K-means clustering algorithm: algorithm AS 136. Appl Statist 28:126–130

Hedberg S (1993) New knowledge tools. Byte (July 1993):106–111

Hennig W (1966) Phylogenetic systematics. University of Illinois Press, Urbana

Jun Y, Bridge PD, Evans HC (1991) An integrated approach to the taxonomy of the genus *Verticillium*. J Gen Microbiol 137:1437–1444

Korf RP, Zhuang WY (1985) A synoptic key to species of *Lambertella* (Sclerotiniaceae), with comments on a version prepared for TaxDat, Anderegg's computer program. Mycotaxon 24:361–386

Krebs CJ (1989) Ecological methodology. Harper Collins, New York

Kruskal JB (1964a) Multidimensional scaling by optimizing goodness of fit to a nonmetric hypothesis. Psychometrika 29:1–27

Kruskal JB (1964b) Nonmetric multidimensional scaling: a numerical method. Psychometrika 29:115–129

Kuhls K, Lieckfeldt E, Samuels GJ, Kovacs W, Meyer W, Petrini O, Gams W, Börner T, Kubicek CP (1996) Molecular evidence that the asexual industrial fungus *Trichoderma reesei* is a clonal derivative of the ascomycete *Hypocrea jecorina*. Proc Natl Acad Sci USA 93:7755–7760

Leuchtmann A, Petrini O, Samuels GJ (1996) Isozyme subgroups in *Trichoderma* section *Longibrachiatum*. Mycologia 88:384–394

Lieckfeldt E, Samuels GJ, Nirenberg HI, Petrini O (1999). A morphological and molecular perspective of *Trichoderma viride*: Is it one or two species? Appl Environm Microbiol 65:2418–2428

LoBuglio KF, Pitt JI, Taylor JW (1993) Phylogenetic analysis of two ribosomal DNA regions indicates multiple independent losses of a sexual *Talaromyces* state among asexual *Penicillium* species in subgenus *Biverticillium*. Mycologia 85:592–604

Ludwig JA, Reynolds JF (1988) Statistical ecology; a primer on method and computing. John Wiley, New York

Lutzoni F, Vilgalys R (1995) Integration of morphological and molecular data sets in estimating fungal phylogenies. Can J Bot 73 (Suppl 1):S649–S659

Nei M (1972) Genetic distance between populations. Am Nat 106:283–292

Nei M (1978) Estimation of average heterozygosity and genetic distance from a small number of individuals. Genetics 89:583–590

Nei M, Li WH (1979) Mathematical model for studying genetic variation in terms of restriction endonucleases. Proc Natl Acad Sci USA 76:5269–5273

Oliver D, Ernst W, Fincher G (1996) Netscape 2 unleashed. Sams. Net Publishing, Indianapolis

Pankhurst RJ (1991) Practical taxonomic computing. Cambridge University Press, Cambridge, UK

Parsaye K, Chignell M, Khoshafian S, Wong H (1989) Intelligent databases. Object-oriented, deductive hypermedia technologies. Wiley, New York

Petrini LE (1992) *Rosellinia* species of the temperate zones. Sydowia 44:169–281

Petrini LE, Petrini O, Sieber TN (1988) Host specificity of *Hypoxylon fuscum*: a statistical approach to the problem. Sydowia 40:227–234

Petrini LE, Petrini O, Francis SM (1989) On *Rosellinia mammaeformis* and other related species. Sydowia 41:257–276

Petrini O, Petrini LE (1996) Polyphasische Taxonomie. Probleme und Methoden. Mycol Helv 8(2):83–90

Petrini O, Petrinic LE, Laflamme G, Ouellette GB (1989) Taxonomic position of *Gremmeniella abietina* and related species: a reapprisal. Can J Bot 67:2805–2814

Petrini O, Rusca CV, Szabo I (1990) ASCUS: an error-tolerant mycological classification system. Sydowia 42:273–285

Petrunak DM, Christ BJ (1992) Isozyme variability in *Alternaria solani* and *A. alternata*. Phytopathology 82:1343–1347

Samuels GJ, Petrini O, Kuhls K, Lieckfeldt E, Kubicek CP (1998) The *Hypocrea schweinitzii* complex and *Trichoderma* sect. *Longibrachiatum*. Stud Mycol 41:1–54

Sieber-Canavesi F, Petrini O, Sieber TN (1991) Endophytic *Leptostroma* species on *Picea abies*, *Abies alba*, and *Abies balsamea*: a cultural, biochemical, and numerical study. Mycologia 83:89–96

Sieber TN, Petrini O, Greenacre M (1997) Correspondence analysis as a powerful tool in fungal taxonomy. Syst Appl Microbiol 21:433–441

Sneath PHA (1989) Analysis and interpretation of sequence data for bacterial systematics: the view of a numerical taxonomist. Syst Appl Microbiol 12:15–31

Sneath PHA, Sokal RR (1973) Numerical taxonomy. WH Freeman, San Francisco

Sokal RR, Rohlf FJ (1981) Biostatistics, 2nd edn. Freeman, San Francisco

Sokal RR, Sneath PHA (1963) Principles of numerical taxonomy. Freeman, San Francisco

Sreenivasaprasad S, Mills PR, Meehan BM, Brown AE (1996) Phylogeny and systematics of 18 *Colletotrichum* species based on ribosomal DNA spacer sequences. Genome 39:499–512

Svendsen A, Frisvad JC (1994) A chemotaxonomic study of the terverticillate *Penicillia* based on high performance liquid chromatography of secondary metabolites. Mycol Res 98:1317–1328

Swofford DL (1993) PAUP: Phylogenetic analysis using parsimony. Version 3.0s + 4. Illinois Natural History Survey, Champaign, Illinois

Swofford DL, Olsen GJ (1990) Phylogeny reconstruction. In: Hillis DM, Moritz C (eds) Molecular systematics. Sinauer, Sunderland, MA, pp 411–501

Swofford DL, Olsen GJ, Waddell PJ, Hillis DM (1996) Phylogenetic inference. In: Hillis DM, Moritz C, Mable BK (eds) Molecular systematics, 2nd edn. Sinauer Associates, Sunderland, MA, pp 407–514

Systat (1996). Statistics, version 8 edition. SPSS Inc., Chicago, IL

Thrane U (1990) Grouping *Fusarium* section *Discolor* isolates by statistical analysis of quantitative high performance liquid chromatographic data on secondary metabolite production. J Microbiol Meth 12:23–39

Trigiano RN, Caetano-Anollés G, Bassam BJ, Windham MT (1995) DNA amplification fingerprinting provides evidence that *Discula destructiva*, the cause of dogwood anthracnose in North America, is an introduced pathogen. Mycologia 87:490–500

Tukey JW (1977) Exploratory data analysis. Addison-Wesley, Reading, MA

Whalley AJS, Whalley MA (1977) Stromal pigments and taxonomy of *Hypoxylon*. Mycopathologia 61:99–103

Wolfaardt JF, Wingfield MJ, Kendrick WB (1992) Synoptic key and computer database for identification of species of *Ceratocystis sensu lato*. S Afr J Bot 58:277–285

Xu CQ, Wang TZ, Hao H, Wu LJ, Wei JH (1991) Application of fuzzy cluster analysis to epidemic dynamics of grey leaf spot of poplars. For Pest Dis 1, 6-8:33–35

Evolution and Speciation

9 Speciation Phenomena

Paul A. Lemke†

CONTENTS

I.	Introduction	219
II.	Incompatibility Systems	219
A.	Homogenic Incompatibility	219
B.	Heterogenic Incompatibility	220
III.	Other Factors that influence Gene Flow	220
A.	Isolating Mechanisms	220
	1. Homothallism (Self-Fertility)	221
	2. Ecological Barriers	222
	3. Apomixis or Loss of Fertility	223
B.	Hybridization Potential	223
IV.	Conclusion	225
	References	225

I. Introduction

Nature abhors categories. Of all categories, the one most difficult to reconcile with nature is the biological species. The fungal species is no exception, and, in certain respects, the fungi pose special problems for species delimitation. These problems are the subject of this chapter.

The most objective and widely accepted definition of the species is based on genetic homology. Simply defined, the *species* is an integral system for genetic recombination, and members of a given species are expected to share either directly or tangentially in a common gene pool.

It is apparent from what has been repeatedly observed that not all species of fungi conform with this definition. Gene flow within a fungal species may be discontinuous or even negligible. Moreover, related species of fungi may hybridize and thereby lose their specific identity. Sterility barriers within species and hybridization between species undermine the concept of the species as a well-integrated, reproductively isolated system for gene flow.

Nonconformity with the common-gene-pool concept of the species has long been recognized in nonfungal life-forms, principally in the higher plants (Stebbins 1950), and is considered to be indicative evidence that speciation is the result of divergent and convergent evolution of existing populations through change in gene frequency. Clearly, in fungi, as in other organisms, the population, not the species, is the unit of evolution. Populations that constitute a species embody change, and that change is brought about principally through three phenomena: mutation, selection, and recombination.

In considering these phenomena, Esser and Hoffmann (1977) proposed a modification of the species concept: *populations* (*races*) belong to different species when the failure to interbreed and to produce viable offspring in nature is not caused by genetic parameters operating in the completion of the sexual cycle.

Mutation induction and selective pressures are largely extrinsic phenomena–influences of the environment. The characteristic intrinsic feature of a species is its potential for recombination of genetic material. In this regard, fungi display considerable diversity ranging from sexual dimorphism to apomixis.

II. Incompatibility Systems

A. Homogenic Incompatibility

Recombination in most fungi is under strict genetic control through systems of incompatibility. Systems of incompatibility determine the breeding potential of a species in the absence of sexual

---

Department of Botany and Microbiology, Auburn University, Auburn, Alabama 36849-5407, USA
† After finishing the first draft of this manuscript unfortunately Paul A. Lemke passed away on November 18, 1995. Since this chapter needed only editorial assistance, I have refrained from any updating in order to honor the merits of Paul Lemke as coeditor of *The Mycota*. Thus, this chapter has to be understood as Paul's posthumous contribution to mycology, a discipline of science to which he dedicated his scientific efforts. Karl Esser, Series Editor

dimorphism. Two functional types of incompatibility are now recognized in fungi, and these should not be confused, since one promotes outbreeding and the other promotes inbreeding. The first type is common-factor or homogenic incompatibility, commonly referred to as heterothallism (Blakeslee 1904, 1906; Whitehouse 1949a,b; Esser and Raper 1965). Homogenic incompatibility favors recombination through association of dissimilar alleles. The mating-type factors of bipolar and tetrapolar fungi are the elements of homogenic incompatibility. Homogenic incompatibility is analogous to sexuality only insofar as it comfers self-sterility and cross-fertility upon individuals in a population. *Schizophyllum commune* demonstrates homogenic incompatibility at its best (Raper, 1966). This fungus with a two-factor, multiple-allelic system of incompatibility enjoys considerable potential for outbreeding. *Schizophyllum* fits well the concept of the species as an integral system for gene flow, since allopatric isolates of *S. commune* are panmictic, and incompatibility in that fungus is predictable on the basis of common mating-type factors. *Schizophyllum commune* and *Coprinus cinereus* the most extensively studied of fungi in this regard, may not be typical of fungi with regard to their potential for recombination. Ecological barriers and/or genetic factors operate in these fungi as isolating mechanisms to delimit gene flow. Although fungal populations have not been studied extensively in this regard, there is already sufficient evidence to indicate that isolating mechanisms act not only between species of fungi but within species of fungi as well.

B. Heterogenic Incompatibility

This brings us to a second functional type of incompatibility prevalent in fungi (Biggs 1937; Burnett and Boulter 1963; Esser 1966, 1971; Grindle 1963; Mounce and Macrae 1938). This type of incompatibility is heterogenic (Esser 1966; Esser and Blaich 1994). Heterogenic incompatibility, in contrast with homogenic incompatibility, restricts recombination when alleles differ. It therefore promotes inbreeding and homozygosity. Many of the sterility barriers recognized in species of fungi apparently have a genetic basis in heterogenic incompatibility. Heterogenic incompatibility may be superimposed on homogenic incompatibility as it is in *Podospora anserina* (Esser 1956, 1959a,b; Rizet and Esser 1953). As Esser has pointed out, allelic differences at several loci are involved in producing partial or complete intersterility among isolates of *P. anserina*. Heterogenic incompatibility operates at different stages of the life cycle. It may influence heterokaryon formation or any of several sexual events leading to the formation of a zygote. Heterogenic incompatibility may function in the absence of homogenic incompatibility, as it does in certain homothallic and anamorphic (imperfect) species of *Aspergillus* (Caten 1971; Grindle 1963; Jinks et al. 1966).

Reduced potential for gene flow in fungi can be influenced by genetic factors other than those underlying heterogenic incompatibility. Homothallism (self-fertility) in its various forms restricts recombination. Secondary or heterokaryotic homothallism promotes inbreeding in heterothallic (self-incompatible) fungi. Primary or homokaryotic homothallism is de facto loss of sexual competence, since no recombination is effected through meiosis from a homozygote. The normal absence of sexuality in the imperfect fungi represent considerable, if not complete, loss of potential for recombination. Parasexuality (see Sect. III. A.3) in natural populations of these fungi doubtfully compensates for such loss (Caten 1971).

Cytogenetic mechanisms, principally inversions and translocations, are known to control genetic recombination in higher plants and animals (for review see Stebbins 1950; Ehrlich and Holm 1963). Fungal populations are poorly understood in this regard. Since chromosomal abberations operate as underlying mechanisms for postzygotic isolation in other organisms, such mechanisms conceivably could also fractionate natural populations of fungal species. Perkins (1972) has shown that a translocation involving the mating-type locus of *Neurospora crassa* can lead to inviable progeny or to progeny inhibited for growth. The latter progeny are heterozygous for mating-type and contain duplications of the translocated region. The former progeny represent corresponding lethal deficiencies of the translocated region.

III. Other Factors that Influence Gene Flow

A. Isolating Mechanisms

Prezygotic isolation, however, appears to be more characteristic of fungi and is directed frequently toward the formation and maintenance of heterokaryons. Regulation of heterokaryosis provides fungi with a unique opportunity to experiment with genetic isolation, much as behavioral patterns in animals enforce species recognition and prevent gene wastage (Mayr 1970). Heterogenic controls of heterokaryosis and of sexual events subsequent to heterokaryosis clearly operate not only between species but also within certain species of fungi. In *Podospora anserina* such controls are polygenic. It is reasonable to assume that incompatibility between species of fungi is also heterogenic and determined through numerous genetic loci. Fungi, relative to higher plants, have a limited potential for interspecific hybridization. As a rule, restrictions to heterokaryon formation between fungal species are stringent. Confirmed examples of interspecific hybridization in the fungi, although indeed rare,, are available for all major taxonomic groups of the true fungi. Undoubted instances of such interactions have been reported in lower fungi, principally *Allomyces* (Emerson and Wilson 1954); in ascomycetes, both in yeasts (Winge and Roberts 1949) and filamentous genera, *Neurospora* (Dodge 1927; Howe and Haysman 1966) and *Cochliobolus* (Nelson 1963); and finally in basidiomycetes, notably in smuts (Holton 1931; Holton and Fischer 1941; Holton and Kendric 1956), and recently in *Sistotrema* of the homobasidiomycetes (Lemke 1966, 1969; Ullrich and Raper 1975).

Interspecific hybridization, when it does occur in fungi, does not occur with impunity. Hybrid progeny show reduced viability, an indication that postzygotic isolation may operate as well in fungi to maintain separate species.

1. Homothallism (Self-Fertility)

The resupinate basidiomycete *Sistotrema brinkmannii* embodies several of the speciation phenomena. Biggs (1937) recognized six intersterile groups among 14 isolates of this fungus. Two groups were bipolar, three groups were tetrapolar, and the remaining group was homothallic. All six groups were sympatric and morphologically similar on the basis of hymenial structures. Biggs in 1937 suggested that *S. brinkmannii* represented a minimum of three cryptic species, a homothallic species, a bipolar species, and a tetrapolar species, and that sterility barriers were present in the two heterothallic species. Between 1963 and 1966 reinvestigation of this system essentially confirmed Biggs' earlier observations (Lemke 1966, 1969). Sterility barriers for heterokaryon formation exist in all three component species, the homothallic as well as both heterothallic species. The homothallic species exhibits primary or homokaryotic homothallism, and homokaryons are phenotypically dikaryotic with clamp connections. Dikaryosis in heterothallic species is heterokaryotic and controlled through multiple-allelic incompatibility.

Homothallic strains were paired with heterothallic strains, and genetic evidence for heterokaryosis was obtained in one instance and subsequently confirmed by Ullrich and Raper (1975). This nutritionally forced heterokaryon between a homothallic strain and a bipolar strain (mating type A1) proved to be dikaryotic in phenotype and was brought to sporulation. From this cross a sample of 109 germinating basidiospores was isolated. From this sample only eight mycelia developed, and an analysis of these sparse progeny revealed a number of interesting points: 1) seven of the eight progeny were recombinant; 2) only one of eight progeny was homothallic (phenotypically dikaryotic) and this strain was recombinant for nutritional markers; it sporulated and appeared normal in all respects; 3) all seven bipolar (nondikaryotic) progeny were compatible with a nonparental bipolar strain of A2 mating-type, 4) only six of these bipolar progeny were incompatible with the parental bipolar strain A1.

In the later study by Ullrich and Raper (1975), 613 progeny from 9 different homothallic bipolar heterothallic crosses of, nutritionally forced hybrids that fruited also showed many recombinant progeny and low viability of the progeny (11–56% germination). However, the homothallic and bipolar traits segregated one to one with all of the bipolar progeny having the same mating types as their bipolar progenitors.

In higher basidiomycetes the homothallic and heterothallic conditions may not be phylogenetically as distinct as they outwardly appear to be. Raper and coworkers (1965) and Parag (1962) demonstrated that the A and B incompatibility factors in the tetrapolar *Schizophyllum commune* are subject to mutational impairment. The homokaryon of *Schizophyllum* carrying mutations for both factors, A mut B mut, is phenotypically dikaryotic and fertile (Koltin 1970). This doubly mutant homokaryon provides experimental evidence for derivation of a homothallic condition from a heterothallic one. The idea that homothallism evolved from heterothallism in higher basidiomycetes through specific mutations is at least mechanistically feasible.

The homothallic (self-fertile) species of *Sistotrema brinkmannii* has been investigated further through induced mutation (Lemke 1966). Mutations that lead to self-sterility in homokaryons were obtained readily, and these often disrupted dikaryosis. Four distinct phenotypes were observed among self-sterile homokaryons. Hyphae were either 1) simple-septate, 2) irregularly clamped with scattered pseudoclamp-connections, 3) regularly clamped and dikaryotic, or 4) dikaryotic with aborted or immature basidia. Forty of the self-sterile mutations were analyzed through complementation analysis, paired in all possible combinations. In this analysis for cross-fertility, only one of the 40 mutations proved to be dominant, and all crosses involving it exhibited heterokaryon incompatibility. Otherwise, the mutations to self-sterility were recessive and crosses resulted in normal dikaryosis and sporulation in practically all cases. Six of the self-sterility mutations were genetically mapped and are distributed on three linkage groups.

These results with a homothallic strain of *Sistotrema* are comparable to those from studies conducted with homothallic ascomycetes by other investigators (El Ani and Olive 1962; Esser and Straub 1958: Olive 1958; Wheeler 1954). Studies with both groups of homothallic fungi indicate that sexual progression in a homothallic homokaryon comprises a large number of distinct stages subject to mutational impairment. In *Sistotrema* the loci for 40 such mutations are scattered and their number, although uncertain, is a minimum of 6 and a maximum of 36, probably closer to the latter number. It should be emphasized that none of the mutations to self-sterility in *Sistotrema* formed a pattern of heterothallism comparable to bipolarity or tetrapolarity of higher basidiomycetes. These mutations rather constitute a separate order of phenomena leading to self-sterility and cross-fertility. They are unrelated to multiple-allelic heterothallism and most likely represent mutations that modify any of the many structural genes that encode for dikaryosis and sporulation in basidiomycetes.

The relatively simple biallelic form of heterothallism present in the ascomycetes may have been derived from homothallic ancestry through complementary mutations to self-sterility. This hypothesis was proposed independently by Olive (1958) and Wheeler (1954) and is supported principally by studies with *Sordaria fimicola* (El Ani and Olive 1962). In that homothallic fungus, two very closely linked mutations to self-sterility have been obtained which exhibit complementation for cross-fertility. Two complementary, nonrecombinable mutations for such self-sterility would, in essence, constitute the biallelism characteristic of bipolarity in ascomycetes. The suggestion that heterothallism in ascomycetes evolved repeatedly from homothallic forms through intragenic self-sterility mutations is at least plausible.

2. Ecological Barriers

Isolation among fungi can be brought about by ecological factors, microecological as well as macroecological (Kukkonen 1971). The physiological races of rust fungi demonstrate intraspecific isolation imposed through host specialization (Stakman and Harrar 1957). Host and parasite are genetically balanced with respect to resistance and virulence. This balance is restrictive for outbreeding and maintains the physiological race (Flor 1956; Person 1966). Although host-parasite associations are not, strictly speaking, heterokaryotic, they do involve a highly specialized form of genetic interspecific "complementation".

Ecological barriers presumably operate, but are not always apparent, in saprobic species. Intersterile races have been recognized among North American isolates of *Fomes pinicola*, a wood-rotting basidiomycete. Isolates from the same tree may belong to separate races (Mounce and Macrae 1938). An extreme case for the presence of a sterility barrier within a single ecotype of a given species involves *Mycocalia denudata*, a bipolar gasteromycete (Burnett and Boulter 1963). Two genetically isolated races of this fungus were obtained from a single fructification. In this instance, the one apparent fructification encompassed two confluent but reproductively isolated basidiocarps. *Mycocalia denudata* illustrates well a problem inherent in the study of fungal populations: what constitutes an individual? A fungus in nature may represent a genetic mosaic. The intermingling of genetically distinct heterokaryons complicates resolution of fungal populations into individual phenotypes. Any statistical analysis of gene frequencies within fungal populations must take this into consideration. An extreme example of mosaicism has been observed in an "isolate" of *Schizophyllum fasciatum*.

From a confluent collection of 7 fruitbodies of this basidiomycete obtained from a single fence post in Mexico, over 48 incompatible A factors (homogenic) were recovered (Raper 1966). Such heterogeneity confuses a reasonable or rational criterion for an individual in a population. An opposite extreme in this regard is evident from analysis of a population of the mushroom *Armillaria bulbosa* from a Michigan forest (Smith et al. 1992). An isogenic mycelium of this fungus covered more than 15 ha and all individual fruit-bodies were identical for incompatibility components. Collectively, this mycelium may constitute the largest "individual" organism known to date.

Less is known about a rather large group of fungi, the imperfect or anamorphic fungi, and of their potential for speciation. These fungi are considered to be the derived or relic species of perfect or teleomorphic ancestors, and there is ample taxonomic evidence to support this conclusion. The fungi have still retained specific identity in the formal absence of sexuality. The large number of imperfect species and their diversity provide convincing testimony for successful exploitation of prezygotic, or, more correctly, azygotic, isolation by fungi to maintain the integrity of a species.

3. Apomixis or Loss of Fertility

Parasexual or somatic recombination was discovered 40 years ago in *Aspergillus nidulans* (Pontecorvo et al. 1953) and has been recognized experimentally in many other fungi (Bos 1996). Parasexuality offers a recourse for some recombination in the absence of meiosis, and, in view of this, the parasexual process should have special significance among populations of anamorphic fungi. This, however, does not appear to be the case in *Aspergillus* (Caten 1971; Grindle 1963; Jinks et al. 1966). Conspecific isolates of *Aspergillus* have been examined in considerable detail for competence to form heterokaryons. Heterokaryon incompatibility has proven to be rampant among wild-type isolates of a given species. For example, Caten (1971) reported that among 126 combinations involving 21 isolates of *A. versicolor* only 3 combinations or about 2% of the sample formed heterokaryons. Heterokaryon incompatibility effectively precludes parasexual recombination. Thus, parasexuality may simply be an incidental derangement of mitosis with no real significance for genetic recombination in natural populations.

The trend to restrict recombination is not exclusively that of anamorphic fungi. Many fungi with known perfect or telomorph states are essentially asexual species in nature. The incidence of sexual reproduction, even in lower fungi (Mucorales) is recognized to be low because of the poor frequency for germination of zygotes and the common occurrence within the order of sexually neutral strains (Blakeslee et al. 1927). Even in higher fungi, sex is often vestigial. Witness such species as *Emericella (Aspergillus) nidulans*, *Neurospora (Monilia) sitophila*, or *Thanatephorus cucumeris* (= *Rhizoctonia solani*). Although these fungi have the potential to exhibit metagenesis through telomorphy, they are for all intents and purposes imperfect or anamorphic secies.

B. Hybridization Potential

The following diagram (Fig. 1) is an attempt to provide a synoptic outline of speciation phenomena in fungi. Three basic phenomena underlie speciation in fungi, as in other biological systems. These are mutation, selection, and recombination. Although the induction of mutations and the pressures of selection are acknowledged as significant factors in fungal speciation, these phenomena have not been discussed extensively in this chapter, which has been concerned rather with the competence of a fungal population to exchange genetic material.

It has been pointed out that fungal species vary considerably in this ragard. Certain fungi are inbreeders and limit gene flow through any of several mechanisms, i.e., heterogenic incompatibility, homothallism, apomixis, or restrictive ecological adaptations. On the other side of the ledger are fungi that typically outbreed and recombine efficiently through homogenic incompatibility and panmixis. A few fungal species exhibit morphological differentiation as per sexual dimorphism, and even fewer fungal species are known to converge through hybridization. The best-documented case of interspecific hybridization in fungi involves formation of the natural hybrid, *Allomyces javanicus*, in a cross between related species of different ploidy (Emerson and Wilson 1954). Among viable hybrids from the cross were forms intermediate between the two parents, *A. arbuscula x A. macrogynus*, and karyological data confirmed the hybrid nature of these progeny. Other, and far more recent, examples of hybrid karyotype analysis based on molecular analysis, i.e., restriction polymorphism of DNA sequences

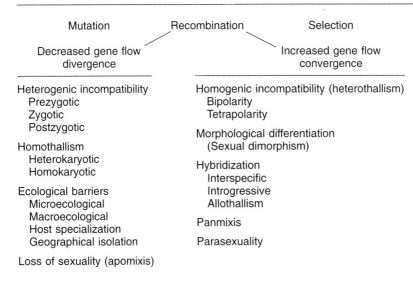

Fig. 1. Synoptical scheme of speciation in fungi. (Lemke 1973)

and/or chromosomal analysis by electrophoresis, are now available.

In another early study of hybridization, Nelson (1963, 1964) crossed 16 species of *Cochliobolus (Helminthosporium)* in all combinations. The majority of crosses were either completely infertile or produced only immature or sterile asci. However, 13 out of 120 crosses yielded progeny, but the viability of ascospores was, in all instances, low. A few of the surviving hybrid progeny were subsequently backcrossed or outcrossed, and, in some instances, second-generation progeny from these crosses exhibited increased viability (Nelson 1964). Improved viability of hybrid progeny through backcrossing or outcrossing has been observed in higher plants and is known as introgressive hybridization (Stebbins 1950).

Some fungi are clearly paradoxical with regard to their potential for gene flow, as populations often combine genetic systems that have opposite effects on recombination. Secondary or heterokaryotic homothallism is often superimposed on homogenic incompatibility. In the bipolar *Mycocalia denudata*, secondary homothallism is determined by a dominant allele (*Pd*) for precocious mitotic division of the four meiotic products in the basidium (Burnett and Boulter 1963). The resultant eight nuclei are distributed at random into four basidiospores. Thus, 50% of the spores are heterokaryotic with respect to mating-type factors. In the absence of the dominant allele for precocious division (*pd*), basidia of *M. denudata* regularly contain four nuclei, and basidiospores are uninucleate upon their inception. Basidiospores at maturity are binucleate but homokaryotic. In *Coprinus bisporus* secondary homothallism is brought about in yet another way – by reduction of spore number per basidium. The selective pressures for secondary homothallism in this species may be related to genetic restriction upon hyphal anastomosis and nuclear migration (Kemp 1971).

As mentioned earlier, heterogenic incompatibility and homogenic incompatibility can coexist in the same species. *Podospora anserina* demonstrates this paradoxical association, and gene flow in this species is further complicated by secondary homothallism.

In view of the diversity among fungi for the control of recombination, it is indeed difficult to generalize as to the significance of gene flow in fungal speciation. Much has been said in the past about the importance of sexuality and of recombination in fungal evolution (Kniep 1928; Hartman 1943; Raper 1966; Whitehouse 1949a,b), but there has been relatively little discussion concerning selective pressures for asexuality and a propensity for nonrecombination in fungi. Sex and recombination are clearly dispensable commodities in a great many fungi, and the selective advantage for their dispensation is not now apparent.

In the absence of recombination, speciation should be brought about principally through the interplay of mutation and selection. Fungi are predominantly haploid organisms, and mutant genotypes in haploid populations can be conserved or eliminated directly through selection.

Thus, fungi, relative to higher diploid organisms, should be more readily susceptible to the effects of mutation and selection (Raper 1968).

Sixty years ago, H.J. Muller (1932) suggested that there was no basic biological reason why evolution, especially in haploid forms, could not go on indefinitely without sexuality. In this opinion, "Sex is not an absolute necessity, it is a luxury. It is necessary only in a relativistic sense, for sexless beings, although often at a temporary advantage, cannot keep up the pace of evolution set by sexual beings. In an evolutionary race between competitive species, the sexless must eventually lose out. "Stebbins (1950) extends this dialectic with two further generalizations. First, "in rapidly reproducing organisms the genetic system that operates is usually one which favors fitness at the expense of flexibility." Secondly, "That genetic system most strongly promoting immediate fitness at the expense of flexibility is one in which sex is absent." Organisms committed to immediate fitness are prone to comprise recombination for the safety of numbers and resort to proliferous asexual multiplication.

Spores and vegetative propagules of several types are formed by fungi in great profusion. These cells represent a vast collection of haploid genotypes which, subject to mutation and selection, could be channelled into a wide variety of specialized ecological situations. Divergent speciation could thus occur in haploid organisms without recombination. However, species generated in this fashion would probably be highly specialized, isolated entities, filling extremely narrow ecological niches. The physiological races of parasitic fungi and the heterogenic races of anamorphic as well as telomorphic fungi conceivably have arisen through such divergence and may represent incipient species.

IV. Conclusion

What then are biological processes that influence fungal speciation? Three aspects of fungal biology discussed herein deverve to be reemphasized: 1) the vast majority of fungi are haploid organisms; 2) fungi have expended a considerable amount of genetic energy to control the formation and maintenance of specific heterokaryons; and 3) the fungi, taken as a group, are paradoxical; against a background of extravagant and varied sexual cycles, fungi have a tendency to restrict gene flow and recombination.

The prevalence of haploidy, the strict control of heterokaryosis, and an irregular but definite trend toward apomixis are attributes of fungal biology. These phenomena undoubtedly influence the operational details of fungal evolution and must be accommodated by any model which intends to explain the origin of fungal species.

Virtually nothing is known about rates of evolution in fungal populations. However, at least three patterns are now evident among fungi with regard to their potential for change in gene frequency through time. The first pattern is that exhibited by imperfect or anamorphic fungi. Evolution in these fungi would be dependent largely upon mutation and selection with, at best, some parasexual activity but generally impaired gene flow. The second pattern is typified by *Schizophyllum commune*. This pattern includes sexuality; gene flow is continuous and controlled through homogenic incompatibility. The third pattern is recognized in *Podospora anserina* as well as in several basidiomycetes, *Sistotrema brinkmannii*, *Mycocalia denudata*, and *Fomes pinicola*. These species combine opposite effects for recombination, and gene flow within the species is discontinuous. Evolution in species of the third pattern should occur in quantum jumps rather than as a continuum. Sewall Wright (1931, 1932) in considering rates of evolution among populations has suggested that most rapid evolution might progress through subpopulations partially isolated which only occasionally or indirectly exchange genetic material. We simply do not now have sufficient data from fungal populations to integrate the fungi into such evolutionary theory.

References

Biggs R (1937) The species Concept in *Corticium coronilla*. Mycologia 29:686–706
Blakeslee AF (1904) Zygospore formation a sexual process. Science 19:864–866
Blakeslee AF (1906) II. Differentiation of sex in thallus, gametophyte and sporophyte. Bot Gaz 42:161–178
Blakeslee AF, Cartledge JL, Welch DS, Bergner AD (1927) Sexual dimorphism in Mucorales. I. Intraspecific reactions. Bot Gaz 84:27–50
Bos CJ (1996) Somatic recombination. In: Bos CJ (ed) Fungal genetics, principles and practice. Marcel Dekker, New York
Burnett JH, Boulter ME (1963) The mating systems of fungi. II. Mating systems of the gasteromycetes

Mycocalia denudata and *M. duriaeana*. New Phytol 62:217–236

Caten CE (1971) Heterokaryon incompatibility in imperfect species of *Aspergillus*. Heredity 26:299–312

Dodge BO (1927) Nuclear phenomena associated with heterothallism and homothallism in the ascomycete *Neurospora*. J Agric Res 35:289–305

El Ani AS, Olive LS (1962) The induction of balanced heterothallism in *Sordaria fimicola*. Proc Natl Acad Sci USA 48:17–19

Emerson R, Wilson CM (1954) Interspecific hybrids and the cytogenetics and cytotaxonomy of *Euallomyces*. Mycologia 46:393–434

Ehrlich P, Holm RW (1963) The process of evolution. New York

Esser K (1956) Die Incompatibilitätsbeziehungen zwischen geographischen Rassen von *Podospora anserina*. (CES) REHM I. Genetische Analyse der Semi-Incompatibilität. Z Indukt Abstammungs-Vererbungsl 87:595–624

Esser K (1959a) Die Incompatibilitätsbeziehungen zwischen geographischen Rassen von *Podospora anserina*. (CES) REHM II. Die Wirkungsweise der Semi-Incompatibilitäts-Gene. Z Vererbungsl 90:29–52

Esser K (1959b) Die Incompatibilitätsbeziehungen zwischen geographischen Rassen von *Podospora anserina*. (CES) REHM III. Untersuchungen zur Genphysiologie der Barragebildung und der Semi-Incompatibilität. Z Vererbungsl 90:445–456

Esser K (1966) Incompatibility. In: Ainsworth GC, Sussmann AS (eds) The Fungi, Vol 2. Academic Press, New York, pp 661–676

Esser K (1971) Breeding systems in fungi and their significance for genetic recombination. Mol Gen Genet 110:86–100

Esser K, Blaich R (1994) Heterogenic incompatibility. In: Esser K, Lemke PA (eds) The Mycota, Vol I (eds: Wessels J, Meinhardt F) Growth, differentiation and sexuality. Springer, Berlin Heidelberg New York, pp 211–232

Esser K, Hoffmann P (1977) Genetic basis for speciation in higher basidiomycetes with special reference to the genus *Polyporus*. In: Clemencon H (ed) The species concept in hymenomycetes. Cramer, Vaduz, pp 189–214

Esser K, Raper JR (1965) Heterogenic incompatibility. Proc X Int Congr Bot, Edinburgh 1964, p 45 (abstr). Original in: Esser K, Raper JR (eds) Incompatibility in fungi. Springer, Berlin Heidelberg New York, pp 6–13

Esser K, Straub J (1958) Genetische Untersuchungen an *Sordaria macrospora* AUERSW, Kompensation und Induktion bei genbedingten Entwicklungsdefekten. Z Vererbungsl 89:729–746

Flor HH (1956) The complementary genic system in flax and flax rust. Adv Genet 8:29–54

Grindle M (1963) Heterokaryon compatibility of unrelated strains in the *Aspergillus nidulans* group. Heredity 18:191–204

Hartman M (1943) Die Sexualität. Fischer, Jena

Holton CS (1931) Hybridization and segretation in the oat smuts. Phytopathology 21:835–842

Holton CS, Fischer GW (1941) Hybridization between *Ustilago avenae* and *U. prennans*. J Agric Res 62:121–128

Holton CS, Kendric EL (1956) Problems of delimitation of species of *Tilletia* occurring on wheat. Res Stud State Coll Wash 24:318–325

Howe HB, Haysman P (1966) Linkage group establishment in *Neurospora tetrasperma* by interspecific hybridization with *N. crassa*. Genetics 54:293–302

Jinks JL, Caten EC, Simchen G, Croft JC (1966) Heterokaryon incompatibility and variation in wild populations of *Aspergillus nidulans*. Heredity 21:227–239

Kemp RFU (1971) Breeding systems, speciation and taxonomy of the genus *Coprinus* (abstract). In: Ainsworth GC, Webster J (eds) First International Mycological Congress, Exeter, UK, Sept 1971:50.

Kniep H (1928) Die Sexualität der niederen Pflanzen. Fischer, Jena

Koltin Y (1970) Development of the *A mut B mut* strain of *Schizophyllum commune*. Arch Microbiol 74:123–128

Kukkonen I (1971) Micro- and macroecological factors in the speciation of obligate parasites (abstract). In: Ainsworth GC, Webster J (eds) 1st Int Mycological Congr, Exeter, UK, Sept 1971:54

Lemke PA (1966) The genetics of dikaryosis in a homothallic basidiomycete, *Sistorema brinkmanni*. PhD Thesis, Harward University, Cambridge, MA

Lemke PA (1969) A reevaluation of homothallism, heterothallism, and the species concept in *Sistotrema brinkmanni*. Mycologia 61:57–76

Lemke PA (1973) Isolating mechanisms in Fungi – Prezygotic, puttzygotic and azygitic persounia 71:249–260

Mayr E (1970) Populations, species, and evolution: an abridgement of animal species and evolution. Cambridge, USA

Mounce I, Macrae R (1938) Interfertility phenomena in *Fomes pinicola*. Can J Res (Bot Sci) 16:364–376

Muller HJ (1932) Some genetic aspects of sex. Am Nat 92:233–251

Nelson RR (1963) Interspecific hybridization in the fungi. Mycologia 55:104–123

Nelson RR (1964) Bridging interspecific incompatibility in the ascomycetous genus *Cochliobolus*. Evolution 18:700–704

Olive LS (1958) On the evolution of heterothallism in fungi. Am Nat 92:233–251

Parag Y (1962) Mutation in the *B* incompatibility factor of *Schizophyllum commune*. Proc Natl Acad Sci USA 48:743–750

Perkins DD (1972) An insertional translocation in *Neurospora* that generates duplications heterozygous for mating type. Genetics 71:25–51

Person C (1966) Genetic polymorphism in parasitic systems. Nature (Lond) 212:266–267

Pontecorvo G, Roper JA, Hemmons LM, MacDonald KD, Bufton AWJ (1953) The genetics of *Aspergillus nidulans*. Adv Genet 5:141–237

Raper JR (1966) Genetics of sexuality in higher fungi. Ronald Press, New York

Raper JR (1968) On the evolution of fungi. In: Ainsworth GC, Sussman As (eds) The Fungi, Vol. 3. Academic Press, New York, pp 677–693

Raper JR, Boyd DH, Raper CA (1965) Primary and secondary mutations at the incompatibility loci in *Schizophyllum*. Proc Natl Acad Sci USA 53:1324–1332

Rizet G, Esser K (1953) Sur des phénomènes d'incompatibilité entre souches d'origines differentes chez *Podospora anserina*. C R Acad Sci Paris 237:760–761

Smith ML, Bruhn JN, Anderson JB (1992) The fungus *Armillaria bulbosa* is among the largest and oldest living organisms. Nature 356:428–431

Stakman EC, Harrar JG (1957) Principles of plant pathology. New York

Stebbins GL (1950) Variation and evolution in plants. New York
Ullrich RC, Raper JR (1975) Primary homothallsim – relation to heterothallism in the regulation of sexual morphogenesis in *Sistotrema*. Genetics 80:311–321
Wheeler HE (1954) Genetics and evolution of heterothallism in *Glomerella*. Phytopathology 44:342–345
Whitehouse HLK (1949a) Multiple allelomorph heterothallism in the fungi. New Phytol 48:212–244
Whitehouse HLK (1949b) Heterothallism and sex in the fungi. Biol Rev 24:411–447
Winge Ö, Robberts C (1949) A gene for diploidization in yeasts. CR Trav Lab Carlsberg (Physiol) 24:341–346
Wright S (1931) Evolution in mendelian populations. Genetics 16:97–159
Wright S (1932) The roles of mutation, inbreeding, crossbreeding and selection in evolution. Proc 6th Int Congr Genet 1:356–366

10 Fungal Molecular Evolution: Gene Trees and Geologic Time

Mary L. Berbee[1] and John W. Taylor[2]

CONTENTS

I. Introduction	229
II. Genes for Phylogenies: Congruence and Conflict	229
A. Nuclear Ribosomal Repeat Regions	229
B. When are Ribosomal Genes not Enough?	230
C. Protein-Coding Genes	231
III. Inferring Phylogeny and Timing of Fungal Divergences Using 18S rRNA Gene Sequences	232
A. Fungal Time Tree Under Molecular Clock Assumptions	233
B. Origin of the Fungi	239
1. Flagella and the Basal Fungi	240
2. Colonization of Land	240
C. Radiation of the Terrestrial Fungi: Origin of the Ascomycota	241
1. Euascomycetes	241
D. Basidiomycete Radiation	243
IV. Summary	243
References	243

I. Introduction

Fungal phylogenetics has always been based on characters, but technological and intellectual advances are introducing new kinds of characters and new ways of thinking about them. First light microscopy, then electron microscopy, and now DNA sequencing successively upset previous views of fungal relationships. Phenetics, cladistics, and computerized data analysis and phylogenetic tree generation are now changing the intellectual rules for taxonomy and phylogenetics. The combination of new characters and new analytical tools have supported some taxonomic groups, established some new ones, and demolished a few old ones.

In this chapter, we do not discuss methods of phylogenetic analysis, but we will discuss some of the ribosomal and protein-coding DNA characters that are increasingly important in fungal phylogenetics. For a discussion of phylogenetic methods applied to fungi, see Berbee and Taylor (1999). To show the strengths and limitations of 18S rRNA sequences for phylogenetic analysis, we analyze and discuss trees from 49 diverse species of fungi and 3 animal outgroups. We consider the implications of the tree and nucleotide substitution rates for the timing of fungal divergences.

II. Genes for Phylogenies: Congruence and Conflict

A. Nuclear Ribosomal Repeat Regions

The most commonly used DNA for fungal phylogenetic studies is located in the ribosomal DNA repeat regions of the nuclei. The ribosomal repeats have been regions of choice for several good reasons. Fungal ribosomal gene clusters are arranged in roughly 200 tandem repeats (Butler and Metzenberg 1989). Because each nucleus contains 200 or so identical copies of the region, at least one intact copy for molecular analysis can usually be recovered, even from low-quality DNA preparations. The rDNA regions permit phylogenetic comparisons at many taxonomic levels (Bruns et al. 1991). Within the rDNA repeats are three coding genes (18S, 5.8S, and 28S) and two spacers (ITS 1 and ITS 2) that are transcribed together as a single unit and then cleaved into separate rRNA products. The rRNAs encoded by the 18S, 5.8S, and 28S genes become structural parts of the ribosomes and are essential for protein synthesis. The genes are highly conserved and are universally present, permitting phylogenetic comparisons among domains and kingdoms as well as within the more restricted group, the Eumycota. Compared with the genes,

---

[1] Department of Botany, 6270 University Blvd., University of British Columbia, Vancouver, British, Columbia, V6T 1Z4 Canada, berbee@unixg.ubc.ca
[2] Department of Plant and Microbial Biology, 111 Koshland Hall, University of California, Berkeley, California, 94720-3102, USA jtaylor@socrates.berkeley.edn

the spacers ITS1 and ITS2 are much freer to vary because their transcripts are excised from the rRNA and discarded instead of becoming part of the ribosome. The ITS regions are useful for taxonomic comparisons of closely related species and genera. A third spacer, the intergenic spacer or IGS, is not transcribed. It lies between pairs of repeats and is highly variable. In basidiomycetes, an additional RNA gene, the 5S rRNA gene, is within the IGS. In other fungi, the 5S gene is usually elsewhere in the genome (Metzenberg et al. 1985).

Why have phylogenetic studies of major fungal taxa based on rRNA been successful? Part of the reason lies in the ease of primer design for the rRNA regions. Nucleotides defining rRNA function seem to be distributed in highly conserved patches. Patches of 20 or more consecutive, highly conserved nucleotides are ideal for designing primers that will amplify DNAs from diverse organisms. Universal 18S rRNA gene primers (White et al. 1990), for example, amplify DNAs from insects as well as from red algae.

Also in favor of ribosomal genes for tree building is the fact that gene trees and species trees are likely to be the same, at least for distantly related species. Heterogeneity within species is usually slight. Among the ITS regions of 25 clinical isolates of *Coccidioides immitis*, for example, Burt et al. (1996) found a single polymorphic site. The ribosomal genes undergo concerted evolution so that sequence heterogeneity among repeats within a nucleus is rare. Horizontal transfer of ribosomal genes is unknown.

Sometimes the most compelling reason for choosing rRNA genes for a phylogenetic study is that the many sequences already in the molecular genetic databases can be reused to answer new questions. Sequencing ribosomal genes for just a few ingroup taxa while drawing the remaining ingroups and outgroups from national databases can answer many questions with minimum effort. In October 1997, GenBank contained over 5400 fungal rRNA sequences, accessible through the National Center for Biotechnology Computing (http://www.ncbi.nlm.nih.gov/). We took advantage of 52 of these sequences, from the 1997 GenBank pool to infer the trees in Figs. 2–5, discussed later in this chapter.

B. When are Ribosomal Genes not Enough?

Resolution of relationships among the fungi might be better if complete sequences were available from all three rRNA genes from the ribosomal repeats. However, this situation is likely to remain hypothetical for some time because DNA sequencing is time-consuming and expensive, so that complete sequences are generally available for the shorter 5.8S and 18S genes, but not the longer 28S genes.

Even if complete rRNA sequences were available for all fungi, the ribosomal repeats would not answer all questions about relationships. Although the number of nucleotides in the ribosomal repeats is substantial (about 9000), the number of possible characters useful for a specific question is often small. Imagine that the repeats are being used for a question about an ancient divergence in the fungi. Of the 9000 nucleotides in a repeat, the 18S gene accounts for about 1800bp, the 5.8S gene for about 120bp, and the 28S gene for about 3200bp. The remaining nucleotides in the IGS and ITS are highly variable and change too quickly to record ancient divergences. That leaves about 5100 conserved positions in the RNA genes. Of the 5100 or so conserved positions, most do not vary among the fungi, presumably because they are functionally important and mutations in these positions are lethal. From the 18S data set we describe later in this chapter, PAUP 4.0d55 estimated that 56% of the nucleotide positions were not free to vary and so could not have undergone substitutions useful in phylogenetic reconstruction. Assuming that the 28S and 5.8S genes have similar substitution rates, this leaves about 2225 nucleotides in the rRNA genes that could potentially record ancient fungal divergences. Of course, over short periods of time between divergences in a rapid radiation, only a few of these 2225 positions will record the event by substituting nucleotides (Philippe et al. 1994). Therefore, there is every reason to look beyond the ribosomal genes and spacers to the rest of the genome. Fortunately, the whole genome of a fungus like *Neurospora crassa* contains about 47000000 nucleotides (Orbach et al. 1988). If lack of resolution of branching order is due to insufficient data, there is a good chance of finding a record of radiation outside the ribosomal repeats.

Worse than lacking information, genes including the rRNA genes can be sources of phylogenetic misinformation. Occasionally, ribosomal gene sequences evolve at very different rates in different species. If the rate differences is extreme, long branches "attract" one another and quickly evolving taxa group together or with distantly related outgroups rather than with their slowly evolving relatives (Felsenstein 1978). A clear demonstration of the phenomenon comes from the 18S sequences of the insect world. Flies are close relatives of fleas and scorpionflies based on strong evidence from comparative morphology. However, because their sequences evolved unusually rapidly, flies group with outgroups or with long branched ingroups instead of with their flea and scorpionfly relatives (Carmean and Crespie 1995; Carmean et al. 1992; Huelsenbeck 1997).

Long branch attraction might also explain conflicting positions of slime molds and microsporidia in 18S rRNA and protein coding gene trees. Is it long branch attraction or the true phylogeny that positions the divergence of the slime molds down among the first protists in 18S rRNA trees (Cavalier-Smith 1993)? Actin and tubulin gene trees point to a more recent origin for the slime molds, near the time of origin of animals and plants (Baldauf and Palmer 1993). Although in the 18S gene trees the microsporidia appeared to be among the first eukaryotes to diverge, this was probably an artifact of long branch attraction. Trees from three other genes, from α- and β-tubulin (Keeling and Doolittle 1996; Keeling et al. 2000) and from the large subunit of RNA polymerase II (Hirt et al. 1999) all link microsporidia to the fungi.

As in morphological phylogenetics, comparing homologous characters is important in molecular phylogenetics. Because the rRNA genes are prone to insertions and deletions, the homologous patches can be difficult to recognize in distantly related species. The secondary structure of the rRNA is more highly conserved than the nucleotide sequence and can be a useful guide to identifying and aligning homologous regions. The rRNA World Wide Web (WWW) server (http://rrna.uia.ac.be/index.html) provides a universal alignment, based on secondary structure, for all available complete 18S and 28S sequences (Van de Peer et al. 1997). As an example from the universal alignment showing the effect of the history of added and subtracted nucleotides, the 18S sequence of *Alternaria alternata* required twice as many gaps as nucleotides to fit in with the universal alignment of diverse prokaryotes and eukaryotes.

Even with secondary structure as a guide, decisions about alignment can be subjective. Authors should make their alignments available to other researchers who can then challenge the initial results with a new alignment and fresh analysis. TreeBASE (http://www.herbaria.harvard.edu/treebase/index.html), not limited to rRNA, is being developed as a repository for alignments and trees from published reports. Since 1998, the journal *Mycologia* has required that authors of phylogenetic studies submit alignments and trees to TreeBASE before publication.

In general, one of the advantages to working with ribosomal genes over protein genes is that all the copies of a ribosomal gene from the same nucleus are identical or nearly identical. However, as occasionally reported for 5S and ITS regions, copies of the repeat can differ significantly in the same organism (O'Donnell 1992; O'Donnell and Cigelnik 1997). If the genes are polymorphic within an individual, then the same complications could arise as when inferring phylogenies from paralogous protein-coding genes.

C. Protein-Coding Genes

Protein sequences offer several advantages over rRNA gene sequences for phylogenetic analysis. Homology and convergence are easier to recognize in protein sequences made up of the 20 amino acids than in DNA sequences of the four nucleotides. Length changes are infrequent in protein-coding genes because insertions and deletions often lead to fatal frame shifts and elimination through natural selection. Eukaryotic genomes offer a wide range of protein-coding genes for phylogenetic analysis. However, studies of fungal phylogenetics based on protein-coding genes must surmount some minor obstacles beginning with primer design. Because selection acts mainly at the amino acid level in protein-coding genes, different DNA triplets can code for the same amino acid. A primer that works for one fungus may fail with its close relatives due to substitutions that destroy the priming site in the DNA without changing the amino acid sequence. The priming problem can sometimes be solved by using degenerate primers consisting of sequence mixtures that complement several different target template sequences, and by designing the primers so that their 3' ends match the more conservative first and second codon positions.

Like nuclear rRNA sequences, mitochondria and their protein-coding genes are present in multiple copies in most cells. Given the right primers, mitochondrial genes can be easy to amplify, even from low-quality starting DNA. However, to date, relatively few fungal phylogenetic studies use mitochondrial protein-coding genes. Paquin et al. (1995a, b) inferred plausible trees from the mitochondrial genes coding for subunits 1 and 3 of cytochrome oxidase and for subunit 5 of NADH dehydrogenase. More recently, Paquin et al. (1997) used subunits 1–3 and cytochrome b in a phylogenetic analysis to infer strong support for animals and fungi as sister taxa. Within the fungi, their analysis shows Spizellomycetales rather than Blastocladiales at the base of the fungi, and relationships at the base of the Ascomycota are no better resolved with the mtDNA genes than those inferred from ribosomal genes. Other analyses by Paquin et al. (1997) of entire mtDNA sequences give a fascinating view into the evolution of mt introns, codon usage, and the interplay of nuclear and mitochondrial genes.

Nuclear protein-coding genes are usually present in one copy per nucleus, so recovering enough intact copies to amplify requires good quality genomic DNA. Only a few studies of fungal relationships used single copy nuclear genes (Smith 1989; Radford 1993; Keeling et al. 2000; Liu et al. 1999). Baldauf and Palmer (1993) used highly conserved nuclear genes for actin, α- and β-tubulin, histones, and for the translation elongation factor Ef-1α to show that animals and fungi appear as sister groups (Baldauf and Palmer 1993). Analyzing a total of 53 different enzyme encoding genes from the international genetic data bases, Doolittle et al. (1996) inferred that fungi diverged from animals about 965 Ma ago.

The genes used in the above studies were chosen in part because they did not appear to have undergone horizontal transfer or frequent gene duplication. Genes that do duplicate to form gene families require extra sequencing effort. The phylogenetic analysis of chitin synthase genes by Bowen et al. (1992) provides a good example of the technical steps needed to survey members of a gene family from several fungi. A number of genes code for chitin synthase in the Euascomycetes (Bowen et al. 1992; Din et al. 1996). Bowen et al. (1992) designed degenerate primers to amplify chitin synthase genes belonging to three gene families from a wide range of Ascomycota. The amplification product from a single fungus was usually a mixture of chitin synthases from two or three of the families. To separate genes from the three families, the polymerase chain reaction (PCR) products were cloned and sorted based on similarity. Clones representing each family were sequenced and a tree of the sequences was inferred using UPGMA. The chitin synthase gene must have duplicated twice before the Euascomycetes radiated, because several of the Euascomycetes species had three chitin synthase genes and, within each gene family, the same phylogeny was repeated (Fig. 1). In this case, interpreting organismal evolution from the gene tree simply involved following the branching pattern within each chitin synthase gene family. Interpreting organismal evolution from gene trees can be much more difficult when the pattern of gene duplication and loss is more complex.

Gene trees and organismal trees conflict when horizontal transfer occurs and unrelated organisms exchange genetic material. As an example of the consequences of horizontal transfer, flowering plant genes for the beta subunit of the F1-ATPase are polyphyletic even though flowering plants clearly have a single origin (Iwabe et al. 1989). In the F1-ATPase tree, the carrot clusters with the bread mold *Neurospora* but the tobacco and spinach cluster with bacteria. The explanation? The *Neurospora* and carrot genes were from the nucleus and had been vertically transmitted. The tobacco and spinach genes were from chloroplasts. Plants gained their chloroplast genomes through a horizontal transfer event involving symbiosis. An ancestral plant captured and enslaved the cyanobacterium that became the first chloroplast. The chloroplast genes retain the traces of their cyanobacterial origin, clustering with *E. coli* in gene trees rather than grouping with eukaryotes. The phylogeny showing the polyphyletic plants is correct for the F1-ATPase gene but would be wrong if generalized to the whole organisms.

III. Inferring Phylogeny and Timing of Fungal Divergences Using 18S rRNA Gene Sequences

Which fungal relationships can be resolved with confidence using 18S rRNA sequence data? Which regions of the phylogeny of major taxa might benefit most from investigation with other genes? To provide a sense of the strengths and limits of tree building from the 18S genes, we inferred the phylogeny for 49 fungi and 3 outgroups (Table 1) using parsimony and the computer program PAUP 4.0d55. We added sequences from a broad selection of species from the Ascomycota and Basidiomycota, from the few chytrids available in the databases and from zygomycetes including

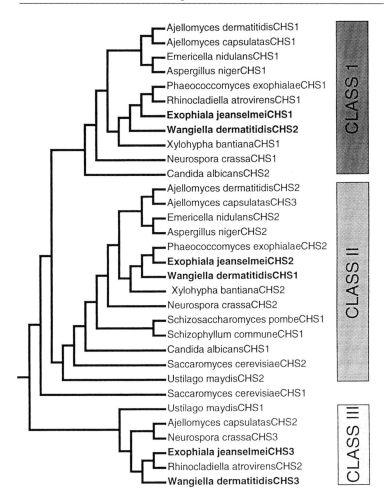

Fig. 1. Phylogenetic analysis of three families or classes of chitin synthase genes. The branching order for the species in the Euascomycetes is similar in the subtree from each of the three classes. As an example, note that chitin synthase genes from black yeasts, including *Wangiella dermatitidis* and *Exophiala jeanselmei*, group together in all three subtrees. The repetition of the phylogeny in each subtree suggests that the ancestral chitin synthase gene duplicated at least twice, before the radiation of the Euascomycetes, to form the three gene families. (After Bowen et al. 1992)

endomycorrhizal species and animal-associated trichomycetes and Entomophthorales (Table 1). The outgroups include one collar flagellate (Choanoflagellate) (Wainright et al. 1993) and two members of a group of fish parasitic protists (Ragan et al. 1996; Spanggaard et al. 1996). Each of our 50 replicated heuristic searches with random addition of taxa and tree bisection and reconnection branch swapping found the same 45 equally parsimonious trees of length 1209 (Fig. 2). The trees differed in branching order mainly in the Ascomycota. The branching order of the Archiascomycetes (see Kurtzman and Sugiyama, Chap. Vol. VII, Part A) varied from tree to tree and the branching order of taxa within the Euascomycetes was not consistent (not illustrated). For a rough estimate of the support from the data for the branching order, we performed 500 neighbor-joining bootstrap replicates, calculating the distance matrix using a Kimura correction for multiple hits, assuming that half of the sites in the alignment were not variable and that the transition to transversion ratio was two. The bootstrap numbers over 50% are shown in Fig. 3. In general, branches with more than 50% bootstrap support were also present in all 45 of the equally parsimonious trees (not illustrated). Weakly supported branches can also be discovered by determining which branches are lost in the consensus of trees that are one step longer than the most parsimonious tree (Bremer 1988). We next used PAUP 4.0d55 to find all trees one step longer than the most parsimonious tree in a single heuristic search with tree bisection and reconnection; PAUP found 531 trees. The consensus of the 531 trees is shown in Fig. 3.

A. Fungal Time Tree Under Molecular Clock Assumptions

If the rate of nucleotide substitution is approximately constant for all lineages and if fossil evi-

Table 1. Species used for phylogenetic analysis and the accession number of their 18S rRNA genes

Group	Name	Accession no.[a]
Ascomycota	*Blumeria graminis*	L26253
Ascomycota	*Candida albicans*	X53497
Ascomycota	*Capronia pilosella*	U42473
Ascomycota	*Dipodascopsis uninucleata*	U00969
Ascomycota	*Eremascus albus*	M83258
Ascomycota	*Eurotium rubrum*	U00970
Ascomycota	*Galactomyces geotrichum*	U00974
Ascomycota	*Herpotrichia juniperi*	U42483
Ascomycota	*Hypomyces chrysospermus*	M89993
Ascomycota	*Lecanora dispersa*	L37734
Ascomycota	*Morchella elata*	L37537
Ascomycota	*Neurospora crassa*	X04971
Ascomycota	*Ophiostoma ulmi*	M83261
Ascomycota	*Peziza badia*	L37539
Ascomycota	*Pleospora rudis*	U00975
Ascomycota	*Pneumocystis carinii*	X12708
Ascomycota	*Porpidia crustulata*	L37735
Ascomycota	*Saccharomyces cerevisiae*	J01353
Ascomycota	*Schizosaccharomyces pombe*	X54866
Ascomycota	*Sclerotinia sclerotiorum*	X69850
Ascomycota	*Talaromyces flavus*	M83262
Ascomycota	*Taphrina deformans*	U00971
Basidiomycota	*Boletus satanas*	M94337
Basidiomycota	*Chroogomphus vinicolor*	M90822
Basidiomycota	*Coprinus cinereus*	M92991
Basidiomycota	*Cronartium ribicola*	M94338
Basidiomycota	*Leucosporidium scottii*	X53499
Basidiomycota	*Russula compacta*	U59093
Basidiomycota	*Spongipellis unicolor*	M59760
Basidiomycota	*Suillus cavipes*	M90828
Basidiomycota	*Tilletia caries*	U00972
Basidiomycota	*Tremella globospora*	U00976
Basidiomycota	*Ustilago hordii*	U00973
Chytridiales (Chytridiomycota)	*Chytridium confervae*	M59758
Blastocladiales (Chytridiomycota)	*Blastocladiella emersonii*	M54937
Neocallimasticales (Chytridiomycota)	*Caecomyces (Shaeromonas) communis*	M62707
Neocallimasticales (Chytridiomycota)	*Neocallimastix frontalis*	X80341
Neocallimasticales (Chytridiomycota)	*Neocallimastix joynii*	M62705
Neocallimasticales (Chytridiomycota)	*Neocallimastix* sp.	M59761
Neocallimasticales (Chytridiomycota)	*Piromyces (Piromonas) communis*	M62706
Spizellomycetales (Chytridiomycota)	*Spizellomyces acuminatus*	M59759
Endogonales (Zygomycota)	*Endogone pisiformis*	X58724
Entomophthorales (Zygomycota)	*Basidiobolus ranarum*	D29946
Entomophthorales (Zygomycota)	*Conidiobolus coronatus*	D29947
Entomophthorales (Zygomycota)	*Entomophthora muscae*	D29948
Glomales (Zygomycota)	*Gigaspora margarita*	X58726
Glomales (Zygomycota)	*Glomus intraradices*	X58725
Glomales (Zygomycota)	*Glomus mosseae*	Z14007
Trichomycetes (Zygomycota)	*Smittium culisetae*	D29950
Basal animal	*Dermocystidium salmonis*	U21337
Basal animal	*Ichthyophonus hoferi*	U25637
Collar flagellate	*Diaphanoeca grandis*	L10824

[a] GenBank database accession numbers (http://www3.ncbi.nlm.nih.gov/Entrez/).

dence is available to calibrate the rate of nucleotide substitutions, then the percentage substitution between pairs of species can be used to estimate their times of divergence. This approach has been used for bacteria (Ochman and Wilson 1987), animals including humans (Hasegawa et al. 1993), plants (Wolfe et al. 1989), and fungi (Berbee and Taylor 1993; Simon et al. 1993). Our initial estimate of the divergence times for major groups of fungi was based on 18S rRNA gene sequences

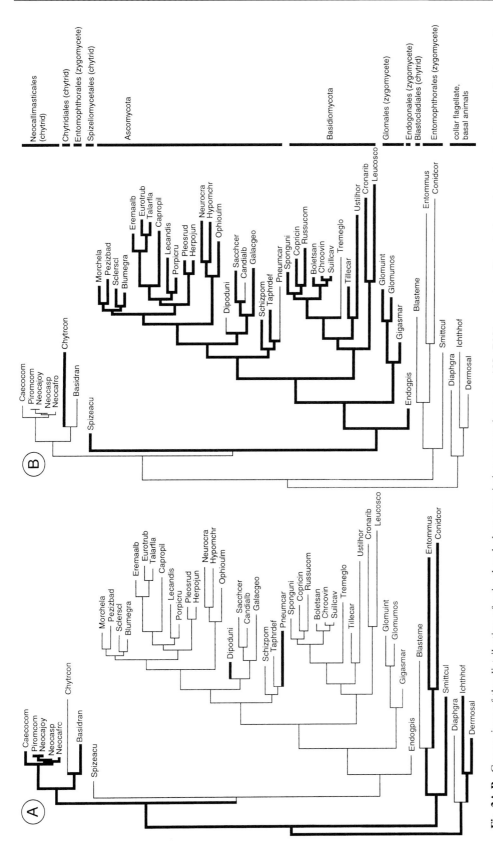

Fig. 2A,B. Comparison of the distribution of animal and plant association mapped onto an 18S rRNA gene tree. Tree **A** shows branches leading to animal parasites or commensals in bold. Tree **B** shows branches leading to plant associated fungi, including endosymbionts, pathogens and saprophytes, *in bold*. Animal association appears at the *base of the tree* while association with plants characterizes most members of the monophyletic group including the Ascomycota, the Basidiomycota, and endomycorrhizal zygomycetes (Glomales). The tree, with a length of 1209 steps, was one of 45 equally parsimonious trees found in 50 replicated heuristic parsimony searches with PAUP 4.0d55. *Species names are abbreviated with the first five letters of the genus followed by the first three letters of the specific epithet.* Table 1 gives full names

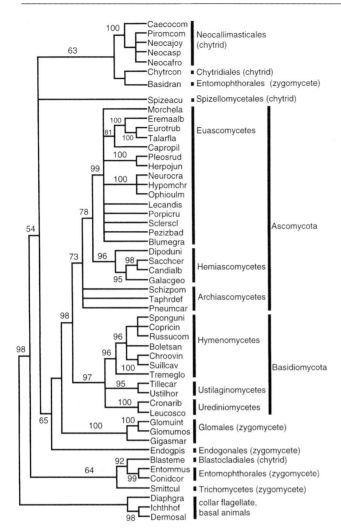

Fig. 3. The many polytomies in this tree indicate areas where the 18S rRNA gene provides little support for resolution of branching order. This is a strict consensus of 531 trees of length 1210 (one step longer than the most parsimonious tree). The *numbers on the branches* are bootstrap percentages from neighbor joining analysis of 500 boot-strapped data sets. *Species names are abbreviated with the first five letters of the genus followed by the first three letters of the specific epithet.* Table 1 gives full names

(Berbee and Taylor 1993). Although based on single gene and although the fossil data available for calibration of the tree are few, our estimates have been generally supported by fossil finds over the past few years. We estimated, for example, that basidiomycetous mushrooms radiated after the Cretaceous. Hibbett et al. (1995) reported finding the oldest mushroom known from amber dated at 90–94 Ma. We suggested that ectomycorrhizae in the fossil record should be good indicators for the radiation of the mushroom-forming basidiomycetes. LePage et al. (1997) found 50-Ma fossilized ectomycorrhizae typical of fungi in the suilloid group. Based on our analysis, Blastocladiales should have diverged well before the Devonian, and one could expect to find their fossils in the Devonian. Taylor et al. (1994) subsequently described a well-preserved blastocladiaceous fossil from the Devonian, showing how static fungal morphology can be over even 400 Ma.

An intriguing exception to the predictions of our hypothesis was Taylor's (1999) discovery of a complex, fruiting body producing, ascomycetous fossil fungus from the 400 Ma Rhynie chert. We estimated that the Ascomycota diverged from the Basidiomycota about 400 Ma, and that the Euascomycetes, with complex fruiting bodies, did not appear for another 100 million years or so.

Since our initial estimate of the timing of fungal origins, Nagahama et al. (1995) have released 18S rRNA gene sequences for some of the zygomycetes that radiated early in fungal history. Choanoflagellates and the fish parasitic protists *Dermocystidium* and *Ichthyophonus* were among the first organisms on the branch leading to animals to diverge and their 18S sequences have recently been released. These sequences allowed us to update our time tree, looking for evidence of how the very first fungi colonized land. We repeated our 1993 estimate using the phylogenetic data set described above and one of the 45 equally parsimonious trees from that data set (Fig. 2).

We then used maximum likelihood in PAUP 4.0d55 to calculate branch lengths for the tree (Fig. 4) assuming that (1) that a molecular clock was operating and (2) the rate of change varied from site to site in the data set. We assumed that the substitution rates showed a gamma distribution with shape parameter 0.5. An estimated 56% of sites were invariable. The variable sites were distributed into three classes of substitution rates.

To calibrate the molecular clock, we established a correlation between points in the geological time scale and percentage substitutions for fungal divergences. As can be seen in Fig. 5, we estimated the overall relationship between geologic time and nucleotide substitution to be 1.26% per 100Ma, compared to the 1.0% estimated previously (Berbee and Taylor 1993). The new, higher substitution rate probably results from using a less conservative alignment with a higher proportion of variable characters. In our current estimate, age estimates for divergences are generally higher. The Glomales, for example, diverged from the Ascomycota and Basidiomycota 600vs. 500Ma in our previous estimate. The older age estimate results in part from using the 965Ma date for the divergence of animals from fungi as a calibration point. Molecular clock estimates are approximations and the differences among estimates may also be due to differences in taxon sampling or to variations in DNA substitution rates among lineages.

Our calibration points include:

1. The divergence of animals from fungi was estimated at 965 Ma, plus or minus about 140 Ma, by Doolittle et al. (1996). Based on 18S rRNA sequence data, some of the first "animals" to diverge are protists, including the collar flagellate, *Dermocystidium*, and *Ichthyophonus* (Wainright et al. 1993; Ragan et al. 1996; Spanggaard et al. 1996). The corresponding percentage substitution per lineage was 11.1% for ca. 1 billion years, or 1.1% per 100 million years.

2. Most Glomales, Endogonales, ascomycetes, and basidiomycetes are associated with terrestrial plants. The most parsimonious assumption is that radiation of these fungi followed the origin of land plants. While the date of origin of the first terrestrial plants is uncertain, microfossils from 460Ma (Gray 1985) have been attributed to terrestrial plants. Conservatively placing the origin of land plants at 600Ma, 140Ma earlier than their appearance as fossils, provides an earliest possible date for terrestrial fungus radiation. The corresponding percentage substitution for the first terrestrial fungus radiation (*Endogone* from the rest) is 7.6 or 1.26% per 100Ma.

3. Fossil spores and arbuscules from about 390Ma (Kidston and Lang 1921; Remy et al. 1994) represent the most recent possible date for the origin of the Glomales. The divergence of the Glomales from the ascomycete/basidiomycete clade, corresponding to 7.4% substitution, must have happened before the 390-Ma fossils were produced. The divergence of the Ascomycota from the Basidiomycota, corresponding to 6.9% substitution came after the origin of the Glomales.

4. Regularly septate hyphae have also been reported from 390-Ma Rhynie chert (Kidston and Lang 1921), and most likely originated earlier. Most ascomycetes and basidiomycetes have regularly septate hyphae, but most of the zygomycetes and chytrids do not. Compared with the origin of regularly septate hyphae, the divergence of Glomales from ascomycete/basidiomycete stem species, again corresponding to 7.4% substitution, was earlier, while the divergence of the Ascomycota from the Basidiomycota, corresponding to 6.9% substitution, was later.

5. A 290-Ma clamp connection (Dennis 1970) provides a most recent possible date for Basidiomycota. Compared with clamp origin, the split of Ascomycota from Basidiomycota, corresponding to 6.9% substitution, must have been earlier, while the radiation of Basidiomycota, corresponding to 6.2% substitution, was probably later.

6. *Caecomyces*, *Piromyces*, and *Neocallimastix* are genera of chytrids found in stomachs of mammals. Chytrid stomach inhabitants probably radiated after the first mammal radiations, following the split of placentals and marsupials, about 150–200 Ma. The divergence of the stomach chytrids from free-living chytrids, corresponding to 3.3% substitution, may have occurred after the host organisms

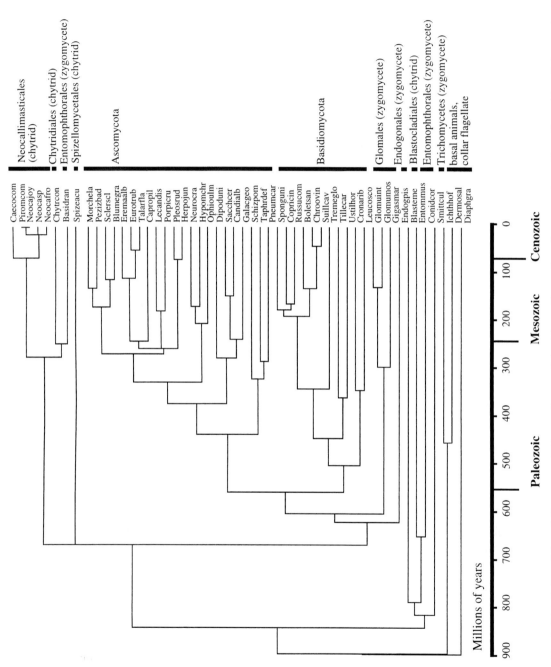

Fig. 4. The timing of fungal divergences. This tree is one of the 45 equally parsimonious trees from 50 replicated heuristic searches with PAUP 4.0d55. The branch lengths were calculated using maximum likelihood, assuming site-to-site rate variation and a molecular clock, and then absolute ages of divergences were estimated using the calibration of 1.26% substitution per lineage per 100 Ma. *Species names are abbreviated with the first five letters of the genus followed by the first three letters of the specific epithet*. Table 1 gives full names

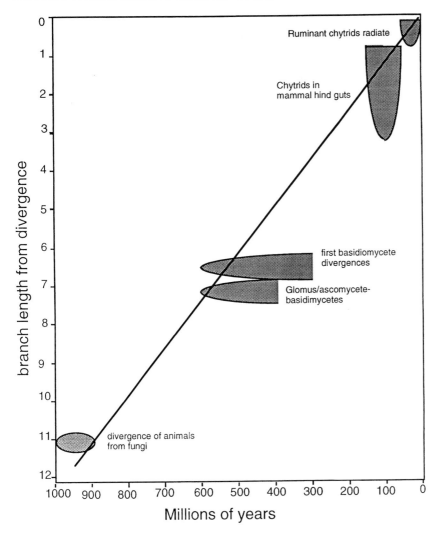

Fig. 5. Calibration of the molecular clock. After establishing the relationship between points in the absolute geological time scale and percent substitution associated with fungal radiations, we estimated that the DNA substitution rate was about 1.26% per 100 Ma

evolved, but before the split between placental and marsupial mammals, given that modern stomach chytrids inhabit both groups.

7. Chytrids in the genus *Neocallimastix* are found only in ruminants. The origin of ruminants, about 40 Ma, provides a most recent likely date for the radiation of these chytrids (associated with 0.2% substitution).

We had to drop one calibration that we used once before (Berbee and Taylor 1993). We used *Phellinites*, originally reported as a 165-Ma fossil polypore (Singer and Archangelsky 1958), as a calibration point for the minimum age for homobasidiomycetes, but Hibbett et al. (1997) demonstrated convincingly that the microscopical anatomy of *Phellinites* is typical for gymnosperm bark.

B. Origin of the Fungi

Trees from 18S rRNA gene sequences show animals as closest relatives to the fungi. Although fungi do not much resemble beetles and great apes, in habit and flagellation they do resemble some of the basal representatives of the animal kingdom. *Dermocystidium* and *Ichthyophonus* are protists that share with the basal fungi the habit of parasitizing fish and invertebrates (Ragan et al.

1996; Spanggaard et al. 1996). Collar flagellates are marine filter feeders that share with the chytrids the feature unusual among the protists of having a single posterior flagellum.

Based on our estimates of substitution rates, the most ancient of the fungal divergences occurred before plants colonized land (Fig. 4). Basal fungal lineages diverging before the likely origin of land plants include zygomycetes in the Trichomycetes and in the Entomophthorales. These fungi are usually obligate parasites or commensals of animals, most commonly of invertebrates (Fig. 2). The search for fungi parasitizing marine arthropods and other invertebrates has not been thorough. Additional search might reveal additional highly divergent fungi on marine hosts.

1. Flagella and the Basal Fungi

Among the fungi, only chytrids have flagella. The chytrid flagellum, being entirely typical for a eukaryote, is probably a feature retained from ancestral protists. Whether flagella were lost once or more than once during the evolution of terrestrial fungi is currently debated, and different gene trees show conflicting patterns. The flagellated chytrids do not appear at the base of the 18S tree. By inference, then, from the 18S data set all the groups of the non-flagellated fungi that cluster below chytrids or as sister taxa to chytrids came from the same flagellated ancestor that gave rise to chytrids. These nonflagellated fungi must have undergone convergent flagellar losses (Nagahama et al. 1995). The nonflagellated trichomycete *Smittium culisetae* clusters below the chytrid *Blastocladiella emersonii*, requiring one loss of the flagellum. *Entomophthora muscae* and *Conidiobolus coronatus* constitute the sister taxon to *B. emersonii*, requiring a second flagellar loss. The zygomycete *Basidiobolus ranarum* is the sister taxon to the *Chytridium confervae*, requiring a third flagellar loss (Fig. 2). This phylogenetic pattern suggests that the zygomycetous fungi colonized the land and lost flagella at least three times in association with animal hosts. The Glomales represent a fourth loss, this time in association with plant hosts.

There are, however, two important qualifications to this interpretation of evolutionary events based on the tree from 18S sequence data. First, taxon sampling for the first fungi is incomplete. Possibly, the preference for association with animals would not remain basal if the diversity of chytrids and zygomycetes was more fully represented in the tree. Second, branch lengths among some of the zygomycetes, including some of the Entomophthorales, are unusually long (Fig. 2). A long branch also leads to the chytrid *B. emersonii*. The lack of monophyly for the chytrids and the zygomycetes in 18S rRNA gene trees may be an artifact of long branch attraction (Felsenstein 1978). The next step in assessing the monophyly of chytrids and zygomycetes involves assessing whether another gene without the long branch problem supports the same topology. The other available gene trees including zygomycetes and chytrids are from β-tubulin (Keeling et al. 2000). Interpretation of β-tubulin trees is complicated by historical gene duplication. One of the two copies of β-tubulin gene in the chytrid *Spizellomyces punctatus* clusters with the zygomycetes (Keeling et al. 2000), which supports convergent loss of flagella. However, other aspects of the zygomycete phylogeny are different in the β-tubulin and 18S gene trees, and more data are needed to resolve branching patterns and ancient fungal history. The mitochondrial protein-coding gene NAD5, for example, might be useful because branches leading to a chytrid, a zygomycete, and other fungi appear to be about the same length (Paquin et al. 1995b).

2. Colonization of Land

While association with animals is particularly common at the base of the fungal tree, association with plants as mycorrhizal symbionts, as parasites, and as saprobes is the most common condition for the monophyletic group of terrestrial fungi that includes the Endogonales, Glomales, and Ascomycota and Basidiomycota (Fig. 2). The endomycorrhizal Endogonales and Glomales comprise the fourth group of zygomycetes that independently lost flagella, possibly during colonization of land in association with terrestrial plants; but were terrestrial plants available as food sources when these fungi were radiating? Some of the earliest fossil evidence for land plants comes in the form of dispersed microfossils of spores and tracheid-like structures, some of them from the Ordovician from 460 Ma. Our most recent tree (Fig. 4) suggests that the first divergences among the terrestrial fungi, the divergences of Endogonales and Glomales, occurred about 600 Ma or even earlier. This is about 140 Ma before any evidence for vascular plants that are now hosts for these obligate endomycorrhizal fungi and is about 100 Ma older than our estimate for the age of the divergences based on our earlier work. The difference between the age estimate of fungi and hosts may reflect error because of rate variation in sequence evolution. It may reflect a greater age for the vascular plants than the fossil record suggests, or it may be that the fungi were initially associated with the ancestors of the vascular plants, following their hosts onto land.

C. Radiation of the Terrestrial Fungi: Origin of the Ascomycota

The tree from 18S rRNA sequence data shows Ascomycota and Basidiomycota diverging from one another in the Paleozoic, about 500Ma (Fig. 4).

In the Ascomycota, an early radiation established the Archiascomycetes. (see Kurtzman and Sugiyama, Chap. 9, Vol. VII, Part A). The Archiascomycetes is a diverse group of fungi with and without hyphal fruiting bodies. It includes the fission yeast *Schizosaccharomyces pombe*, the animal pathogen *Pneumocystis carinii*, the peach leaf curl fungus *Taphrina deformans*, and the discomycete *Neolecta vitellina*. No obvious morphological character unites these taxa, and they do not always cluster together based on 18Ss rRNA sequence data (Fig. 3). Inferring the sequence of evolutionary events leading to their origin will require better resolution of their branching order and an improved understanding of the patterns underlying their morphological diversity. A more complete fossil record would contribute greatly to reconstruction of the first ascomycetes. Possibly, the complex, 400Ma old "pyrenomycete" fossil (Taylor et al. 1999) represents an Archiascomycete, revealing morphological features of a very early ascomycete.

The two other primary groups in the Ascomycota, the yeasts or Hemiascomycetes (also known as Saccharomycetales) and the filamentous ascomycetes or Euascomycetes, are both clearly monophyletic. In our analysis, the Hemiascomycetes received 96% support and the Euascomycetes received 99% bootstrap support (Fig. 3). The two groups appear in trees from other genes as well, including, for example, orotidine 5'-monophosphate decarboxylase (Radford 1993), β-tubulin (Baldauf and Palmer 1993), and the second largest subunit of RNA polymerase II (Liu et al. 1999).

The Hemiascomycetes include budding yeasts like the baker's yeast *Saccharomyces cerevisiae* as well as hyphal "yeasts" like *Dipodascopsis uninucleata* and *Galactomyces geotrichum*. The hyphal yeasts appear to be basal in our trees, suggesting that the budding yeasts may have been derived from hyphal ancestors.

1. Euascomycetes

The Euascomycetes always appear as a monophyletic group (see Barr, Chap. 13, Vol. VII, Part A). However, the branching order along the backbone at the base of the Euascomycetes is unresolved with available data (Fig. 3). Conflicting phylogenies from different laboratories and low bootstrap support for any particular phylogeny characterize this region of the tree. In all of the 45 equally parsimonious trees from our data set, the pyrenomycetes including *Neurospora crassa* appeared at the base of the Euascomycetes but without bootstrap support (Fig. 2). However, the basal group is unresolved in a consensus of 531 trees found in a search for trees up to one step longer (Fig. 3). Using a similar method of analysis but including a different selection of taxa, Gargas et al. (1995) found Pezizales at the base of the Euascomycetes, although again without bootstrap support. Spatafora's (1995) analysis showed a basal divergence between pyrenomycetes plus Pleosporales and the discomycetes plus plectomycetes and allies.

While the divergence order for the lineages of Euascomycetes is unclear, all groups of Euascomycetes have features in common which, by inference, should also be characters of the ancestor to the whole group. The basal members of Euascomycetes lineages share long, slender asci that are arranged in a single layer in a fruiting body and which shoot their spores forcibly into the air. Most groups of Euascomycetes produce conidia (i.e., mitospores). Conidia, along with pollen and other spores, are preserved in the fossil record. Finding evidence of conidial diversification in the fossil record would provide a most recent possible age for radiation of the Euascomycetes. Our time tree suggests that this radiation may have been taking place in the Mesozoic, about 240Ma.

Originating from the backbone of the Euascomycetes are several well-supported, monophyletic groups of taxa, some corresponding to traditional Ascomycota classes. Plectomycetes is a traditional class characterized by cleistothecial fruiting bodies containing thin-walled asci (see Geiser and LoBuglio, Chap. 10, Vol. VII, Part A). The plectomycetes consistently form a monophyletic group in trees from 18S rRNA (Berbee and Taylor 1992a); (Fig. 3) and other genes (Fig. 1); (Bowen et al. 1992). They encompass fungi ranging from false truffles in the Elaphomycetales (Landvik et al. 1996; LoBuglio et al. 1996) through human pathogens in the Onygenales to *Penicillium* species in the Trichocomaceae (Berbee et al. 1995). Similarly, the pyrenomycetes is a traditional class strongly supported by sequence data in all studies (see Samuels and Blackwell, Chap. 11, Vol. VII, Part A). The typical pyrenomycete has long, slim, thin-walled asci arranged in a single layer at the base of a flask-like fruiting body. Ascospores are forcibly discharged at maturity. Suggesting that the fruiting body shape, ascus shape, and ascus disposition are all functionally correlated to forcible ascospore discharge, the fungi that lose forcible discharge often lose the other features too, so that they resemble plectomycetes. The 18S rRNA trees provide clear evidence that *Ophiostoma* and *Ceratocystis*, for example, are descendants of fungi with all the typical pyrenomycete characters (Berbee and Taylor 1992b; Spatafora and Blackwell 1994).

In contrast to the above examples, the discomycetes (cup fungi) is a traditional class that lacks support from most 18S rRNA studies (see Pfister and Kimbrough, Chap. 12, Vol. VII, Part A). Two discomycetes orders, the Leotiales and the Pezizales sometimes cluster together (Fig. 2) but without bootstrap support (Fig. 3). The powdery mildews, or Erysiphales, often cluster within the Leotiales even though they had not previously been considered discomycetes, but again bootstrap support for the connection is weak (Saenz et al. 1994). Lichenized discomycetes in the Lecanorales occupy different positions, not necessarily with other discomycetes, in different phylogenetic analyses. As mentioned above, *Neolecta vitellina* although a discomycete, clusters with the Archiascomycetes (Landvik et al. 1993).

Like the discomycetes, the loculoascomycetes is a traditional ascomycete class without strong support from 18S rRNA studies (see Barr and Huhndorf, Chap. 13, Vol. VII, Part A). Before molecular evidence contradicted morphological evidence, Chaetothyriales were always included in the loculoascomycetes. Like other loculoascomycetes, they have bitunicate asci arranged in a hymenial layer. From these two morphological characters, the Chaetothyriales have little in common with the plectomycetes. However, consistently in 18S rDNA sequence-based trees (Berbee and Taylor 1995; Haase et al. 1995; Spatafora et al. 1995; Silva-Hanlin and Hanlin 1999) and occasionally in trees from RNA polymerase subunit II (Liu et al. 1999), the Chaetothyriales cluster with the plectomycetes. This suggests that the cleistothecial form with irregularly distributed asci lacking forcible discharge in the plectomycetes may have been derived from fruiting bodies with asci arranged in a single layer having forcible ascospore discharge (Berbee 1996).

Although the taxonomic position of the groups of ascomycete lichens is uncertain, the phylogenies do show that ascomycete lichens are polyphyletic. Lichenization appears to have originated independently in at least two orders, the Arthoniales and the Lecanorales (Gargas et al. 1995; Stenroos and DePriest 1998), although uncertainty about the branching order along the backbone of the Euascomycetes makes a single origin difficult to rule out (F. Lutzoni, AIBS oral presentation, August 1997). In our trees, lichenization does not appear as the most ancient habit for the Ascomycota. Lecanorales appear as the earliest group of lichenized ascomycetes and the separation of *Porpidia crustulata* from *Lecanora dispersa* represents an old divergence for the group. If lichenization arose after the divergence of Lecanorales from other Euascomycetes, it could be as old as about 240 Ma and it most parsimoniously arose before the separation of *P. crustulata* from *L. dispersa* at about 180 Ma.

Until recently, Euascomycetes fungi could only be classified once their sexual state was observed and studied. No sexual state is known for over 5000 species of fungi (Rossman 1993), a situation that has confounded classification. For the Ascomycota, sequencing of rDNA regions is becoming increasingly useful in integrating the asexual and sexual species into the same classification system by revealing similarities at the nucleotide level (Berbee and Taylor 1992c; Gaudet et al. 1989; LoBuglio et al. 1993; Reynolds and Taylor 1993; Taylor 1995). Seifert et al. (1997) used 18S and 28S ribosomal gene sequences to show that the asexual *Trichothecium roseum* is probably a member of the pyrenomycete order Hypocreales, and Kuhls et al. (1996) showed that asexual *Trichoderma reesei* is probably a clonal derivative of the sexual species *Hypocrea jecorina*.

D. Basidiomycete Radiation

One of the first molecular phylogenetic analyses of fungi, from 5S rRNA data, divided the Basidiomycota into two groups, one with inflated rims or dolipores surrounding the septa between adjacent hyphal cells, and the other lacking the inflated rims (Walker and Doolittle 1982). Partly because 5S sequences include only about 120 base pairs of DNA, the exact branching order of the smuts, the rust group, and the mushrooms and jelly fungi differed in different analyses (Blanz and Unseld 1987; Walker 1984). 18S and 28S ribosomal RNA gene sequences provided more characters and resolving power. The trees from 18S and 28S sequences show three lineages in the Basidiomycota that diverged early in the history of the group but do not clearly resolve the branching order for the three (Fig. 3). These are the rusts or Uredinomycetes (see Swann and Frieders, Chap. 2, this Vol.), the smuts or Ustilaginomycetes (see Bauer et al., Chap. 3, this Vol.) and the Hymenomycetes (Swann and Taylor 1993; Begerow et al. 1997; Swann et al. 1999); see Wells et al., Chap 4, this Vol.; Hibbett and Thorn, Chap. 5, this Vol.). The clamp connection at the septa that is found in at least some members of all three groups may be a primitive character of the phylum that evolved before the first basidiomycete radiation, by about 440Ma. If this estimate of phylogeny and of timing is correct, then it may be possible to find fossilized clamp connections much older than the 290-Ma clamps found in Pennsylvanian coal (Dennis 1970).

IV. Summary

This chapter has emphasized many phylogenetic questions where sequence data provide clear resolution. Because of the evolutionary constraints limiting horizontal transfer and paralogous evolution that act on the ribosomal genes, they will probably remain valuable phylogenetic indicators for quite some time. In future studies, the ribosomal genes will continue to bring misplaced fungi into their correct phylogenetic positions. As we have emphasized in this chapter, however, for many phylogenetic questions ribosomal gene sequences do not provide clear answers. For further understanding of the deeper branches in fungal tree, sequences from more genes and characters of all other types will be welcome.

References

Baldauf SL, Palmer JD (1993) Animals and fungi are each other's closest relatives: congruent evidence from multiple proteins. Proc Natl Acad Sci USA 90:11558–11562

Begerow D, Bauer R, Oberwinkler F (1997) Phylogenetic studies on nuclear large subunit ribosomal DNA sequences of smut fungi and related taxa. Can J Bot 75:2045–2056

Berbee ML (1996) Loculoascomycete origins and evolution of filamentous ascomycete morphology based on 18S rRNA gene sequence data. Mol Biol Evol 13:462–470

Berbee ML, Taylor JW (1992a) Two ascomycete classes based on fruiting-body characters and ribosomal DNA sequence. Mol Biol Evol 9:278–284

Berbee ML, Taylor JW (1992b) Convergence in ascospore discharge mechanism among pyrenomycete fungi based on 18S ribosomal RNA gene sequence. Mol Phylogenet Evol 1:59–71

Berbee ML, Taylor JW (1992c) 18S ribosomal RNA sequence characters place the human pathogen *Sporothrix shenckii* in the genus *Ophistoma*. Exp Mycol 16:87–91

Berbee ML, Taylor JW (1993) Dating the evolutionary radiations of the true fungi. Can J Bot 71:1114–1127

Berbee ML, Taylor JW (1995) From 18S ribosomal sequence data to evolution of morphology among the fungi. Can J Bot 73 (Suppl 1):S677–S683

Berbee ML, Taylor JW (1999) Fungal phylogeny. In: Oliver RP, Schweizer M (eds) Molecular fungal biology. Cambridge University Press, New York, pp 21–77

Berbee ML, Yoshimura A, Sugiyama J, Taylor JW (1995) Is *Penicillium* monophyletic? An evaluation of phylogeny in the family Trichocomaceae from 18S, 5.8S and ITS ribosomal DNA sequence data. Mycologia 87:210–222

Blanz PA, Unseld M (1987) Ribosomal RNA as a taxonomic tool in mycology. Stud Mycol 30:247–258

Bowen AR, Chen-Wu JL, Momany M, Young R, Szaniszlo PJ, Robbins PW (1992) Classification of fungal chitin synthases. Proc Natl Acad Sci USA 89:519–523

Bremer K (1988) The limits of amino acid sequence data in angiosperm phylogenetic reconstruction. Evolution 42:795–803

Bruns TD, White TJ, Taylor JW (1991) Fungal molecular systematics. Annu Rev Ecol Syst 22:525–564

Burt A, Carter DA, Koenig GL, White TJ, Taylor JW (1996) Molecular markers reveal cryptic sex in the human pathogen *Coccidioides immitis*. Proc Natl Acad Sci USA 93:770–773

Butler DK, Metzenberg RL (1989) Premeiotic change of nucleolus organizer size in *Neurospora*. Genetics 122:783–791

Carmean D, Crespie BJ (1995) Do long branches attract flies? Nature 373:666

Carmean D, Kimsey LS, Berbee ML (1992) 18S rDNA sequences and the holometabolous insects. Mol Phylogenet Evol 1:270–278

Cavalier-Smith T (1993) Kingdom Protozoa and its 18 phyla. Microbiol Rev 57:953–994

Dennis RL (1970) A Middle Pennsylvanian basidiomycete mycelium with clamp connections. Mycologia 62:578–584

Din AB, Specht CA, Robbins PW, Yarden O (1996) Chs-4, a class IV chitin synthase gene from *Neurospora crassa*. Mol Gen Genet 250:214–222

Doolittle RF, Feng D-F, Tsang S, Cho G, Little E (1996) Determining divergence times of the major kingdoms of living organisms with a protein clock. Science 271:470–477

Felsenstein J (1978) Cases in which parsimony or compatibility methods will be positively misleading. Syst Zool 27:401–410

Gargas A, DePriest PT, Grube M, Tehler A (1995) Multiple origins of lichen symbiosis in fungi suggested by SSU rDNA phylogeny. Science 268:1492–1495

Gaudet J, Julien J, Lafay JF, Brygoo Y (1989) Phylogeny of some *Fusarium* species as determined by large-subunit rRNA sequence comparison. Mol Biol Evol 6:227–242

Gray J (1985) The microfossil record of early land plants: advances in understanding of early terrestrialization, 1970–1984. Philos Trans R Soc Lond, Ser B, 309:167–195

Haase G, Sonntag L, Van de Peer Y, Uijthof JMM, Podbielski A, Melzer-Krick B (1995) Phylogenetic analysis of ten black yeast species using nuclear small subunit rRNA gene sequences. Antonie van Leeuwenhoek 68:19–33

Hasegawa M, Di Rienza A, Kocher TD, Wilson AC (1993) Toward a more accurate time scale for the human mitochondrial DNA tree. J Mol Evol 37:347–354

Hibbett DS, Grimaldi D, Donoghue MJ (1995) Cretaceous mushrooms in amber. Nature 377:487

Hibbett DS, Donoghue MJ, Tomlinson PB (1997) Is *Phellinites digiustoi* the oldest homobasidiomycete? Am J Bot 84:1005–1011

Hirt RP, Logsdon Jr. JM, Healy B, Dorey MW, Doolittle WF, Embley TM (1999) Microsporidia are related to Fungi: evidence from the largest subunit of RNA polymerase II and other proteins. Proc Natl Acad Sci USA 96:580–585

Huelsenbeck JP (1997) Is the Felsenstein zone a fly trap? Syst Biol 46:69–74

Iwabe N, Kuma K-I, Hasegawa M, Osawa S, Miyata T (1989) Evolutionary relationships of archaebacteria, eubacteria, and eukaryotes inferred from phylogenetic trees of duplicated genes. Proc Natl Acad Sci USA 86:9355–9359

Keeling PJ, Doolittle WF (1996) Alpha-tubulin from early-diverging eukaryotic lineages and the evolution of the tubulin family. Mol Biol Evol 13:1297–1305

Keeling PJ, Luker MA, Palmer JD (2000) Evidence from beta-tubulin phylogeny that Microsporidia evolved from within the Fungi. Mol Biol Evol 17:23–31

Kidston R, Lang WH (1921) On old red sandstone plants, V. The Thallophyta occurring in the peat-bed; the succession of accumulation and preservation of the deposit. Trans R Soc Edinb 52:855–902

Kuhls K, Lieckfeldt E, Samuels GJ, Kovacs W, Meyer W, Petrini O, Gams W, Börner T, Kubicek CP (1996) Molecular evidence that the asexual industrial fungus *Trichoderma reesei* is a clonal derivative of the ascomycete *Hypocrea jecorina*. Proc Natl Acad Sci USA 93:7755–7760

Landvik S, Eriksson OE, Gargas A, Gustafsson P (1993) Relationships of the genus *Neolecta* (Neolectales ordo nov., Ascomycotina) inferred from 18s rDNA sequences. Syst Ascomyc 11:107–118

Landvik S, Shailer NFJ, Eriksson OE (1996) SSU rDNA sequence support for a close relationship between the Elaphomycetales and the Eurotiales and Onygenales. Mycoscience 37:237–241

LePage BA, Currah RS, Stockey RA, Rothwell GW (1997) Fossil ectomycorrhizae from the middle Eocene. Am J Bot 84:410–412

Liu Y, Whelen S, Hall B (1999) Phylogenetic Relationships among Ascomycetes: evidence from an RNA polymerase II subunit. Mol Biol Evol 16:1799–1808

LoBuglio K, Berbee ML, Taylor JW (1996) Phylogenetic origins of the asexual mycorrhizal symbiont *Cenococcum geophilum* Fr. and other mycorrhizal fungi among the ascomycetes. Mol Phylogenet Evol 6:287–294

LoBuglio KF, Pitt JI, Taylor JW (1993) Phylogenetic analysis of two ribosomal DNA regions indicates multiple independent losses of a sexual *Talaromyces* state among asexual *Penicillium* species in subgenus *Biverticillium*. Mycologia 85:592–604

Metzenberg RL, Stevens JN, Selker EU, Morzycka-Wroblewska E (1985) Identification and chromosomal distribution of 5S rRNA genes in *Neurospora crassa*. Proc Natl Acad Sci USA 82:2067–2071

Nagahama T, Sato H, Shimazu M, Sugiyama J (1995) Phylogenetic divergence of the entomophthoralean fungi: evidence from 18S ribosomal RNA gene sequences. Mycologia 87:203–209

Ochman H, Wilson AC (1987) Evolution in bacteria: evidence for a universal substitution rate in cellular genomes. J Mol Evol 26:74–86

O'Donnell K (1992) Ribosomal DNA internal transcribed spacers are highly divergent in the phytopathogenic ascomycete *Fusarium sambucinum* (*Gibberella pulicaris*). Curr Genet 22:213–220

O'Donnell K, Cigelnik E (1997) Two divergent intragenomic rDNA ITS2 types within a monophyletic lineage of the fungus *Fusarium* are nonorthologous. Mol Phylogenet Evol 7:103–116

Orbach MJ, Vollrath D, Davis RW, Yanofsky C (1988) An electrophoretic karyotype of *Neurospora crassa*. Mol Cell Biol 8:1469–1473

Paquin B, Forget L, Roewer I, Lang BF (1995a) Molecular phylogeny of *Allomyces macrogynus*: congruency between nuclear ribosomal RNA- and mitochondrial protein-based trees. J Mol Evol 41:657–665

Paquin B, Roewer I, Wang Z, Lang BF (1995b) A robust fungal phylogeny using the mitochondrially encoded NAD5 protein sequence. Can J Bot 73 (Suppl 1):S180–S185

Paquin B, Laforest M-J, Forget L, Roewer I, Wang Z, Longcore J, Lang BF (1997) The fungal mitochondrial genome project: evolution of fungal mitochondrial genomes and their gene expression. Curr Genet 31:380–395

Philippe H, Chenuil A, Adoutte A (1994) Can the Cambrian explosion be inferred through molecular phylogeny? Development (Suppl):15–25

Radford A (1993) A fungal phylogeny based upon orotidine 5′-monophosphate decarboxylase. J Mol Evol 36:389–395

Ragan MA, Goggins CL, Cawthorn RJ, Cerenius L, Jamieson AVC, Plourde SM, Rand TG, Söderhäll K, Guttell RR (1996) A novel clade of protistan parasites near the animal-fungal divergence. Proc Natl Acad Sci USA 93:11907–11912

Remy W, Taylor TN, Hass H, Kerp H (1994) Four hundred-million-year-old vesicular arbuscular mycorrhizae. Proc Natl Acad Sci USA 91:11841–11843

Reynolds DR, Taylor JW (eds) (1993) The fungal holomorph: mitotic, meiotic and pleomorphic speciation in fungal systematics. CAB International, Wallingford, 375 pp

Rossman AY (1993) Holomorphic hypocrealean fungi: *Nectria sensu stricto* and teleomorphs of *Fusarium*. In: Reynolds DR, Taylor JW (eds) The fungal holomorph: mitotic, meiotic and pleomorphic speciation in fungal systematics. CAB International, Wallingford, pp 149–160

Saenz GS, Taylor JW, Gargas A (1994) 18S rRNA gene sequences and supraordinal classification of the Erysiphales. Mycologia 86:212–216

Seifert KA, Louis-Seize G, Savard ME (1997) The phylogenetic relationships of two trichothecene-producing hyphomycetes, *Spiecellum roseum* and *Trichothecium roseum*. Mycologia 89:250–257

Silva-Hanlin DMW, Hanlin RT (1999) Small subunit ribosomal RNA gene phylogney of several loculoascomycetes and its taxonomic implications. Mycol Res 103:153–160

Simon L, Bousquet J, Lévesque RC, Lalonde M (1993) Origin and diversification of endomycorrhizal fungi and coincidence with vascular land plants. Nature 363:67–69

Singer R, Archangelsky S (1958) A petrified basidiomycete from Patagonia. Am J Bot 45:194–198

Smith TL (1989) Disparate evolution of yeasts and filamentous fungi indicated by phylogenetic analysis of glyceraldehyde-3-phosphate dehydrogenase genes. Proc Natl Acad Sci USA 86:7063–7066

Spanggaard B, Skouboe P, Rossen L, Taylor JW (1996) Phylogenetic relationships of the intercellar fish pathogen *Ichthyophonus hoferi* and fungi, choanoflagellates and the rosette agent. Mar Biol 126:109–115

Spatafora JW (1995) Ascomal evolution of filamentous ascomycetes: evidence from molecular data. Can J Bot 73 (Suppl 1):S811–S815

Spatafora JW, Blackwell M (1994) The polyphyletic origins of opiostomatoid fungi. Mycol Res 98:1–9

Spatafora JW, Mitchell TG, Vilgalys R (1995) Analysis of genes coding for small-subunit rDNA sequences in studying phylogenetics of dematiaceous fungal pathogens. J Clin Microbiol 33:1322–1326

Stenroos SK, DePriest PT (1998) SSU rDNA phylogeny of cladoniiform lichens. Am J Bot 85:1548–1559

Swann EC, Taylor JW (1993) Higher taxa of basidiomycetes: an 18S rRNA gene perspective. Mycologia 85:923–936

Swann EC, Frieders EM, McLaughlin DJ (1999) *Microbotryum*, *Kriegeria* and the changing paradigm in basidiomycete classification. Mycologia 91:51–66

Taylor JW (1995) Making the Deuteromycota redundant: a practical integration of mitosporic and meiosporic fungi. Can J Bot 73 (Suppl 1):S754–S759

Taylor TN, Hass H, Kerp H (1999) The oldest fossil ascomycetes. Nature 399–648

Taylor TN, Remy W, Hass H (1994) *Allomyces* in the Devonian. Nature 367:601

Van de Peer Y, Jansen J, De Rijk P, De Wachter R (1997) Database on the structure of small ribosomal subunit RNA. Nucleic Acids Res 25:111–116

Wainright PO, Hinkle G, Sogin ML, Stickel SK (1993) Monophyletic origins of the Metazoa: an evolutionary link with Fungi. Science 260:340–342

Walker WF (1984) 5S rRNA sequences from Atractiellales, and basidiomycetous yeasts and Fungi Imperfecti. Syst Appl Microbiol 5:352–359

Walker WF, Doolittle WF (1982) Redividing the Basidiomycetes on the basis of 5S rRNA sequences. Nature 299:723–724

White TJ, Bruns T, Lee S, Taylor J (1990) Amplification and direct sequencing of fungal ribosomal RNA genes for phylogenetics. In: Innis MA, Gelfand DH, Sninsky JJ, White TJ (eds) PCR protocols, a guide to methods and applications. Academic Press, San Diego, California, pp 315–322

Wolfe KH, Gouy M, Yang Y-W, Sharp PM, Li W-H (1989) Date of the monocot-dicot divergence estimated from chloroplast DNA sequence data. Proc Natl Acad Sci USA 86:6201–6205

Subject Index

acetate-malonate pathway 150
aeciospore 41, 51
agriculture 38, 57, 75, 77
amphispore 41
anamorph 3, 6, 19, 37, 60, 67, 69, 70, 75, 77, 78, 109, 112, 146–148, 182, 242
antibiotic 3, 5
antimicrobic 193
apomixis 219, 223
aquatic fungi 39
aquatic plants 78
arthrospore 97, 98
ascocarp 236
ascus 241
asexual reproduction 6, 41, 225
asexual species 37, 50, 51, 242
astaxanthin 4
atrotomentin 149
author citation 176

bacteriolytic ability 155
ballistoconidium 60, 74, 77, 78
ballistospore 6, 76, 87, 92, 99, 104, 108, 109
basidiocarp 13, 40, 49, 86, 92, 94, 95, 98, 99, 101, 102, 104, 106, 109–112, 114, 115, 236
basidiospore 3, 40–42, 52, 60, 62, 75–79, 87, 90, 92, 93, 96, 99, 101, 102, 108–110, 114, 134, 139, 144
  ontogeny 143
basidium 6, 9, 11, 39–42, 52, 60, 79, 95, 101, 113, 139, 143, 144
  auricularioid 10, 25, 42, 99, 106, 110
  furcate 87, 96
  gasteroid 49
  holobasidium 10, 25–27, 41, 42, 62, 74–77, 79, 86, 87, 106, 111, 112
  holometabasidium 7
  metabasidium 19, 26, 41, 42
  morphology 47, 48, 58, 62
  ontogeny 62, 78, 89, 92, 93, 96, 101, 102, 104, 107, 109, 110
  phragmobasidium 7, 10, 22, 41, 62, 74, 75, 77, 87, 93, 99, 106
  phragmometabasidium 25
  probasidium 10, 19, 41, 42, 78, 92, 93, 99
  sphaeropedunculate 86
binomial system 171, 172

biochemical characters 15–18, 25, 27, 37, 43, 48, 52
  pigments 149
biocontrol 38
bioluminescence 151
biosystematics 172
blastospore 49, 87
budding 3, 6, 108, 109, 111, 112
  bipolar 6
  enteroblastic 6
  polar 6
bulbils 148

cell wall 6, 49
  branching 44, 46
  carbohydrates 15, 18, 27, 43, 49, 57, 59
chromatin
  interphase 46
cladistics 204, 210
clamp connections 6, 9, 12, 41, 61, 92, 93, 99, 102, 115, 237, 243
classification xi, 3, 6, 48, 50, 51, 64, 67, 87, 93, 95, 106, 111, 135, 207
cluster analysis 207
clustering 208
coevolution 79, 80
colacosome 4, 16, 39, 45, 50, 51
compatibility systems 86, 92
conidioma 147
conidiogenesis 6
conidium 39, 42, 49, 51, 60, 95, 96, 109, 111, 148, 241
  arthroconidium 3, 6, 8, 26, 147
  ballistoconidium 3, 6, 8, 19, 25, 26
  blastoconidium 3, 6, 27
  microconidium 42, 43, 87, 89, 93, 95, 96, 98, 99, 105
convergence (see homoplasy)
cryopreservation 200
cultural characters 16, 17, 77, 96, 113
culture collections 5, 52, 193, 195, 196
culture media 194
culture methods 52, 115, 193
  ballistospore discharge 5, 9, 52, 114
  growth conditions 195
  sterilization methods 194
  storage 196
  streak plating 52
cystidium 104, 132

database 203
  SPECIES 2000 189
  TreeBASE 231
decomposer 151
Dictionary of the Fungi 48, 172, 175, 182
dikaryon 6, 87
dikaryophysis 97, 100, 103
dimorphic life cycle 87
discharge apparatus 138, 140, 242
diseases and infections 155
  damping off 154
  root rot 154
dispersal 39, 41, 60, 63, 78, 98, 138, 139, 141, 147, 148
DNA typing methods 17, 18

ecological barriers 222
ecology 4, 38, 62, 63, 75, 78, 80, 155
endoplasmic reticulum 46
endospore 6, 43
ER cap 44, 45
evolution xi, 57, 78–80, 224
  basidiospore 139
  ectomycorrhiza 154
  flagellar loss 240
  fungi 241, 243
  hypha 142
  morphological 135
  origin of Fungi 232, 239
  trophic state 151
exploratory data analysis 205, 206
exudate, plant 4, 39

fimbrial glycosylation patterns 49
fission 3
folliculitis 76
food 148
form genera 175
fossil fungi 121, 175, 183, 237, 239, 243
fungal names, indices 187, 188
fungicide 38

gasteroid taxa 39, 42, 48, 52, 74, 79
gene flow 219, 220, 224
gene frequency 219
gene sequences
  internal transcribed spacer 129
  mitochondrial 122, 124, 128, 129, 231
    cytochrome 232
    cytochrome oxidase 232
    NADH dehydrogenase 232

nuclear rRNA 123, 124
  5S 15, 43, 48, 57, 59, 229–231
  18–28S 9, 15, 18, 49, 106,
    112, 122, 128, 129, 229–231,
    234–236, 239
  protein
    actin 231, 232
    chitin synthase 231–233
    EF1-α 231
    F1-ATPase 231, 232
    histones 231, 232
    laccase 152
    NAD5 240
    orotidine 5′-monophosphate
      decarboxylase 241
    RNA polymerase II 231, 241,
      242
    tubulin 231, 232, 240, 241
germination
  basidiospore 41, 78, 87, 89, 92, 96,
    100, 107, 109, 113, 141
gloeocystidium 94, 95, 100, 101, 104

habitat 4, 25, 51, 86, 92, 95, 98, 104,
    110
  aquatic 38, 141
  freshwater 5, 121
  marine 5, 51, 121, 141
  terrestrial 5, 26, 38, 105
haustorium 4, 6, 12, 39, 67, 74, 78,
    79, 108, 109
  tremelloid 107
heterothallic fungi 9, 111, 149
holomorph 182
homobasidiomycetes, numbers of
    121, 128
homoplasy xi, 37, 59, 78, 80, 131,
    139, 142, 144, 146, 147, 149,
    150, 153, 240, 242
homothallic fungi 9, 60, 111, 149,
    221
horizontal gene transfer 232
host range 62–64, 78, 79
host resistance 38
host specificity 62
host-parasite interaction 58, 60, 64,
    67, 70, 74, 77–79
hybridization 221
hypha
  conducting elements 142
  systems 142

incompatibility
  heterogenic 220
  heterokaryon 223
  homogenic 219
incompatibility system 9
industry 121, 152
  brewing 51
  chemical 4
Ingoldian fungi 39
insect associations 38, 39
intergenic spacer (IGS) 230
internal transcribed spacer
  ITS1 229–231
  ITS2 229–231
International Code of Botanical

Nomenclature 172, 174
International Committee on
  Bionomenclature 173
intraspecific isolation 222

Java 212

karotype analysis 17
karyogamy 41, 95, 96
killer toxin 17

lenticular body 16, 45
life cycle 10, 59, 60, 74–79, 86, 92,
    93, 95, 98, 104, 106, 110
  dimorphic 9
light microscopic characters 47, 52,
    71, 95, 98, 124, 139, 141, 143
long branch attraction 230, 231,
    240

marine fungi 39
mating systems 104, 106, 111, 149
mating type
  bipolar 221
  tetrapolar 221
medicine 148
meiosis 9, 44, 45, 49, 92, 93, 95, 96,
    99, 101, 102, 109, 111
  spindle 143
mevalonate pathway 150
microbodies 44, 45
microscala 44, 46, 49, 52
mite infestation 195
mitosis 45
  postmeiotic 143, 144
mitotic fungi (see asexual species)
modified bifactorial compatibility
    85, 86, 111
molecular clock 233, 237, 238
monokaryon 87
morphology 6, 9, 25, 52, 74, 75, 86,
    95, 122, 124, 132, 135, 137, 159,
    204
  basidiocarp 133
multivariate analysis 208
mutation 219
mycoherbicides 38
mycoparasite 4, 13, 16, 39, 45, 49,
    51, 86, 87, 95, 102, 106,
    108–110, 112, 114, 116, 148,
    151, 155
  facultative 155
mycorrhiza 86, 92, 95, 105, 153,
    240
  ectomycorrhiza 128, 153, 155, 197,
    236
  ericoid 154
  orchid 133, 154

necrotrophic parasites 156
nematotoxin 156
nomenclature xi
  changes 173
  codes 185
  conservation 184
  Heterobasidiomycetes 93
  history 171

misapplied names 177
  priority 181
  registration 178
  rules 172, 180
  stability 187
  synonymy 182
  terms 173
  valid publication 177
nomenclatural filter 174, 175
nonparametric tests 205
nuclear division 43, 44, 51
nuclear migration 143
nucleic acid sequences 4, 18, 25, 43,
    49, 51, 52, 58, 64, 70, 72, 73, 75,
    76, 89, 108, 112, 159
  distinguishing species and strains
    27
nutrition 7, 8, 16, 17, 38, 60, 86, 95,
    155, 193, 194

ordination 207

parametric tests 205
parasexuality 220, 223
parenthesome (see septal pore
    apparatus)
pathogens 38, 39
  alga 156
  angiosperm 38
  animal 26, 76, 240
  bryophyte 157
  fern 45, 51
  gymnosperm 38
  human 3, 4, 26, 76, 112
  insect 52
  lichen 39
  moss 38, 41, 45, 51, 52, 157
  plant 20, 22, 49, 51, 57, 59, 76, 77,
    79, 92, 154, 155
  protozoa 95
  sedge 38
peridium 138
pheromone 9, 111
phylogeny 9, 41, 43, 46, 52, 71–73,
    76, 79, 80, 85, 86, 122–124, 128,
    135, 149, 159, 229, 233
  fungi 236, 238, 240
  fungi-animal association 235
  fungi-plant association 235
  monophyletic xi, 37, 47, 49, 50,
    58
  phenetic xi, 48
pityriasis versicolor 76
pleomorphic fungi 182
pneumococcal-type pneumonia 3
polynomial system 171
populations 219
principal components analysis 208
pseudosclerotium 148

recombination 219, 223
root rot 38

saprobe 38, 49, 155, 157
saprotrophy 5, 38, 60, 86, 98, 105, 151
sclerotium 92, 148
seborrheic dermatitis 76

Subject Index

selection 219
septal pore apparatus 7, 8, 13–15, 25, 43–45, 47–49, 58, 59, 64, 65, 67, 68, 70, 85, 89, 90, 92, 93, 96, 99, 101, 106, 108–112, 122, 128, 130, 144, 145, 243
septum 6, 13–15, 26, 27, 43, 68, 96, 101
   dolipore (see septal pore apparatus)
sequence alignment 231
sequence primer design 231
sexual reproduction 9, 10, 37, 225
shikimate-chorismate pathway 149
software 212–214
sorus 60, 61
SPB (see spindle pole body)
special form 175
speciation 219
species concept 63, 64, 219
Species plantarum 172, 181
specific epithet 172
spermatium 41
spermogonium 47
spindle pole body 16, 43–45, 85, 87, 89, 92, 93, 95, 96, 98, 99, 101, 106, 108, 122
spore, secondary 87
statistics 204
sterigma 41, 139
symbiont 156, 157, 159
symbioses 156
symplechosome 46, 49
synanamorph 182
synnema 95
systematics
   history xi, 48, 64, 85, 86, 122, 124
   phlyogenetic xi, 18–28, 46–48, 64, 75, 89, 108, 115, 116, 121, 125

taxonomic hierarchy 174
taxonomic hypotheses 37, 46, 135, 137–139, 159
taxonomic keys
   computer keys 211
   identification keys 210
taxonomy 172
   numerical 203, 204
teleomorph 6, 19, 37, 67, 76, 77, 112, 147, 182, 223

teliospore 6, 7, 10, 11, 19, 20, 22, 25, 26, 41, 49, 60, 61, 74–79, 112
teliospore ball 74–76, 78
telium 40
thelephoric acid 150
transformations, data 207
trophic state 151
typification 178
   cultures 179

ubiquinone 43
ultrastructural characters 7, 8, 13–16, 26, 37, 43, 44, 46, 48–50, 52, 60, 64, 70, 71, 75, 85, 86, 89, 92, 93, 95, 96, 99, 101, 108, 144
urediniospore 41, 51, 52

wood decay 38, 51, 98, 105, 151
   brown rot 152
   white rot 152
Woronin body 13, 45

yeast 76
yeast cell 148
yeast phase 60, 74, 77, 109–113

Biosystematic Index

Abortiporus 125, 148
Acanthophysium 126
Aegerita 148
aeroaquatic fungi 148
agaric 58, 135, 137
Agaricaceae 122, 125, 129–131, 140, 154, 158, 159
Agaricales 122, 124–126, 129–132, 135, 139, 144
agaricoid fungi 134
Agaricostilbaceae 40, 50, 51
Agaricostilbales 38, 50, 51
Agaricostilbomycetidae 46, 50, 51
Agaricostilbum 9, 10, 19, 25, 37, 38, 40, 41, 43, 45, 46, 49–51
 pulcherrimum 13, 40, 42, 51
Agaricus 125, 145, 150, 151, 155
Agrocybe 125, 148
Albatrellus 125, 126, 129, 132, 140, 142, 150, 153, 154
 syringae 129
Aleurocystis 147
Aleurodiscus 126
alga
 green 156, 157
 red 141, 194
Alismataceae 70
Allescheriella 148
Allomyces 221
 arbuscula X macrogynus 223
 javanicus 223
Allomycetes xiii
Alpova 126
Alternaria alternata 231
Amanita 125, 143, 145, 153
Amanitaceae 125, 129, 131
amoeba 95
Amylostereum 126
Anacardiaceae 70
angiosperm 38, 62, 78, 79
animal 76, 232, 234, 235, 237
Annellaria 125
ant, attine 158
Anthracoidea 60, 67, 75, 79
 altiphila 63
Antrodia 125, 148, 153
 carbonica 158
Antrodiella 125
Antromycopsis 147
Aphyllophorales 95, 98, 105, 122, 124–127, 129–132, 135, 140, 144, 150
Apiotrichum 8
Aporpium 106

apple 4
Aquathanatephorus 93
aquatic hyphomycete 37
Araceae 67
Araliaceae 67
Arcangeliella 132
Archaemycota xiii
Archiascomycetes xii, 233, 241
Arecaceae 70
Armillaria 125, 151, 154, 155
 bulbosa 223
 mellea 155
Arrhenia 125
Arrhytidia 99
Arthoniales 242
Arthrinium 109
Arthrosporella 148
ascomycetes 37, 47, 59, 105, 154, 156, 221, 237
Ascomycetes xii, xiii
Ascomycota xii, xiii, 6, 45, 49, 150, 182, 232–236, 240, 241
Ascomycotina xii
Aseroe 127
asparagus 38
Aspergillus 171, 197, 199, 220, 223
 nidulans 197, 223
 versicolor 223
Asteraceae 59
Asterophora 125, 148, 155
 lycoperdoides 149
Astraeus 124, 138, 155
Astrogastraceae 126, 132
Athelia 125, 137, 155
 epiphylla 156
 rolfsii 154
Atractiella 38–43, 46, 47, 49, 50
Atractiellales 38, 39, 43, 45, 46, 49, 50
Atractogloea 50
 stillata 43, 49
Attini 158
Aurantioporus 125
Aurantiosporium 50, 59
Auricularia 86, 89, 101, 105, 114, 115, 143
 auricula-judae 89, 90, 99, 100, 105, 115
 fuscosuccinea 99, 143
 mesenterica 99
Auriculariaceae 48, 85, 99, 105, 116
Auriculariales 45, 47, 48, 52, 86, 87, 89, 93, 99, 101, 104–106, 114, 115, 122, 134, 139, 142, 143, 146, 147, 149, 151–155

Auriculariopsis 125
 ampla 137
Auriculoscypha 50
Auriscalpiaceae 126, 132
Auriscalpium 126, 137, 145
 vulgare 137
Austroboletus 126, 140
Autobasidiomycetes 86

bacteria 193
Baeospora 142
Ballistosporomyces 6
Bankera 131, 132, 145, 150
Bankeraceae 131, 132, 150
barley 4, 57
Basidiobolus ranarum 234, 240
Basidiodendron 101, 102, 105
Basidiodendron cinereum 105
 eyrei 100, 101, 104
basidiomycetes 3, 43, 47, 48, 58, 80, 85, 221, 225, 237
Basidiomycota xii, xiii, 6, 18, 43, 46, 47, 57, 86, 182, 234–236, 240, 241, 243
Basidiomycotina xii
Basidioradulum 126, 133, 145
basswood 115
Battarraea 138
Bauerago 59
bean 77, 194
beetle
 ambrosia 158
 bark 149, 158, 159
 mountain pine 147
Bensingtonia 6, 8, 16, 17, 20, 50
 ingoldii 6
 intermedia 6, 20
 yamatoana 13, 16
Bigyra xiii
bird's nest fungi 130, 139
Bjerkandera 125, 150, 157
Blasia 157
Blastocladiales 232, 234, 236
Blastocladiella emersonii 234, 240
Blastomyces 201
Blumeria graminis 234
Boidinia 126
Bolbitiaceae 125, 130, 131, 140
Bolbitius 125
Boletaceae 126, 131
Boletales 140, 153, 159
bolete 131, 135
Boletellus 131, 140
Boletineae 131, 150

Boletinellus 148
Boletopsis 126, 131, 132, 144, 150
Boletus 126, 137, 145, 150
  astraeicola 155
  parasiticus 155
  satanas 234
Bolomycetes xiii
Bombacaceae 70
Bondarzewia 126, 142, 147, 155
Bondarzewiaceae 126, 132
Botryobasidiales 130, 147
Botryobasidium 126, 133, 134, 137–139, 145, 147, 148
  isabellinum 140
Botryoconis 70
Bourdotia 106
  galzinii 104
  galzinii f. *microcystidiata* 104
Brachybasidiaceae 70, 78, 79
Brachybasidium 70
  pinangae 78
Brassicaceae 59, 67
Brauniellula 126
Bromus 64
bryophyte 156, 157
buckwheat 38
Bulbilomyces 148
Bullera 6, 8, 17, 25, 26
  alba 16, 109
  pseudoalba 26
  unica 26
Bulleromyces 3, 6, 7, 10, 25, 113
  albus 12, 14, 15, 26, 109
Burgoa 148
Burrillia 70, 74, 78
Byssoporia 126, 132

cacao 154
Caecomyces 237, 242
  communis 234
Callitrichaceae 70
Calocera 99
  cornea 98, 115
  viscosa 98
Calostoma 126, 130
Calostomataceae 126, 131, 138
Calvatia 125, 130, 151
Campanulaceae 59, 67
Camptobasidiaceae 50
Camptobasidium 39, 49, 50
  hydrophilum 39, 49
Candelabrochaete 125
Candida 4, 27
  albicans 234
cannon ball fungi 134, 138
Cantharellaceae 122, 126, 133, 153
cantharelloid fungi 134
Cantharellus 126, 133, 134, 139, 145
  cibarius 137
Cantharocybe 124
Capronia pilosella 234
Carcinomyces 112, 113
Carcinomycetaceae 112, 113
carrot 194
Caryophyllaceae 59
celeriac 38
Ceraceomyces 148

Ceraceosorus bombacis 70
Ceratobasidiaceae 86, 90, 93, 99, 125
Ceratobasidiales 85–87, 89, 92, 93, 106, 114, 116, 122, 125, 129, 130, 134, 141, 144, 147
Ceratobasidium 92, 93, 125, 130, 139, 145–148, 154
  cornigerum 89, 92
Ceratorhiza 147
Ceratosebacina 102, 105
Cercozoa xiii
cereal 154
Cerinomyces 149
Cerinomyces 99, 149
  aculeatus 98
  altaicus 89
Cerinomycetaceae
Ceriporia 125, 150
Ceriporiopsis 125, 152
Cerrena 157, 159
  unicolor 157
Chaetoderma 153
Chaetospermum 114
Chaetothyriales 242
Chamonixia 126, 150
Chenopodiaceae 67
Chionosphaera 7, 40, 41, 49–51, 79
  apobasidialis 13
Chionosphaeraceae 38–40, 50, 51
Chlorophyllum 125
choanoflagellate 233, 236
Chondrogaster 127
Christiansenia 112, 113
  pallida 112
Christianseniaceae 108, 113
Christianseniales 87, 106, 108, 113
Chromista xii, xiii, 175, 184
Chroogomphus 126, 150
  vinicolor 234
chrysanthemum 4
Chrysomphalina 151
Chrysorhiza 147
chytrid 182, 237, 239, 240
Chytridiales 234
Chytridiomycetes xii, xiii
Chytridiomycota xii, xiii, 234
Chytridium confervae 234, 240
Cintractia 62, 67, 75
  axicola 63
Citrus 38
Clathraceae 127, 134
Clathrus 127, 145, 146
Clavaria 125, 130, 171
  argillacea 154
  ignicolor 146
Clavariaceae 122, 124–127, 130, 133, 134
Clavariachaete 133
Clavariadelphus 127, 134, 135, 137, 140, 151
  truncatus 137
clavarioid fungi 134
Claviceps 150
Clavicorona 126, 137, 142, 144
Clavulicium 143, 145
Clavulina 126, 133, 134, 137, 143, 156

Clavulinaceae 126, 133
Clavulinopsis 125, 130, 151
  fusiformis 137
Climacocystis 125
Climacodon 125
Clinoconidium 70
  farinosum 79
Clintamra 62, 67
Clitocybe 125
Clitopilus 125, 148
club fungi 135
Coccidiodictyon 50
Coccidioides 201
  immitis 230
Cochliobolus 221, 224
coffee 38, 154
coffee rust 38
Colacogloea 9, 22, 43, 50
  peniophorae 10, 12, 16, 20, 39, 41
Coleosporium 41
  asterum 40, 41
collar flagellate 233, 234, 237, 240
Collybia 112, 125, 147–149
Coltricia 126, 133, 145
Columnocystis 153
Commelinaceae 59, 70
Conferticium 126
Conidiobolus 114
  coronatus 234, 240
Conidiosporomyces 69
  ayresii 61, 67
conifer 62, 115, 152
Coniodictyum 70, 79
Coniophora 126, 131, 137, 140, 145, 150, 153
Coniophoraceae 126, 131
Conocybe 125, 156
Convolvulaceae 69, 80
Coprinaceae 124, 125, 129, 131, 140, 154
Coprinus 125, 145
  bisporus 224
  cinereus 220, 234
  domesticus 147
coral fungi 135
coralloid fungi 134
Cornus 40
Corticiaceae 92, 124–126, 129–133, 139, 153
corticioid fungi 127, 132, 134, 137, 150, 155
Corticium 148
  minnsiae 147
Cortinariaceae 125–127, 140
Cortinarius 125, 141, 150, 153
Cotylidia 125
Craterellus 126, 134, 143, 151
Craterocolla 105
Creolophus 126
Crepidotaceae 130, 140
Crepidotus crocophyllus 130
Crinipellis 125, 159
  perniciosa 154
Cronartium ribicola 234
Crucibulum 125, 139, 147, 152, 157
Crustoderma 125, 148, 153
Cryptobasidiaceae 70, 79

Cryptoccocus 4, 6, 8, 16, 17, 25–27
  albidus 4
  ater 27
  cellulolyticus 4, 26
  curvatus 4
  diffluens 4
  elinovii 4
  feraegula 17
  huempii 17
  humicola 4
  humicolus 4
  laurentii 4, 14, 26
  magnus 27
  neoformans 16–18, 28
  skinneri 26
  terreus 4
  tsukubaensis 4
Cryptomycocolacales 50
Cryptomycocolax 41, 45, 50
  abnorme 16, 51
Cryptoporus 125
  volvatus 158
cucumber 4
cucumber powdery mildew 4
cyanobacteria 156, 157
Cyathus 125, 139, 147
Cyclomyces 133
Cyperaceae 59, 62, 67, 74–76, 79, 80
cyphelloid fungi 137
Cyphomyrmex 148
Cyrenella 148
Cystangium 126
Cystobasidium 3, 9, 39, 50
Cystoderma 125
Cystofilobasidiaceae 112, 113
Cystofilobasidiales 23, 25–27, 112
Cystofilobasidium 7, 9, 10, 17, 25, 26, 112, 113
  bisporidii 17
  capitatum 9, 11, 112
  infirmominiatum 9, 10, 17

Dacrymyces 43, 99, 113, 143
  capitatus 98
  chrysospermus 98
  minor 98
  ovisporus 96
  stillatus 96–98
  unisporus 96
Dacrymycetaceae 86, 99
Dacrymycetales 85–87, 89, 93, 96, 98, 99, 105, 106, 114, 116, 122, 139, 142–144, 146, 149, 151–153
Dacryonaema 99
Dacryopinax 99, 143
  spathularia 96
Daedalea 125, 129, 153
Daedaleopsis 125, 150
  confragosa 137
Datronia 125
Dendrocorticium 125, 137
Dendroctonus 125, 159
  brevicomis 158
  frontalis 148, 158
  ponderosae 147
Dendrophora 126
Dendrosporomyces 148

Dendrothele 148, 155
Dentipellis 126
Dentocorticium 125
Dermatosorus 69
Dermocybe 125, 150
Dermocystidium 236, 237, 239
  salmonis 234
deuteromycetes 37
Diaphanoeca grandis 234
Diaporthe 109
Dicellomyces 70
  gloeosporus 78
Dichostereum 126, 147
dicot 16, 59, 67, 69, 70, 74–76, 79, 80
Dictyonema 125, 145, 156
Dictyonemataceae 125, 130
Dictyopanus 151
Digitatispora marina 141
Dikaryomycota xii
Diliaceae 67
Dipodascopsis uninucleata 234, 241
Dipsacaceae 59
discomycetes 109, 114, 151, 241, 242
Discomycetes xiii
Ditiola 99
Ditiola pezizaeformis 98
Doassansia 70, 74, 78
Doassansiaceae 69
Doassansiales 9, 67, 69, 76, 78, 80
Doassansiopsaceae 67, 74, 75
Doassansiopsis 67, 74, 75, 78, 79
  ticonis 61
Doassinga 70, 78
  callitrichis 67, 68
Drepanoconis 70
Ductifera 105
Duportella 126

earthstar 134, 138, 139
Eballistra 69, 74, 77
Eballistraceae 69, 77
Echinodontiaceae 126, 132
Echinodontium 126, 142, 155
Efibulobasidium 105
  albescens 114
Eichleriella 101, 105
Elaphomycetales 242
Ellula 147
Elmerina 106
  caryae 102, 104
Emericella nidulans 223
Endogonales 234, 237, 240
Endogone 237
  pisiformis 234
Endomycetes xiii
Endoperplexa 102, 105
  septocystidiata 102
Endophyllum sempervivi 41
Enteromycetes xiii
Entoloma 125, 141
  abortivum 155
  parasitica 155
Entolomataceae 125, 130, 131, 140
Entomocorticium 126, 159
  dendroctoni 158
Entomophthora muscae 234, 240
Entomophthorales 233, 234, 240

Entorrhiza 58, 61, 67, 70, 72–74, 79
  casparyana 63, 68
Entorrhizaceae 67
Entorrhizales 67, 68
Entorrhizomycetidae 67, 70
Entyloma 59, 62, 63, 69, 77, 78
  dactylidis 63
  guaraniticum 63
  hieracii 67
  microsporum 61, 77
Entylomataceae 69
Entylomatales 67, 69, 70, 77, 78
Eocronartiaceae 50
Eocronartium 37, 38, 40, 41, 45–47, 50–52
  muscicola 43, 44, 144
Epithele 125
Epulorhiza 147
Eremascus albus 234
Ericaceae 79, 80
Erisyphe graminis 4
Erratomyces 69, 77
  patelii 77
Erysiphales 242
Erythricium salmonicolor 155
Erythrobasidium 7, 10, 19, 25, 46, 50
Erythromyces 154
euagarics 129–131
Euascomycetes xii, 232, 233, 236, 241
Eumycota xii, xiii
Eurotium rubrum 234
Exidia 86, 101, 105, 113, 114, 143
  crenata 101, 105
  glandulosa 100, 101
  grisea 101
  nucleata 143
  pinicola diversa 101
  pithya 115
  recisa 101, 105
  saccharina 101
Exidiaceae 101, 105, 147
Exidiales 105
Exidiopsis 101, 102, 105, 114
  calospora 102, 105, 114
  diversa 101
  fugacissima 102
  gloeophora 102
  longispora 102
  opalea 104
  plumbescens 89, 90
Exobasidiaceae 70, 78, 79
Exobasidiales 64, 70, 76
Exobasidianae 64, 68, 69, 77
Exobasidiellum 70
Exobasidiomycetidae 27, 64, 69, 70, 76, 77, 79
Exobasidium 60, 61, 67, 70, 76, 79
  oxycocci 67
Exophiala jeanselmei 233
Exoteliospora 62, 67
  osmundae 62, 74

Faerberia 125, 127, 129, 137
Fagaceae 69
false truffles 242
Farysia 69

Farysporium 69
Favolaschia 151
Fellomyces 6, 8, 25, 26
Femsjonia 99
fern 38, 45, 51, 62
Fibulobasidium 9, 25, 113
  inconspicuum 111
Fibulomyces 148
Fibulosebacea 105
Fibulostilbum 40, 50, 51
Fibulotaeniella 148
Filobasidiaceae 108, 111–113
Filobasidiales 12, 23, 25–27, 87, 106, 108, 112–114, 116
Filobasidiella 3, 7, 9, 10, 14, 25, 112, 113
  depauperata 14, 15
  lutea 108, 113
  neoformans 3, 9–11, 14, 26, 112
Filobasidium 3, 7, 9, 10, 14, 17, 26, 27, 112, 113
  capsuligenum 4
  elegans 27
  floriforme 4, 27
  uniguttulatum 11
Fistulina 125, 142, 145, 152
Fistulinaceae 125, 129, 130
Flammulina 125
fly 138
Fomes 125, 140, 142, 145
  pinicola 222, 225
Fomitopsis 125, 129, 142, 148, 153, 155
  pinicola 158
Formicidae 158
Franzpetrakia 69
Fulvisporium 50, 59
Fungi xiii, 175
Fusarium 197, 204
  avenaceum 197

Galactomyces geotrichum 234, 241
Galerina 209
  antheliae 209
  arctica 209
  atkinsoniana 209
  calyptrata 209
  clavata 209
  heterocystis 209
  hypnorum 209
  muricellospora 209
  paludosa 157
  pseudotundrae 209
  rubiginosa 209
  stagnina 209
  tundrae 209
  vittaeformis 209
Galzinia 125
  culmigna 92
Ganoderma 125, 129, 140, 142, 145, 154, 155
Ganodermataceae 124, 125, 129
gasteroid fungi 135
Gasteromycetes 122, 124–127, 129–131, 134, 135, 138, 142, 144
gasteromycetes 222
Gastrosuillus 126

Gautieria 127, 134, 135, 151, 153
Gautieriaceae 127
Gautieriales 127, 134
Geastraceae 127, 138
Geastrum 127, 134, 139, 141, 145, 146
  saccatum 137
Gelimycetes xiii
Geminago 69
Gentianaceae 59
Geomycetes xiii
Geopetalum 127
Georgefischeria 69
  riveae 63
Georgefischeriaceae 69, 77
Georgefischeriales 67–70, 74, 76, 78, 80
Gerronema 125, 151
  marchantiae 157
Gigaspora margarita 234
Gloeocantharellus 127, 134, 137
Gloeocystidiellum 126, 142
Gloeocystidium 125, 158
  ipidophilum 130, 158
Gloeodontia 126
Gloeophyllum 125, 129, 153
Gloeoporus 125
Gloeosynnema 147
Gloeotulasnella 95, 96
  cystidiophora 93, 94
Gloiocephala aquatica 141
Gloiodon 126
Glomales 234, 235, 237, 240
Glomomycetes xiii
Glomosporiaceae 64, 67, 75
Glomosporium 64, 67
  leptideum 75
Glomus intraradices 234
  mosseae 234
Glutinoagger 147
Gomphaceae 127, 134, 135, 151, 153
Gomphidiaceae 126, 130, 131, 159
Gomphidius 126, 137, 150
Gomphus 127, 134, 137, 151
Graphiola 61, 64, 70, 79
  phoenicis 60, 63
Graphiolaceae 70, 79
grass 64, 77, 154
gray mold 1
Gremmeniella 204
Guepiniopsis 96, 99
Gymnoconia peckiana 41
Gymnomyces 126
Gymnopilus 150
gymnosperm 38
Gyrodon 140, 150
Gyrodontium 133
Gyroporus 126, 140

Haasiella 151
Halocyphina villosa 141
Haloragaceae 59
hamamelids 77
Hanseniaspora 9
Haplotrichum 148
Hebeloma 125, 153

Heimiomyces 137
Helicobasidium 37, 40, 45, 46, 48, 50, 51, 147
  brebissonii 44
  mompa 13, 46
Helicogloea 38–41, 45–50
  intermedia 42
  lagerheimii 44
  variabilis 44
Heliconiaceae 70
Heliocybe 152, 153
Helminthosporium 224
Hemiascomycetes xii, 241
Hemibasidii 86
Hericiaceae 126, 132
Hericium 126, 145, 155
Herpobasidium 38, 40, 45, 46, 50, 51
Herpotrichia juniperi 234
Heteroacanthella 106
Heterobasidiomycetes xii, xiii, 85, 122, 123, 126, 130, 143, 144, 149, 151, 152
heterobasidiomycetes 41, 43, 46, 47, 52
Heterobasidiomycetidae 86, 87, 114, 116
Heterobasidion 126, 132, 140, 142, 147, 155
Heterocephalacria 112, 113
Heterochaete 101, 105
Heterochaetella 106
Heterodoassansia 70, 74, 78
Heterogastridiales 20, 50
Heterogastridium 40, 50
  pycnidioideum 13, 14, 16, 20, 25, 39
Heterokonta xii
Heteroscypha 106
Heterotestus 98
Heterotextus 99
Heterotolyposporium 69
Hirneola 105
Histoplasma 201
Hoehnelomycetaceae 49, 50
Hohenbuehelia 125, 156
Holtermania 9
Holtermannia 109, 113
  corniformis 26
homobasidiomycete clades 124, 125, 135, 137, 140, 142–147, 149, 152–155, 157, 159
Homobasidiomycetes xii, xiii, 46, 47, 121
homobasidiomycetes 86, 89, 92, 104, 106, 221
Horakia 124, 132, 133
Hyaloria 106
Hyaloriaceae 86, 99, 102, 104, 105, 114, 116
Hydnaceae 122, 125, 126, 129, 133, 153
Hydnangium 125, 130
Hydnellum 126, 131, 132, 137, 144, 145, 150
hydnoid fungi 134
Hydnum 126, 133, 134, 143, 156

Hydrocharitaceae 70
Hydropus 142
Hygrocybe 125, 150
Hygrophoraceae 125, 130
Hygrophoropsis 130, 131, 148, 150
Hygrophorus 125, 139, 153, 154
Hymenoascomycetes xii, xiii
Hymenochaetaceae 126, 133, 150
Hymenochaete 126, 145, 150
Hymenogaster 126
Hymenogastraceae 126, 127, 131
Hymenogastrales 124, 126, 127, 132, 134
Hymenomycetes xii, 19, 22, 23, 47, 57, 122, 125–127, 138, 243
Hymenomycetidae 25
Hyphelia 148
Hyphochytrea xiii
Hyphochytriomycetes xii, xiii
Hyphochytriomycota xii, xiii
Hyphoderma 125, 147, 156
Hyphodontia 126, 133, 145
 sambuci 137
Hypholoma 125, 147, 148, 150
hyphomycetes 74
Hypocrea jecorina 243
Hypocreales 150, 243
Hypomyces 150
 chrysospermus 234
Hypoxylon 204
 fuscum 204
Hypsizygus 152
Hyrophoropsis 126, 140, 153
Hysterangiaceae 127, 134
Hysterangium 127

Ichthyophonus 236, 237, 239
 hoferi 234
imperfect fungi 220, 223
Ingoldiella 141, 148
Ingoldiomyces 69
Inocybe 125, 153
Inonotus 126, 133, 148, 150, 153
insect 38, 40, 50, 52, 138, 148, 157
Insolibasidium 38, 45, 50, 51
 deformans 41
Ips typographus 158
Irpex 125
Itersonilia 8, 12, 15
 perplexans 4, 12, 15, 16

Jamesdicksonia 69, 74, 77
jelly fungi 3, 37, 85, 86, 243
Jola 38, 41, 45–48, 50–52
 cf. *javensis* 40, 42, 44
Juhnghunia 125
Juncaceae 59, 67, 74, 80

Kavinia 127, 134
Kjeldsenia 127
Kockovaella 6, 8, 25, 26
Kondoa 7, 19, 46, 50, 51
 aeria 10
 malvinella 9, 43
Kordyana 70, 78
Kriegeria 9, 22, 39, 40, 41, 43, 45, 50
 eriophori 14, 20, 38, 42

Krieglsteinera 50
Krieglsteineraceae 50
Kryptastrina 50
Kuhneromyces 125
Kuntzeomyces 69
Kurtzmanomyces 6, 8, 19, 25, 50
 insolitus 25

Labyrinthulomycetes xiii
Laccaria 125, 130, 153, 154
 trullisata 148
Lachnocladiaceae 126, 132, 143
Lacrimaria 125
Lactarius 126, 132, 142, 156
Laetiporus 125, 127, 142, 147, 153
Laetisaria fuciformis 155
Lamiaceae 59
Lampteromyces 125, 130, 150, 151, 159
Langermannia 157
Lauraceae 70
laurel 79
Laurilia 147
Laurobasidium 70, 79
 lauri 79
Laxitextum 126, 145
leaf blight of coffee 154
leaf blight of maize 154
Lecanora dispersa 234, 242
Lecanorales 242
Leguminosae 77, 88
Lemnaceae 70
Lentaria 127, 134
Lentibulariaceae 59
Lentinellus 126, 132, 137, 142
Lentinula 125, 145, 152, 154, 158
Lentinus 125, 127, 129, 140, 142, 152, 153, 159
 tigrinus 133
Lenzites 125, 157
 betulina 155
Lenzitopsis 132, 133
Leotiales 242
Lepiota 125
Lepista 125
Leptoporus 125
Leptosporomyces 148
Leratia 125, 131
Leucoagaricus 125, 158
Leucocintractia 69
Leucocoprinus 125, 148, 158
Leucogyrophana 148, 150
Leucopaxillus 150
Leucosporidium 3, 5, 7, 9, 10, 16, 19, 20, 22, 39, 45, 49, 50, 86
 antarcticum 9
 fellii 9, 11, 20
 scottii 9, 11, 13, 16, 234
lichen 58, 110, 171, 182–185, 242
Licrostroma 147
Liliaceae 67
Limacella 125
Limonomyces 148
 roseipellis 155
Lindtneria 138
Liroa 50, 59
liverwort 157

loculoascomycetes 242
Loculoascomycetes xii, xiii
Loculomycetes xiii
Lonicera 38
Lopharia 125
Loweporus 154
Lycoperdaceae 125, 129, 138, 154
Lycoperdales 125, 127, 130, 131, 134
Lycoperdon 125, 141, 145
lycophyte 38
Lyophylleae 159
Lyophyllum 125, 159
 palustre 159
Lysurus 127

Macalpinomyces 69
Macowanites 126, 132
Macrocystidia 140
Macrotermitinae 159
Macrotyphula 125
Magnolia 38
maize 57, 154
Malassezia 3, 5, 6, 8, 17, 27, 60, 69, 76
 furfur 17, 18
Malasseziales 27, 69, 76
mammal 237
mangrove 5, 141
Marasmiellus 125, 154
 affixus 156
Marasmius 112, 125, 149, 154, 158
Marchantia 157
Martellia 126
Mastigobasidium 10, 20, 22
 intermedium 20
Matula 147
Melaniella 62, 69, 78
Melaniellaceae 69
Melanogaster 126, 131
Melanogastraceae 126, 131
Melanogastrales 126
Melanoleuca 125
Melanopsichium 69
Melanotaeniaceae 67, 68, 74
Melanotaenium 63, 67, 74
 endogenum 74
 euphorbiae 74
Meripilus 125, 142
 giganteus 158
Meruliacae 122
Merulius 150
Metabourdotia 105
Michenera 147
Microbotryaceae 20, 50, 59
Microbotryales 20, 25, 38, 43, 45, 48–51, 59, 60
Microbotryomycetidae 19, 39, 46, 49, 50
Microbotryum 9, 19, 22, 25, 37, 43, 50, 59
 reticulatum 42
 violaceum 20
Microcollybia 148
Microporus 141, 154
Microsebacina 102, 105
Microsporea xiii
Microsporidea xiii
Microsporidia 231

Microstroma 9, 64, 69, 77
　juglandis 27, 77
Microstromataceae 69
Microstromatales 9, 27, 67–70, 77
Minisporea xiii
mite 193, 195
Mixia 46, 49, 50
　osmundae 49
Mixiaceae 50
Moesziomyces 69
Monilia sitophila 223
Moniliopsis 147
　solani 147
monocot 62, 67, 70, 74–76, 78–80
Monosporonella 106
Montagnea 125, 131, 142
Morchella elata 234
Moreaua 69
moss 40
Mrakia 5, 7, 15, 17, 25–27, 112, 113
　frigidia 9, 27
　gelida 27
mulberry 38
Multiclavula 126, 133, 134, 137, 139, 143, 144, 146, 156
Mundkurella 67, 74, 79
Muribasidiospora 70, 79
　triumfetticola 77
mushroom 25, 171
Mutatoderma 147
Mycaureola dilseae 141
Mycena 125, 148, 151
　citricolor 154
Mycetozoa 184
Mycoacia 125
　denudata 222, 224, 225
Mycogloea 9, 39, 41, 50
Mycosyringaceae 67, 75
Mycosyrinx 67, 75, 79
　cissi 63, 68, 75
Mylitta 148
Mylittopsis 105
Myriococcum 148
Myxariaceae 105
Myxarium 106
　mesomorphum 104
　nucleatum 102, 115
Myxomycota 184

Naiadella 50
　fluitans 37, 39
Nannfeldtiomyces 70, 74, 78
Naohidea 9, 39, 43, 46, 50
　sebacea 10, 12, 51
Narasimhania 70, 74, 78
Necator 147
nematode 156
nematode-trapping fungi 156
Neocallimastigales 234
Neocallimastix 234, 237, 239
　frontalis 234
　joynii 234
Neolecta vitellina 241, 242
Neolentinus 142, 152, 153
　lepideus 133
Neovossia 69
Neurospora 199, 221, 232

　crassa 197, 220, 230, 234, 241
　sitophila 223
Nia vibrissa 141
Nidularia 139
Nidulariaceae 125, 129, 139, 141
Nidulariales 125, 127, 130, 134
Nothoclavulina 148
Nyctalis 148, 155
Nymphaeaceae 70

oat 4, 115
Oberwinkleria 69
Occultifur 9, 19, 25, 50
　externus 10, 12
　internus 10, 11, 43
Oligoporus 125, 158, 153
Oliveonia 92, 93
　atrata 91, 92
Oliveroniaceae 92, 93
Omphalina 125, 156
Omphalotus 125, 130, 150, 151, 153, 159
Onagraceae 59
Oncobasidium 93
Onnia 145, 150
Onygenales 242
Oomycetes xiii, 182, 184, 185
Oomycota xii, xiii, 175, 184
Ophiostoma 242
　ulmi 234
orchid 95, 105, 154
Orchidaceae 62
Ordonia 50
Orphanomyces 62, 69
Osmundaceae 67
Ossicaulis 152
Oxalidaceae 59
Oxyporus 126, 133, 155
Ozonium 147

Pachnocybaceae 40, 50, 52
Pachnocybe 37, 39–41, 45, 46, 50–52, 79
　ferruginea 38, 39, 42, 52
Pachykytospora 124
Pachyma 148
palm 51, 79, 80
Panaeolus 148
Panellus 125, 139, 151
Paneolina 125, 156
Paneolus 125
Panus 125, 127, 142, 148, 152, 153
　fulvus 148
Paracoccidioides 201
Paragyrodon 126
Paraphelaria 105
parsnip 4
Patouillardina 106
　cinerea 100, 101
Patouillardinaceae 101, 106
Paullicorticium pearsonii 146
Paxillaceae 125, 126, 130, 131, 150
Paxillus 126, 130, 131, 140, 151, 158
peach leaf curl fungus 241
Penicillium 197, 199, 242
Peniophora 39, 126, 132, 140, 151, 158

Perenniporia 125
Pericladium 69
Peronosporomycetes xii, xiii
Peziza badia 234
Pezizales 241, 242
Phaeolus 125, 127, 145, 150, 153, 155
Phaeotellus 125
Phaeotrametes decipiens 147
Phaffia 4, 8, 25
　rhodozyma 4, 17, 27, 28
Phallaceae 127, 134
Phallales 124, 127, 134, 138, 139, 151, 159
Phallus 127, 171
　tenuis 137
Phanerochaete 125, 137, 145, 147
　chrysosporium 152
　cremea 112
　salmonicolor 147
　sordida 143
Phellinites 239
Phellinus 126, 133, 141, 142, 145, 154, 155
Phellodon 131, 132, 150
Phlebia 125, 137, 139, 147, 148, 150
Phlebiopsis 125
　gigantea 158
Phleogena 37–40, 46, 47, 49, 50
Phleogenaceae 49, 50, 86, 116
Phlogiotis 106, 143
　helvelloides 104
Pholiota 125, 144, 147, 150
Phragmotaenium 69, 74
Phragmoxenidiaceae 106, 110, 113
Phragmoxenidium 113
　mycophilum 106, 110
Phylloporia 126
Phylloporus 126, 131, 137, 150
Phyllotopsis 125, 151
Physisporinus 125
Phytomyxea xiii
Pietraia 148
Piloderma 125, 130
　croceum 130
pink disease 155
Piptoporus 125, 129, 153, 155
Piromonas communis 234
Piromyces 237
　communis 199, 234
Pisolithus 126, 148, 150
Planetella 69
plant 232, 235, 237
Plasmodiophoromycetes xii, xiii
Plasmodiophoromycota xii, xiii
Platycarpa 50
Platygloea 37–39, 48, 50
　disciformis 51, 52
　effusa 43, 47
Platygloeaceae 50
Platygloeales 20, 47, 48, 51, 52
Platygloeales *sensu stricto* 50
Platypodidae 158
Plectomycetes xii, xiii, 241, 242
Pleospora rudis 234
Pleosporales 241
Pleurocybella 151

Pleurogala igapoensis 156
Pleurotaceae 129, 154
Pleurotus 125, 142, 145, 147, 148, 156, 158
  noctilucens 151
  tuberregium 148
Pluteaceae 125, 140
Pluteales 140
Pluteus 125
Pneumocystis carinii 234, 241
Poaceae 59, 62, 63, 70, 75, 76, 79, 80
Podabrella 159
Podaxaceae 125
Podaxales 125
Podaxis 125, 131, 142
Podohydnangium 125
Podospora anserina 220, 221, 224, 225
Polygonaceae 59
Polyozellus 150
Polyporaceae 122, 124–126, 129, 132, 133, 150
polypore 58, 76, 127, 135, 141, 150, 159, 239
Polyporoletus 125, 129
Polyporus 125, 141, 145, 150, 159
  tuberaster 148
poplar 115
Porogramme 125
Porpidia crustulata 234, 242
Portulacaceae 59
Postia 125, 153, 155
potato 194
powdery mildews 242
Proliferobasidium 70
  heliconiae 78
Protobasidiomycetes 86
Protococcus 157
Protoctista 184
Protodaedalea 106
Protodontia 106
  uda 102
Protohydnum 106
  cartilagineum 104
Protomerulius 106
Protomycetales 59
Protophallaceae 127
Protozoa xiii
Protubera 127
Psathyrella 125, 155
  epimyces 155
Pseudocolus 127, 134
Pseudodermatosorus 70
Pseudodermatosporus 78
Pseudodoassansia 70, 74, 78
Pseudohydnum 106
  gelatinosum 104
Pseudomycota xii, xiii
Pseudostypella 105
Pseudotomentella 126, 131, 150
Pseudotracya 70, 78
Pseudotulasnella 96
Pseudoxenasma 126
Pseudozyma 4, 8, 27, 60, 69, 75
  antarctica 4
  flocculosa 4

  rugulosa 4
  tsukubaensis 4
Psilocybe 125, 147, 148, 156
Ptechetelium 50
Pterula 125, 142
Ptychogaster 148
Puccinia 171
  lagenophorae 38
  malvacearum 44
puffball 25, 131, 137, 141
Pulcherricium 125, 139, 158
Pulveroboletus 145
Punctularia 125, 148, 150
Pycnoporellus 147, 153
Pycnoporus 125
pyrenomycete 114, 241, 242
Pyrenomycetes xiii
Pyrus 38

Quercus 38

Radulomyces confluens 146
Ramaria 127, 134, 137, 144, 145, 151
  ignicolor 146
Ramaricium 133, 134
Renatobasidium 105
  notabile 102
Reniforma 6, 8, 20
  strues 17, 25
Repetobasidiellum 102
Repetobasidium 102, 139
Resinicium 147
  bicolor 156
resupinate fungi 135
Resupinatus 125, 156
Rhamnaceae 70
Rhamphospora 9, 62, 70, 77, 78
  nymphaeae 61, 63, 78
Rhamphosporaceae 70, 78
Rhizoctonia 105, 147
  solani 223
Rhizopogon 126, 131, 137, 141, 150, 159
Rhizopogonaceae 126
Rhodocybe 150
Rhodosporidium 3, 7, 10, 16, 19, 25, 37, 39, 49–51
  diobovatum 9
  fluviale 9
  kratochvilovae 9
  lusitaniae 5, 9
  paludigenum 9
  sphaerocarpum 9, 13
  toruloides 4, 9, 13, 16
Rhodotorula 3, 4, 6, 8, 17, 20, 27, 50, 148
  acheniorum 27
  fujisanensis 22
  glutinis 4, 25
  gracilis 4
  graminis 4
  hinnulea 27
  mucilaginosa 4
  phylloplana 27
  rubra 4
  yarrowii 16

Rhynchogastrema 113
  coronata 112
Rhynchogastremaceae 87, 108, 112, 113
Richoniella 141
Rickenella pseudogrisella 157
Riessia 147
Rigidoporus 125
Ripartites 125, 130
Rosaceae 67
rose 4
rose powdery mildew 4
Rosellinia 204, 206
  abscondita 206
  chusqueae 206
  etrusca 206
  mammaeformis 204
  nectrioides 206
  subsimilis 206
Russula 126, 132, 142, 150
  compacta 234
Russulaceae 126, 130, 132, 153, 155
rust 37, 38, 41, 45, 47, 48, 51, 52, 61, 79, 80, 85, 116, 198, 243
  endocyclic 41

Saccharomyces cerevisiae 198, 234, 241
Saccharomycetales 241
Saccharomycetes xii, xiii
Saccoblastia 46, 47, 49, 50
Sakaguchia 7, 10, 19, 25, 46, 50, 51
  dacryoidea 9, 12
Sarcodon 126, 131, 132, 144, 150
Sarcodontia 150
scale insect 39
Schizonella 61, 69
Schizophyllaceae 125, 129, 130
Schizophyllum 125, 145, 156, 220, 221
  ampla 137
  commune 220, 221, 225
  fasciatum 222
Schizopora 126, 133, 145
Schizosaccharomyces pombe 234, 241
Schroeteria 59
Scirpus 141
Scleroderma 126, 141, 150, 155
Sclerodermataceae 126, 131, 138, 153
Sclerodermatales 126, 138
Sclerostilbum 147
Sclerotinia sclerotiorum 234
Sclerotium 148
Scolytidae 158
Scopuloides 125
Scutigeraceae 153
Scytinostroma 126, 145
Sebacina 102, 105, 106, 134, 143, 147
  crozalsii 102–104
  deminuta 101
  epigaea 104
  fugacissima 102
  incrustans 104
  podlachica 104
  subhyalina 99, 102, 104

Sebacinaceae 104, 106, 114, 116
Sebacinella 93
secotioid fungi 130, 131, 138, 141
Selaginella 38
Selaginellaceae 69
Semiomphalina leptoglossoides 156
Senecio vulgaris 38
Septobasidiaceae 50, 52
Septobasidiales 37, 38, 50–52
Septobasidium 9, 38, 40, 43, 46, 48, 50, 52
   carestianum 40
Septomycetes xiii
Serendipita 102, 105
   vermifera 99, 101, 105
Serpula 126, 131, 137, 140, 145, 153, 156
Shaeromonas communis 234
Sirex 157
Siricidae 157
Sirobasidiaceae 26, 86, 87, 106, 110, 111, 113
Sirobasidium 9, 109, 110, 113
   intermedium 26
   magnum 26, 107, 111, 115
Sirotrema 9, 109, 113
Sistotrema 125, 139, 143, 148, 149, 221, 222
   brinkmannii 92, 157, 221, 222, 225
Skeletocutis 125
slime mold xiii, 182, 184, 231
Smittium culisetae 234, 240
smut 3, 37, 38, 49, 57, 58, 85, 86, 221, 243
   urediniomycetous 42, 52
snow molds of cereals 154
snow molds of turf 154
Sordaria fimicola 222
Sparassidaceae 124, 125, 129
Sparassis 125, 127, 145, 153, 155
Sparganiaceae 70
Sphacelotheca 9, 22, 37, 50, 59
   cinnamomi 70
   fagopyri 38
Sphaerobolaceae 127, 141
Sphaerobolus 127, 134, 138, 139, 150
Sphaerotheca fuliginea 4
   pannosa var. *rosae* 4
Sphagnum capillaceum 157
Spiculogloea 50, 52
   occulta 41
spikemoss 62, 78
Spiniger 147
Spizellomyces acuminatus 234
   punctatus 240
Spizellomycetales 232, 234
Spongipellis 125
   unicolor 234
Sporidiaceae 20
Sporidiales 20, 47, 48
Sporidiobolaceae 19, 20, 43, 50, 51
Sporidiobolus 3, 6, 7, 10, 16, 17, 19, 25, 39, 43, 50
   johnsonii 9, 13, 25
   ruineniae 9, 10, 11, 13, 25
   salmonicolor 9, 10
Sporisorium 69, 79

Sporobolomyces 3, 4, 6, 8, 9, 19, 20, 25, 27, 50, 51, 114
   lactophilus 25
   roseus 4
Sporothrix 4, 41
Sporotrichum 147
Squamanita 148, 155
Steccherinum 125
Stegocintractia 69
Stephanoascus 4, 27
Stephanospora 138
Sterculiaceae 69
Stereaceae 126, 132, 153
Stereum 126, 140, 145, 151, 157
Sterigmatomyces 6, 8, 17, 19, 25, 27, 50, 51
Sterigmatosporidium 7, 26
Stigmatolemma 156
Stilbotulasnella 96, 115, 145, 146
   conidiophora 95, 115
Stilbum 39, 40, 49–51
   vulgare 13, 38, 51
stinkhorn 134, 137–139
Stipitochaete 133
Stramenipila (see Stramenopila)
Stramenopila xii
Strobilomyces 126, 140, 150
Stropharia 125, 145, 148
Strophariaceae 125, 129, 131, 140, 142, 147, 150, 154
Stylina 70
Stypella 102, 106
   grilletii 102
   subgelatinosa 102, 104
Subulicystidium 148
   longisporum 146
Suillus 126, 131, 141, 150
   cavipes 234
Sympodiomycopsis 6, 8, 60
   paphiopedili 27, 70
Synechococcus 157
Syzygospora 112
   alba 111

Taeniospora 148
Talaromyces flavus 234
Taphrina 49
   deformans 234, 241
Taphrinales 49
Taphrinomycetes xiii
Tapinella 126, 130, 153
Teliomycetes 52
Tephrocybe palustre 157
termite 159
Termitomyces 125, 147
Termitomyceteae 159
Testicularia 69
Tetragoniomyces 109, 113
   uliginosus 106
Tetragoniomycetaceae 106, 110, 113
Thanatephorus 92, 93, 130, 147, 148, 159
   cucumeris 92, 93, 130, 147, 154, 223
   praticola 130
Thanatophyton 147

Thaxterogaster 141
Thea 38
Thecaphora 64, 67, 75
   haumanii 75
Thelephora 126, 131, 132, 145, 150
Thelephoraceae 122, 126, 131, 132, 150, 153
Tilachlidiopsis 147
Tiliaceae 77
Tilletia 27, 57, 62, 69, 77
   barclayana 68
   caries 58, 64, 77, 234
   controversa 57, 64, 77, 199
   setaricola 63
   tritici 199
Tilletiaceae 64, 69, 77, 79
Tilletiales 64, 67–69, 76, 77, 80
Tilletiaria 27, 69
   anomala 13, 14, 60, 63, 76
Tilletiariaceae 69
Tilletiopsis 4, 8, 14, 60, 77, 78, 114
   albescens 17, 70
   cremea 70
   flava 69
   fulvescens 69
   lilacina 70
   minor 4, 70
   pallescens 17, 70
   washingtonensis 70
Tofispora 90, 92
Tolyposporella 69, 76
Tolyposporium 69
Tomentella 126, 131, 132, 137, 145
toothed fungi 135
Tothiella 67
   thlaspeos 75
Tracya 70, 74, 78
Trametes 125, 150, 157
   conchifer 147
   versicolor 155
Tranzscheliella 69
Trappea 127
Trechispora alnicola 155
Tremella 9, 10, 26, 43, 85, 108–110, 113–115
   aurantia 109
   brasiliensis 111
   fuciformis 111
   globospora 89, 90, 108, 109, 111, 115, 234
   grilletii 102
   indecorata 111
   lutescens 111
   mesenterica 12, 107, 108, 110, 111, 115
   moriformis 111
   mycophaga var. *obscura* 109
   polyporina 109
Tremellaceae 26, 86, 87, 106, 111, 113
Tremellales 9, 12, 13, 22, 25, 26, 39, 47, 48, 85–87, 106–108, 110, 112, 113, 115, 122, 139, 142, 144, 146, 148, 149, 151, 153
Tremellodendron 105, 106
   candidum 104
Tremellodendropsidaceae 104, 106

Tremellodendropsis 87, 99, 104, 106, 114
  flagelliformis 104
  pusio 104
  transpusio 104
  tuberosa 99, 104
Tremellomycetidae 16, 25, 86, 87, 89, 106, 110, 114–116
Tremelloscypha 106
  gelatinosa 104
Tremellostereum 106
  dichroum 104
Tremex 157
Trichaptum 126, 133, 145
Trichocintractia 69
Trichocomaceae 242
Trichoderma reesei 243
Tricholoma 125, 150, 153, 154
Tricholomataceae 124, 125, 129, 130, 140, 155, 158, 159
Tricholomatales 150
Trichomycetes xii, xiii, 234, 240
Trichosporon 4, 6, 8, 9, 14, 15, 25–27
  beigelii 4
  brassicae 14
  cutaneum 4
  dulcitum 4
  inkin 14, 15
  laibachii 14
  moniliiforme 4
  pullulans 4, 9, 14, 25–27
Trichosporonales 22, 25
Trichothecium roseum 243
Tricladiomyces 148
Trimorphomyces 9, 109, 113
  papilionaceus 109, 112
Trogia 142
truffle, false 130, 131, 134, 137–139, 141, 151
Truncocolumella 126
Tsuchiyaea 6, 8, 25
Tulasnella 89, 95, 96, 104, 122, 126, 134, 137–139, 143–145, 147, 154
  araneosa 95
  pinicola 96
  pruinosa 95
  zooctonia 95
Tulasnellaceae 86, 95, 96, 116, 126
Tulasnellales 85–87, 89, 93, 95, 96, 101, 106, 114, 116, 122, 126, 134, 141, 145, 146

Tulostoma 125, 130, 139
Tulostomataceae 125, 129, 138, 142
Tulostomatales 125, 126, 130, 131
Tylopilus 126
Typhula 125, 148, 154, 157
  phacorhiza 130
  uncialis 146
Tyromyces 125

Udeniomyces 6, 8, 17, 25, 27
Uleiella 62, 69
Ulmaceae 70
Uredinales 37, 38, 41, 43, 45, 47, 48, 50–52, 116
Uredinella 50
Urediniomycetes xii, xiii, 3, 9, 15–21, 25, 37, 57, 59, 141, 143, 144, 147, 243
Urediniomycetidae 45, 46, 50, 51
urediniomycetous yeast 12, 17
Uredo farinosa 79
Urocerus 157
Urocystaceae 67, 74, 75
Urocystales 64, 67, 74, 77
Urocystis 62, 63, 67, 74, 77
  ranunculi 63
Uromyces dianthi 43
Ustacystis 67, 77
  waldsteiniae 67, 68, 79
Ustanciosporium 69
Ustilaginaceae 64, 75, 76, 79
Ustilaginales 9, 27, 47, 48, 64, 67, 68, 75
Ustilaginomycetes xii, xiii, 3, 9, 15, 18, 19, 27, 43, 46, 48, 49, 57, 141, 143, 144, 149, 243
Ustilaginomycetidae 20, 27, 67, 74, 76, 79
Ustilago 9, 17, 43, 57, 62, 69, 75, 79, 86
  avenae 16
  hordei 58, 75, 234
  maydis 16, 57, 58, 60, 61, 63, 75, 144
  onumae 70
  oxalidis 67
  speculariae 74
  tritici 75
Ustilentyloma 50, 59
  fluitans 20
Ustilentylomataceae 50, 59
Ustomycetes xiii
Ustomycota 47, 86

Uthatobasidium 92, 93
  fusisporum 110

Vaccinium 154
Valsa 109
Vararia 126
Veluticeps 153
Verrucospora 132
Vesiculomyces 126
Vitaceae 75, 80
Volvariella 155
  surrecta 155
Volvocisporiaceae 69
Volvocisporium 69, 77
  triumfetticola 77
Vuilleminia 125

Waitea 122, 125, 130, 145–147
  circinata 92, 130
Wangiella dermatitidis 233
Weraroa 125, 131
wheat 57, 77
witches' broom of cacao 154
Wolfiporia 125, 148, 153
woodwasp 157

Xanthophyllomyces 113
  dendrorhous 4, 27, 28
Xenasma 139
Xenolachne 9, 109, 113
Xerocomus 126
Xeromphalina 125
Xerulaceae 142
Xyleborus 125, 159
  dispar 158

yeast 3, 37, 39, 43, 45, 48, 51, 197, 241
  ballistosporic 38, 52
  basidiomycetous 46, 48, 52
  fission 241
  teliospore forming 38, 48
  tremellaceous 14, 17
  urediniomycetous 41
  ustilaginaceous 17, 27

Zelleromyces 126, 145
Zoomycetes xiii
Zundeliomyces 59
Zygogloea 50
  gemellipara 43
zygomycetes 182, 236, 237, 240
Zygomycetes xii, xii
Zygomycota xii, xiii, 234

Printing and Binding: Stürtz AG, Würzburg